| VIII | | | IB | IIB | IIIA | IVA | | | | A |

OF THE ELEMENTS

									003
									He
									2

				10.81	12.011	14.007	15.999	18.998	20.179
				B	C	N	O	F	Ne
				5	6	7	8	9	10

				26.982	28.086	30.9738	32.06	35.453	39.948
				Al	Si	P	S	Cl	Ar
				13	14	15	16	17	18

58.933	58.71	63.546	65.37	69.72	72.59	74.922	78.96	79.904	83.80
Co	Ni	Cu	Zn	Ga	Ge	As	Se	Br	Kr
27	28	29	30	31	32	33	34	35	36

102.906	106.4	107.868	112.40	114.82	118.69	121.75	127.60	126.904	131.30
Rh	Pd	Ag	Cd	In	Sn	Sb	Te	I	Xe
45	46	47	48	49	50	51	52	53	54

192.22	195.09	196.967	200.59	204.37	207.2	208.981	(209)	(210)	(222)
Ir	Pt	Au	Hg	Tl	Pb	Bi	Po	At	Rn
77	78	79	80	81	82	83	84	85	86

151.96	157.25	158.925	162.50	164.930	167.26	168.934	173.04	174.97
Eu	Gd	Tb	Dy	Ho	Er	Tm	Yb	Lu
63	64	65	66	67	68	69	70	71

(243)	(247)	(247)	(251)	(254)	(253)	(256)	(253)	(257)
Am	Cm	Bk	Cf	Es	Fm	Md	No	Lr
95	96	97	98	99	100	101	102	103

Chemistry and
Our Changing World

Alan Sherman
Sharon J. Sherman
Middlesex County College, Edison, New Jersey

Chemistry and Our Changing World

Prentice-Hall, Inc., Englewood Cliffs, New Jersey 07632

Library of Congress Cataloging in Publication Data

Sherman, Alan.
   Chemistry and our changing world.

   Includes index.
   1. Chemistry. I. Sherman, Sharon II. Title.
QD31.2.S483   1983   540   82-13217
ISBN 0-13-129361-3

Editorial/production and interior design
by Zita de Schauensee
Chapter-opening and front matter design by Judith A. Matz
Cover design by Photo Plus Art
Manufacturing buyer: John Hall

Printed in the United States of America

10 9 8 7 6 5 4 3 2

The following figures are taken from Alan Sherman and Sharon J. Sherman,
Elements of Life, copyright © 1980 by Prentice-Hall, Inc.: 1-1, 2-1 to 2-5,
3-17, 3-18, 3-27, 4-1 to 4-19, 5-1 to 5-6, 7-1, 7-3 to 7-18, 8-1 to 8-13, 11-2 to
11-5, 11-10, 11-14, 12-1 to 12-10, 12-12, 15-2, 15-4 to 15-6, 15-8 to 15-13,
16-1 to 16-6, 16-8, 16-10 to 16-14, 16-16 to 16-21. Used by permission.

ISBN 0-13-129361-3

Prentice-Hall International, Inc., *London*
Prentice-Hall of Australia Pty. Limited, *Sydney*
Editora Prentice-Hall do Brasil, Ltda, *Rio de Janeiro*
Prentice-Hall Canada Inc., *Toronto*
Prentice-Hall of India Private Limited, *New Delhi*
Prentice-Hall of Japan, Inc., *Tokyo*
Prentice-Hall of Southeast Asia Pte. Ltd., *Singapore*
Whitehall Books Limited, *Wellington, New Zealand*

To Michael and Robert

Contents

Chapter 2

Atoms, Molecules, Elements, and Compounds 17

Some Things You Should Know After Reading This Chapter 17

Chapter 3

Atomic Theory 35

Some Things You Should Know After Reading This Chapter 35

Chapter 4

Chemical Bonding 68

Chapter 5

Chemical Reactions 92

Chapter 6

Energy: Where Does It Come from and Where Does It Go? 111

Some Things You Should Know After Reading This Chapter 111

Chapter 7

The Three States of Matter: Solids, Liquids, and Gases 133

Some Things You Should Know After Reading This Chapter 133

Chapter 8

Solutions of Acids, Bases, and Salts   159

Some Things You Should Know After Reading This Chapter   159

Chapter 9

Water and Water Pollution 185

Chapter 10

Air Pollution 206

Chapter 11

Nuclear Power 232

Chapter 12

Organic Chemistry 258

Part 1 The Basics 259

Chapter 13

Plastics and Polymers 298

Chapter 14

Agricultural Chemistry 325

Chapter 15

Food: A Chemical Collection 339

Chapter 16

Biochemistry: The Molecules of Life 371

Chapter 17

Pharmaceuticals and Drugs 400

Some Things You Should Know After Reading This Chapter 400

Chapter 18

The Chemistry of Home Care and Personal Products 438

Chapter 19

Chemistry and Outer Space 463

Appendix A

Basic Mathematics 487

Some Things You Should Know After Reading This Appendix 487

Appendix B

The Metric System and Measurement 499

Some Things You Should Know After Reading This Appendix 499

Appendix C

Expanded Rules for Writing and Naming Chemical Compounds 517

Important Tables 527

Glossary 537

Answers to Selected Exercises 552

Index 561

Preface

Chemistry and Our Changing World is a text written for the nonscience major. The purpose of the text is to introduce the nonscience student to the world of chemistry and to show this student how chemistry relates to the everyday world. The topics found in the table of contents reflect this goal.

To make this text more appealing and easy to learn from, a number of features have been included. Some of the outstanding features of the text are:

1. The use of *simple language*. The reader does not need to sit by a dictionary to comprehend the written material.

2. Clear presentation, usually in a *conversational style*.

3. Visual illustration through *numerous cartoons, figures, photographs, and tables*. The cartoons are used to explain difficult concepts and to add humor. The photographs and figures reinforce the written material, and the tables summarize the information given in the text.

4. *Learning goals* listed at the beginning of each chapter. The student will learn from these what he or she is expected to know after completing the chapter.

5. A chapter *summary* at the end of each chapter. These summaries reinforce the learning goals.

6. Numerous *worked-out examples* in each chapter. Some of the worked-out examples show step by step how to solve a quantitative problem; the strategy as well as all the mathematics is shown in the solution of the problem. Other worked-out examples supply sample test questions, with the complete solution following each question.

7. An abundant supply of *self-test exercises* at the end of each chapter. The exercises relate back to the learning goals.

8. An extremely useful set of *supplements*, including a glossary, an appendix on basic mathematics, an appendix on the metric system and measurement, an appendix on expanded nomenclature rules, and a set of important tables.

We have also prepared an instructor's guide to accompany the text.

Our text is designed for use in a one-semester or two-quarter course. The basic core of nine chapters can be covered in as many weeks. The instructor then has the freedom to pick and choose from the remaining 10 chapters to fit the needs of the course and the students.

Our text has undergone two years of class testing, as well as review by many reviewers. These reviewers are all professors in various colleges (two-year and four-year) and universities. In fine-tuning the text, we tried to incorporate as many of the suggestions from the reviewers as possible. This is, of course, a difficult task because the liberal arts (nonscience majors) chemistry course is viewed differently by each instructor. Some instructors want basic theory and mathematics to be a part of the course. Other instructors think applications should be stressed and that theory and mathematics should be kept at a minimum or omitted entirely. We've tried to strike a balance in this area. We realize that our text can't be all things to all people, but we have tried to keep the needs of instructors and students in mind as we wrote each chapter.

We are always open to new ideas. The methods we use in the text work well with our students, but we welcome any and all constructive criticism from the experience of others with these materials, so that we can make future editions of this text even more effective and useful.

We would like to thank some of our friends and relatives whose efforts and encouragement helped us to complete this project. Our sincere appreciation goes to Robert Schulman and Gale Dillman. We would also like to thank Professors Mary Ann Miller, Mary Mullen, Dominic Macchia, John Murray, Diane Trainor, and Linda Burns, all of Middlesex County College. Also, a very special thanks to Lois Bertha for her help in preparing the manuscript.

We would like to express our appreciation to several people who contributed significantly to the development of this text: Professor Jon M. Bellama of the University of Maryland, Professor Allen A. Denio of the University of Wisconsin–Eau Claire, Professor Emily P. Dudek of Brandeis University, Professor Richard H. Hanson of the University of Arkansas at Little Rock,

Professor John Healey of Chabot College, Professor Russell H. Johnsen of Florida State University, Professor Leo E. Kallan of El Camino College, Professor James L. McAtee, Jr. of Baylor University, Professor Charles R. Ward of the University of North Carolina at Wilmington, Professor John A. Weyh of Western Washington University, and Professor Charles M. Wynn of Eastern Connecticut State College. To all of you we give our sincere thanks.

Finally, our sincere gratitude to the staff of Prentice-Hall, who are true professionals in their work. A very special thanks to Betsy Perry, our editor, whose constant interest in our project was a real help to us. Also, a very special thank you to Zita de Schauensee, who handled beautifully all aspects of production.

ALAN SHERMAN
SHARON J. SHERMAN

North Brunswick, New Jersey

Chapter 1

Science, Technology, Man, and Environment: An Important Interplay

Some Things You Should Know After Reading This Chapter

You should be able to:

1. State the differences between science and technology.
2. State some of the ways in which human beings have shaped their environment.
3. Describe the six steps of the scientific method.
4. Discuss the interrelationships among science, technology, man, and environment.
5. State the relationship between chemistry and other sciences.
6. State the relationship between chemistry and today's society.
7. State some of the contributions of the Egyptians to early chemistry.
8. State some of the contributions of the Greeks to early science.
9. State some of the contributions of the alchemists to modern science and medicine.
10. State some of the merits of chemistry today.
11. Discuss the case of the herpes virus.
12. Explain how chemical research has helped save millions of Norwegian spruce trees.
13. Discuss the case of food additives and hyperactive behavior in children.
14. Explain how chemists have found a way to lower the acidity of some U.S. wines.

INTRODUCTION

The decades of the 1980's and 1990's: will they signal a bright new beginning for humankind or will they signal the beginning of the end? In some ways, the choice is ours. Will science and technology be used to improve the quality of life on this planet for all people or will they prove to be the tools of our own destruction? These questions are not new. Many scientists, educators, and futurists have been asking these questions over the past 30 years. However, many of these questions will be answered before the year 2000. Either we will find a way to solve our future energy problems soon, or our technological society will grind to a halt. Either we will find a way to produce sufficient food to feed the billions of people on our planet or millions will die of starvation.

In order to understand and discuss the numerous problems that we will be facing in the coming years, we first have to understand how science and technology have brought us to where we are now. It is the purpose of this chapter to briefly review the interrelationships among science, technology, man, and environment. However, before we do this, let us take an imaginary journey into the future and look at some "way out" possibilities.

SCENARIO

THE DAY WE LOST NEW JERSEY

It's a bright sunny spring day in New Jersey on April 5, 1990. New Jersey has grown through the years to become a major industrial state in the Northeast; a state that consumes a tremendous amount of industrial power. Almost 80% of this power is supplied by 10 nuclear power plants located either in the state or surrounding it. Two of these power plants are off the coast of Atlantic City, built on human-made islands in the Atlantic Ocean. New Jersey is not a state that is prone to many natural hazards. Tornados are infrequent; so are hurricanes and earthquakes. That's why it came as such a complete surprise when on this day a major earthquake hit the state, sending shock-waves from Boston to Virginia Beach. The quake was as strong as the one that devastated the City of San Francisco in the early 1900's. However, in New Jersey, the results were more devastating. The earthquake demolished numerous buildings and tore up many roads. However, the quake was centered on the central Jersey coast, so not many people were killed by the initial devastation. The problems arose with the four nuclear power plants located just off the coast or on the coast of the Jersey shore.

Although these plants were supposed to be earthquake-proof, they weren't strong enough to withstand the power of this quake. Sections of these plants broke away. Cooling systems were lost and the atomic cores heated to meltdown. Radioactive clouds of steam were dispersed into the atmosphere. Offshore breezes blew the clouds across the state. Tens of thousands of New Jersey residents died of the effects of this radiation, and tens of thousands more were to die of radiation-linked diseases.

SCENARIO

NEW BODIES FOR OLD

The year is 2050. During the first 50 years of the twenty-first century, science and technology were used by the governments of the earth to improve the quality of life on this planet for all people. It was the year 2000 when the major powers on this planet agreed to stop spending billions of dollars on weapons of war and destruction and decided instead to pool their knowledge and wealth to improve the human condition. Many of the problems that plagued human society in the late 1900's were brought under control by 2025. Programs to stabilize population growth and produce adequate food for those living on the planet were becoming effective. Medical science as well as the natural sciences like biology, chemistry, and physics reached new frontiers.

However, it was the discovery in this year of 2050 that was to bring startling news to the scientific community. Scientists and engineers had transplanted the brain of an individual into an artificial body. This new body had the appearance of a human being and functioned in a completely humanlike manner. The transplant had been a success. The ramifications of this event were awesome. This new body would not age and would last 200 years.

Some interesting possibilities? Perhaps, but for now they're only stories. Actually, most scientists don't like to contemplate such occurrences. Even science writers tend to be more conservative about predictions of the future. However, we just couldn't resist the opportunity of postulating a world where all of us could collect social security for over 200 years! But now let's turn our attention to more serious matters.

SCIENCE AND TECHNOLOGY

What Is Science?

Science is a way of looking at the natural world. It is a way of trying to explain how the world around us operates. Observations are made, then information is gathered and examined. Patterns or regularities are searched for and these patterns help the scientist make general statements about the phenomena being observed (Fig. 1-1). Finally, some conclusions about the observations are drawn that enable the scientist to formulate a theory. Then predictions are made based on the theory. Next, the scientist tests the theory by performing more experiments, gathering more information, and analyzing this information.

Figure 1-1 Based on their observations and experiments, scientists attempt to make *decisions* about various phenomena.

The entire process, called the *scientific method*, repeats itself. The new information gathered either backs up the theory or disputes it.

Scientists deal with *variables*, which are factors in an experiment that can be changed. Usually, the scientist controls certain variables in an experiment and changes only one variable at a time to test its effect. In other words, the scientist uses a systematic approach to examine a particular phenomenon. Once again, the process is called the scientific method. It is not a procedure that every scientist must go through every time an experiment is performed. It is, however, a general description of how the body of knowledge that we call science has been developed, and it shows how science is constantly undergoing scrutiny and change.

The scientific method essentially consists of these phases:

1. Observation and description

2. Searching for patterns and regularities

3. Making generalizations

4. Formulating theories or hypotheses

5. Making predictions based on theory

6. Testing the predictions by experimentation

Step 6 leads us back to step 1. When the new experiments are performed to see if the theory is correct, new observations are made. The new information that is gathered is analyzed by the various steps in the scientific method. Thus the body of knowledge that we call science grows.

Let's see how the scientific method works by using the following example.

Why do some things burn whereas others don't? In other words, what is the nature of *combustion*? According to the ancient Greeks, something that could burn contained within itself the element *fire*—which would leave the substance under the proper conditions. In 1687, this notion was described in a more modern form by the German physician and chemist Georg Ernst Stahl (1660–1734), who said that objects which burn contain *phlogiston* (from the Greek word meaning "to set on fire").

According to Stahl, substances that were able to burn were rich in phlogiston. Upon burning, the phlogiston was released into the air, so the material left behind was poor in phlogiston and could not burn.

Stahl used his theory to explain a number of observations. For example, wood, which burned readily, was supposed to be rich in phlogiston. But ash, which was left behind after the wood burned, was supposed to be poor in phlogiston (which of course was the reason why the ash would not burn any further).

Stahl explained the rusting of metals in a similar manner, because he envisioned the process of rusting as being analogous to the burning of wood. According to the phlogiston theory, a metal was rich in phlogiston, whereas its "rust" was not. With this in mind, consider how Stahl explained the *refining process*, in which a *rocky ore* is turned into a pure *metal*. According to Stahl, the rocky ore is poor in phlogiston. The ore is heated with charcoal, which is rich in phlogiston. The phlogiston passes from the charcoal into the ore and in the process the charcoal turns into ash, which is low in phlogiston. Simultaneously, the phlogiston-poor ore is turned into the phlogiston-rich metal.

Stahl considered air to be only passively useful for combustion. The air simply acted as the carrier of the phlogiston as it left the metal or wood to pass on to something else (if that something else were available).

The phlogiston theory seemed to explain the process of combustion very nicely. By 1780 it was almost universally accepted by chemists. It had withstood the test of the scientific method for almost 100 years. Over that 100-year period, scientists were able to use the phlogiston theory to explain observations and make predictions.

But all theories are always subject to new data and new experiments. With the arrival of the nineteenth century, careful measurement became part of every chemical investigation. A problem with the phlogiston theory (a problem that was known to eighteenth-century chemists but never addressed) soon became very apparent to the nineteenth-century chemists as experiments using careful measurement were performed and data were collected. The problem was that when certain objects, such as wood and paper, underwent combustion, the resulting products (the ashes) weighed *less* than the original material. This was to be expected since the phlogiston left the material as the substance burned. But when metals rusted the products formed weighed *more* than the original materials. Yet they too should lose phlogiston and weigh less.

These facts, coupled with the studies of the French chemist Antoine Lavoisier (1743–1794), led to the downfall of the phlogiston theory, and gave rise to a totally new theory of combustion in which the element oxygen had a central role.

According to Lavoisier, if a substance burned it combined with oxygen (that was part of the air). If the combustion of the substance plus oxygen gave rise to other gases, such as carbon dioxide, the original substance would appear to lose weight as it was turned into ash. This was the case in the burning of wood and paper. On the other hand, if the combination of the substance plus oxygen gave rise to a nongaseous product, the original substance would appear to gain weight as the reaction proceeded. Such was the case in the rusting of metals. These experimental results led to the development of Lavoisier's new theory.

The overthrow of the seemingly well entrenched phlogiston theory by the modern theory of combustion was a triumph of the scientific method. This simple episode illustrates what a scientist does. The scientist makes general statements and tests these statements by performing experiments. The general statements lead us to ask "Why?". The answers lead us to develop theories. Theories explain why certain facts of nature are so. If a theory doesn't agree with a particular fact of nature, it must be changed. This is how science works. Using this method, scientists have developed the theories and ideas needed by technologists to put human beings on the moon.

What Is Technology?

Technology produces material objects for use by people. People bring about technology and then consume the products that are developed to sustain life and to make life more comfortable. What motivates technological developments? It is the desire to produce material objects in greater quantity and of better quality.

You might ask how *science* and *technology* differ. Basically, science is *knowledge* and technology is the *application of science in the production of material things*. Some technological ideas are outgrowths of scientific breakthroughs. Some technology has developed throughout history independent of scientific knowledge, simply by trial and error. In many cases the existing technology is the basis upon which technological change occurs. In other cases scientific knowledge is used as a basis for new technological developments. We might say that technology builds on two things: existing technology and science.

Let's look at some examples of the interrelationships of science and technology. Thomas Edison was probably one of the greatest innovators in American history. In 1890, he wrote that his accomplishments arose from "long and patient labor, and were the results of countless experiments, all directed toward some well-defined object," for example, the electric light bulb. In general, his experiments were *not* based on scientific facts.

In 1900, the General Electric Company established a research laboratory to carry out basic scientific research. Irving Langmuir was a scientist working in this laboratory. His research studied the effects produced by introducing different gases into a high-vacuum lamp containing a tungsten filament. He found that when nitrogen gas was introduced into the bulb, the lifetime of the tungsten filament increased. This scientific discovery was the basis for the production of a more modern incandescent light bulb. Here is an example of a technological advance based on scientific knowledge. As a matter of fact, Langmuir won the Nobel Prize in Chemistry for his research.

SCIENCE, TECHNOLOGY, AND THE ARTS AND HUMANITIES

There probably isn't a facet of life today that hasn't been affected by science and technology. Take music, for example. The development of such devices as the Moog synthesizer has given us the ability to produce music electronically. Guitars and other stringed instruments, as well, are commonly connected to electronic amplifiers (Fig. 1-2). Just about everyone has access to stereo

Figure 1-2 Technology has allowed us to produce music electronically. (The Wurlitzer Company, DeKalb, Illinois)

recordings. These are all examples of how technology, based on scientific knowledge, has affected music.

Then think of the field of drama and movie production. Special effects in such movies as "Star Wars," "Close Encounters of the Third Kind," "The Empire Strikes Back," "Clash of the Titans," "Superman," "Star Trek," and "Tron" are accomplished by use of technological advancements in electronics.

Even art has been touched upon by science and technology. It is possible to examine art objects for authenticity by the use of *x-ray technology*. Older paintings use paint that absorbs different amounts of radiation than newer paints. This has made it possible to distinguish original works of art from frauds in many cases.

Use of spectrophotometric analysis (which is the science of using an optical instrument for producing and examining the spectrum of light or radiation from any source) has even helped to verify historical events.

So we see that science and technology are two interrelated fields. The union of these two fields has affected almost all areas of life today.

HUMANKIND AND THE ENVIRONMENT

Through the application of science and technology, humankind, like no other organism, has been able to change the environment drastically. Many of these changes have bettered the condition of humankind. However, some of these changes have also produced adverse effects for all of us living on this planet. We shall look at some specific examples later in this text.

Part 2   Chemistry Through the Ages

SCENARIO

NEW GENES FOR THE SICK

It's May 27, 2030, and Mrs. Jones has given birth to a beautiful baby boy. He is examined and found to have *sickle cell anemia*. This means that he has been affected by an error in his genes. He is admitted to the pediatric wing of the hospital and doctors treat him to correct the genetic error. Years ago this condition could have cost him his life (Fig. 1-3).

Mr. Smith is getting a new heart this morning. Years ago doctors would have worried about his body rejecting the new heart. But today he is given an injection of a blocking antibody which helps protect the transplanted organ

(a)   (b)

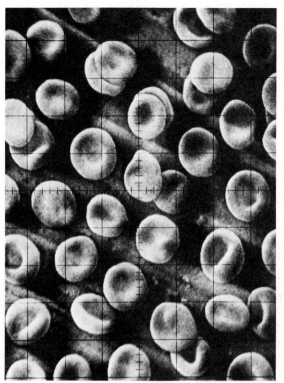 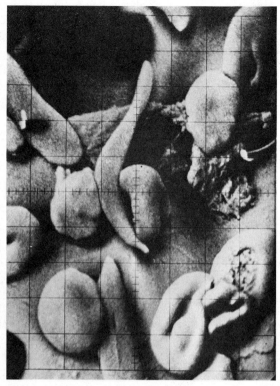

Figure 1-3 Electron micrograph of normal red blood cells (a) and sickle cells (b). The magnification is about 5000. (Philips Electronic Instruments)

from attack by Mr. Smith's immune system. His body accepts the new heart well.

Last winter we were endangered by a flu epidemic, caused by a new strain of flu virus, which could have taken millions of lives. However, millions of people received *passive immunization* against the new virus by means of wholesale production of flu vaccine, and millions of lives were saved.

Cancer is no longer a disease that threatens us. There are injections of specific antigens that immunize us against the formation of cancerous tumors. Medical science is also able to regulate cell growth, which, when uncontrolled, led to cancer.

This scenario deals with the benefits of recombinant DNA research. The early 1970's saw the emergence of a powerful new biological tool. Molecular biologists learned how to remove small bits of genetic material, called DNA, from various organisms. The DNA was then inserted into bacteria in such a way that the new DNA became part of the genetic material of the bacteria. As the bacteria reproduced, the new genetic information was also reproduced. With this technology specific genes could be isolated and the basic mechanism of genetics in all organisms could be studied. Today there is a debate going on as to whether DNA should be recombined. There are potential hazards to this type of research which we'll discuss later in more detail. However, our scenario takes us into the future, where we assume that such research has been successfully carried out to achieve the episodes we discussed.

In this section of the chapter we look at chemistry through the ages and see what's going on today.

OUR BEGINNINGS

In Part 1 of this chapter we looked at the role of science and technology in society, but how does chemistry fit into this picture?

From our early beginnings, humankind has always felt a need to know. We were constantly exploring the environment around us and proposing answers for the many natural but seemingly mysterious phenomena.

It is with this attitude that the early Egyptian and Greek civilizations came on the scene. The Egyptians, as long ago as perhaps 1000 B.C., were practicing a sort of mystical chemistry called "khemia." It was an applied chemistry of sorts and involved the formation of embalming fluids and other potions. It also involved the mining and refining of metals and the extraction of pigments and juices from various plants. Some of these plant extracts had powerful healing and pain-killing properties. In Egypt it was the priests who practiced "khemia" and much of their time was spent in trying to understand the processes of life and death.

The Greeks, on the other hand, were more interested in the fundamental composition of things. They wanted to know if there was one thing in the universe from which all other things can be derived. By about 600 B.C. the Greek philosopher Thales thought he had the answer—water! By 546 B.C. the Greek philosopher Anaximenes thought he had the answer—air! By 450 B.C. the Greek philosopher Empedocles thought he had the answers—earth, air, fire, and water! Today, we think we have the answers, but we'll hold off on that for a while.

The Chinese and Arabs were also practicing their own form of "khemia." In fact, the Egyptian word "khemia" became the Arabic word "al-kimiya," and today we use the word "alchemy" to define everything that happened in chemistry between A.D. 300 and 1600. The alchemists employed the techniques and the knowledge of both the Greeks and Egyptians.

One of the goals of the alchemists was to find the "elixir of life." This elixir would cure all diseases and prolong life indefinitely. Unfortunately, this goal was never realized. Just think of some of the consequences if it was! Another goal of the alchemists was to turn unworthy metals (metals with little value like lead) into gold. This goal was also never realized. But the work of the alchemists, especially the latter ones (A.D. 500 to 1000) did prove important in setting the stage for modern chemistry and medicine. Many substances that chemists use today were discovered by the alchemists. Indeed, in the year 1000, the alchemist Avicenna published his "Book of the Remedy" and in 1330, the alchemist Bonus published his "Introduction to the Arts of Alchemy." Both of these texts proved valuable to later scientists and may have even helped Van Helmont in 1620 to lay the foundations of chemical physiology.

CHEMISTRY TODAY: WHAT CAN IT DO FOR US?

Chemistry affects just about every area of our daily lives. Not only can it stand on its own merits, but it is important in other disciplines as well. Chemistry applied to the science of biology has brought about many new advances. It has helped us to understand better how living organisms function (Fig. 1-4). Many of our environmental problems, such as cleaning up the air and water and treating our solid wastes, are being solved by the use of chemical knowledge.

(a)

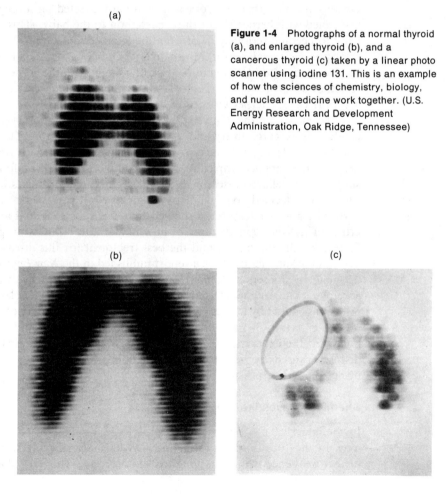

Figure 1-4 Photographs of a normal thyroid (a), and enlarged thyroid (b), and a cancerous thyroid (c) taken by a linear photo scanner using iodine 131. This is an example of how the sciences of chemistry, biology, and nuclear medicine work together. (U.S. Energy Research and Development Administration, Oak Ridge, Tennessee)

(b)

(c)

People are living longer and healthier lives through research, which has improved our standard of living and developed drugs to treat our diseases, and fertilizers and pesticides to help us grow better foods. Let's see how chemistry continues to help us.

Chemistry Against Herpes

A serious epidemic of a new venereal disease is capturing public attention. This new venereal disease is caused by the genital herpes II virus and the common cold sore virus, herpes I. Although herpes I is not venereal, it is capable of causing venereal disease. Herpes II is now more common than syphilis and is found more often in teenagers than in older people. This is because there seems to be some cross-immunity between herpes I and herpes II and the older you are, the more likely you are to have had cold sores and therefore you have built up some immunity to herpes I which helps protect you against herpes II. What's so bad about herpes II anyway?

Women who are infected with herpes II virus have higher rates of spontaneous abortion than do women who are not infected. If a pregnant woman is infected with herpes II, it can be passed on to the baby. This can cause the baby to become seriously ill or even die from the infection. Because of this, some physicians will perform the delivery of a baby from such infected women by Caesarian section, to avoid the risk of contact with the herpes virus and the resultant 50% mortality. There is also a link between cervical cancer and herpes II virus. Herpes virus can't be cured by antibiotics.

Scientists at the University of Pennsylvania School of Medicine are conducting research to try and find a cure for herpes virus. They have found that the simple sugar 2-deoxy-D-glucose is highly effective in treating genital herpes when applied topically. The group of researchers found that this simple sugar relieved discomfort and also cured most patients in early trials. The group decided to try 2-deoxy-D-glucose because work done 20 years earlier by another researcher showed that this substance had antiviral properties. This is a good example of how new research builds on past research. Research will continue to find the best treatment for this disease.

In the summer of 1981, a report published in the *New England Journal of Medicine* by doctors at Johns Hopkins Medical School tells of a drug called acyclovir. The drug was tested on patients who faced a high risk of developing infections caused by herpes virus. The drug was extremely effective in preventing the growth of herpes virus in this particular patient population. Doctors hope that the experimental drug will also prove effective against a variety of diseases caused by herpes viruses, including chicken pox, shingles, mononucleosis, and herpes encephalitis.

Chemistry Saves the Norwegian Spruce

In the late 1960's strong winds toppled large numbers of Norwegian spruce trees in the forests of Norway and Sweden. Shortly afterward a drought further weakened many surviving trees in Norway. This caused a tremendous rise in the *Ips beetle* population that became noticeable in the 1970's. These beetles, also called *bark beetles*, locate Norwegian spruce trees. They use these trees to mate and propagate (and in so doing destroy the trees). In 1979 the *Ips* beetles destroyed 5 million trees in Norway, which represented a loss of more than $20 million in the monetary value of the wood. About 1 million trees were destroyed in Sweden. To avert further heavy damage to the spruce trees, the governments of Sweden and Norway joined forces to cut down the number of beetles so that there would be just enough of them to maintain their proper balance in the ecosystem. How did they do this?

Pheromones were used to trap the beetles. Pheromones, more commonly known as sex attractants, are chemicals which provide a means of communication between insects. Let's see how pheromones help beetles communicate. The *Ips* beetles live under the soil during the winter and scouts emerge in early summer to look for trees that will serve as breeding places. These beetles then emit chemicals called pheromones which attract the male population to the tree. The males then make tunnels behind the bark in preparation for mating. When everything is ready, the males emit the pheromone once again. This time the pheromone attracts the females to the trees, where mating occurs. The eggs hatch, the larvae grow and develop behind the bark, and the new generation leaves the trees and burrows in the soil, as did the generation before.

Robert M. Silverstein of the State University of New York at Syracuse and J. Otto Rodin of SRI International of California identified the components of the pheromone of one type of beetle in 1966. David L. Wood of University of California at Berkeley showed that a mixture of these components were effective in attracting the *Ips* beetle. Lars Skattebøl, a chemistry professor at the University of Oslo, identified the components of the pheromone of the particular type of beetle plaguing the Norwegian spruce. He and his associates supplied Alf Bakke, head of the entomology department at the Norwegian Forest Research Institute, with the synthetic (human-made) pheromone and along with the help of Borregaard Company of Sarpsborg, Norway, traps were constructed and enough of the pheromone was synthesized to limit the Ips beetle population. Research is still going on to determine if the pheromone-baited traps are the best method of beetle population control. Concern for maintaining the proper balance of organisms in the ecosystem was exercised throughout when a solution to this problem was sought.

Chemicals in Our Foods:
Can They Produce Hyperactive Children?

Did you ever read the label on some of your favorite foods? You might see a list of ingredients that seems to be a mile long. You might be afraid to put these foods in your stomach. You may well be right in exercising caution. Some food additives have been causing great concern among food scientists lately. There is even thought that food additives can cause hyperactivity in children. In 1973, Ben Feingold, a California pediatric allergist, proposed that salicylates, which are naturally occurring substances found in many foods, such as fruits and vegetables, as well as artificial colors and artificial flavors are the cause of hyperactivity. Scientific tests have *not* been able to relate diet to hyperactivity in children. Nevertheless, many parents have reported a tremendous improvement in their childrens' behavior when all manufactured baked goods, luncheon meats, ice cream, powdered puddings, candies, soft drinks, punch, teas, coffee, margarine, and condiments are excluded from the diet, as suggested by Feingold. Such nonfood items as mouthwash, toothpaste, cough drops, perfumes, and some over-the-counter drugs are also prohibited on the Feingold diet. Experiments will continue to see if, indeed, there is some scientific basis or validity to Feingold's claims.

Chemistry Improves Wine

Most U.S. wines, with the exception of those from California, are highly acidic because they contain a strong substance called malic acid. Robert Beelman of Pennsylvania State University has developed a new strain of bacteria, called PSU-1, which may give winemakers a way to reduce the acidity of U.S. wines. The bacterium, which can survive and reproduce quickly in highly acidic grape juice, will convert the strong malic acid into weaker lactic acid. This will reduce the acidity of the wine and retain its quality. So you can see that scientists are also trying to help "the little old winemakers."

SCIENCE AND CHEMISTRY: DO THEY NEED A NEW DIRECTION?

So far we've taken a look at some of the ways in which the branch of science called chemistry functions. There are those who feel that science is too involved in destructive ends such as defense and arms, and not involved enough in fulfilling human needs. Some feel that science should play a bigger role in improving health, in providing more food and better shelter for more people. One such group is called the *Club of Rome*, which consists of a group of academic thinkers and industrialists who have an eye toward predicting the future. In their reports they characteristically forecast that there is big trouble ahead. Their seventh report, called "The Human Gap," says that although things are getting worse, they could improve if we shifted our educational system from a process of "maintenance learning" to one of more "innovative learning." The authors of the report say that for them "learning means an approach, both to knowledge and to life that emphasizes human initiative. It encompasses the acquisition and practice of new methodologies, new skills, new attitudes and new values necessary to live in a world of change." They feel that we aren't learning at the "levels, intensities, and speeds which are needed to cope with life today." They say that we can solve the problem by redirecting the workings of science and technology in areas where they are "needed most—in health, food, shelter and education." Science should be "redirected in an innovative way toward human needs," the authors say.

In closing the authors ask: "Will humanity be taught by shocks, whose lessons entail prohibitive costs, or will people learn how to shape those events which, with intelligence and will power, could be controlled?"

SUMMARY

The purpose of this chapter was to set the stage for you as you begin your study of chemistry in our changing world. You should now understand how science operates and be able to explain the differences between science and technology.

We also surveyed chemistry through the ages. We showed the relationship between chemistry and other sciences, and we looked into history and saw the contributions of the Egyptians and Greeks to early chemistry. We also saw how the alchemists contributed to modern science and medicine. Then we jumped to today's society and looked at a few of the projects being undertaken by scientists. We looked at the cases of herpes virus, the Norwegian spruce trees, food additives, and the production of U.S. wines. Finally, we discussed some of the shortcomings of science today in a report by the Club of Rome.

At the end of each chapter you will find a number of self-test exercises containing questions and/or problems that should be answered. In this way you can fully test your knowledge of the chapter.

Good luck to you as you begin this course. We hope that you will find some exciting discoveries and challenges awaiting you.

EXERCISES

1. What is science?

2. What is technology?

3. Distinguish between science and technology.

4. List the six steps of the scientific method.

5. Discuss several areas of your life that have been affected by science and technology.

6. Lavoisier maintained that mass was never created or destroyed in a chemical reaction, but merely shifted from one substance to another. This statement, known as the Law of Conservation of Mass, was based on the results of his combustion experiments. How would you arrange such an experiment to show that this was true?

7. Write your own scenario of the future in the next 100 years.

8. Classify each of the following as being science or technology.
   (a) A doctor performs a kidney transplant.
   (b) A doctor searches for viruses that produce cancer.
   (c) A communications satellite is sent into space.

(d) A television set is developed that is the size of a pack of cigarettes.

(e) The process of nuclear fusion is discovered.

9. Cite at least five examples of how humankind has changed the environment through the use of science and technology.

10. Make a list of the positive aspects of science and technology on our civilization. Do the same thing for the negative aspects. Compare the two lists; which list is longer? Compare your list with your friends' lists. What do you find in terms of the total number of positive aspects versus negative aspects of science and technology on our civilization?

11. Which group of people were involved in the formation of embalming fluids and other potions?

12. What is meant by the term "khemia"?

13. Which group of people were involved in determining the fundamental composition of the universe?

14. What is meant by the term "alchemy"?

15. What are some of the benefits that we reap from chemistry today?

16. Explain why herpes II virus is so dangerous. Discuss how one group of scientists is trying to deal with it.

17. What are your feelings about the idea of limiting the Ips beetle population by the use of pheromones?

18. Discuss your thoughts about limiting chemicals added to foods.

19. Now that you've read about the Club of Rome report "The Human Gap," how do you feel about the role that science is playing in society?

20. Do you feel that scientists should have a greater role in the application of scientific knowledge than they presently do? Do you feel that politicians are sufficiently prepared to make decisions involving the application of scientific discoveries?

21. Write a short scenario showing a danger of recombinant DNA research.

22. According to Empedocles, the elements earth, air, fire, and water compose all matter. If this were true, what elements would compose paper?

23. What was the "elixir of life"?

24. A very well known chemical corporation has stated: "No chemical is safe all the time, but without chemicals, life itself would be impossible." Comment on this statement.

25. If you know a hyperactive child, see if you can find out from the child's parents what the child's diet is. Compare this diet with the list of foods that Feingold says should be avoided by hyperactive children.

Chapter 2

Atoms, Molecules, Elements, and Compounds

Some Things You Should Know After Reading This Chapter

You should be able to:

1. Describe the use of the scientific method and deductive reasoning in explaining the nature of our universe.
2. Classify matter according to Figure 2-4.
3. Define the words heterogeneous, homogeneous, mixture, solution, element, and compound.
4. Define the words atom and molecule.
5. Write the chemical symbol of an element using either the periodic table or an alphabetical listing of the elements, and write the name of the element when given the symbol (for the more common elements).
6. Say what is meant by a relative system of atomic weights.
7. Define what is meant by a mole and give an example.
8. Determine the formula weight of a compound, using the periodic table to look up atomic weights.
9. Determine the number of moles of a compound when given the number of grams of compound and its formula weight.
10. Determine the number of moles of an element when given the number of grams of the element and its atomic weight.
11. Determine the number of grams of a compound when given the number of moles and the formula weight of the compound.
12. Determine the number of grams of an element when given the number of moles and the atomic weight of the element.

Figure 2-1 The philosopher-scientist.

INTRODUCTION

In the previous chapter we reviewed briefly the history of chemistry and science and some of its applications in today's society. Although the ancient Egyptian and Greek civilizations laid the foundations of modern science, the methods of scientists from the mid-1600's on differed greatly from the methods used by the "ancients." For example, the Greek philosophers never bothered to subject their ideas to experiment to see if they were true. In other words, they never bothered to test the many substances that they knew about to see if they were composed of only four elements: earth, air, fire, and water (Fig. 2-1). However, beginning with Robert Boyle (1627–1691), scientists did subject their ideas to experimental proof. They used an approach that came to be called the *scientific method* (which we discussed in Chap. 1). The scientific method uses *deductive reasoning* backed up by *experimental proof* in order to get at the heart of a matter. The famous practitioner, best known for his use of deductive reasoning, was Sherlock Holmes.

SCENARIO

THE AMAZING SHERLOCK HOLMES

The following is an excerpt from Nicholas Meyer's The Seven-Per-Cent Solution.*
Holmes had been tricked by his friend Watson into traveling to Vienna to see Sigmund Freud. Holmes is unaware of the identity of the individual to whom he has just been introduced.

"Who am I and why should your friends be so eager to have us meet?" Holmes eyed him coldly.
"Beyond the fact that you are a brilliant Jewish physician who was born in Hungary and studied for a time in Paris, and that some radical theories of yours have alienated the respectable medical community so that you have severed your connection with various hospitals and branches of the medical fraternity—

---

beyond the fact that you have ceased to practise medicine as a result, I can deduce little. You are married, possess a sense of honour, and enjoy playing cards and reading Shakespeare and a Russian author whose name I am unable to pronounce. I can say little besides that will be of interest to you."

Freud stared at Holmes for a moment in utter shock. Then, suddenly, he broke into a smile—and this came as another surprise to me [Watson], for it was a child-like expression of awe and pleasure.

"But this is wonderful!" he exclaimed.

"Commonplace," was the reply. "I am still awaiting an explanation for this intolerable ruse, if ruse it was. Dr. Watson may tell you that it is very dangerous for me to leave London for any length of time. It generates in the criminal classes an unhealthy excitement when my absence is discovered."

"Still," Freud insisted, smiling with fascination, "I should very much like to know how you guessed the details of my life with such uncanny accuracy."

"I never guess," Holmes corrected smoothly. "It is an appalling habit, destructive to the logical faculty." He rose, and though he tried not to show it, I suspected a thaw was creeping into his replies. Holmes could be as vain as a girl about his gifts, and there was nothing patronizing or insincere in the Viennese doctor's admiration. He now prepared to forget or ignore the danger he supposed he was in, and to enjoy his last moments to the fullest.

"A private study is an ideal place for observing facets of a man's character," he began in a familiar tone, reminiscent of an anatomy professor explicating the intricacies of a skeleton before a class. "That the study belongs to you, exclusively, is evident from the dust. Not even the maid is permitted here, else she would hardly have ventured to let matters come to this pass," and he swept a finger over some nearby bindings, accumulating soot on the tip.

"Go on," Freud requested, clearly delighted.

"Very well. Now when a man is interested in religion and possesses a well-stocked library, he generally keeps all books on such a subject in one place. Yet your editions of the Koran, the King James Bible, the Book of Mormon, and various other works of a similar nature are separate—across the room, in fact—from your handsomely bound copy of the Talmud and a Hebrew Bible. These, therefore, do not enter into your studies merely, but constitute some special importance of their own. And what could that be, save that you are yourself of the Jewish faith? The nine branched candelabrum on your desk confirms my interpretation. It is called a Menorah, is it not?"

"Your studies in France are to be inferred from the great many medical works you possess in French, including a number by someone named Charcot. Medicine is complex enough already and not to be studied in a foreign language for one's private amusement. Then, too, the well-worn appearance of these volumes speaks plainly of the many hours you have spent poring over them. And where else should a German student read French medical texts but in France? It is a longer shot, but the particularly dog-eared appearance of those works of Charcot—whose name seems to have a contemporary ring—makes me venture to suggest that he was your own teacher; either that, or his writing had some special appeal for you, connected with the development of your own ideas. It can be taken for granted," Holmes went on with the same didactic formality, "that only a brilliant mind could penetrate the mysteries of medicine in a foreign tongue, to say nothing of concerning itself with the wide range of subject matter covered by the books in this library."

He walked about the room as if it were a laboratory and nothing more, paying us only the most cursory attention as he continued his lecture.

Freud watched, leaning back with his fingers interlaced across his waistcoat. He was unable to stop smiling.

"That you read Shakespeare is to be deduced from the fact that the book has been replaced upside-down. You can scarcely miss it here amidst the English literature, but the fact that you have not adjusted the volume suggests to my mind that you no doubt intended pulling it out again as in the near future, which leads me to believe that you are fond of reading it. As for the Russian author—"

"Dostoievski," Freud prompted.

"Dostoievski . . . the lack of dust on that volume—also lacking on the Shakespeare, incidentally—proclaims your consistent interest in it. That you are a physician is obvious to me when I glance at your medical degree on the wall. That you no longer practise medicine is evident by your presence here at home in the middle of the day, with no apparent anxiety on your part about a schedule to keep. Your separation from various societies is indicated by those spaces on the wall, clearly meant to display additional certificates. The color of the paint is there somewhat paler in small rectangles, and an outline of dust shows me where they used to be. Now, what can it be that forces a man to remove such testimonials to his success? Why, only that he ceased to affiliate himself with those various societies, hospitals, and so forth. And why should he do this, since once he troubled himself to join them all? It is possible that he became disillusioned with one or two, but not likely that disenchantment with the lot of them set in, and all at once. Therefore I conclude that it is they who became disenchanted with you, doctor, and asked that you resign your membership in each of them. And why should they do this—and in a body, from the look of the wall? You are still living placidly enough in the same city where this has all taken place, so some position you have taken— evidently a professional one—has discredited you in their eyes and they have in response—all of them—asked you to leave. What can this position be? I have no real idea, but your library, as I noted earlier, is evidence of a far-ranging, enquiring, and brilliant mind. Therefore I take the liberty of postu-lating some sort of radical theory, too advanced or too shocking to gain ready acceptance in current medical thinking. Possibly the theory is connected with the work of the M. Charcot who seems to have been such an influence on you. That is not certain. Your marriage is, however. It is plainly blazoned forth on the finger of your left hand, and your Balkanized accent hints at Hungary or Moravia. I do not know that I have omitted anything of importance in my conclusions."

"You said that I possessed a sense of honour," the other reminded him.

"I am hoping that you do," Holmes replied. "I inferred it from the fact that you bothered to remove the plaques and testimonials of those societies which have ceased to recognise you. In the privacy of your own home you might have permitted them to remain and made what discreet capital out of them that you would, and no one the wiser."

*From Nicholas Meyer, *The Seven-Per-Cent Solution*, Ballantine Books, New York. Copyright © 1974 by Nicholas Meyer. Used by permission.

Holmes was a master of deductive reasoning. But we have to take our leave of Holmes, Watson, and Freud at this point and return to the world of chemistry. (However, if you're interested as to what happens next in this Holmes–Freud encounter, pick up a copy of *The Seven-Per-Cent Solution*.)

The use of deductive reasoning played an important part in helping scientists understand our universe and its composition. Let's see what deductive reasoning and experimentation have yielded with respect to our understanding of nature.

MATTER AND ENERGY: THAT'S ALL THERE IS!

Classification has always been associated with scientists. It goes hand in hand with using the scientific method. More important, it allows us to take a great amount of information and arrange it in a convenient form (Fig. 2-2). For example, let's say that we want to classify everything in our universe. That's a pretty tall order! But if we want to answer the question of the Greeks—"what is the one substance of which all other substances are composed"—devising a classification system might help. Let's try to keep it simple and make use of our general knowledge of our environment.

Figure 2-2 Classification—it sure helps! (After a cartoon by Marchant in the January 1973 I²R calendar. Used by permission of Instruments for Research and Industry.)

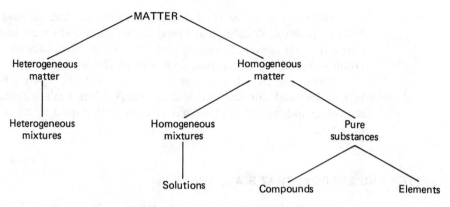

Figure 2-3 The classification of matter.

All that exists in our universe falls into two broad categories: matter and energy. *Matter is anything that occupies space and has mass.* Some examples are clothes, furniture, water, and you. *Energy is often defined as the ability to do work.* It comes in a variety of forms: heat, light, electrical, and chemical energy, to mention a few. We talk more about energy (especially energy shortages) later, but for now, let's turn our attention to matter.

WHAT IS MATTER?

Our classification scheme at this point is quite simple, and based on our common experience. But just to say that the universe is composed of matter and energy is not very helpful. A classification system should be simple, but not to the point where the categories are too general. There are many different types of matter, so we may want to devise a few more categories for our classification system. One method would be to say that matter exists in three states: solid, liquid, and gas. However, a more useful classification system is shown in Figure 2-3.

THE TYPES OF MATTER EXPLAINED

Let's briefly discuss our classification system.

1. *Heterogeneous matter* Heterogeneous means having different parts with different properties.

   Heterogeneous mixtures This is the common type of mixture, such as salt and sand. It is a mixture because it consists of two or more substances "tossed" together. It's a heterogeneous mixture because each part of the mixture retains its own characteristic properties. Another example of a heterogeneous mixture is a tossed salad. (That tomato certainly tastes and looks different from that piece of lettuce!)

2. *Homogeneous matter*   Homogeneous means having similar consistency throughout.

(a) *Solutions*   These are homogeneous mixtures. For example, salt and water mix to form a solution. The saltwater solution is said to be homogeneous because there is uniformity throughout. All parts of the solution have the same properties.

(b) *Elements*   These are the basic building blocks of all matter. This is in keeping with the definition formulated by the Greeks. Remember, they thought that there were only four elements: earth, air, fire, and water. However, today we know of more than 105 elements (and earth, air, fire, and water are not even on the list, since they aren't really elements). What are some of the elements? Examples are gold, silver, oxygen, nitrogen, and calcium. Each element has its own specific properties and has been given a chemical symbol. The elements have been arranged in what is known as the periodic table of the elements; see the inside front cover of the book. There is also an alphabetical listing of the elements on the inside back cover of the book.

Elements may be classified into three major groups: metals, metalloids, and nonmetals. Examples of *metallic elements* are sodium (which has the symbol Na), calcium (Ca), iron (Fe), cobalt (Co), and silver (Ag). The reason that these elements are classified as metals is that they have certain properties, among which are: they have luster (they are shiny), they conduct electricity well, and they conduct heat well.

Some examples of *nonmetals* are chlorine (which has the symbol Cl), oxygen (O), carbon (C), and iodine (I). Nonmetals don't shine, don't conduct electricity well, and don't conduct heat well.

The *metalloids* fall halfway between the metals and nonmetals. Metalloids have some properties which are like those of metals and other properties which are like those of nonmetals. Some examples are arsenic (As), germanium (Ge), and silicon (Si).

(c) *Compounds*   Compounds are substances composed of two or more elements, chemically combined in a definite proportion by mass or weight. Unlike mixtures, which can be physically tossed together with varying composition, compounds have a definite composition. Water, for example, is composed of the elements hydrogen and oxygen in the ratio of 11.1% hydrogen to 88.9% oxygen, by weight. Regardless of the source of the water, it always has the same composition of hydrogen and oxygen.

Compounds can be broken down into their elements by various chemical means. For example, you can use an electric current to decompose water into hydrogen and oxygen. Approximately 4 million compounds have been reported to date, and there are possibilities for millions more

Figure 2-4 Compounds are combinations of elements.

to be discovered. Remember that compounds are combinations of elements, and there are many ways to combine over 100 elements (Fig. 2-4). Perhaps some of the new compounds, yet to be discovered, will hold the cure to some of our most dreaded diseases or will have the property to slow the aging process.

The Atom

Elements may be the basic building blocks of matter, but what—if anything—makes up the elements? In other words, what would be the result of taking an element, a piece of gold, for example, and cutting it in half, and in half again, ad infinitum. We would soon reach the point of having such a small piece of gold that it would be beyond our ability to cut it. It is at times like these when scientists must use their knowledge about how elements react to continue the experiment in their minds. Scientists have done just that and have agreed that if they could continue to cut a piece of gold in half, they would eventually reach a particle called the *atom* (in this case, an atom of gold). *The atom is the smallest part of an element that retains the chemical properties of the element.* One gold atom is so small that billions of them are required to make a tiny speck of gold that can be seen with a microscope. The atom, therefore, is the basic particle which constitutes the elements. Gold is composed of gold atoms, iron of iron atoms, and oxygen of oxygen atoms. Each element is composed of its own type of atoms.

At this point there are usually some skeptics in the room. They ask: "How do you know that atoms exist, if no one has ever seen an atom?" The answer is by indirect evidence. We postulate the existence of atoms based on how elements react to form compounds. Much of what we know about chemi-

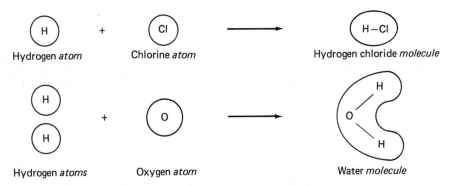

Figure 2-5 Molecules are composed of combinations of atoms. (This figure is not intended to describe a chemical reaction. It shows only that molecules are composed of combinations of atoms.)

cal reactions is easily explained by using the concept of the atom. (However, it's not quite true that no one has ever seen an atom. In a manner of speaking, atoms have been seen. In 1970, Albert Crewe and his staff at the University of Chicago's Enrico Fermi Institute took the first black-and-white pictures of single atoms using a special type of electron microscope. In late 1978, Crewe and his staff reported taking the first color time-lapse moving pictures of individual atoms with the aid of a high-resolution scanning electron microscope. It was uranium atoms that were observed.)

The Molecule

We have just seen what happens when we continually subdivide a sample of an element; we eventually reach an atom. What happens when we continually subdivide a compound? Consider the compound sugar. As we continually divide a sugar grain, we come to a point where further breakdown of the sugar will result in a loss of its physical and chemical properties. This ultimate particle of a compound is called a *molecule*. Molecules, too, are extremely small, but with the aid of an *electron microscope*, molecules of some of the larger and more complex compounds have been observed (Fig. 2-5).

SYMBOLS AND FORMULAS OF ELEMENTS AND COMPOUNDS

Scientists have developed a chemical shorthand to abbreviate the names of the elements. You don't have to write the word *oxygen*; you can just use the symbol *O*. The names of most elements are derived from Latin or English, and their chemical symbols may be the first letter or first two letters of their names. The first letter in the symbol is always a capital letter. The symbol for hydrogen is *H*, the symbol for neon is *Ne*, and the symbol for cobalt is *Co*. (Note this: CO does not represent cobalt but the elements carbon and oxygen, which is a compound, carbon monoxide. We'll discuss compounds in a moment.) Some symbols come from the Latin names of the elements; iron is *Fe* (from the Latin *ferrum*), lead is *Pb* (from the Latin *plumbum*). These chemical symbols extend beyond international boundaries and are used by scientists throughout the world. (Just ask a Soviet scientist, the next time you see one, what the symbol U stands for; see what he or she has to say!) A chemistry article written in Russian would of course be in the Cyrillic alphabet. But the chemical equations would be written in the symbols that we use in the alphabet for the English language.

Since a *chemical compound* is composed of two or more elements, we combine symbols of the particular elements to represent the *chemical formula* of the compound. For example, hydrogen chloride contains one atom of hydrogen and one atom of chlorine. The chemical formula for hydrogen chloride is HCl. Water is another example: the water molecule contains two atoms of hydrogen and one atom of oxygen, so we write H_2O. The number 2 in the formula is called a *subscript*. It indicates the number of atoms of a particular element present in the molecule. A molecule of ethyl alcohol contains two atoms of carbon, one atom of oxygen, and six atoms of hydrogen: C_2H_6O.

ATOMIC WEIGHT

If it were possible to assemble one atom of each of the 105 known elements and compare their weights, what would we find? We would find that hydrogen was the lightest atom and that a helium atom weighed 4 times as much as the hydrogen atom, the oxygen 16 times as much, and the gold atom, 197 times as much. Based on this information we could establish a relative atomic weight scale. We could let hydrogen be one weight unit, and let everything else be a multiple of this unit. The atomic weight of each element is shown above its symbol in the periodic table (see the inside front cover). A method similar to the one just discussed was used by scientists to determine the atomic weights of the elements. [Actually, the atomic weight scale is based on an isotope of the element carbon. This isotope called carbon-12 is assigned a weight of 12 units (sometimes called 12 atomic mass units, or 12 amu for short). If you don't know what an isotope is, see Chap. 3.] The fact that atoms of different elements had different atomic weights played an important part in helping scientists understand the nature of matter. We make use of this atomic weight scale later in this chapter.

GRAM ATOMIC WEIGHT AND THE MOLE

It's all well and good to talk about atoms and relative atomic weights, but scientists work with large amounts of materials, not a few atoms or molecules. What we need is a way to relate the numbers of atoms of an element (or molecules of a compound) to a weight, let's say in grams. Fortunately, there is a way to do this; it's by defining a quantity called the "mole."

What's a *mole*? It is simply a number of items, such as a dozen (12 items) or a gross (144 items). How many items are there in a mole? A huge number.

$$602{,}300{,}000{,}000{,}000{,}000{,}000{,}000$$

items, which can be more conveniently written as 6.023×10^{23}. (This number is written in *scientific notation*, which is discussed in Appendix A at the end of the text.) You can see why we normally wouldn't use the mole when we talk about purchasing eggs, apples, or oranges. But it is a very convenient number for the chemist. Why? Because by definition, *the atomic weight of an element in grams is one mole of that element*. In other words, 1 gram of hydrogen is 1 mole of H atoms (remember that hydrogen has an atomic weight of 1). Sixteen grams of oxygen is 1 mole of O atoms (remember that O has an atomic weight of 16) and 238 g of uranium is 1 mole of U atoms (remember that U has an atomic weight of 238).

The mole concept is very important in chemistry. It gives the chemist a convenient way to describe a large number of atoms or molecules and to relate this number of atoms or molecules to a weight (usually in grams).

We will now learn to convert moles of atoms of an element to its corresponding weight in grams, and vice versa. This will be helpful to us later, and it's really easy. It's like converting inches to feet and feet to inches. Let's try some examples.

Example 2-1   Do the following conversions:

(a) 48 inches = ? feet

(b) 3.0 feet = ? inches

Solution   In this text all math problems will be solved by the *factor-unit method*. This method involves setting up the problem in the following manner.

What I want to know = (*quantity given*)(*factor unit*)

The idea is that when you multiply the "quantity given" by the "factor unit," the proper units cancel to give you the desired quantity. Let's see how this works.

(a) 48 inches = ? feet. In this problem the *factor unit* is 12 inches = 1 foot, which can be written in two ways:

$$\frac{12 \text{ inches}}{1 \text{ foot}} \quad \text{or} \quad \frac{1 \text{ foot}}{12 \text{ inches}}$$

Our job is to choose the proper *factor unit* so that our answer comes out in feet.

What I want to know = (quantity given)(factor unit)

$$? \text{ feet} = (48 \text{ inches}) \left(\frac{1 \text{ foot}}{12 \text{ inches}} \right)$$

$$? \text{ feet} = 4.0 \text{ feet}$$

Notice how the term "inches" cancels to give "feet." If we had chosen the other factor unit, the terms wouldn't have canceled. Let's do it so that you can see for yourself.

$$? \text{ feet} = (48 \text{ inches}) \left(\frac{12 \text{ inches}}{1 \text{ foot}} \right)$$

$$? \text{ feet} = 576 \frac{\text{inches}^2}{\text{foot}}$$

The units don't cancel and we don't get the answer feet!

(b) 3.0 feet = ? inches

What I want to know = (quantity given)(factor unit)

$$? \text{ inches} = (3.0 \text{ feet}) \left(\frac{2 \text{ inches}}{1 \text{ foot}} \right)$$

$$? \text{ inches} = 36 \text{ inches}$$

Notice that in this problem we use the factor unit

$$\frac{12 \text{ inches}}{1 \text{ foot}}$$

in order for the proper terms to cancel.

Now let's apply our knowledge of moles, atomic weight, and the factor-unit method to do some chemical arithmetic. (By the way, throughout this text we will use the rules for significant figures in rounding off our answers. If you are interested in learning these rules, you can read about them in Appendix B of this text.)

Example 2-2   Determine the moles of atoms for each of the following elements.

(a) 36 g of C   (b) 64 g of O

(c) 6.4 g of O   (d) 23.8 g of U

Solution   (a) The atomic weight of C is 12. This gives us the factor unit

$$\frac{12 \text{ g}}{1 \text{ mole}} \text{ or } \frac{1 \text{ mole}}{12 \text{ g}}.$$

What I want to know = (quantity given)(factor unit)

$$? \text{ moles} = (36 \text{ g}) \left(\frac{1 \text{ mole}}{12 \text{ g}} \right)$$

$$? \text{ moles} = 3.0 \text{ moles}$$

$$\left(\text{Do you see why we chose the factor unit } \frac{1 \text{ mole}}{12 \text{ g}}? \right)$$

(b) The atomic weight of O = 16. This gives us the factor unit

$$\frac{16 \text{ g}}{1 \text{ mole}} \text{ or } \frac{1 \text{ mole}}{16 \text{ g}}.$$

What I want to know = (quantity given)(factor unit)

$$? \text{ moles} = (64 \text{ g}) \left(\frac{1 \text{ mole}}{16 \text{ g}} \right)$$

$$? \text{ moles} = 4.0 \text{ moles}$$

(c) The atomic weight of O is 16. The same factor unit applies here as in part (b).

What I want to know = (quantity given)(factor unit)

$$? \text{ moles} = (6.4 \text{ g}) \left(\frac{1 \text{ mole}}{16 \text{ g}} \right)$$

$$? \text{ moles} = 0.40 \text{ mole}$$

(d) The atomic weight of U is 238. This gives us the factor unit

$$\frac{238 \text{ g}}{1 \text{ mole}} \text{ or } \frac{1 \text{ mole}}{238 \text{ g}}.$$

What I want to know = (quantity given)(factor unit)

$$? \text{ moles} = (23.8 \text{ g}) \left(\frac{1 \text{ mole}}{238 \text{ g}} \right)$$

$$? \text{ moles} = 0.100 \text{ mole}$$

Example 2-3   Determine the grams of each element in

(a) 3.0 moles of He   (b) 0.10 mole of C

(c) 2.0 moles of K   (d) 0.100 mole of U

Solution   (a) The atomic weight of He is 4. This gives us the factor unit

$$\frac{4 \text{ g}}{1 \text{ mole}} \text{ or } \frac{1 \text{ mole}}{4 \text{ g}}.$$

What I want to know = (quantity given)(factor unit)

$$? \text{ g} = (3.0 \text{ moles}) \left(\frac{4 \text{ g}}{1 \text{ mole}} \right)$$

$$? \text{ g} = 12 \text{ g} \cdot$$

(b) The atomic weight of C is 12. This gives us the factor unit

$$\frac{12 \text{ g}}{1 \text{ mole}} \quad \text{or} \quad \frac{1 \text{ mole}}{12 \text{ g}}.$$

What I want to know = (quantity given)(factor unit)

$$? \text{ g} = (0.10 \text{ mole}) \left(\frac{12 \text{ g}}{1 \text{ mole}} \right)$$

$$? \text{ g} = 1.2 \text{ g}$$

(c) The atomic weight of K is 39. This gives us the factor unit

$$\frac{39 \text{ g}}{1 \text{ mole}} \quad \text{or} \quad \frac{1 \text{ mole}}{39 \text{ g}}.$$

What I want to know = (quantity given)(factor unit)

$$? \text{ g} = (2.0 \text{ moles}) \left(\frac{39 \text{ g}}{1 \text{ mole}} \right)$$

$$? \text{ g} = 78 \text{ g}$$

(d) The atomic weight of U is 238. This gives us the factor unit

$$\frac{238 \text{ g}}{1 \text{ mole}} \quad \text{or} \quad \frac{1 \text{ mole}}{238 \text{ g}}.$$

What I want to know = (quantity given)(factor unit)

$$? \text{ g} = (0.100 \text{ mole}) \left(\frac{238 \text{ g}}{1 \text{ mole}} \right)$$

$$? \text{ g} = 23.8 \text{ g}$$

THE FORMULA WEIGHT OF A COMPOUND

The formula of water is H_2O, and that for table salt is NaCl. What is the formula weight of these compounds? The formula weight is simply the *sum* of the atomic weights of the elements that compose the compound.

For H_2O, the atomic weight of H = 1, and O = 16.

Formula weight = (2 × atomic wt. of H) + (1 × atomic wt. of O)

Formula weight $= (2 \times 1) + (1 \times 16) = 2 + 16 = 18$

Atoms per
molecule

Atomic
weight of
hydrogen

For NaCl, the atomic weight of Na $= 23$ and Cl $= 35.5$; therefore, the formula weight of NaCl is calculated as follows:

Formula weight $= (1 \times 23) + (1 \times 35.5) = 23 + 35.5 = 58.5$

Example 2-4   Determine the formula weights of the following compounds.

(a) Fe_2O_3 (b) CO_2

(c) Na_3PO_4 (d) $Cu_3(PO_4)_2$

Solution   (a) The atomic weight of Fe $= 56$ and O $= 16$.

$$\text{Formula weight} = (2 \times 56) + (3 \times 16)$$
$$= 112 + 48$$
$$= 160$$

(b) The atomic weight of C $= 12$ and O $= 16$.

$$\text{Formula weight} = (1 \times 12) + (2 \times 16)$$
$$= 12 + 32$$
$$= 44$$

(c) The atomic weight of Na $= 23$, P $= 31$, and O $= 16$.

$$\text{Formula weight} = (3 \times 23) + (1 \times 31) + (4 \times 16)$$
$$= 69 + 31 + 64$$
$$= 164$$

(d) The atomic weight of Cu $= 64$, P $= 31$, and O $= 16$.

$$\text{Formula weight} = (3 \times 64) + (2 \times 31) + (8 \times 16)$$
$$= 192 + 62 + 128$$
$$= 382$$

COMPOUNDS AND MOLES

As you may have guessed, the formula weight of a compound in grams is the weight of 1 mole of molecules. Therefore, 18 g of H_2O (remember that the formula weight of water is 18) is 1 mole of water molecules. And 44 g of CO_2 (remember that the formula weight of carbon dioxide is 44) is 1 mole of carbon dioxide molecules. Let's apply this knowledge to the following problems.

Example 2-5 Determine the moles of molecules in each weight of the following compounds.

(a) 500 g of $CaCO_3$ (b) 13.2 g of $(NH_4)_2SO_4$

Solution (a) First determine the formula weight of $CaCO_3$. Atomic weight of $Ca = 40$, $C = 12$, and $O = 16$.

$$\text{Formula weight} = (1 \times 40) + (1 \times 12) + (3 \times 16)$$
$$= \quad 40 \quad + \quad 12 \quad + \quad 48$$
$$= 100$$

This gives us the factor unit $\dfrac{100 \text{ g}}{1 \text{ mole}}$ or $\dfrac{1 \text{ mole}}{100 \text{ g}}$.

What I want to know = (quantity given)(factor unit)

$$? \text{ moles} = (500 \text{ g})\left(\frac{1 \text{ mole}}{100 \text{ g}}\right)$$

$$? \text{ moles} = 5 \text{ moles}$$

(b) First determine the formula weight of $(NH_4)_2SO_4$. Atomic weight of $N = 14$, $H = 1$, $S = 32$, and $O = 16$.

$$\text{Formula weight} = (2 \times 14) + (8 \times 1) + (1 \times 32) + (4 \times 16)$$
$$= \quad 28 \quad + \quad 8 \quad + \quad 32 \quad + \quad 64$$
$$= 132$$

This gives us the factor unit $\dfrac{132 \text{ g}}{1 \text{ mole}}$ or $\dfrac{1 \text{ mole}}{132 \text{ g}}$.

What I want to know = (quantity given)(factor unit)

$$? \text{ moles} = (13.2 \text{ g})\left(\frac{1 \text{ mole}}{132 \text{ g}}\right)$$

$$? \text{ moles} = 0.100 \text{ mole}$$

Example 2-6 Determine the grams of each compound in

(a) 2.5 moles of $Ca(NO_3)_2$

(b) 0.050 mole of MgO

Solution (a) First determine the formula weight of $Ca(NO_3)_2$. Atomic weight of $Ca = 40$, $N = 14$, and $O = 16$.

$$\text{Formula weight} = (1 \times 40) + (2 \times 14) + (6 \times 16)$$
$$= \quad 40 \quad + \quad 28 \quad + \quad 96$$
$$= 164$$

This gives us the factor unit $\dfrac{164 \text{ g}}{1 \text{ mole}}$ or $\dfrac{1 \text{ mole}}{164 \text{ g}}$.

What I want to know = (quantity given)(factor unit)

$$? \text{ g} = (2.5 \text{ moles})\left(\dfrac{164 \text{ g}}{1 \text{ mole}}\right)$$

$$? \text{ g} = 410 \text{ g}$$

(b) First determine the formula weight of MgO. Atomic weight of Mg = 24 and O = 16.

$$\text{Formula weight} = (1 \times 24) + (1 \times 16)$$
$$= \quad 24 \quad + \quad 16$$
$$= 40$$

This gives us the factor unit $\dfrac{40 \text{ g}}{1 \text{ mole}}$ or $\dfrac{1 \text{ mole}}{40 \text{ g}}$.

What I want to know = (quantity given)(factor unit)

$$? \text{ g} = (0.050 \text{ mole})\left(\dfrac{40 \text{ g}}{1 \text{ mole}}\right)$$

$$? \text{ g} = 2.0 \text{ g}$$

SUMMARY

In this chapter we discussed the scientific method. We talked about how scientists use this method together with deductive reasoning to try and understand our universe. We also studied the different kinds of matter and talked about atoms, molecules, chemical symbols, and moles. What we have tried to do is give you a basic understanding of important chemical principles. We will use these principles in subsequent chapters. Before you proceed, however, try to answer the questions and problems that follow.

EXERCISES

1. Give an example of how the following people might use the scientific method.
   (a) lawyer   (b) physician
   (c) stock market analyst (d) department store manager

2. State whether each of the following is an element, compound, or mixture.
   (a) milk   (b) glass   (c) gold   (d) calcium   (e) wood
   (f) sugar   (g) paper   (h) salt   (i) iron   (j) beach sand

3. Choose the word "atom" or "molecule" for each of the following statements.
   (a) The smallest part of an element that can enter into a chemical reaction is called a(n)_____.
   (b) The smallest part of a compound that can enter into a chemical reaction is called a(n)_____.

4. Look up the symbol and write the name of the first 20 elements in the periodic table.

5. If the periodic table was set up so that calcium had an atomic weight of 1, what would be the weight of
   (a) bromine?   (b) neon?

6. If you had 1 mole of dollar bills and divided it equally among the 4 billion people in the world, how long would it take you to spend your share, if you spent 1 million dollars a day? (Assume no return on your money.)

7. Determine the formula weights of the following compounds. (You may round off the atomic weights in the periodic table to whole numbers.)
   (a) Br_2 (b) OF_2 (c) H_3PO_4
   (d) K_2SO_4 (e) $(NH_4)_3PO_4$ (f) $(NH_4)_2SO_4$

8. Determine the number of grams of each of the following.
   (a) 2.0 moles of Na   (b) 0.20 mole of Na
   (c) 0.50 mole of NO_2 (d) 4.0 moles of H_2

9. Determine the number of moles of each of the following.
   (a) 78 g of K   (b) 7.8 g of K
   (c) 640 g of SO_2 (d) 150 g of $CaCO_3$

10. Cholesterol is a compound suspected of causing hardening of the arteries. The formula for cholesterol is $C_{27}H_{46}O$. Determine the formula weight of cholesterol. How many moles is 3.86 g?

11. Vitamin C may be a cure for colds or perhaps a preventative. The formula for vitamin C is $C_6H_8O_6$. How many moles is 8.8 g of vitamin C?

12. Put yourself in the place of Sherlock Holmes. You have just been given an unknown substance. This substance is either a compound or an element. What might you do to determine whether it is a compound or an element? Be specific.

Chapter 3

Atomic Theory

Some Things You Should Know After Reading This Chapter

You should be able to:

1. Discuss the events that led scientists to believe that atoms were divisible.
2. Give an example of the following terms or define them: electrolysis, electrode, anode, cathode, ion, anion, cation.
3. Reproduce the following table from memory.

PARTICLE	CHARGE	APPROX. MASS (amu)
Proton	+1	1
Electron	−1	Negligible
Neutron	0	1

4. Discuss the historical development of the electron, proton, and neutron.
5. Describe the Thomson model of the atom.
6. Discuss the historical development of the periodic table.
7. Discuss some of the important features of the periodic table.
8. Discuss the important developments that led to the discovery of radiation.
9. Discuss the major differences among alpha, beta, and gamma radiation.
10. Describe the Rutherford gold foil experiment.
11. Define and give an example of atomic number, atomic weight, and isotope.
12. Explain the major features of the Bohr model of the atom.
13. Determine the number of protons, electrons, and neutrons in an atom when given its mass number and atomic number, or vice versa.
14. Explain the difference between a continuous and a line spectrum.
15. Diagram or write the electron configurations for the first 18 elements.
16. Describe the major features of the quantum mechanical model of the atom.

INTRODUCTION

> Matter matter everywhere,
> But what are we to think?
> What it is all composed of,
> Is essentially the link.

In Chapter 2 we talked about atoms. We said that elements were composed of atoms, each being composed of its own type of atoms. The element helium is composed of helium atoms, and the element lithium is composed of lithium atoms. But what, if anything, are atoms composed of? Up until the mid-1800's most scientists thought that atoms were indivisible. They thought that atoms were solid through and through. After all, this is what John Dalton had postulated in his atomic theory of the early 1800's. Dalton's theory assumed that:

1. *All elements are composed of tiny indivisible particles called atoms.* This idea was similar to the idea of Democritus about 460 B.C. Both men felt that there was an ultimate particle of which all matter was composed. They felt that all matter was made up of tiny indestructible and indivisible spheres. Democritus chose the sphere because the Greeks felt that the sphere was the most perfect three-dimensional shape, and because he thought that his atoms should be the perfection of matter.

2. *All matter is composed of combinations of these atoms.* Dalton suggested that the atoms of different elements combine to form molecules of compounds. For example, two atoms of hydrogen and one atom of oxygen combine to form a water molecule.

3. *Atoms of a particular element are identical. Atoms of different elements are different.* Dalton felt that gold was different from silver because somehow the atoms of gold are different from the atoms of silver.

4. *Atoms of the same element have the same size, mass, and form.* Dalton felt that all gold atoms are the same as all other gold atoms in every respect. This is why, if you look at a piece of gold, it appears homogeneous. The same holds true for the atoms of other elements.

Most of what Dalton theorized is still believed today. But his basic idea about the *indivisibility* of the atom was eventually shown to be incorrect, as was his idea that all atoms of the same element have the same size, mass, and form. This led to revisions of his theory. Let's look at the historic events that led to our present atomic theory.

THE DIVISIBLE ATOM: A HISTORICAL REVIEW

Even as Dalton was proposing his theory of the indivisible atom, scientists were performing experiments which indicated that this may not be the case. Much of the evidence was based on the premise that matter was electrical

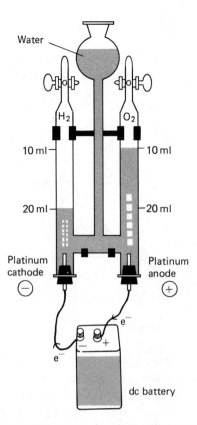

Figure 3-1 Decomposing water into its two elements, hydrogen and oxygen.

in nature. In 1800, Nicholson and Carlisle passed an electric current through water and decomposed it into its two elements, *hydrogen* and *oxygen* (Fig. 3-1). In 1807, the English chemist Sir Humphry Davy (1778–1829) used a powerful battery to pass electric current through various molten salts. Two of the salts that Davy used were molten potash (now known as potassium carbonate) and molten soda (now known as sodium carbonate). In this manner, Davy discovered two elements, *potassium* and *sodium*. Using this technique, Davy also obtained the pure metals of *magnesium, calcium, strontium,* and *barium*.

Michael Faraday (1791–1867), who served as Davy's assistant and protégé, continued his teacher's work. Faraday discovered that many compounds could be decomposed into their elements by the use of electricity. He called this process *electrolysis*, and he named the compound or solution that could carry an electric current, an *electrolyte*. The metal rods inserted into the molten salt (or solution) were called *electrodes*. The electrode carrying the positive charge was called the *anode*, and the one carrying the negative charge, the *cathode* (Fig. 3-2). Faraday hypothesized that the current was carried through the melt or solution by entities called *ions* (Greek for wanderers). The ions that traveled toward the anode were called *anions* and those that

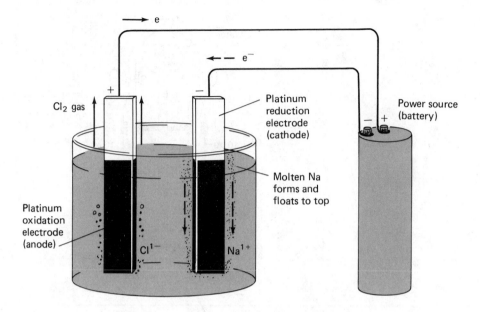

Figure 3-2 Electrolysis of molten sodium chloride (NaCl).

traveled toward the cathode were called *cations*. However, it wasn't until 1884 that the Swedish chemist Svante August Arrhenius (1859–1927) explained Faraday's ions. In his Ph.D. thesis, Arrhenius suggested that Faraday's ions were really simple atoms carrying a positive or negative charge. But it was Faraday's work in the mid-1800's that confirmed the electrical nature of matter.

The Electron

Throughout the 1800's, studies on the electrical nature of matter continued. Scientists attempted to drive electric current through various materials. They also attempted to drive an electric current across a vacuum. Faraday made one attempt, but failed for lack of a good enough vacuum. However, by 1875 the English physicist William Crookes (1832–1919) devised a tube (which came to be called the *Crookes tube*) in which an electric current could be driven across a vacuum. The tube contained two pieces of metal, called electrodes. Each electrode was attached by a wire to the source of an electric current. The source had two terminals, positive and negative. The electrode attached to the positive terminal was called the *anode*; the electrode attached to the negative terminal was called the *cathode*. Crookes showed that when the current was turned on, a beam moved from the cathode to the anode; in other words, the beam moved from the negative to the positive terminal (Fig. 3-3). Therefore, the beam had to be negative in nature.

What was this beam? Was it made of particles or waves? Did it come from the electricity or from the metal electrodes? Physicists in Crookes' time were not sure about the answers to these questions, but they did make guesses.

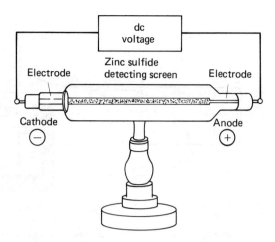

Figure 3-3 A Crookes or cathode ray tube.

Whatever this beam was, it traveled in straight lines (they knew this because it cast sharp shadows). In 1876, Eugen Goldstein (1850–1930), a German physicist, proposed that until it was decided what this beam was composed of, it be thought of as radiation and called *cathode rays*.

The German physicists in Crookes' time favored the wave theory of cathode rays, because the beam traveled in straight lines, like water waves. But the English physicists favored the particle theory. They said that the beam was composed of tiny particles which moved very quickly—so quickly that they were hardly influenced by gravity. That was why the particles moved in a straight path. Notice how one experimental observation led to two different theories.

Crookes proposed a method to solve this dilemma. If the beam was composed of negative particles, a magnet would deflect them. But if the beam was a wave, a magnet would cause almost no deflection. Particles should also be deflected by an electric field. In 1897, the British physicist J. J. Thomson (1856–1940) used both these techniques—magnetic and electric—to show that these rays were composed of particles (Fig. 3-4). A name had to be chosen for these subatomic particles. Back in 1891, six years before Thomson's classic experiment, an Irish physicist, George Johnstone Stoney (1826–1911), suggested that the fundamental unit of electricity, whether it was a particle or something else, be called an *electron*. Now here was this fundamental unit that Thomson's experiment had uncovered, so these particles became known as electrons. Electrons were arbitrarily said to have a unit charge of minus one (-1). In 1906, Thomson won the Nobel Prize in Physics for his amazing work.

In 1911, a young American physicist named Robert Andrews Millikan (1868–1953) determined the mass of an electron and found it to be 9.11×10^{-28} gram. (To get an idea of how small this is, notice that minus sign up there in the exponent, and think of all the zeros we would have to put before the "9" if we wrote the entire number as a decimal.)

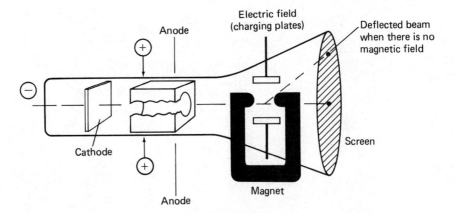

Figure 3-4 The Thomson experiment: (1) A beam of electrons (dashed line) moves from the cathode toward the anodes. (2) Some electrons pass between the anodes. (3) The electric field causes the beam of electrons to bend. This is visible on the screen. (4) But the experimenter can add a magnet to counteract the deflection caused by the electric field and make the beam follow a straight path. (5) One can then measure the strengths of the electric field and the magnetic field and calculate the charge-to-mass ratio (e/m) of the electron.

Next, someone had to prove that the electrons weren't coming from the electricity, but were being given off by the metal electrodes. Proof that metals do give off electrons came from the laboratories of Philipp Lenard (1862–1947), a German physicist. In 1902, he showed that *ultraviolet light* directed onto a metal makes it send out, or emit, electrons. From this evidence it was safe to assume that metal atoms—and the atoms of all other elements—contain electrons.

The Proton

The discovery of the electron and the fact that it is a negative particle of electricity was very important, but it left many unanswered questions. Since electrons are part of all atoms, and since atoms in their normal states are electrically neutral, what else is there in the atom that balances the negative electron? If matter had a negative charge, you would get a shock every time you touched anything. It seems that there must be some other sort of particle that has the same degree of charge as the electron, only positive in nature. This reasoning led scientists to suppose that there were positive particles in the atom.

Actually what happened was that in 1886, Eugen Goldstein, the German physicist who gave cathode rays their name, was experimenting with Crookes tubes that had perforated cathodes. Goldstein noticed that when the cathode rays were given off in one direction (toward the anode), other rays found their way through holes in the cathode and sped off in the opposite direction (Fig. 3-5).

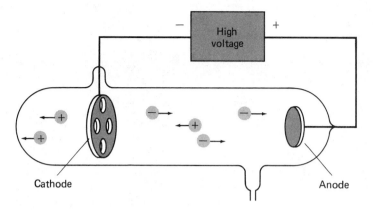

Figure 3-5 Goldstein used a cathode ray tube with a perforated cathode to study positive particles emanating from the tube.

It appeared that these rays must be positive in character. However, it wasn't until 1907 that this hypothesis was confirmed when studies were performed on these rays in a magnetic field. These positive rays, as J. J. Thomson called them in 1907, were different from electrons. All electrons had the same mass, but these positive rays or actually particles, came in different masses (depending on what trace gases were present in the Crookes tube). The lightest positive particle had the mass of a hydrogen atom. It was Ernest Rutherford (1871–1937) who suggested in 1914 that this lightest positive particle be accepted as the unit of positive charge. Further experiments confirmed Rutherford's hypothesis. In 1920, Rutherford suggested the name *proton* for this positive particle. The proton therefore had a unit charge of plus one (+1). In other words, it had the same magnitude of charge as an electron, but opposite to that of the electron. However, in terms of mass the proton was very much different from the electron. One proton weighed 1837 times more than one electron.

The Thomson Model of the Atom

It seemed pretty certain at this point that atoms had internal structure. At least two subatomic particles, protons and electrons, composed atoms. This fact, however, led to another question: *What was the arrangement of protons and electrons in atoms?* It was J. J. Thomson who proposed one solution.

Thomson said that we should think of the atom as a sphere made up of positive electricity in which electrons were embedded (Fig. 3-6); he called this the *plum-pudding theory of the atom*, because he analogized it to plums stuck in a pudding. However, a better analogy for us Americans would be the *chocolate-chip ice cream theory of the atom*. Think of a scoop of chocolate-chip ice cream. The chocolate chips represent the electrons. The ice cream represents a sea of positive electricity (composed of protons). Each element is different because each has a different number of electrons and protons arranged in a

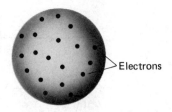

Figure 3-6 The Thomson model of the atom.

different way. It's like different scoops of ice cream with different numbers of chocolate chips. For example, hydrogen atoms have one electron and one proton. Helium atoms have two electrons and two protons. Thomson seemed to have given scientists an idea that they could agree on, but his theory still needed experimental proof.

THE DEVELOPMENT OF THE PERIODIC TABLE

Before we complete our discussion on the nature of the atom, we must first look at an important parallel development that was taking place in the nineteenth century. For you see, while many scientists were investigating the structure of the atom, a number of other scientists were investigating the properties of the various known elements, seeking to find some order among them. Little did anyone realize at the time that these two areas of research would be brought together in the twentieth century to complete a basic understanding of the atom and the nature of matter. Here's what happened.

From the time of the ancient Greeks up to 1700, only 14 elements were known. Then in a short span of 10 years, (1800–1810) 14 more elements were discovered. By 1830, 45 elements were known. Chemists began to wonder how many elements actually existed and if these elements had any relationship to one another. A number of chemists began to investigate the relationships between elements.

The first chemist to notice some order among the elements was the German scientist Johann Döbereiner (1780–1849), who published an account of his findings in 1829. It occurred to Döbereiner that bromine had chemical and physical properties somewhere between those of chlorine and iodine, and that bromine's atomic weight was almost midway between those of chlorine and iodine (Fig. 3-7). Could this be coincidence or was there really some order to the elements? Döbereiner found two more groups of similar elements: the first was comprised of calcium (Ca), strontium (Sr), and barium (Ba); the second, sulfur (S), selenium (Se), and tellurium (Te) (Fig. 3-8). He called these groups *triads* (threes). However, Döbereiner could not find any additional triads from among the known elements, so his results left much doubt in the minds of many chemists.

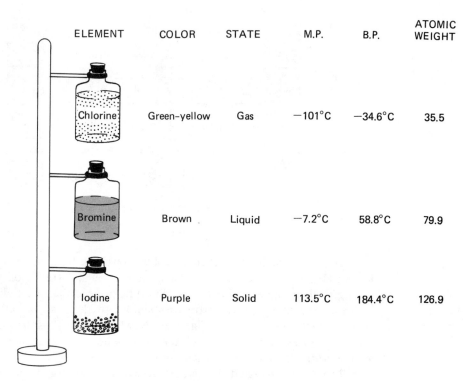

ELEMENT	COLOR	STATE	M.P.	B.P.	ATOMIC WEIGHT
Chlorine	Green–yellow	Gas	−101°C	−34.6°C	35.5
Bromine	Brown	Liquid	−7.2°C	58.8°C	79.9
Iodine	Purple	Solid	113.5°C	184.4°C	126.9

Figure 3-7 A Döbereiner triad.

Figure 3-8 More Döbereiner triads.

	COLOR	STATE	M.P. (°C)	B.P. (°C)	ATOMIC WEIGHT amu
Ca	Silver–white	Solid	842.8	1487	40.08
Sr	Silver–white to pale yellow	Solid	769	1384	87.62
Ba	Yellow–silvery	Solid	725	1140	137.34
S	Yellow	Solid	113	444	32.06
Se	Blue–gray	Solid	217	685	98.96
Te	Silver–white	Solid	452	1390	127.6

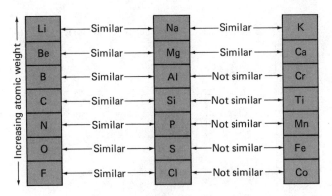

Figure 3-9 Newland's Law of Octaves.

In 1864, the English chemist John Newlands tried to arrange all the known elements in order of increasing atomic weight. He hit on the idea of arranging them in vertical columns. Since he noticed that the eighth element had chemical and physical properties similar to the first, he let the eighth element start a new column (Fig. 3-9). Newlands called his arrangement the *Law of Octaves*. But there were many places in his arrangement where dissimilar elements were next to each other.

In 1869, the Russian chemist Dmitri Mendeleev (pronounced "Men-duh-lay-eff") (1834–1907) arranged elements so that their order depended on the similarity of their chemical properties. Mendeleev established horizontal rows or *periods*. Hydrogen, by itself made up the first period. The next two periods each contained seven elements. The periods after that contained more than seven elements. This is the point at which Mendeleev's table differed from Newlands' table.

In arranging his table, Mendeleev occasionally put heavier elements before lighter ones in order to keep elements with the same chemical properties in the same vertical column (*group*) (Fig. 3-10). Sometimes he left open spaces in the table, where he reasoned that unknown elements would go. (This was a bold prediction in 1869 and many chemists looked at Mendeleev's predictions with great skepticism.) For example, in 1869 the element gallium was unknown; however, Mendeleev predicted its existence and properties. He did this based on the properties of aluminum (which appeared directly above the space left open for this unknown element). Mendeleev even went so far as to predict the melting point, boiling point, and atomic weight of the then-unknown gallium, which he called eka-aluminum. Then, six years later, while analyzing zinc ore, the French chemist Lecoq de Boisbaudran (1838–1912) discovered the element gallium. Its properties were identical to those Mendeleev had predicted six years earlier (Fig. 3-11).

The periodic table of today is similar to Mendeleev's; however, many new elements have been discovered since 1869. Today's table consists of seven horizontal rows called *periods* and a number of vertical columns called *groups*

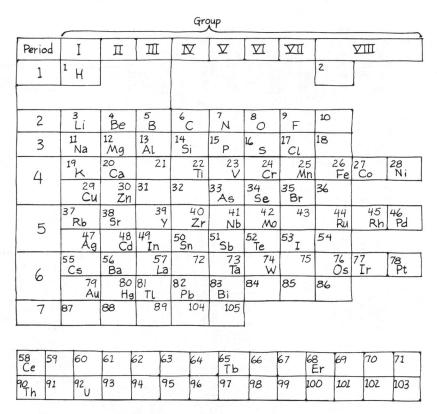

Figure 3-10 Mendeleev's periodic table in modern form.

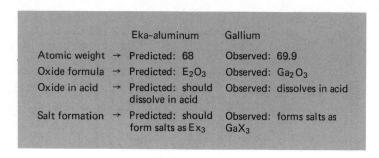

Figure 3-11 The predicted properties of eka-aluminum and the actual properties of gallium.

or (*families*) (Table 3-1). Those elements which have similar chemical properties appear in the same group. The question chemists were asking in the late 1800's was: Why did elements in the same group have similar chemical properties? Could it have something to do with the structure of the atom? Let's get back to our discussion of atomic theory and complete our story.

TABLE 3-1 *Periodic table of the elements*

Group

Period	IA	IIA	IIIB	IVB	VB	VIB	VIIB	VIII	VIII	VIII	IB	IIB	IIIA	IVA	VA	VIA	VIIA	VIIIA
1	1.008 H 1																	4.003 He 2
2	6.941 Li 3	9.012 Be 4											10.81 B 5	12.011 C 6	14.007 N 7	15.999 O 8	18.998 F 9	20.179 Ne 10
3	22.990 Na 11	24.305 Mg 12											26.982 Al 13	28.086 Si 14	30.9738 P 15	32.06 S 16	35.453 Cl 17	39.948 Ar 18
4	39.0983 K 19	40.08 Ca 20	44.956 Sc 21	47.90 Ti 22	50.941 V 23	51.996 Cr 24	54.938 Mn 25	55.847 Fe 26	58.933 Co 27	58.70 Ni 28	63.546 Cu 29	65.38 Zn 30	69.72 Ga 31	72.59 Ge 32	74.922 As 33	78.96 Se 34	79.904 Br 35	83.80 Kr 36
5	85.468 Rb 37	87.62 Sr 38	88.906 Y 39	91.22 Zr 40	92.9064 Nb 41	95.94 Mo 42	(97) Tc 43	101.07 Ru 44	102.906 Rh 45	106.4 Pd 46	107.868 Ag 47	112.41 Cd 48	114.82 In 49	118.69 Sn 50	121.75 Sb 51	127.60 Te 52	126.905 I 53	131.30 Xe 54
6	132.905 Cs 55	137.33 Ba 56	La–Lu 57–71	178.49 Hf 72	180.948 Ta 73	183.85 W 74	186.2 Re 75	190.2 Os 76	192.22 Ir 77	195.09 Pt 78	196.967 Au 79	200.59 Hg 80	204.37 Tl 81	207.2 Pb 82	208.980 Bi 83	(209) Po 84	(210) At 85	(222) Rn 86
7	(223) Fr 87	226.025 Ra 88	Ac–Lr 89–103	(260) [Rf] 104	(260) [Ha] 105	(263) 106												

Transition elements

Lanthanides:

138.906 La 57	140.12 Ce 58	140.908 Pr 59	144.24 Nd 60	(145) Pm 61	150.4 Sm 62	151.96 Eu 63	157.25 Gd 64	158.925 Tb 65	162.50 Dy 66	164.930 Ho 67	167.26 Er 68	168.934 Tm 69	173.04 Yb 70	174.97 Lu 71

Actinides:

(227) Ac 89	232.038 Th 90	231.036 Pa 91	238.029 U 92	237.048 Np 93	(244) Pu 94	(243) Am 95	(247) Cm 96	(247) Bk 97	(251) Cf 98	(254) Es 99	(257) Fm 100	(258) Md 101	(259) No 102	(260) Lr 103

85.468 Rb 37
← Atomic weight
← Atomic number

Metal   Metalloid
Nonmetal   Inert

THE DISCOVERY OF RADIATION

The discovery of radioactivity played an important part in deciphering the mystery of the atom. It all began in 1895 when the German physicist Wilhelm Konrad Röntgen (1845–1923) was investigating the ability of cathode rays to make certain chemicals glow. To do this work he usually worked in a darkened room. On one such occasion while working with his cathode ray tubes Röntgen noticed a flash of light some distance from the tube. The light was coming from a chemically coated paper. The paper glowed only when the cathode rays were in action. Even when Röntgen took the chemically coated paper into the next room, it glowed when the cathode ray tube was turned on. In other words, something given off by the cathode ray tube was able to penetrate through walls and strike the chemically coated paper. Röntgen called these penetrating rays *x rays*. This discovery immediately got the attention of many chemists and physicists, who turned their research efforts in this direction.

One such physicist was Antoine Becquerel (1852–1908). He was working on fluorescent substances, substances that have the ability to glow on their own. (Strictly speaking, fluorescent substances do not have the ability to glow on their own but instead are reemitting radiation that was absorbed a very short time prior to this reemission.) Becquerel wanted to know if fluorescent substances gave off x rays. In 1896, Becquerel took a sample of the fluorescent substance uranium and placed it on a piece of photographic film that was still wrapped in its black paper. Becquerel hypothesized that if the uranium was simply giving off ordinary light, it would not pass through the black paper to expose the film. However, if x rays were present, they would pass through the paper and fog the film. (We should point out that this experiment was done in sunlight, which was needed to make the uranium fluoresce.) Sure enough, the film was exposed. However, even when this experiment was not done in sunlight, the film was exposed. In other words, the uranium crystals were giving off penetrating radiation at all times (Fig. 3-12).

At the same time there was in Paris a young Polish chemist, Marie Curie (1867–1934), who with her husband Pierre was working in the laboratories

Figure 3-12 Uranium is constantly emitting radiation.

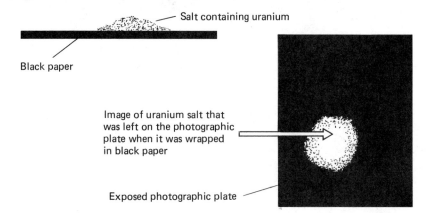

Salt containing uranium

Black paper

Image of uranium salt that was left on the photographic plate when it was wrapped in black paper

Exposed photographic plate

at Sorbonne. The Curies became interested in Becquerel's problem. It was, in fact, Marie Curie who defined the ability of a substance to produce penetrating rays as *radioactivity*. The Curies found that a radioactive substance seems to keep on and on, year after year, emitting these powerful penetrating rays. In 1898, they discovered that the element thorium is radioactive, and so is polonium, and so, most of all, is radium.

It is interesting to note that for their pioneering work, Pierre and Marie Curie, together with Antoine Becquerel, won the 1903 Nobel Prize in Physics. Three years later, Pierre was killed in a traffic accident. His wife continued to work in the field of radioactivity and won the Nobel Prize for Chemistry in 1911. She died in 1934 of pernicious anemia, perhaps caused by the very radioactive chemicals to which she devoted her life's work.

Although early experimenters first thought that radioactive materials produced only x rays, they soon discovered that the situation was more complex. For example, when they allowed the radiation produced by uranium to pass through a magnetic field, they could detect three types of radiation.

The English physicist Ernest Rutherford called these three types of radiation *alpha rays*, *beta rays*, and *gamma rays*, from the first three letters of the Greek alphabet, α, β, and γ (Fig. 3-13). We have more to say about these different types of radiation in Chapter 11, but for now it is important for you to know that alpha rays seemed to have a *positive charge*, whereas beta rays seemed to have a negative charge, and gamma rays seemed to have no charge. Also, it was discovered that certain radioactive elements would give off only one type of radiation. In other words, they would be pure alpha, beta, or gamma emitters.

Figure 3-13 The radiation from uranium can be separated into three types: alpha, beta, and gamma.

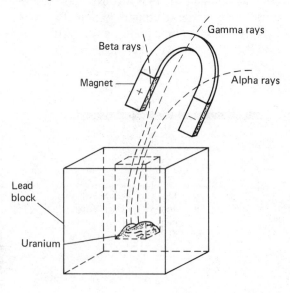

This important discovery allowed experimenters to use radioactive materials as *particle guns*. This is how it worked. The experimenter would take a bit of the radioactive material and place it in a lead-lined box. (Lead absorbs radiation.) However, the lead-lined box would have a small hole in it, so the radiation could leave the box through the hole. In this way an experimenter could have a gunlike device that would shoot a bulletlike stream of radioactive particles at a target. It was Rutherford who made use of such a particle gun in testing the Thomson model of the atom. Let's see what happened.

THE RUTHERFORD MODEL OF THE ATOM: A FIRST FOR THE TWENTIETH CENTURY

Rutherford was a student of Thomson and was interested in testing his teacher's model of the atom. Rutherford set up a particle gun using an element that was a pure alpha emitter. In other words, he had a "gun" that would be able to shoot positive particles at a target. Rutherford reasoned that since Thomson pictured the atom as a *neutral* symmetrical unit—meaning that the positive and negative charges balanced each other—the positive beam would not be scattered as it passed through the atom (Fig. 3-14).

Figure 3-14 According to the Thomson model of the atom, positive particles ought to pass through a metal foil without being deflected.

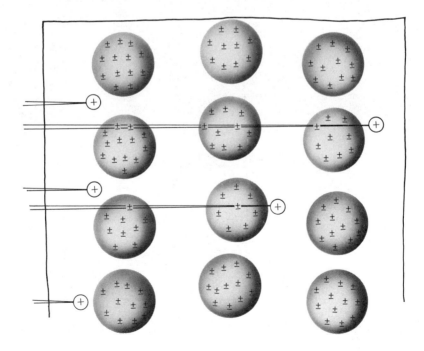

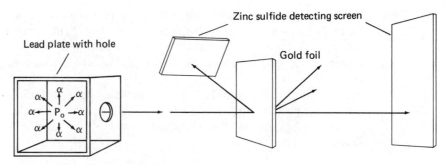

Figure 3-15 Rutherford's gold foil experiment.

Rutherford decided to use the element gold as his target. This was because the gold could be hammered into a target that was only 2000 atoms thick. The experimental apparatus was set up so that as the alpha particles passed through the atoms of gold they (the alpha particles) would be detected by a zinc sulfide screen (Fig. 3-15). Of course, Rutherford expected that the alpha rays would just pass right through the gold atoms, which would prove that the Thomson model of the atom was correct. But instead of getting the expected result, Rutherford found that about one alpha particle out of 20,000 ricocheted or bounced back toward the hole in the lead box. As Rutherford said: "It was

Figure 3-16 The actual way that positive particles pass through metal foil as shown by Rutherford.

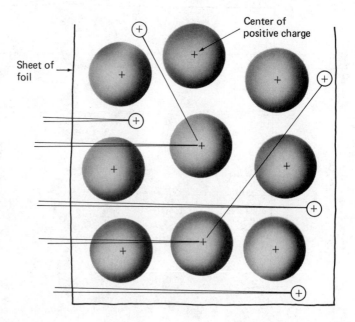

almost as incredible as if you fired a fifteen-inch shell at a piece of tissue paper and it came back and hit you."

To explain the results of his experiment, Rutherford made up a new model of the atom. Let's see what he said and examine his reasoning. Rutherford said that an atom must have a *center of positive charge*, where all the protons must be located. He called this center of positive charge the *nucleus* of the atom. Rutherford felt that the nucleus of the atom must be very small in comparison with the overall size of the atom. He also said that the nucleus must be where the mass of the atom is concentrated. Rutherford's reasoning was that the alpha particles were deflected because they were being repelled by a high concentration of positive charge with an immovable mass (Fig. 3-16). He also decided that the electrons were located around the nucleus instead of in it, and that the mass of the electrons was very small compared with the mass of the protons. His reasoning here was that for 99% of the particles to go through the foil undeflected, the electrons had to be scattered in a relatively large space of the atom outside the nucleus.

Later experiments showed that the diameter of the nucleus of an atom is approximately 10^{-13} cm, whereas the diameter of the whole atom is 100,000 times as great. Rutherford's model of the atom, based on his new experimental evidence, made the Thomson model of the atom obsolete.

THE NEUTRON: ANOTHER SUBATOMIC PARTICLE

Before 1920, scientists knew about two basic subatomic particles: electrons, which are negatively charged and have very small mass, and protons, which are positively charged, and contain most of the atom's mass. But a problem came up. Scientists knew that a hydrogen atom had a relative mass of one unit. We sometimes refer to this as one *atomic mass unit*, abbreviated amu. Scientists also knew that the proton had a mass of 1.0072766 amu, and the electron had a much smaller mass of 0.0005486 amu. These numbers all seemed fine at first— the mass of hydrogen, which is made up of one proton and one electron, was very close to the sum of the masses of a proton and an electron. (Remember that protons account for just about all of the atom's mass; when we are adding up the mass of an atom we can usually ignore the mass of the electrons.) The problem arose when scientists considered helium. The helium atom has two electrons and two protons, so the scientists thought it should have a mass of about 2 amu. But, in fact, its mass is 4 amu. And when they looked at other elements, more puzzles appeared. Carbon has six electrons and six protons, but a mass of 12 amu. What was the answer to this problem? Where was the missing mass?

To account for the extra mass in the atom, another particle was proposed, one that had no charge, but a mass equal to that of the proton. In 1932, a British physicist, James Chadwick (1891–1974), was able to prove the existence

TABLE 3-2   *The particles in the atom*

PARTICLE	SYMBOL	CHARGE	MASS (amu)
Proton	p	+1	1.0072766
Electron	e	−1	0.0005486
Neutron	n	0	1.008665

of a new subatomic particle, called the *neutron*, with a mass of 1.008665 amu. If this particle were also present in the nucleus, it would add to the mass but not affect the charge. This helped to clarify matters (Table 3-2). Scientists could now picture helium as being composed of two electrons, two protons, and two neutrons, so that they could understand helium's mass of 4 amu.

ATOMIC NUMBER

The atoms of different elements have different numbers of electrons and protons. This is, of course, what makes the atoms of one element different from another element, just as Thomson had suspected. The number of electrons or protons in a *neutral* atom is called the *atomic number* of that element. If we look at the periodic table inside the front cover, we see the atomic number below the symbol of each element. An atom of sodium (Na), which has the atomic number 11, has 11 electrons and 11 protons (Fig. 3-17). An atom of magnesium (Mg), which has the atomic number 12, has 12 electrons and 12 protons; and an atom of uranium (U), which has the atomic number 92, has 92 electrons and 92 protons. We can see that the elements in the periodic table are arranged in order of their atomic numbers. In other words, Mendeleev, without realizing it, had arranged his periodic table in order of increasing *atomic number*.

Figure 3-17   In this text the atomic number of an element appears beneath its symbol.

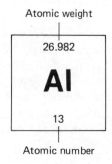

Atomic weight

26.982

Al

13

Atomic number

ISOTOPES

If you look at the periodic table you'll see that the atomic weights of most elements are not whole numbers; for example, sodium (Na) has a weight of 22.990, titanium (Ti) a weight of 47.90, and rubidium (Rb) a weight of 85.468. Why is this so? If the protons and neutrons each give a mass of 1 amu to the atom, the atomic weights of the elements certainly ought to be whole numbers, or very close to whole numbers.

The answer is that most elements exist in more than one *form*. When John Dalton said that the atoms of an element must be identical in size, mass, and shape, he did not know that *isotopes* of the elements existed. What are isotopes?

Isotopes are atoms of an element that have the same number of electrons and protons, but different numbers of neutrons.

From this definition you can see that, although the negative and positive charges of isotopes are identical, their masses are not. Hydrogen, for example, exists in three isotopic forms (Fig. 3-18). The common form of hydrogen (called *protium*) contains one electron and one proton. Another form of hydrogen (called *deuterium*) contains one electron, one proton, and one neutron. A third form of hydrogen (called *tritium*) contains one electron, one proton, and two neutrons.

Sometimes we want to represent an isotope of an element not only by its symbol, but also by its mass number and atomic number. (The mass number is the mass of the isotope. It is the sum of the protons plus neutrons.) A standard notation has been developed for doing this. It consists of writing the symbol of the element, then writing the mass number of the isotope as a *superscript* (above the line) and its atomic number as a *subscript* (below the line).

$$^{16}_{8}O \qquad ^{12}_{6}C \qquad ^{35}_{17}Cl$$

⌐ Mass number
⌐ Atomic number

Figure 3-18 The three isotopes of hydrogen.

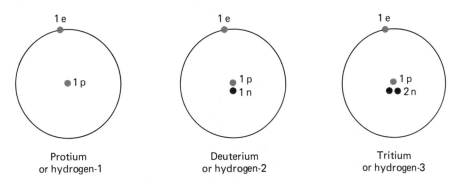

Protium or hydrogen-1	Deuterium or hydrogen-2	Tritium or hydrogen-3

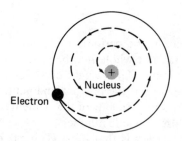

Figure 3-19 What would happen if electrons followed classical physical laws and fell into the nucleus?

With this notation, we can immediately tell the composition of the isotope in terms of protons, electrons, and neutrons. The number of protons (or, in the neutral atom, electrons) is simply the atomic number of the element. The number of neutrons is the difference between the mass number and the atomic number. For example, $^{235}_{92}U$ has 92 protons, 92 electrons, and 143 neutrons $(235 - 92 = 143)$. We have more to say about the isotopes of elements in Chapter 11.

THE BOHR MODEL OF THE ATOM

The scientists of the early 1900's gradually got a mental picture of the atom as having protons and neutrons in the center (the nucleus) and electrons in the outer area. But physicists of the day were bothered to some extent, because the only way they could imagine the electrons remaining outside the nucleus of the atom was to assume that they were in constant motion around the nucleus. But if this were so, why didn't the electrons lose energy as they traveled around the nucleus and get pulled into the nucleus? (Remember, opposite charges attract.) If this did occur, atoms would destroy themselves. However, this does not occur (Fig. 3-19). Why not? An answer was suggested by the Danish physicist Niels Bohr (1885–1962). But before we can review his work, some background information is necessary.

SPECTRA

In the mid-1600's, the English genius Sir Isaac Newton allowed a beam of sunlight to pass through a glass prism, and found that it separated into a whole series of different colors, which we call the *visible spectrum* (Fig. 3-20). A rainbow is pretty much the same sort of thing.

Later experiments showed that a specific amount of energy could be associated with each color of light. In other words, light is a form of energy. But this should come as no surprise to us. Just think of what the sun's light

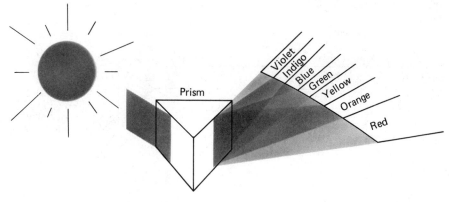

Figure 3-20 When sunlight passes through a prism it is broken up into its component colors.

rays can do to us if we sunbathe too long! The visible spectrum produced by sunlight is only a small part of what is called the *electromagnetic spectrum* (Fig. 3-21).

In the 1850's, the German physicist Robert Kirchhoff (1824–1887) and the German chemist Robert Bunsen (1811–1899) developed an instrument called the *spectroscope*. Its main parts were a heat source (the Bunsen burner),

Figure 3-21 The electromagnetic spectrum.

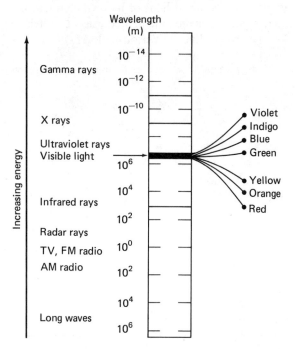

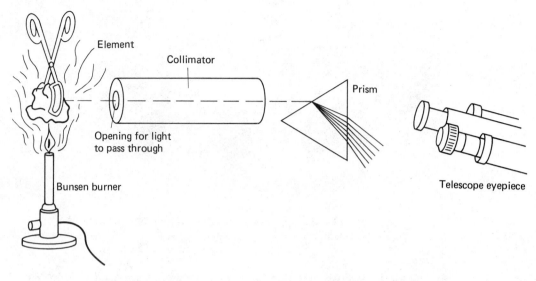

Figure 3-22 A spectroscope: The element is heated in the flame of the Bunsen burner. Some of the light from the heated element passes through the collimator, which is simply a brass tube with a narrow slit at both ends. The collimator directs a fine beam of light onto the prism. When the light passes through the prism, it is broken into its different parts—a spectrum—and observers can look at the spectrum through the telescope.

a prism, and a telescope (Fig. 3-22). Kirchhoff and Bunsen heated samples of various elements. In each case, as an element got hotter and hotter and began to glow, it produced light of its own characteristic color. If they allowed the light from the heated element to pass through the prism, it separated into a series of bright lines of various colors.

Figure 3-23 A continuous spectrum and a line spectrum produced by hydrogen.

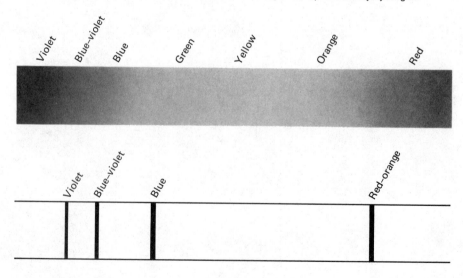

The spectrum produced by sunlight is called a *continuous spectrum*. This means that one color merges into the next. The spectrum found by Bunsen and Kirchhoff is called a *line* spectrum (Fig. 3-23), because it is a series of bright lines separated by dark bands. Each element produces its own kind of line spectrum.

Now how do we use this information in a practical way? Scientists found these line spectra very useful for identifying elements in unknown samples. For example, chemists have discovered that Napoleon died of arsenic poisoning on St. Helena, because they analyzed the line spectra of samples of his hair—in this century, 150 years after his death—and found proof that the poison was given to him on a fairly regular basis. Line spectra were also useful when scientists were studying the sun and other extraterrestrial bodies. For example, the study of the line spectra of the sun produced the first identification of the element helium. But until 1913 no one realized that these line spectra held the key to the structure of the atom.

LIGHT AND ENERGY

As we mentioned before, light is a form of energy. Most objects, when you put them in sunlight or in the light from some other type of source, begin to get warm. The reason is that they absorb heat energy. But some objects react differently than others. Consider watches with luminescent dials. To get the watch dial to glow, you first have to expose the watch to a strong light source and then turn off the light. The watch dial glows. What happened? When you exposed the dial to the light source, it absorbed energy from the light. After you took away the light source, the dial emitted this energy in the form of light.

Think what this experiment tells us about the line spectra of the elements. Each element, when heated, absorbs energy. After the element has heated up, it emits this energy in the form of light, and this is what causes the spectrum.

After scientists became aware of line spectra, they began to ask themselves questions. Why did elements have line spectra and not continuous spectra? What subatomic particles (electrons, protons, or neutrons) caused these line spectra?

THE BOHR THEORY

In 1913, Niels Bohr suggested answers to these puzzling questions. He said that electrons were responsible for line spectra. He proposed the ideas of shells, or imaginary spherical surfaces centered about the nucleus; each shell was associated with a particular *energy level*. According to Bohr, electrons traveled around the nucleus of the atom in specific shells, in the same way that planets revolve around the sun. But electrons, unlike planets, are able to jump from one shell (at one energy level) to another (at a higher energy level) if they are given enough energy from an outside source to do it (Fig. 3-24). They can also fall back to their original shells by emitting this energy.

57 *Atomic Theory*

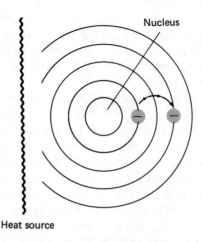

Heat source

Figure 3-24 Electrons in atoms can "jump" from one shell to another if an outside source gives them enough energy to do so.

Figure 3-25 A hydrogen atom in its ground state.

Let's look at the hydrogen atom (Fig. 3-25). Here we have one electron in a shell moving around the nucleus. (This shell is the one closest to the nucleus.) This is called the *ground state* of the hydrogen atom. Niels Bohr postulated that as long as the electron remains in this ground state, no energy is lost or gained. The negative electron doesn't fall into the positive nucleus because it has enough energy of motion to counterbalance the attraction of the nucleus; and as long as there is no outside influence, it never loses energy. Bohr was assuming that electrons do not have to obey classical physical laws. Now we have the answer to the question that we posed several pages ago: Why is it that negative electrons don't spiral into the positive nucleus? Because these very small particles do not obey classical physical laws.

Suppose that we heat a hydrogen atom. According to Bohr, the electron, when it absorbs heat energy, jumps to a higher shell (a higher energy level) farther away from the nucleus (Fig. 3-26). If we supply still more energy to the atom, the electron might jump into a still higher shell (a still higher energy level). There are a number of possible shells into which the electron can jump. When an electron is at an energy level higher than its ground state, we say that the atom is in an *excited state*. But if the atom is given too much energy, the electron can jump completely out of it (this process is called *ionization*).

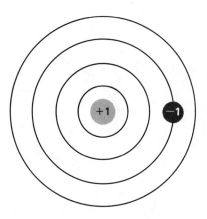

Figure 3-26 An excited hydrogen atom.

What eventually happens to the atom in its excited state? It falls back to its ground state again, and emits exactly the same amount of energy it absorbed to get to the excited state in the first place. We should point out that when the electron moves from one energy level to the other, it moves instantaneously; in other words, there is no spiraling. This is what causes the line spectra. If electrons spiraled from one energy level to the other, they would emit a continuous stream of energy and light. However, since they jump or hop back and forth, they emit energy in bursts, and the separate bursts show up as line spectra.

It's hard to think of electrons moving instantaneously from one energy level to another. To help you visualize it, think of a marble rolling down a flight of stairs. The marble can rest on a step, but it can never pause or stop between steps. As the marble falls (jumps) down the steps, it also loses (emits) energy in bursts.

Energy Levels

The energy levels in atoms have been given letter names (and numbers) (Fig. 3-27). The farther away the level is from the nucleus, the higher the energy of an electron in that level. The K or first energy level has electrons with the lowest energy, and the Q or seventh energy level has electrons with the highest energy. Each energy level can hold only a certain number of electrons at any one time. The theoretical *maximum number* of electrons for each level (or shell) is given in Table 3-3. [By the way, an easy way to find the maximum electron number for an energy level is to use the formula

$$\text{maximum number of electrons} = 2(n)^2$$

where n is the number of the energy level. For example, $n = 1$ for the K level and $n = 2$ for the L level, and so on. Therefore the maximum number of electrons for the K level is $2(1)^2 = 2$, and the maximum number of electrons allowed in the L level is $2(2)^2 = 8$.]

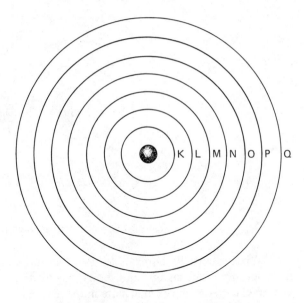

Figure 3-27 Names of the energy levels in atoms.

TABLE 3-3 *Theoretical maximum number of electrons for each energy level*

ENERGY LEVEL	MAXIMUM NUMBER OF ELECTRONS
K	2
L	8
M	18
N	32
O	50
P	72
Q	98

Ground-State Electron Configurations

When we use the Bohr model of the atom we can picture the internal structure of the atom more clearly than we can with earlier models. Let's look at the electron configuration of the atoms of various elements in their ground states (Fig. 3-28). Electron configuration refers to the way the electrons fill the various energy levels of the atom. For example, hydrogen has its one electron in the *K* level. This is expected, since the *K* or first level has the lowest energy. Both of helium's two electrons are also in the *K* energy level. Lithium, however, has three electrons, so two of them go into the *K* energy level and one into the *L* or second energy level. (Remember, the *K* level can only hold two electrons.) If we look at Table 3-4 we can see that—up to element 18—the electrons fill

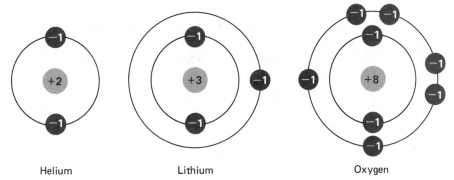

Helium · Lithium · Oxygen

Figure 3-28 Some ground-state electron configurations.

TABLE 3-4 *Ground-State electron configurations of the first 21 elements*

ELEMENT	ATOMIC NUMBER	ELECTRON CONFIGURATION			
		K	L	M	N
Hydrogen	1	1			
Helium	2	2			
Lithium	3	2	1		
Beryllium	4	2	2		
Boron	5	2	3		
Carbon	6	2	4		
Nitrogen	7	2	5		
Oxygen	8	2	6		
Fluorine	9	2	7		
Neon	10	2	8		
Sodium	11	2	8	1	
Magnesium	12	2	8	2	
Aluminum	13	2	8	3	
Silicon	14	2	8	4	
Phosphorus	15	2	8	5	
Sulfur	16	2	8	6	
Chlorine	17	2	8	7	
Argon	18	2	8	8	
Potassium	19	2	8	8	1
Calcium	20	2	8	8	2
Scandium	21	2	8	9	2

the shells or levels in the order that we might predict. Each shell is filled before any electrons are added to the next. But when we write the electron configuration of potassium, other factors come into play; the last electron in potassium enters the N or fourth energy level, even though the M or third energy level is not filled to capacity. This phenomenon gives rise to a concept known as the octet rule.

The Octet Rule

An atom whose highest energy level has eight electrons is very stable. (An exception to this is helium, which is extremely stable even though it has only two electrons. The reason is that a *completed* energy level makes for a very stable element.) Experiments with neon and argon—each of which has eight electrons in its outermost shell—show them to be *inert*; that is, they don't react with other elements (*at least, not in the usual sense, although some of these inert elements have been made to react under what might be termed unusual conditions*). It seems that once an atom has eight electrons in an energy level, it tends to start filling the next-higher level. This tendency is called the *octet rule*. Although the octet rule is not obeyed in all instances, it does help to explain the electron configuration of many elements.

The Importance of Electron Configuration

When we know the electron configuration of elements, we can predict how elements will react with each other. It is the number of electrons in the outermost shell that controls the chemical properties of an element. Elements with similar *outer electron configurations* behave in very similar ways. For instance, Ne, Ar, Kr, Xe, and Rn tend to be very unreactive. They all have eight electrons in their outermost energy levels. On the other hand, the elements Li, Na, K, Rb, Cs, and Fr all react violently with water; they all have one electron in their outermost energy levels.

Now we know why Mendeleev's scheme for the periodic table worked so well. Without realizing it, Mendeleev had placed elements with the same number of outermost electrons in the same *vertical* columns of his table. So the Bohr model of the atom accounts for the similarity of elements in a chemical *family*.

The electron configuration of the elements also helps us to predict how some elements will bond to each other. We discuss this in Chapter 4.

THE MODERN QUANTUM MECHANICAL VIEW OF THE ATOM

In the Bohr model of the atom we looked at electrons as if they were in definite energy levels, which had discretely defined orbital paths. For most of our work in this course, this picture of the atom is the one that best suits our needs.

However, a more modern theory of the atom, the *quantum mechanical model*, has developed since Bohr's day. This is a mathematical model of the atom, according to which we are asked *not* to think of electrons as being in discretely defined orbits, but instead as occupying a volume of space. This volume of space is called an orbital. The orbital represents a region of space in which the electron can be found with 95% probability. Each orbital can hold a maximum of two electrons within its space, which can be shown to have a specific shape. The shape of the *s* orbital is a sphere in which a maximum of two electrons may be found. Figure 3-29 shows the relationship of the 1*s*, 2*s*, and 3*s* orbitals to each other. All *s* orbitals have the same spherical shape; however, they occupy different regions of space around the nucleus of the atom.

The *p* orbitals are dumbbell shaped. There are three *p* orbitals, p_x, p_y, and p_z, which correspond to the three axes in space. Each of these orbitals can hold a maximum of two electrons (Fig. 3-30).

To clarify the idea about an electron being found in a specific orbital with a probability of 95%, consider the following analogy. A bank teller, who we'll call George, works at the First National Bank between the hours of 10:00 A.M. and 3:00 P.M. The probability (or chance) of finding George in the bank between 10:00 A.M. and 3:00 P.M. on a given day is extremely high, let's say 95%. (It's not 100% because there's always a chance that George may be sick on a given day, or he might have taken the day off.) George works at window 3 in the bank. Therefore, the best chance of finding George is not just to look in the bank, but to look specifically at window 3 in the bank.

Let's say that we make a map of the inside of the bank, and mark on it George's position every 10 minutes during the course of the day. For most of the day, we would find him at window 3. This is the region of highest probability. This is where most of our marks would be on the map. However, there would also be marks on the map that would show George in other parts of the bank. Perhaps he took a coffee break, or left his window to get some

Figure 3-29 The relationship of the 1*s*, 2*s*, and 3*s* orbitals to each other.

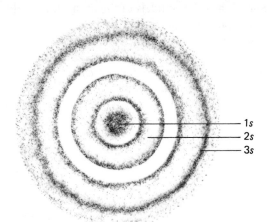

1*s*
2*s*
3*s*

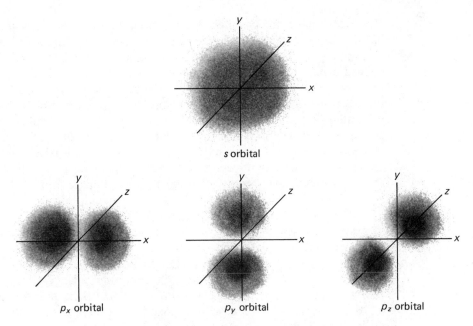

Figure 3-30 The shape of the s, p_x, p_y, and p_z orbitals. The shape of the orbital defines the region of space where an electron may be found, with a probability of 95%. The darker areas of the clouds are the regions of higher probability.

information from the bank manager. The shading in our orbital pictures is just like our marked-up map. It tells us where the electron is expected to be most of the time. On our map, a lot of marks in a particular spot correspond to a greater probability of finding George. Just as on the orbital pictures, the darker regions at the edge of each orbital show the region of highest probability, or the most likely place to look, for finding an electron. However, the entire orbital is the region of finding the electron with a probability of 95%.

Although the quantum mechanical model of the atom is very useful for many scientists in certain fields of study, it is not the perfect model. In fact, some of the other models of the atom that we discussed in this chapter explain certain phenomena better than the quantum mechanical picture. (Perhaps we should say that the other models of the atom offer an easier explanation than the quantum mechanical model.) But that's perfectly fine because scientists are still looking to refine their model of the atom to explain more and more of the mysteries of matter.

In subsequent chapters we use some of the atomic models discussed here to explain various phenomena.

SUMMARY

In this chapter we studied the history and development of atomic theory. We began this chapter with John Dalton proposing that atoms were indivisible spheres. Next, we reviewed the events that led scientists to believe that, indeed, the atom was divisible. We read about Crookes' experiments that gave rise to the existence of electrons. We also reviewed the experiments that gave rise to the existence of protons and neutrons. We learned about the Thomson model of the atom, and how Rutherford, a few years later, proved it to be incorrect. Rutherford then went on to suggest a new model of his own.

We also examined a number of parallel developments that occured in the nineteenth century in the search for the structure of the atom: for example, the development of the periodic table and the discovery of radiation. These developments led Niels Bohr to propose his model of the atom.

Finally, we learned about the quantum mechanical model of the atom. This highly sophisticated, mathematical model represents the present state of the art with regard to atomic theory.

This chapter was quite lengthy and filled with a lot of information. But it was our aim to give you a basic introduction to atomic theory. We hope that you found it to be an interesting story.

EXERCISES

1. Discuss how the experiments of the early 1800's led scientists to believe that atoms were divisible. Cite at least two major experiments.

2. Match the word on the left with its definition on the right.
   (1) electrolysis
   (2) electrode
   (3) anode
   (4) cathode
   (5) ion
   (6) anion
   (7) cation

   (a) In an electrolysis device these are usually metal rods inserted into a molten salt.
   (b) Greek for wanderers, these entities, according to Faraday, carried electric current.
   (c) The electrode carrying the positive charge.
   (d) Ions that travel to the cathode.
   (e) The decomposition of compounds into their elements by electricity.
   (f) Ions that travel to the anode.
   (g) The electrode carrying the negative charge.

3. Complete the following table:

PARTICLE	CHARGE	APPROX. MASS (amu)
	−1	
Neutron		
	+1	

4. Explain the importance of the Crookes tube in the development of the structure of the atom.

5. Discuss some of the evidence that led to the discovery of the proton.

6. Who discovered the neutron? Why was it such an important discovery?

7. Match the model of the atom with its discoverer.
   (1) Dalton   (a) the nuclear atom
   (2) Thomson   (b) the indivisible atom
   (3) Rutherford   (c) the "plum-pudding model"
   (4) Bohr   (d) the energy-level (shell) atom

8. Discuss some of the important features of the periodic table.

9. Match the scientist with a description of his work.
   (1) Döbereiner   (a) the Law of Octaves
   (2) Newlands   (b) periods of different lengths
   (3) Mendeleev   (c) triads

10. In the periodic table, elements in the same chemical family (group) have similar chemical properties. How was this explained by Niels Bohr?

11. How was the discovery of radiation important in developing the structure of the atom?

12. Match the type of radiation with its description.
    (1) alpha   (a) no charge
    (2) beta   (b) negative charge
    (3) gamma   (c) positive charge

13. Determine the number of protons, electrons, and neutrons in neutral atoms of the following isotopes.

    (a) $^{238}_{92}U$ (b) $^{3}_{1}H$ (c) $^{81}_{35}Br$ (d) $^{40}_{20}Ca$

14. What is a continuous spectrum? What is a line spectrum?

15. In the periodic table, a horizontal row is called a _____ . A vertical column is called a _____ .

16. Write the symbol for each of the following isotopes.
    (a) $7p, 7e, 8n$ (b) $6p, 6e, 8n$

17. List some of the unique features of the Bohr model of the atom.

18. An easy way to find the maximum electron number for an energy level is to use the formula

$$\text{maximum number of electrons} = 2(n)^2$$

where n is the number of the energy level (for example, $K = 1, L = 2$, etc.). Use the $2n^2$ rule to calculate the maximum number of electrons that the Q energy level can hold.

19. Discuss the events that led to the discovery of radiation.

20. How did the results of Rutherford's gold foil experiment suggest that an atom had a nucleus?

21. How many electrons are in each energy level for the following elements?
   (a) $_6$C   (b) $_7$N   (c) $_8$O   (d) $_{18}$Ar

22. What is the octet rule?

23. What similarity exists in the electron configurations of elements in group VIA?

24. Describe the major features of the quantum mechanical model of the atom.

25. The radiation that results from particles being ejected from the nucleus of an atom is alpha, beta, and gamma radiation. (This is why alpha, beta, and gamma radiation are known as nuclear radiation.) However, x rays do not come from the nucleus of an atom. Suggest a source of x radiation and an explanation as to how it occurs.

Chapter 4

Chemical Bonding

Some Things You Should Know After Reading This Chapter

You should be able to:

1. Write Lewis dot structures for the A-group elements.
2. Define and give examples of ionic and covalent bonds.
3. Write bonding structures for various covalent compounds.
4. Define single covalent bond, double covalent bond, and triple covalent bond.
5. Use the concept of electronegativity to decide whether a compound is bonded ionically or covalently.
6. Define and give an example of a polar covalent bond.
7. Define and give an example of a nonpolar covalent bond.
8. Write the formula of a chemical compound given its name.
9. Write the name of a chemical compound given its formula.
10. Decide whether a molecule is polar or nonpolar when given its three-dimensional shape.

ELEMENTS TO COMPOUNDS

The time is set about 5 billion years ago. Imagine that you are there. We are sitting in the void of space. But look, our sun, Sol, is not there, and neither is the earth or any of the other planets in our present solar system. What do we see? It looks like a giant star in the distance about to explode.

There is no noise in space, but we see a tremendous explosion that seems to occur all around us. The black void is filled with blinding light. It is our solar system forming before our very eyes—out of the remnants of an exploding star! All of the basic elements that we're familiar with in the periodic table are present. The most abundant element is hydrogen. A tremendous amount of hydrogen forms our sun. Other elements that are present begin to coalesce and form planets. Many of these elements combine to form compounds.

On one particular planet the conditions are just right, so that over the course of the next 5 billion years the simple compounds that are formed react with each other to form more complex compounds. Eventually, these compounds form living organisms of a very complex nature. This planet is the third planet from the sun, and it is called earth, by us, those complex living organisms.

INTRODUCTION

In Chapter 3 we discussed the elements that compose our world and our universe. Yet if you were to look for these elements in the earth, you would not find many of them in their pure elemental form. Of course, you would find some of them, such as gold and platinum. But most of the elements would be found as parts of chemical compounds. For example, you would not find pure sodium metal occurring naturally on the earth, but you would find plenty of sodium contained in the compound sodium chloride (table salt). How come? Why are most of the elements found in compounds and not as pure elements? The answer must have something to do with the stability of elements in compounds. In other words, many elements must be more stable in the form of compounds than as the pure elements. This is a good thing for us! After all, we biological species are composed of compounds.

In this chapter we look at the various ways that elements get together to form compounds.

LEWIS DOT STRUCTURES

In Chapter 3 we learned that elements in the same column of the periodic table have similar chemical properties because they have *the same number of electrons in the outermost energy levels*. Chemists have discovered that it is these outermost electrons which play a major role in determining how the atoms of these elements bond to form molecules of compounds. In this chapter we describe the bonding of some compounds. It will therefore be a great help to

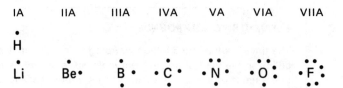

Figure 4-1 Electron dot diagrams of some A-group elements.

us to have a way to depict this bonding, in other words, a standard notation. Fortunately, such a notation exists. It was developed by the famous physical chemist G. N. Lewis in 1916.

The Lewis dot notation shows the symbol of the element and the number of outermost electrons that an atom of the element contains (Fig. 4-1). The rules for writing the Lewis dot notation for elements are as follows:

RULE 1.   Write the symbol for the element.

RULE 2.   Consider the symbol as having four sides. Use a dot to represent each outer electron, beginning on top and going clockwise. (You don't actually have to begin on top and go clockwise, but it is a good bookkeeping procedure.)

RULE 3.   Don't put two dots next to each other until you have distributed the the first four dots.

The nice thing about the Lewis dot notation is that all elements in the same group of the periodic table have the same notation. This is because they all have the same number of outermost electrons. Also, for the A-group elements, the group number tells you the number of electrons in the outermost energy level. (For the B-group elements, this is not the case in most instances, so we'll confine our discussion to the A-group elements.) For example, the notation for a sodium atom (group IA) is the same as that for a lithium atom (group IA).

$$\dot{\text{Na}} \qquad \dot{\text{Li}}$$

And the notation for an oxygen atom (group VIA) is the same as that for an atom of selenium (group VIA).

$$\cdot \ddot{\text{O}} \colon \qquad \cdot \ddot{\text{Se}} \colon$$

Example 4-1   Write the Lewis dot notation for the following A-group elements.

(a) gallium   (b) cesium   (c) iodine   (d) antimony

Solution   Look up the symbol of the element. Then find its position in the periodic table. The group number tells you the number of electrons in the outermost energy level.

(a) Gallium is in group IIIA: $\overset{\cdot}{\underset{\cdot}{Ga}}\cdot$

(b) Cesium is in group IA: $\overset{\cdot}{Cs}$

(c) Iodine is in group VIIA: $\cdot\overset{\cdot\cdot}{\underset{\cdot\cdot}{I}}\colon$

(d) Antimony is in group VA: $\cdot\overset{\cdot\cdot}{Sb}\cdot$

Now let's see how we use the Lewis dot notation to show how the atoms of elements bond to form molecules of compounds.

COVALENT BONDING

Scientists have long known that elements in group VIIIA of the periodic table (the noble gases) tend to be unreactive. In other words, they tend not to react with other elements to form compounds. The interesting thing about these group VIIIA elements is that their atoms all have *eight* electrons in their outermost energy level. We say that they have an *octet* of electrons. (An exception to this statement is helium, whose atoms have two electrons in the outermost energy level.) It appears that when atoms have an octet of electrons in their outermost energy level (or in the case of helium, a filled first energy level) they become very stable. The question is: How can atoms that don't have eight electrons in their outermost energy level obtain an octet? The answer is that they can attain eight electrons in their outermost energy level by sharing electrons with other atoms that also want an octet. In fact, chemists have found that this is one of the main reasons why elements bond to form compounds— to get eight electrons in their outermost energy level. This observation has come to be known as the *octet rule*. Let's see how this is done in *covalent bonding*. Remember that *covalent bonding* is a sharing of outermost electrons between two atoms, so that each can attain an octet of electrons. (An exception to this rule is hydrogen, which needs only two electrons in its outermost energy level. In this manner it will have an electron configuration similar to the noble gas helium.)

The Diatomic Elements:
Some Nice Examples of Covalent Bonding

A number of elements are known to exist naturally in diatomic form. This means that there are two atoms to a molecule. The elements hydrogen, oxygen, nitrogen, chlorine, bromine, iodine, and fluorine all exist naturally as diatomic elements. This means that they exist as H_2, O_2, N_2, Cl_2, Br_2, I_2, and F_2 (Fig. 4-2). You can remember these diatomic elements by simply remembering

Figure 4-2 A diatomic molecule may be pictured in this manner.

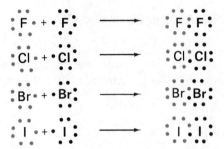

Figure 4-3 Covalent bonding of the group VIIA diatomic elements.

the name HONClBrIF (pronounced "honkelbrif"). Why do these elements exist as diatomics? Because by doing so they can form *covalent bonds* and obtain an octet of electrons. (In the case of hydrogen, it would be a duet of electrons.) Let's see what we mean by this (Fig. 4-3). Notice that in Figure 4-3, each fluorine atom has seven electrons in its outermost energy level. But by sharing the "lone" electron, each fluorine atom can have eight electrons in its outermost energy level. The same is true for a molecule of Cl_2, Br_2, and I_2.

In hydrogen the situation is similar, except that each H atom has one electron, and by the process of sharing, each H atom can have two electrons (Fig. 4-4). (Remember what we said about hydrogen wanting only two electrons in its outermost energy level.)

The type of covalent bond we have discussed so far is called a *single bond*. In a single bond a pair of electrons is shared by the two atoms forming the bond. In our example each atom has donated one electron to form the bond.

In a molecule of oxygen, we see that the bond is slightly different from the previous examples (Fig. 4-5). In order for each oxygen atom to have an octet of electrons, it is necessary that each oxygen atom use two electrons to

Figure 4-4 Covalent bonding in diatomic hydrogen.

H • + • H ⟶ H : H

Figure 4-5 Covalent bonding in diatomic oxygen.

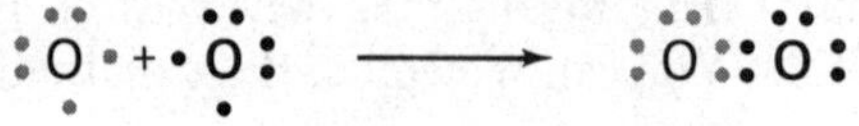

: O • + • O : ⟶ : O :: O :

$$\overset{\bullet\bullet}{\underset{\bullet}{N}} \overset{\bullet}{} + \overset{\bullet\bullet}{\underset{\bullet}{N}} \overset{\bullet}{} \longrightarrow \quad \overset{\bullet\bullet}{N} \overset{\bullet\bullet}{::} \overset{\bullet\bullet}{N}$$

Figure 4-6 Covalent bonding in diatomic nitrogen.

form the bond. This type of covalent bond is called a *double bond*. A double bond occurs when four electrons (two electron pairs) are shared by the two atoms forming the bond. In our example each oxygen atom uses two electrons to form the bond. Notice that this is the only structure that will give each oxygen atom an octet. (We should point out that although the structure in Figure 4-5 is not entirely accurate for the oxygen molecule, it does serve our needs and gets across the general idea.) There are many compounds that have double bonds. We will look at a few of them in the forthcoming pages.

In the molecule nitrogen, we see another type of covalent bond (Fig. 4-6). Notice that in this molecule, each nitrogen atom uses three electrons to form the bond. This type of covalent bond is called a *triple bond*. A triple bond occurs when six electrons (three electron pairs) are shared by the two atoms forming the bond. In our example each nitrogen atom uses three electrons to form the bond. Again, this is the only structure that we can show so that each nitrogen atom has an octet of electrons. There are many compounds that have triple bonds. We shall look at a few in a moment.

Other Molecules Having Covalent Bonds

Up to this point we've looked at covalent bonds formed between diatomic elements. Let's now turn our attention to covalent bonds formed between the atoms of different elements.

One of the most important compounds to biological organisms is water (H_2O). The water molecule bonds in the following manner:

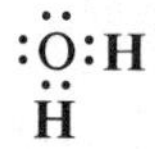

$$\overset{\bullet\bullet}{\underset{\bullet\bullet}{O}} \!:\! H$$
$$H$$

(By the way, although this is the bonding arrangement for water, it is not the actual shape of the water molecule. The Lewis dot diagram shows only how the electrons are shared. The actual bond angles are determined by experiments using x rays—the method is called x-ray diffraction spectroscopy. For water, these experiments show that the actual bond angle is about $105°$.)

$$\overset{\displaystyle O}{H \underset{105°}{\frown} H}$$

Notice, that by sharing electrons in the manner shown, the oxygen atom has eight electrons, and each hydrogen atom has two electrons. (Remember that the hydrogen atom needs only two electrons.)

The compound chloroform ($CHCl_3$), aside from its use in spy stories, was used by the medical profession as an anesthetic. The bonding picture for

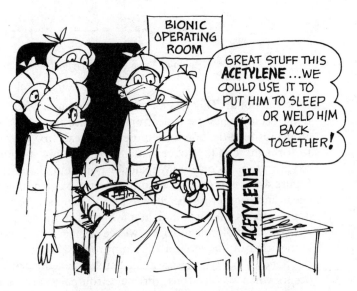

Figure 4-7 Acetylene, a very useful compound.

chloroform looks like this:

$$\ddot{\underset{\cdot\cdot}{\text{Cl}}}$$
$$:\!\overset{\cdot\cdot}{\underset{\cdot\cdot}{\text{Cl}}}\!:\!\overset{\cdot\cdot}{\text{C}}\!:\!\text{H}$$
$$\overset{\cdot\cdot}{\underset{\cdot\cdot}{\text{Cl}}}\!:$$

As you probably know, we breathe in oxygen gas (O_2), whose bonding we described earlier, and we exhale carbon dioxide (CO_2). This process is known as *respiration*. The bonding structure for carbon dioxide is as follows:

$$:\!\overset{\cdot\cdot}{O}\!::\!C\!::\!\overset{\cdot\cdot}{O}\!:$$

Notice that there are two double bonds in this compound.

The compound acetylene (C_2H_2), which is used today for welding (for example, as the fuel in an oxyacetylene torch), was at one time used in hospitals as a surgical anesthetic (Fig. 4-7). The bonding in a molecule of acetylene looks like this:

$$\text{H}\!:\!\text{C}\!::\!\text{C}\!:\!\text{H}$$

Notice that this is the only way that each carbon atom can have an octet of electrons and each hydrogen atom a duet.

Now, see if you can write the bonding structures for some covalent compounds in the example that follows.

Example 4-2 Write electron dot pictures for the following covalent compounds.

(a) H_2Se (Both hydrogens are bonded to the selenium.)

(b) CH_2O (This is the compound formaldehyde. The oxygen and each hydrogen are bonded to the carbon.)

(c) NH_3 (This is ammonia gas. Each hydrogen is bonded to the nitrogen.)

(d) HCN (This is hydrogen cyanide. The hydrogen and nitrogen are both bonded to the carbon.)

Solution

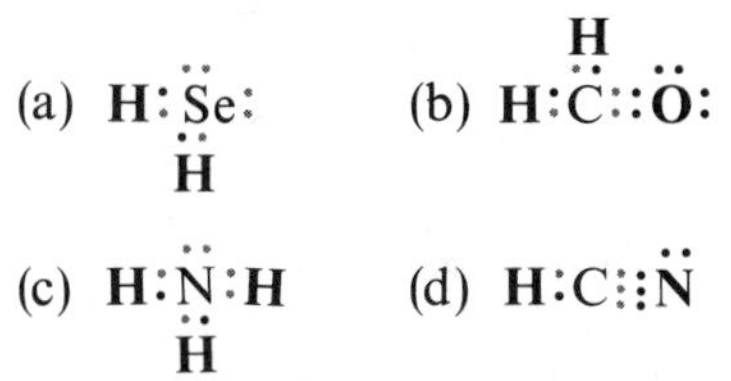

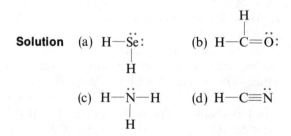

Before we leave the topic of covalent bonding, we should mention the use of the *dash*. The dash (—) is used to simplify the writing of electron dot pictures. The dash represents a bond [in other words, a shared pair of electrons (:)]. Let's see how this works.

Example 4-3 Write the bonding structures in Example 4-2 using the dash notation.

Solution (a) H—S̈e: (b) H—C̈=Ö:
                   | |
                   H   H

(c) H—N̈—H   (d) H—C≡N̈
         |
         H

IONIC BONDING

Another way that atoms can attain an octet is by transferring electrons among each other. Consider the compound sodium chloride, table salt (NaCl). The electron dot notation for a sodium atom is Ṅa and the electron dot notation for a chlorine atom is

$$\cdot \ddot{C}l\!:$$

A sharing of electrons might help the chlorine atom attain an octet, but it certainly wouldn't help the sodium atom. Therefore, this compound cannot bond covalently. However, if the sodium atom transfers its lone electron to the chlorine atom, both atoms can have an octet (Fig. 4-8). Do you see why this is so? For the chlorine atom it's easy to see. But how does the sodium

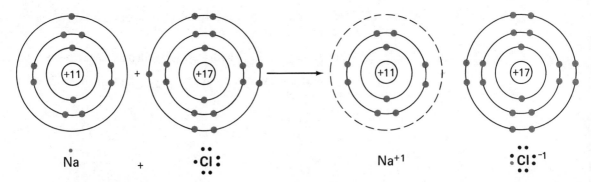

Figure 4-8 Ionic bonding in sodium chloride.

atom gain an octet by giving an electron away? Remember, the electron configuration of a sodium atom is as follows. The K level has two electrons, L level has eight electrons, and M level has one electron. If the sodium atom gives away its M-level electron, the L level becomes the outermost level, and this level has eight electrons.

But what holds the sodium and chlorine atoms together to form the bond? If you remember, the sodium atom has given up one of its electrons. It's no longer a neutral sodium atom. It now contains 11 protons but only 10 electrons. We say that the sodium atom has become a *positively charged sodium ion* (Na^{+1}). The chlorine, on the other hand, has gained an electron. It now has 17 protons and 18 electrons. We say that the chlorine atom has become a *negatively charged chloride ion* (Cl^{-1}).

The word "ion" is used because it indicates that we are talking about atoms that have *either gained or lost electrons*, and therefore have a *positive or negative charge*. By the way, notice that we said chlor*ide* ion, not chlor*ine* ion. We'll explain why the negative ion has an *ide* ending when we discuss the topic of chemical naming later in this chapter. The important thing is that because the sodium ion and chloride ion are oppositely charged, they are attracted to each other. It is this positive–negative attraction which is the *ionic bond*. Many compounds bond ionically. Some examples are given in Figure 4-9.

Figure 4-9 Ionic bonding in magnesium oxide (MgO) and magnesium chloride (MgCl$_2$).

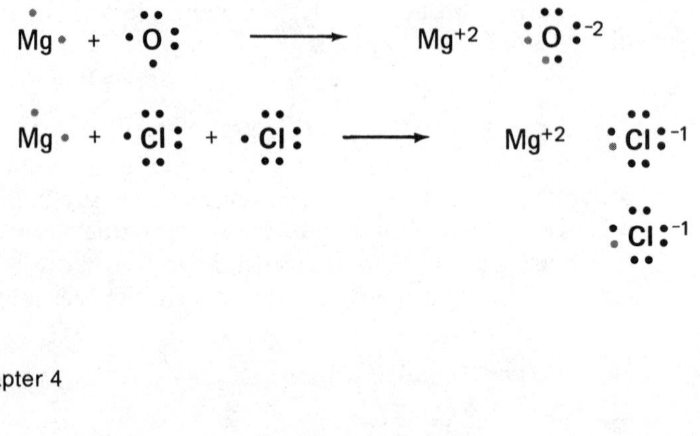

$$Mg\cdot \ + \ \cdot \ddot{\underset{\cdot}{O}}: \ \longrightarrow \ Mg^{+2} \ :\ddot{\underset{\cdot\cdot}{O}}:^{-2}$$

$$Mg\cdot \ + \ \cdot \ddot{\underset{\cdot\cdot}{Cl}}: \ + \ \cdot \ddot{\underset{\cdot\cdot}{Cl}}: \ \longrightarrow \ Mg^{+2} \ :\ddot{\underset{\cdot\cdot}{Cl}}:^{-1}$$

$$:\ddot{\underset{\cdot\cdot}{Cl}}:^{-1}$$

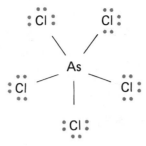

Figure 4-10   Covalent bonding in arsenic pentachloride.

Figure 4-11   Covalent bonding in boron trifluoride.

SOME EXCEPTIONS TO THE OCTET RULE

We have now looked at two ways that atoms bond to form molecules. We have also seen that the major force behind this bonding is the attaining of an octet of electrons in the outermost energy level. However, it is only fair to point out that there are exceptions to this octet phenomenon (Fig. 4-10). There are molecules that exist in which an atom has six electrons in its outermost energy level or even 10 electrons in its outermost energy level (Fig. 4-11). But for most compounds, we find that the octet rule leads to the proper bonding structure.

ELECTRONEGATIVITY:
A WAY TO DIAGNOSE IONIC OR COVALENT BONDS

At this point you may be asking: "How can I tell whether the atoms of two elements bond ionically or covalently?" The answer is simple. We use the concept of *electronegativity*, which is the attraction that an atom has for the electrons it is sharing with another atom. The electronegativities of the elements are shown in Table 4-1. The concept of electronegativity was devised by Linus Pauling, the recipient of two Nobel Prizes (you may know him as the "vitamin C man"). Pauling assigned fluorine an electronegativity value of 4.0 because it attracts electrons more strongly than any other element. The other elements have various values, depending on their ability to attract electrons.

TABLE 4-1 *Periodic table of electronegativities*

Period	IA	IIA	IIIB	IVB	VB	VIB	VIIB	VIII			IB	IIB	IIIA	IVA	VA	VIA	VIIA	VIIIA
1	2.1 H 1																	He 2
2	1.0 Li 3	1.5 Be 4				Transition elements							2.0 B 5	2.5 C 6	3.0 N 7	3.5 O 8	4.0 F 9	Ne 10
3	0.9 Na 11	1.2 Mg 12											1.5 Al 13	1.8 Si 14	2.1 P 15	2.5 S 16	3.0 Cl 17	Ar 18
4	0.8 K 19	1.0 Ca 20	1.3 Sc 21	1.5 Ti 22	1.6 V 23	1.6 Cr 24	1.5 Mn 25	1.8 Fe 26	1.8 Co 27	1.8 Ni 28	1.9 Cu 29	1.6 Zn 30	1.6 Ga 31	1.8 Ge 32	2.0 As 33	2.4 Se 34	2.8 Br 35	Kr 36
5	0.8 Rb 37	1.0 Sr 38	1.2 Y 39	1.4 Zr 40	1.6 Nb 41	1.8 Mo 42	1.9 Tc 43	2.2 Ru 44	2.2 Rh 45	2.2 Pd 46	1.9 Ag 47	1.7 Cd 48	1.7 In 49	1.8 Sn 50	1.9 Sb 51	2.1 Te 52	2.5 I 53	Xe 54
6	0.7 Cs 55	0.9 Ba 56	La–Lu 57–71	1.3 Hf 72	1.5 Ta 73	1.7 W 74	1.9 Re 75	2.2 Os 76	2.2 Ir 77	2.2 Pt 78	2.4 Au 79	1.9 Hg 80	1.8 Tl 81	1.8 Pb 82	1.9 Bi 83	2.0 Po 84	2.2 At 85	Rn 86
7	0.7 Fr 87	0.9 Ra 88	Ac–Lr 89–103	[Rf] 104	[Ha] 105													

1.1 La 57	1.1 Ce 58	1.1 Pr 59	1.1 Nd 60	1.1 Pm 61	1.1 Sm 62	1.1 Eu 63	1.1 Gd 64	1.1 Tb 65	1.1 Dy 66	1.1 Ho 67	1.1 Er 68	1.1 Tm 69	1.1 Yb 70	1.2 Lu 71
1.1 Ac 89	1.3 Th 90	1.5 Pa 91	1.7 U 92	1.3 Np 93	1.3 Pu 94	1.3 Am 95	1.3 Cm 96	1.3 Bk 97	1.3 Cf 98	1.3 Es 99	1.3 Fm 100	1.3 Md 101	1.3 No 102	Lr 103

2.5 ← Electronegativity
C
6 ← Atomic number

Source: Alan Sherman, Sharon Sherman, and Leonard Russikoff, *Basic Concepts of Chemistry*, Houghton Mifflin Company, Boston, 1976. By permission of the publisher.

You can use the electronegativity table to help you decide whether a bond is ionic or covalent. All you need to do is calculate the difference in electronegativity values between the two elements forming the bond. Once you obtain this difference, you can check Table 4-2. This table will help you convert your electronegativity difference into a percentage covalent and percentage ionic character of the bond. If the electronegativity difference is between zero and 1.6, the percentage covalent character is greater than the percentage ionic. The bond may therefore be considered to be covalent. If the electronegativity difference is greater than 1.6, the percentage ionic character is greater than the percentage covalent. The bond may therefore be considered to be ionic. Let's try some examples to see how this works.

TABLE 4-2 *The relationship between electronegativity difference and the ionic percentage and covalent percentage of a chemical bond*

DIFFERENCE IN ELECTRONEGATIVITY	IONIC PERCENTAGE	COVALENT PERCENTAGE
0.0	0.0	100
0.1	0.5	99.5
0.2	1.0	99.0
0.3	2.0	98.0
0.4	4.0	96.0
0.5	6.0	94.0
0.6	9.0	91.0
0.7	12.0	88.0
0.8	15.0	85.0
0.9	19.0	81.0
1.0	22.0	78.0
1.1	26.0	74.0
1.2	30.0	70.0
1.3	34.0	66.0
1.4	39.0	61.0
1.5	43.0	57.0
1.6	47.0	53.0
1.7	51.0	49.0
1.8	55.0	45.0
1.9	59.0	41.0
2.0	63.0	37.0
2.1	67.0	33.0
2.2	70.0	30.0
2.3	74.0	26.0
2.4	76.0	24.0
2.5	79.0	21.0
2.6	82.0	18.0
2.7	84.0	16.0
2.8	86.0	14.0
2.9	88.0	12.0
3.0	89.0	11.0
3.1	91.0	9.0
3.2	92.0	8.0

Source: Alan Sherman, Sharon Sherman, and Leonard Russikoff, *Basic Concepts of Chemistry*, Houghton Mifflin Company, Boston, 1976. By permission of the publisher.

Example 4-4 Determine whether the bonds in the following compounds are ionic or covalent.

(a) HBr (b) KCl (c) O_2 (d) CO_2 (e) NaF

Solution
(a) We use Table 4-1 to find the electronegativity values of hydrogen and bromine.
H is 2.1 and Br is 2.8.
The electronegativity difference is $2.8 - 2.0 = 0.7$.
Table 4-2 tells us that an electronegativity difference of 0.7 is a bond that is 12% ionic and 88% covalent. The bond is therefore considered to be covalent.

(b) The electronegativity value of K is 0.8.
The electronegativity value of Cl is 3.0.
The electronegativity difference is $3.0 - 0.8 = 2.2$.
Table 4-2 tells us that an electronegativity difference of 2.2 is a bond that is 70% ionic and 30% covalent.
The bond is therefore considered to be ionic.

(c) The electronegativity of oxygen is 3.5; however the bond in question is between two O atoms, so the electronegativity difference is zero.
Table 4-2 tells us that this bond has zero percent ionic character and 100% covalent character.

(d) The electronegativity of C is 2.5.
The electronegativity of O is 3.5.
The electronegativity difference is $3.5 - 2.5 = 1.0$.
Table 4-2 tells us that an electronegativity difference of 1.0 is a bond that is 22% ionic and 70% covalent. The bond is therefore considered to be covalent.

(e) The electronegativity of Na is 0.9.
The electronegativity of F is 4.0.
The electronegativity difference is $4.0 - 0.9 = 3.1$.
Table 4-2 tells us that an electronegativity difference of 3.1 is a bond that is 91% ionic and 9% covalent. The bond is therefore considered to be ionic.

THE POLAR AND NONPOLAR COVALENT BOND

Although a covalent bond is defined as a sharing of electrons between two atoms, the sharing is usually not done on an equal basis. This is because the electronegativity of the two elements will probably not be the same (Fig. 4-12). Only if there is no electronegativity difference between the two atoms forming the covalent bond will there be equal sharing, for example in the diatomic elements (Fig. 4-13). Such a bond is said to be a *nonpolar covalent bond*.

Figure 4-12 The electronegativity difference between H and Br gives rise to a polar covalent bond in the compound hydrogen bromide (HBr).

Figure 4-13 The diatomic molecule hydrogen (H_2) has no electronegativity difference. This gives rise to a nonpolar covalent bond.

Whenever there is an electronegativity difference between two atoms forming a covalent bond, we say that the bond is a *polar covalent bond*. The greater the difference, the more polar the bond. (Of course, an electronegativity difference of greater than 1.6 means that the bond is ionic.) The fact that there are polar and nonpolar covalent bonds is extremely important to life on this planet. A case in point is the water molecule, whose bonds are polar covalent. Partly because of this, water has some unique properties: for example, its unusually high boiling point. If it wasn't for polar covalent bonding, water might have turned out to be a gas, at the temperatures that normally exist on our planet, and life, as we know it, would have never evolved.

Example 4-5 Rank the following bonds in order of increasing polarity:

(a) Br—Br   (b) S—O   (c) H—O   (d) C—Cl

Solution Use Table 4-1 to determine the electronegativity difference between the elements, then rank them in order of least difference to most difference.

(a) The electronegativity difference between the two bromines is zero.

(b) The electronegativity difference between S and O is 1.0.

(c) The electronegativity difference between the H and O is 1.4.

(d) The electronegativity difference between C and Cl is 0.5.

Therefore, the ranking is:

Br—Br   C—Cl   S—O   H—O
Least polar   Most polar
bond   bond

HOW TO WRITE A CHEMICAL FORMULA

Chemists who have studied how elements combine to form compounds have realized that there are certain trends. It seems that elements tend to form ions with specific charges, or tend to form only a certain number of covalent bonds. To describe this phenomenon, chemists have devised a system to indicate how elements combine to form compounds. This system involves the assignment of *oxidation numbers* to the substances involved in forming the compound. The *oxidation number* expresses the charge of the element or polyatomic ion in a particular compound. The oxidation number can be positive or negative, depending on whether the element tends to attract electrons strongly or give them up. In carbon tetrachloride (CCl_4), for example, the carbon has an oxidation number of $+4$ and the chlorine has an oxidation number of -1. Elements with high electronegativity values usually have negative oxidation numbers, and elements with low electronegativity values usually have positive oxidation numbers.

Table 4-3 lists the oxidation numbers of some important ions. Also shown in Table 4-3 are some complex ions (also called *polyatomic ions*). These *polyatomic ions*, for example nitrate ion, $(NO_3)^{-1}$, and sulfate ion, $(SO_4)^{-2}$, are covalently bonded, and are found in nature as parts of numerous compounds. For example, if you take vitamins containing iron, you're not really consuming the element iron but the compound iron(II) sulfate, $FeSO_4$. Notice that sulfate group, $(SO_4)^{-2}$, is part of the compound. And if you're suffering from indigestion and consume an antacid tablet, you might be taking calcium carbonate, $CaCO_3$. Notice that carbonate ion, $(CO_3)^{-2}$, is part of the compound.

By using the table of oxidation numbers, we can easily learn to write chemical formulas by following a few simple rules.

RULE 1. Write the symbol of each ion, together with its charge. For example, if we want to write the formula of calcium chloride, we first write

$$Ca^{+2}Cl^{-1}$$

RULE 2. Because chemical compounds are electrically neutral, we must choose proper subscripts to balance the positive and negative charge of each ion. A simple way to do this is to crisscross the numbers.

$$Ca_1^{+2}Cl_2^{-1}$$

Notice that the subscript numbers are written without charge, in other words, without a plus or minus sign.

RULE 3. Now rewrite the formula in a more professional-looking manner. To do this, don't show the oxidation numbers, just

TABLE 4-3 *Ions frequently used in chemistry*

+1		+2		+3	
Hydrogen	H^{+1}	Calcium	Ca^{+2}	Iron(III)	Fe^{+3}
Lithium	Li^{+1}	Magnesium	Mg^{+2}	Aluminium	Al^{+3}
Sodium	Na^{+1}	Barium	Ba^{+2}		
Potassium	K^{+1}	Zinc	Zn^{+2}		
Mercury(I)	Hg^{+1}	Mercury(II)	Hg^{+2}		
Copper(I)	Cu^{+1}	Tin(II)	Sn^{+2}		
Ammonium	$(NH_4)^{+1}$	Iron(II)	Fe^{+2}		
Silver	Ag^{+1}	Lead(II)	Pb^{+2}		
		Copper(II)	Cu^{+2}		

−1		−2		−3	
Fluoride	F^{-1}	Oxide	O^{-2}	Nitride	N^{-3}
Chloride	Cl^{-1}	Sulfide	S^{-2}	Phosphate	$(PO_4)^{-3}$
Hydroxide	$(OH)^{-1}$	Sulfite	$(SO_3)^{-2}$	Arsenate	$(AsO_4)^{-3}$
Nitrite	$(NO_2)^{-1}$	Sulfate	$(SO_4)^{-2}$		
Nitrate	$(NO_3)^{-1}$	Carbonate	$(CO_3)^{-2}$		
Acetate	$(C_2H_3O_2)^{-1}$	Chromate	$(CrO_4)^{-2}$		

the subscript numbers. Also, the number 1 is not shown. Therefore,

$$Ca_1{}^{+2}Cl_2{}^{-1} \quad \text{becomes} \quad CaCl_2$$

RULE 4. Be sure that the subscript numbers are written in least-common-denominator form. This means that they must be reduced to lowest terms. For example, if you were asked to write the formula for aluminum nitride, you would do the following:

$$Al^{+3}N^{-3} \qquad \text{(Rule 1)}$$

$$Al_3{}^{+3}N_3{}^{-3} \qquad \text{(Rule 2)}$$

$$Al_3N_3 \qquad \text{(Rule 3)}$$

But since the subscripts are divisible by *three*, we simplify the formula to

$$AlN \qquad \text{(Rule 4)}$$

Example 4-6 Write the formulas for the following compounds.

(a) aluminum oxide (b) sodium chloride

(c) potassium sulfate (d) ammonium phosphate

(e) barium arsenate (f) ammonium sulfite

Solution Use Table 4-3 to find the charge of the ions in each compound, then follow the rules we've just learned.

(a) $Al^{+3}O^{-2}$ $Al_2^{+3}O_3^{-2}$ which becomes Al_2O_3

(b) $Na^{+1}Cl^{-1}$ $Na_1^{+1}Cl_1^{-1}$ which becomes NaCl

(c) $K^{+1}(SO_4)^{-2}$ $K_2^{+1}(SO_4)_1^{-2}$ which becomes K_2SO_4

Notice that we handle the polyatomic ion by simply keeping parentheses around it. The parentheses were dropped in the final step because the subscript outside the parentheses was 1.

(d) $(NH_4)^{+1}(PO_4)^{-3}$ $(NH_4)_3^{+1}(PO_4)_1^{-3}$ which becomes $(NH_4)_3PO_4$

Notice that the ammonium ion must have the parentheses because the subscript outside the parentheses is 3. However, the parentheses around the phosphate group are unnecessary because the subscript outside is 1.

(e) $Ba^{+2}(AsO_4)^{-3}$ $Ba_3^{+2}(AsO_4)_2^{-3}$ which becomes $Ba_3(AsO_4)_2$

(f) $(NH_4)^{+1}(SO_3)^{-2}$ $(NH_4)_2^{+1}(SO_3)_1^{-2}$ which becomes $(NH_4)_2SO_3$

What would you do if we asked you to write the formula of iron oxide? The first thing you would have to find out is which iron oxide we mean: iron(II) oxide or iron(III) oxide. Each is a distinct chemical compound with its own properties. *The Roman numeral indicates the charge of the iron in the compound.* (The rules for writing and naming chemical compounds tell us to use a Roman numeral to indicate the charge of transition metal elements and other elements that may have variable oxidation numbers. We have more to say about this in the next section.) The formula of iron(II) oxide would be determined as follows:

$$Fe^{+2}O^{-2} \qquad Fe_2^{+2}O_2^{-2} \qquad \text{which becomes FeO}$$

The formula of iron(III) oxide would be determined as follows:

$$Fe^{+3}O^{-2} \qquad Fe_2^{+3}O_3^{-2} \qquad \text{which becomes } Fe_2O_3$$

Example 4-7 Write the formulas for the following compounds.

(a) copper(I) carbonate (b) tin(II) phosphide

(c) iron(III) acetate (d) lead(II) phosphate

Solution (a) $Cu^{+1}(CO_3)^{-2}$ $Cu_2^{+1}(CO_3)_1^{-2}$ which becomes
Cu_2CO_3

(b) $Sn^{+2}P^{-3}$ $Sn_3^{+2}P_2^{-3}$ which becomes Sn_3P_2

(c) $Fe^{+3}(C_2H_3O_2)^{-1}$ $Fe_1^{+3}(C_2H_3O_2)_3^{-1}$ which becomes
$Fe(C_2H_3O_2)_3$

(d) $Pb^{+2}(PO_4)^{-3}$ $Pb_3^{+2}(PO_4)_2^{-3}$ which becomes
$Pb_3(PO_4)_2$

Whenever you are dealing with chemistry, you need to be able to write chemical formulas. Writing them is much easier and faster if you have the information in Table 4-3 right at your fingertips. So spend some time mulling over this table and getting acquainted with the charges of the most common elements and polyatomic ions. You can predict the charges of most A-group elements, just keep in mind the following:

Group IA is $+1$.

Group IIA is $+2$.

Group IIIA is $+3$.

Group IVA is ± 4.

Group VA is usually ± 3 or $+5$.

Group VIA is usually -2.

Group VIIA is usually -1.

However, it's a good idea to memorize the charges of the polyatomic ions and transition metals in Table 4-3.

HOW TO WRITE A CHEMICAL NAME

Naming chemical compounds is a vast, but systematic, area of chemistry. Chemists from all over the world get together regularly and agree on rules for naming compounds. This is done at meetings of the IUPAC (International Union of Pure and Applied Chemistry). Some of the more important rules are:

RULE 1. The positive ion is always written first: for example, NaCl, not ClNa. (If the compound contains a metal and a nonmetal, the name of the metal comes first.)

RULE 2. The name of the compound consists of the names of both *ions*. (Once the elements join together to form compounds, they are not the same elements any more, but ions.) For example, NaCl

TABLE 4-4 *Greek prefixes*

mono = 1	di = 2	tri = 3	tetra = 4	penta = 5
hexa = 6	hepta = 7	octa = 8	nona = 9	deca = 10

is sodium chlor*ide*, not sodium chlor*ine*. Names of compounds made up of only two elements always end in *-ide*. (Keep in mind that the free elements sodium and chlorine have very different properties from the compound sodium chloride. Sodium is a very reactive metal which is poisonous if injested. Chlorine is a very reactive gas which can be toxic if inhaled. Sodium chloride, on the other hand, is a fairly inert solid which many of us consume each day.)

RULE 3. Compounds made up of elements with variable oxidation numbers must include the Roman numeral as part of the compound name. For example, $Fe(NO_3)_3$ is called iron(III) nitrate, not just iron nitrate.

RULE 4. For compounds made up of oxide ions, we sometimes use a Greek prefix in the compound name (Table 4-4). For example: CO is carbon *mon*oxide, CO_2 is carbon *di*oxide, and SO_3 is sulfur *tri*oxide.

RULE 5. With some compounds, most notably compounds composed of two nonmetals, you have a choice; you can use either the Greek prefix or the Roman numeral. For example, PCl_5 can be called either phosphorus(V) chloride or phosphorus penta-chloride.

Example 4-8 Name the following compounds.

(a) Fe_2O_3 (b) CuO

(c) Cu_2O (d) Ag_2SO_4

(e) $Sr(NO_3)_2$ (f) Cs_2S

(g) $Hg(NO_3)_2$ (h) $Zn(C_2H_3O_2)_2$

(i) SO_2 (j) UF_6

Solution (a) iron(III) oxide   (b) copper(II) oxide

(c) copper(I) oxide   (d) silver sulfate

(e) strontium nitrate   (f) cesium sulfide

(g) mercury(II) nitrate   (h) zinc acetate

(i) sulfur dioxide

(j) uranium hexafluoride or uranium(VI) fluoride

THE THREE-DIMENSIONAL CHARACTERISTICS OF MOLECULES:
SOME ARE POLAR, SOME ARE NOT

Up to this point, we have neglected the fact that molecules are three-dimensional. This fact, however, is very important to the understanding of how some molecules behave. Let's look at the three-dimensional shape of some molecules to see what properties appear.

If we could enlarge a single molecule of any diatomic element (for instance, H_2, O_2, or N_2), it would seem to be linear (in a line) (Fig. 4-14). This would also be true for covalently bonded compounds consisting of two *un*like atoms, such as HCl (Fig. 4-15).

But if we could enlarge a single molecule of water, we would see a bent molecule (Fig. 4-16). Experiments have shown that the angle between the hydrogen atoms is about 105°.

Figure 4-14   Three-dimensional models of some diatomic elements.

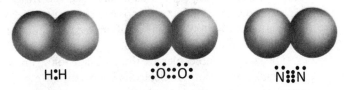

H:H   :O::O: N⦂⦂N

Figure 4-15   A three-dimensional model of HCl.

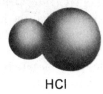

HCl

Figure 4-16   A three-dimensional model of H_2O.

H_2O

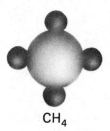

CH$_4$

Figure 4-17 A three-dimensional model of CH$_4$.

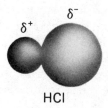

HCl

Figure 4-18 The polar molecule HCl, showing the positive center on the hydrogen and the negative center on the chlorine.

If we were to look at a molecule of methane (CH$_4$), we would see a pyramid-shaped symmetrical molecule (Fig. 4-17). The angle between neighboring hydrogen atoms in methane has been shown to be 109.5°.

Molecules of other compounds exist in many other shapes. But remember when you are discussing shapes of molecules that these shapes are always three-dimensional.

The fact that molecules are three-dimensional, and the fact that in covalent compounds the sharing of electrons is not always equal, gives some molecules the property of being *polar* and other molecules the property of being *nonpolar*. Let's see what this means.

In a molecule of HCl (Fig. 4-18) the hydrogen represents the positive center of the molecule. (We say that the hydrogen appears to have a partial positive charge, δ^+.) The chloride ion represents the negative center of the molecule. (We say that the chloride ion appears to have a partial negative charge, δ^-.) This is because the electronegativity of Cl is greater than that of H. Because the two centers of charge do not coincide, the molecule is said to be *polar*.

In the compound CCl$_4$ (Fig. 4-19) there are four carbon–chlorine bonds, each of which is a polar covalent bond. This is because the electronegativity of C and Cl are different. Because of this electronegativity difference, the carbon appears to have a partial positive charge, and each chlorine atom a partial negative charge. But the molecule is *nonpolar*! This is because the *center of positive charge* (on the carbon atom), coincides with the *center of negative charges* from each of the chloride ions (also on the carbon atom). This is due to the symmetrical shape of CCl$_4$, which is tetrahedral (a triangular pyramid).

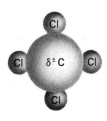

Figure 4-19 The nonpolar molecule carbon tetrachloride (CCl_4) showing that the centers of positive and negative charges coincide.

Example 4-9 Determine (1) if the bonds in each of the following molecules are polar or nonpolar and (2) if the molecule is polar or nonpolar.

(a) Br_2 (shape is linear)

(b) CH_4 (shape is pyramidal)

(c) CO_2 (shape is linear, with each oxygen bonded to the carbon)

(d) H_2O (shape is bent, as in Fig. 4-16)

Solution (a) Bromine is a diatomic element; therefore, the Br—Br bond is non-polar and the molecule is nonpolar.

(b) The bonding in CH_4 is similar to the bonding in CCl_4. Each bond in methane (CH_4) is a polar covalent bond. But the molecule is nonpolar since the centers of positive and negative charges coincide.

(c) Each C=O bond in CO_2 is polar, but the molecule is nonpolar. This is because the centers of positive and negative charge coincide.

(d) Each O—H bond in H_2O is polar. The molecule is also polar since the centers of positive and negative charge do not coincide.

SUMMARY

In this chapter we learned how the chemical elements bond to form compounds. We saw that there are two basic types of bonds: ionic and covalent. And we learned that by using the concept of electronegativity we could tell whether a bond in a compound was ionic or covalent. We found out that there are three types of covalent bonds: single, double, and triple bonds. We also found out that these covalent bonds may be polar or nonpolar.

Our study of bonding led us to the idea of combining capacity of elements, which we called oxidation numbers. Using oxidation numbers we learned how to write the formulas of chemical compounds. We also learned how to name these compounds from their formulas.

Finally, we looked at the three-dimensional characteristics of some covalently bonded molecules. We found that there are two types: polar molecules and nonpolar molecules. In subsequent chapters we will see how polar and nonpolar molecules play an important role in our world.

EXERCISES

1. Write the Lewis dot notation for the following elements.
   (a) barium   (b) fluorine   (c) rubidium   (d) tin

2. Write the Lewis dot notation for
   (a) chloride ion   (b) sulfide ion   (c) phosphide ion
   (*Hint*: Notice that we asked for the *ions* Cl^{-1}, S^{-2}, and P^{-3}, not the elements!)

3. With the exception of helium, all the other group VIIIA elements have _____ electrons in their outermost energy level.

4. Write electron dot diagrams for the following covalent compounds.
   (a) Br_2 (b) HBr   (c) C_2H_4 (d) PH_3 (e) C_2Cl_2

5. Name the diatomic elements from memory.

6. Stability, associated with eight electrons in the outermost energy level, has come to be known as the _____ rule. An exception to this rule is the element hydrogen, which needs only _____ electrons in its outermost energy level.

7. (a) When each atom uses two electrons to form the bond, it is called a _____ covalent bond.
   (b) When each atom uses three electrons to form the bond, it is called a _____ covalent bond.
   (c) When each atom uses one electron to form the bond, it is called a _____ covalent bond.

8. In the formation of the compound magnesium chloride ($MgCl_2$), the magnesium ion has given up _____ electron(s), and each chloride ion has gained _____ electron(s).

9. Determine whether the bonds in the following compounds are ionic or covalent.
   (a) HI   (b) $SrCl_2$ (c) CO   (d) MgF_2 (e) AsH_3

10. List the following compounds in order of increasing polarity of their bonds: H_2Se, H_2O, H_2S, H_2Te.

11. Define the following terms.
    (a) ionic bond   (b) covalent bond
    (c) polar covalent bond   (d) nonpolar covalent bond
    (e) polar molecule   (f) nonpolar molecule

12. Write the formulas for the following compounds.
    (a) sodium arsenate
    (b) potassium sulfite
    (c) barium chloride
    (d) iron(II) nitrate
    (e) copper(II) sulfate
    (f) aluminum sulfide
    (g) rubidium oxide
    (h) cobalt(II) nitrite
    (i) carbon tetrachloride
    (j) sulfur dioxide
    (k) hydrogen acetate
        (also known as acetic acid)
    (l) phosphorus pentachloride

13. Write the names of the following compounds.
    (a) $AgBr$
    (b) Cu_2S
    (c) $Fe(NO_3)_2$
    (d) $Fe(NO_3)_3$
    (e) $Ca(OH)_2$
    (f) $Al_2(SO_3)_3$
    (g) Zn_3P_2
    (h) $CuCrO_4$
    (i) $Hg_3(PO_4)_2$
    (j) $CoCl_2$
    (k) OsO_4
    (l) Cu_3N

14. Determine the oxidation number of the underlined element in each compound.
    (a) $\underline{Ni}Cl_2$
    (b) $\underline{Os}O_4$
    (c) $\underline{V}_2O_5$
    (d) $\underline{U}F_6$

15. Determine whether the bond in each compound is polar or nonpolar, and then whether the molecule is polar or nonpolar.

    Molecule   *Shape*

    (a) NH_3 H—N(—H)—H with N at top, H below

    (b) H_2S S with H—H

    (c) N_2 N≡N

    (d)   HCN   H—C≡N

16. Write the formulas for the following compounds.
    (a) sodium sulfite
    (b) potassium arsenate
    (c) barium nitrate
    (d) iron(II) chloride
    (e) copper(II) sulfide
    (f) aluminum sulfate
    (g) rubidium chloride
    (h) cobalt(II) oxide
    (i) carbon dioxide
    (j) sulfur trioxide
    (k) uranium(VI) oxide
    (l) phosphorus trichloride

17. Write the names of the following compounds.
    (a) Ag_2S
    (b) $CuBr$
    (c) $Hg(NO_3)_2$
    (d) $Fe(NO_2)_3$
    (e) $CaSO_3$
    (f) $Al(OH)_3$
    (g) $ZnCrO_4$
    (h) Cu_3P_2
    (i) HgS
    (j) $Co_3(PO_4)_2$
    (k) Cu_3P
    (l) V_2O_5

Chapter 5

Chemical Reactions

Some Things You Should Know After Reading This Chapter

You should be able to:

1. Write a formula equation from a word equation.
2. Balance a formula equation.
3. Name the four major types of inorganic reactions: combination, decomposition, single replacement, and double replacement.
4. Give an example of each type of reaction mentioned in objective 3.
5. Use a solubility table to predict if a substance is water insoluble.
6. Use the activity series to predict the products of a single replacement reaction.
7. Recognize an oxidation–reduction reaction.
8. Define oxidation as the loss of electrons by a substance undergoing a chemical reaction.
9. Define reduction as the gain of electrons by a substance undergoing a chemical reaction.
10. Define reducing agent as a substance that causes another substance to be reduced, and oxidizing agent as a substance that causes another to be oxidized.

INTRODUCTION

Within our own bodies and without, chemical reactions are taking place all the time (Fig. 5-1). As we breathe oxygen gas (O_2) into our lungs at this moment, some previously inhaled oxygen gas is returned to the atmosphere as carbon dioxide (CO_2). This is a result of a series of chemical reactions known as respiration. In our stomachs, small intestines, and large intestines, food that we've consumed over the past several hours is being broken down into basic nutrients by a series of chemical reactions known as digestion. Numerous other chemical reactions are taking place in our bodies all the time. There are reactions that build new cells and reactions in which old cells are broken down. There are reactions that store energy, and there are reactions that produce energy so that we may move around.

Outside our bodies countless other reactions are taking place. Gasoline is reacting with oxygen, allowing automobiles to run. Plants are absorbing carbon dioxide from the atmosphere and producing oxygen gas in a series of reactions known as photosynthesis. There are metals, such as iron, reacting with oxygen in the atmosphere to produce rust. There are also chemists in numerous laboratories around the world who are reacting various elements and compounds for the purpose of producing the many products that we use in our daily lives. Some of the reactions carried out by chemists are for the purpose of producing pharmaceutical compounds, for use by health practitioners in their never-ending fight against disease and illness.

As you can see, chemical reactions are *what's happening*! Even the production of sunlight is due to a chemical reaction of sorts. And although some chemical reactions occur instantaneously, whereas others occur over a long period of time, all of these reactions have a profound effect on our daily lives.

In this chapter we look at some very basic types of chemical reactions. Most of these reactions are of the inorganic type. However, later in the text we look at more complex reactions, specifically those which are involved with living organisms. To study chemical reactions, we must learn how to represent these reactions using chemical equations. The first part of the chapter is devoted to this task.

FORMULA EQUATIONS FROM WORD EQUATIONS: THE CHEMIST'S SHORTHAND

Just about any chemical reaction can be put down in words. For example:

$$\text{hydrogen gas} + \text{oxygen gas} \xrightarrow[\text{spark}]{\text{electric}} \text{water}$$

The "$+$" sign means *and*, and the "$\rightarrow$" means *yields*. The words *electric spark* over the yields sign indicates the necessary conditions for the reaction to occur. This *word equation* tells us that water can be produced from its elements hydrogen and oxygen. In this equation, the hydrogen and oxygen gas are called the *reactants*. This means that they are the starting substances. The water is called the *product*. This means that it is the substance produced.

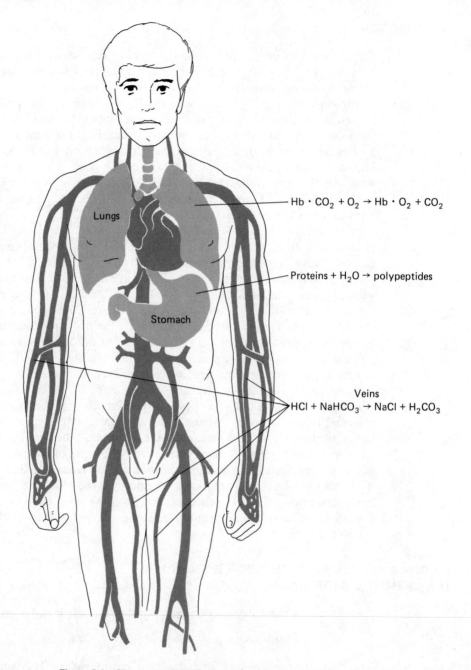

Figure 5-1 Chemical reactions are taking place in our bodies all the time.

However, word equations tell us nothing about the chemistry of the reaction. We can remedy this situation by writing the formula of each substance in place of the words. The information we learned in Chapter 4 will help us with that.

$$\text{Hydrogen gas} + \text{oxygen gas} \xrightarrow[\text{spark}]{\text{electric}} \text{water}$$

$$\text{H}_2 + \text{O}_2 \xrightarrow[\text{spark}]{\text{electric}} \text{H}_2\text{O}$$

We have written the symbols in place of the words. However, we must perform one additional task, and that is to balance the equation. What does this mean? It means that we must be sure that the same number of atoms of each element appear on both sides of the equation—the reactant side and the product side. This is because chemists have found that atoms are not lost or gained in a chemical reaction. This is what has come to be known as the *Law of Conservation of Mass*. (We should point out that atoms are not lost or gained in *common* types of chemical reactions. These are all types of chemical reactions *except* those involving *nuclear fission* or *fusion*. In nuclear fission and fusion, atoms can be lost or gained by being partially turned into energy. However, we'll save further discussion of this topic for Chapter 11.)

In the equation for the formation of water that we've just written, you will see that there are two hydrogen atoms and two oxygen atoms on the left side of the equation, and two hydrogen atoms and one oxygen atom on the right side of the equation. This equation is unbalanced. There is one less oxygen atom on the right side of the equation as opposed to the left. The way we balance a chemical equation is to use *coefficients*, that is, numbers that we place in front of the formulas of elements and compounds (Fig. 5-2).

Figure 5-2 Coefficients are used to balance chemical equations.

$$2\,H_2 \quad + \quad O_2 \quad \rightarrow \quad 2\,H_2O$$

Figure 5-3 Two molecules of hydrogen gas plus one molecule of oxygen gas produce two molecules of water.

We choose the numbers by trial and error, until we have the same number of atoms of each element on both sides of the equation. By the way, we try using the *smallest whole numbers possible*. In our example for water, we do the following:

$$2\,H_2 + 1O_2 \xrightarrow[\text{spark}]{\text{electric}} 2\,H_2O \quad \text{(balanced)}$$

This equation is now balanced. There are four hydrogen atoms and two oxygen atoms on the left side of the equation, and four hydrogen atoms and two oxygen atoms on the right side of the equation. The balanced equation tells us that two molecules of hydrogen gas plus one molecule of oxygen gas produce two molecules of water (Fig. 5-3). The previous equation may also be written in the following manner:

$$2\,H_2 + O_2 \xrightarrow[\text{spark}]{\text{electric}} 2\,H_2O$$

The coefficient in front of the oxygen doesn't have to be shown if it is the number 1.

When you attempt to balance a chemical equation remember that

1. The formulas of the reactants and products must be written properly and *never* changed.

2. The coefficients must be chosen so that the same number of atoms of each element appear on both sides of the equation.

Example 5-1 Write a balanced chemical equation for

$$\text{sodium chloride} \xrightarrow{\text{electricity}} \text{sodium metal} + \text{chlorine gas}$$

Solution Write the formulas for the reactants and products.

$$NaCl \xrightarrow{\text{electricity}} Na + Cl_2$$

There is one chlorine atom on the left side of the equation but two chlorine atoms on the right side. We can balance the chlorines by placing a 2 in front of the NaCl.

$$2 \text{ NaCl } \xrightarrow{\text{electricity}} \text{ Na } + \text{ Cl}_2$$

Now the chlorines are balanced, but the sodiums are not. We can balance the sodiums by placing a 2 in front of the Na atom on the right. Now we have

$$2 \text{ NaCl } \xrightarrow{\text{electricity}} 2 \text{ Na } + \text{ Cl}_2 \quad \text{(balanced)}$$

A very important thing to remember in balancing a chemical equation is never change or use subscripts. For example, you may *not* balance the sodium chloride equation by taking the following steps:

1. $\text{NaCl } \xrightarrow{\text{electricity}} \text{ Na } + \text{ Cl}_2$

2. $\text{NaCl}_2 \xrightarrow{\text{electricity}} \text{ Na } + \text{ Cl}_2 \quad \text{(balanced)}$

What's wrong in step 2? The answer is that the formula for sodium chloride is NaCl, *not* NaCl_2. Also, never insert a coefficient between the atoms in a molecule. Coefficients must be placed in front of the symbol for the molecule. Remember that we choose coefficients by trial and error. Do one element at a time; then after you think the equation is balanced, recheck your results.

Example 5-2 Write a balanced equation for

$$\text{zinc } + \text{ phosphoric acid } \longrightarrow \text{ zinc phosphate } + \text{ hydrogen gas}$$

Solution Write the formulas for the reactants and products.

$$\text{Zn } + \text{ H}_3\text{PO}_4 \longrightarrow \text{ Zn}_3(\text{PO}_4)_2 + \text{ H}_2$$

There is one zinc atom on the left side of the equation and three on the right. Therefore, place the coefficient 3 in front of the zinc atom on the left side of the equation.

$$3 \text{ Zn } + \text{ H}_3\text{PO}_4 \longrightarrow \text{ Zn}_3(\text{PO}_4)_2 + \text{ H}_2$$

The zinc atoms are balanced. However, there is one phosphate group on the left side of the equation and two phosphate groups on the right. Therefore, place the coefficient 2 in front of the phosphoric acid on the left side of the equation.

$$3 \text{ Zn } + 2 \text{ H}_3\text{PO}_4 \longrightarrow \text{ Zn}_3(\text{PO}_4)_2 + \text{ H}_2$$

The zinc atoms and phosphate groups are balanced. However, there are six hydrogen atoms on the left side of the equation and only two on the right. Therefore, place the coefficient 3 in front of the hydrogen on the right side of the equation.

$$3\,Zn + 2\,H_3PO_4 \longrightarrow Zn_3(PO_4)_2 + 3\,H_2 \quad \text{(balanced)}$$

The equation is now balanced. There are the *same number of atoms of each element on both sides of the yield sign.*

What follows are some word equations. See if you can balance them yourself before looking at the solutions that follow.

Example 5-3   Write a balanced equation for each of the following chemical reactions.

(a) potassium oxide + water → potassium hydroxide

(b) hydrogen gas + fluorine gas → hydrogen fluoride

(c) silver nitrate + barium chloride → silver chloride + barium nitrate

(d) calcium hydroxide + nitric acid → calcium nitrate + water

(e) aluminum + sulfuric acid → aluminum sulfate + hydrogen gas

Solution   Write the formulas for the reactants and products. Then balance the equation.

(a) $K_2O + H_2O \rightarrow KOH$
$K_2O + H_2O \rightarrow 2\,KOH$ (balanced)

(b) $H_2 + F_2 \rightarrow HF$
$H_2 + F_2 \rightarrow 2\,HF$ (balanced)

(c) $AgNO_3 + BaCl_2 \rightarrow AgCl + Ba(NO_3)_2$
$2\,AgNO_3 + BaCl_2 \rightarrow 2\,AgCl + Ba(NO_3)_2$ (balanced)

(d) $Ca(OH)_2 + HNO_3 \rightarrow Ca(NO_3)_2 + H_2O$
$Ca(OH)_2 + 2\,HNO_3 \rightarrow Ca(NO_3)_2 + 2\,H_2O$ (balanced)

(e) $Al + H_2SO_4 \rightarrow Al_2(SO_4)_3 + H_2$
$2\,Al + 3\,H_2SO_4 \rightarrow Al_2(SO_4)_3 + 3\,H_2$ (balanced)

SOME MAJOR TYPES OF CHEMICAL REACTIONS

There are tens of thousands of different chemical reactions that take place both inside and outside the body. Because there are so many chemical reactions known, chemists have tried to organize these reactions into a small number of groups or types. In the remainder of this chapter we will study four major types of chemical reactions that are important to inorganic chemistry.

In Chapters 12 to 17 we will study some additional types of reactions important to organic chemistry and biochemistry.

The four types of inorganic reactions we are going to look at are:

1. Combination reactions

2. Decomposition reactions

3. Single replacement reactions

4. Double replacement reactions

Combination Reactions

Combination reactions are those in which two or more substances combine to form a more complex substance. For example:

$$A + B \longrightarrow AB$$

A common type of combination reaction involves two elements forming a compound. For example: Sulfur + oxygen gas *yields* sulfur dioxide.

$$S + O_2 \longrightarrow SO_2 \quad \text{(balanced)}$$

Sulfur dioxide is one of the air pollutants that is responsible for causing such respiratory diseases as *chronic bronchitis* and *pulmonary emphysema* (Fig. 5-4).

Figure 5-4 Phenomena of typical bronchitis.

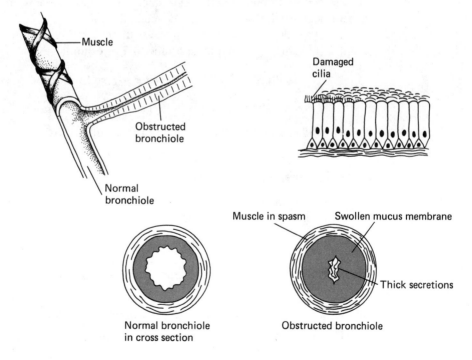

Muscle

Obstructed bronchiole

Normal bronchiole

Damaged cilia

Normal bronchiole in cross section

Muscle in spasm

Swollen mucus membrane

Thick secretions

Obstructed bronchiole

A combination reaction may also involve two simple compounds forming a more complex compound. For example:

$$\text{sodium oxide} + \text{water} \longrightarrow \text{sodium hydroxide}$$

$$\text{Na}_2\text{O} + \text{H}_2\text{O} \longrightarrow 2\,\text{NaOH} \quad \text{(balanced)}$$

In this reaction a metal oxide reacts with water and forms a base (a hydroxide compound). The compound formed here, sodium hydroxide, is commonly known as *household lye*.

Decomposition Reactions

A decomposition reaction is the reverse of a combination reaction. It involves the breakdown of a complex substance into simpler substances. For example:

$$\text{AB} \longrightarrow \text{A} + \text{B}$$

A common type of decomposition reaction involves a compound decomposing into its elements. For example:

$$\text{Mercury(II) oxide} \xrightarrow{\text{heat}} \text{mercury metal} + \text{oxygen gas}$$

$$2\,\text{HgO} \xrightarrow{\text{heat}} 2\,\text{Hg} + \text{O}_2 \quad \text{(balanced)}$$

Mercury(II) oxide has been used as a topical antiseptic, but it has also been used for diluting pigments for painting on porcelain. However, care must be taken in the use of mercury(II) oxide since it is even more poisonous than the element mercury, into which it can be made to decompose.

A decomposition reaction may also involve the breakdown of a more complex substance into simpler substances. For example:

$$\text{sulfurous acid} \xrightarrow{\text{heat}} \text{water} + \text{sulfur dioxide}$$

$$\text{H}_2\text{SO}_3 \xrightarrow{\text{heat}} \text{H}_2\text{O} + \text{SO}_2 \quad \text{(balanced)}$$

We've already mentioned the health hazards of sulfur dioxide.

Single Replacement Reactions

Single replacement reactions are those in which an uncombined element replaces another element from its compound.

$$\text{A} + \text{BC} \longrightarrow \text{AC} + \text{B}$$

A common type of single replacement reaction involves metals and acids. For example:

$$\text{aluminum} + \text{hydrochloric acid} \longrightarrow \text{aluminum chloride} + \text{hydrogen gas}$$

$$2\,\text{Al} + 6\,\text{HCl} \longrightarrow 2\,\text{AlCl}_3 + 3\,\text{H}_2 \quad \text{(balanced)}$$

This reaction shows you why it's not a good idea to store hydrochloric acid in an aluminum container; the container will be turned into a powdery substance, aluminum chloride.

Another type of single replacement reaction involves an uncombined metal replacing another metal which is already in a compound. For example:

$$\text{zinc} + \text{copper(II) sulfate} \longrightarrow \text{zinc sulfate} + \text{copper}$$

$$\text{Zn} + \text{CuSO}_4 \longrightarrow \text{ZnSO}_4 + \text{Cu (balanced)}$$

Copper sulfate solutions have long been used topically as fungicides, and zinc sulfate has been used medically as an astringent.

Double Replacement Reactions

Double replacement reactions are those in which two compounds exchange ions with each other. For example:

$$A^+B^- + C^+D^- \longrightarrow A^+D^- + C^+B^-$$

A common type of double replacement reaction is one that involves an acid and a base. These types of reactions are known as *acid–base reactions*. For example:

$$\text{hydrochloric acid} + \text{magnesium hydroxide} \longrightarrow \text{magnesium chloride} + \text{water}$$

$$2\,\text{HCl} + \text{Mg(OH)}_2 \longrightarrow \text{MgCl}_2 + 2\,\text{H}_2\text{O (balanced)}$$

Hydrochloric acid is produced in the stomach during the process of digestion, and is sometimes called stomach acid. Magnesium hydroxide is commonly called *milk of magnesia*. Its major uses are pharmaceutical. In low dosages of a few hundred milligrams, magnesium hydroxide acts as an antacid. In large dosages of 2 to 4 g, magnesium hydroxide acts as a laxative (Fig. 5-5).

Figure 5-5 In small doses magnesium hydroxide is an antacid; in large doses, a laxative.

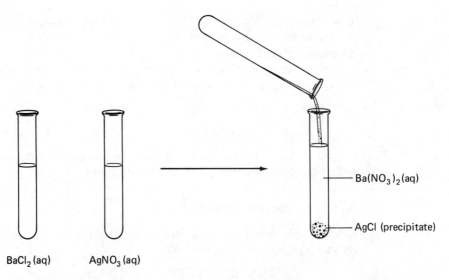

Figure 5-6 The reaction between aqueous barium chloride and silver nitrate produces the soluble salt barium nitrate and the insoluble salt silver chloride.

$BaCl_2$ (aq) $AgNO_3$ (aq)

Magnesium hydroxide neutralizes the excess stomach acid by reacting with the hydrochloric acid and producing the salt, magnesium chloride, and water. Another type of double replacement reaction involves two salts which are soluble in water. (*Soluble* means that the salt can be dissolved in the water.) The salts are usually dissolved in water separately, and then combined. A reaction occurs if one of the products (which is also a salt) is insoluble in water (Fig. 5-6). (*Insoluble* means that the salt is not able to dissolve in the water.) An example of such a reaction is

barium chloride + silver nitrate $\longrightarrow$ barium nitrate + silver chloride

$BaCl_2(aq)$ + 2 $AgNO_3(aq)$ $\longrightarrow$ $Ba(NO_3)_2(aq)$ + 2 $AgCl(s)$ (balanced)

The symbol (*aq*) means aqueous; in other words, the salt is soluble in water. The symbol (*s*) means solid and indicates that the substance is insoluble in water. Double replacement reactions of this type are important to us as biological species because these reactions represent one way that soluble ions can be incorporated into our bodies and then complexed as insoluble compounds. For example, one can imagine soluble compounds containing calcium ions and phosphate ions being consumed in our diets, and then eventually being combined in our bones in the form of calcium phosphate (a relatively insoluble substance). In order to determine whether a substance is soluble or insoluble in water, you may consult a solubility table. Such a table may be found in Appendix C.

Example 5-4 Predict the products of the following reactions, then write the correct formulas and balance the equation.

Combination reactions

(a) 2K + Cl$_2$ →   2KCl

(b) SO$_3$ + H$_2$O →   H$_2$SO$_4$

(c) H$_2$ + I$_2$ →   2HI

Decomposition reactions

(d) 2NH$_3$ →

(e) MgCO$_3$ →   MgO

(f) H$_2$O →

Single replacement reactions

(g) zinc + hydrochloric acid →

(h) magnesium + copper(II) nitrate →

(i) magnesium + phosphoric acid →

Double replacement reactions

(j) phosphoric acid + calcium hydroxide →

(k) AgNO$_3$ + BaCl$_2$ →

(l) H$_2$SO$_4$ + NaOH →

Solution   *Combination reactions*

(a) K + Cl$_2$ → KCl (It's KCl because in compounds K is +1 and Cl is −1.)

   2 K + Cl$_2$ → 2 KCl   (balanced)

(b) SO$_3$ + H$_2$O → H$_2$SO$_4$ (balanced)

(c) H$_2$ + I$_2$ → HI

   H$_2$ + I$_2$ → 2 HI   (balanced)

Decomposition reactions

(d) NH$_3$ → N$_2$ + H$_2$ (Remember, nitrogen and hydrogen are diatomic elements in their free states.)

   2 NH$_3$ → N$_2$ + 3 H$_2$ (balanced)

(e) MgCO$_3$ → MgO + CO$_2$ (balanced)

(f) H$_2$O → H$_2$ + O$_2$

   2 H$_2$O → 2 H$_2$ + O$_2$ (balanced)

Single replacement reactions

(g) $Zn + HCl \rightarrow ZnCl_2 + H_2$

$Zn + 2 HCl \rightarrow ZnCl_2 + H_2$ (balanced)

(h) $Mg + Cu(NO_3)_2 \rightarrow Mg(NO_3)_2 + Cu$ (balanced)

(i) $Mg + H_3PO_4 \rightarrow Mg_3(PO_4)_2 + H_2$

$3 Mg + 2 H_3PO_4 \rightarrow Mg_3(PO_4)_2 + 3 H_2$ (balanced)

Double replacement reactions

(j) $H_3PO_4 + Ca(OH)_2 \rightarrow Ca_3(PO_4)_2 + H_2O$

$2 H_3PO_4 + 3 Ca(OH)_2 \rightarrow Ca_3(PO_4)_2 + 6 H_2O$ (balanced)

(k) $AgNO_3 + BaCl_2 \rightarrow AgCl + Ba(NO_3)_2$

$2 AgNO_3 + BaCl_2 \rightarrow 2 AgCl + Ba(NO_3)_2$ (balanced)

(l) $H_2SO_4 + NaOH \rightarrow Na_2SO_4 + H_2O$

$H_2SO_4 + 2 NaOH \rightarrow Na_2SO_4 + 2 H_2O$ (balanced)

OXIDATION-REDUCTION REACTIONS: ANOTHER WAY TO CLASSIFY REACTIONS

We've just finished learning about four types of inorganic reactions: combination, decomposition, single replacement, and double replacement. Now here we are again with a fifth type—the *oxidation-reduction reaction*. Actually, however, this is not really a fifth type of reaction, because some of the reactions we've already studied fall into this category of reaction. Let's just say that this is simply another way to classify the reactions we've just studied. In other words, reactions may be classified as oxidation–reduction reactions (called redox reactions for short), or non-redox reactions. You'll see what we mean in a moment, but first let's understand what the terms "oxidation" and "reduction" mean.

Oxidation and Reduction Defined

In the past, oxidation was defined as the combination of a substance with oxygen. For example, glucose in our cells reacts with oxygen to produce carbon dioxide and water.

$$C_6H_{12}O_6 + 6 O_2 \longrightarrow 6 CO_2 + 6 H_2O \quad \text{(balanced)}$$

Another example is the rusting of an iron pipe.

$$4 Fe + 3 O_2 \longrightarrow 2 Fe_2O_3 \quad \text{(balanced)}$$

However, even though this is a useful definition, the concept of oxidation has been expanded to include reactions where oxygen is not involved. As presently defined,

oxidation is the loss of electrons by a substance undergoing a chemical reaction or, in other words, the change in oxidation number of a substance to a more positive oxidation number.

For example, in the reaction of iron plus oxygen, the oxidation number of iron changes from zero to positive 3, because each iron atom gives up three electrons in the reaction. (Remember, the oxidation number of an element in its uncombined state is zero.)

$$4 \, Fe^0 + 3 \, O_2 \longrightarrow 2 \, Fe_2^{+3}O_3$$

By our definition this means that the iron is oxidized. But what happens to the oxygen? We say that the oxygen has been reduced.

Reduction is the gain of electrons by a substance undergoing a chemical reaction or, in other words, the change in oxidation number of a substance to a more negative oxidation number.

In the reaction of iron plus oxygen, the oxidation number of the oxygen changes from zero to minus 2, because each oxygen atom gains two electrons from the iron.

$$4 \, Fe + 3 \, O_2{}^0 \longrightarrow 2 \, Fe_2O_3{}^{-2}$$

Notice that in this reaction, one substance (the iron) gets oxidized, while the other substance (oxygen) gets reduced. The processes of oxidation and reduction always occur together. In other words, you can't have oxidation in a reaction unless there is also reduction. And you can't have reduction in a reaction unless there is also oxidation.

Example 5-5   For each reaction, see if you can determine which substance gets oxidized and which substance gets reduced.

(a) $2 \, Na + Cl_2 \rightarrow 2 \, NaCl$

(b) $Zn + 2 \, HCl \rightarrow ZnCl_2 + H_2$

(c) $C + O_2 \rightarrow CO_2$

(d) $2 \, HgO \rightarrow 2 \, Hg + O_2$

Solution   Use your knowledge of oxidation numbers from Chapter 4. Also, remember that the oxidation number of an element in its uncombined state is zero.

(a) $2\,Na^0 + Cl_2{}^0 \longrightarrow 2\,Na^{+1}Cl^{-1}$

The oxidation number of the sodium changes from 0 to $+1$. This means that the sodium is oxidized. The oxidation number of the chlorine changes from 0 to -1. This means that the chlorine is reduced.

(b) $Zn^0 + 2\,H^{+1}Cl^{-1} \longrightarrow Zn^{+2}Cl_2{}^{-1} + H_2{}^0$

The oxidation number of the zinc changes from 0 to $+2$. This means that the zinc is oxidized. The oxidation number of the hydrogen changes from $+1$ to 0. This means that the hydrogen is reduced. The oxidation number of the chlorine does not change. It is neither oxidized nor reduced.

(c) $C^0 + O_2{}^0 \longrightarrow C^{+4}O_2{}^{-2}$

The oxidation number of the carbon changes from 0 to $+4$. This means that the carbon is oxidized. The oxidation number of the oxygen changes from 0 to -2. This means that the oxygen is reduced.

(d) $2\,Hg^{+2}O^{-2} \longrightarrow 2\,Hg^0 + O_2{}^0$

The oxidation number of the oxygen changes from -2 to 0. This means that the oxide ion is oxidized. The oxidation number of the mercury changes from $+2$ to 0. This means that the mercury(II) ion is reduced.

Oxidizing Agents and Reducing Agents

It is important for you to become familiar with the terms "oxidizing agent" and "reducing agent." Simply defined,

an oxidizing agent is a substance that causes something else to be oxidized.
and a reducing agent is a substance that causes something else to be reduced.

As things work out, the oxidizing agent is the substance being reduced. And the reducing agent is the substance being oxidized. Let's take, for example, the reaction involving iron and oxygen to form iron(III) oxide.

$$4\,Fe + 3\,O_2 \qquad 2\,Fe_2O_3$$

The iron is the substance being oxidized, and the oxygen is the substance being reduced. Therefore, the iron is the reducing agent, and the oxygen is the oxidizing agent.

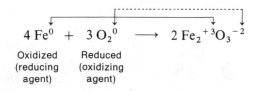

$$4\,Fe^0 + 3\,O_2{}^0 \longrightarrow 2\,Fe_2{}^{+3}O_3{}^{-2}$$

Oxidized   Reduced
(reducing   (oxidizing
agent)   agent)

Example 5-6 Using the reactions in Example 5-5, state which substance is the oxidizing agent, and which substance is the reducing agent.

Solution Just remember that the substance oxidized is the reducing agent, and the substance reduced is the oxidizing agent.

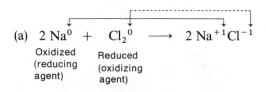

(a) $2\,Na^0 + Cl_2{}^0 \longrightarrow 2\,Na^{+1}Cl^{-1}$

Oxidized   Reduced
(reducing   (oxidizing
agent)   agent)

(b) $Zn^0 + 2\,H^{+1}Cl^{-1} \longrightarrow Zn^{+2}Cl_2{}^{-1} + H_2{}^0$

Oxidized   Reduced
(reducing   (oxidizing
agent)   agent)

Actually, the compound HCl is considered to be the oxidizing agent.

(c) $C^0 + O_2{}^0 \longrightarrow C^{+4}O_2{}^{-2}$

Oxidized   Reduced
(reducing   (oxidizing
agent)   agent)

(d) $2\,Hg^{+2}O^{-2} \rightarrow 2\,Hg^0 + O_2{}^0$

The mercury(II) ion is reduced. The oxide ion is oxidized. The compound, mercury(II) oxide, acts as both the oxidizing agent and reducing agent.

THE ACTIVITY SERIES

After many experiments, chemists have been able to arrange a list of elements according to their reactivities (Table 5-1). In this table, called the *activity series*, the most active (or readily reactive) elements appear at the top. As we move from the top to the bottom of the table, the reactivity of the elements decreases. Each element in the list can replace any element below it, in a compound; that is, by a single replacement reaction, it can take the other element's place.

TABLE 5-1 *The activity series*

METALS	NONMETALS
Lithium	Fluorine
Potassium	Chlorine
Calcium	Bromine
Sodium	Iodine
Magnesium	
Aluminum	
Zinc	
Chromium	
Iron	
Nickel	
Tin	
Lead	
Hydrogen[a]	
Copper	
Mercury	
Silver	
Platinum	
Gold	

[a] Hydrogen is in italic type because the activities of the other elements were calculated relative to hydrogen.

For example,

$$2\ Al + 3\ Zn(NO_3)_2 \longrightarrow 2\ Al(NO_3)_3 + 3\ Zn$$

Aluminum replaces zinc in the compound zinc nitrate, since aluminum is higher than zinc in the activity series.

The activity series helps us determine whether a reaction can take place. It is especially helpful when we are dealing with single replacement reactions.

Example 5-7 Predict whether the following reactions can occur.

(a) $Cu + 2\ HCl \xrightarrow{\text{heat}} CuCl_2 + H_2$

(b) $H_2 + CuO \xrightarrow{\text{heat}} H_2O + Cu$

Solution (a) When we look at Table 5-1, we see that copper is below hydrogen in the activity series. Therefore, copper cannot replace hydrogen in the compound HCl.

$$Cu + HCl \longrightarrow \text{no reaction}$$

(b) Looking at Table 5-1, we see that hydrogen is above copper in the activity series. Therefore, this reaction can occur since hydrogen can replace copper in the compound CuO.

$$H_2 + CuO \longrightarrow H_2O + Cu$$

SUMMARY

In this chapter we learned how to write a formula equation from a chemical word equation. We also learned how to balance chemical equations using a trial-and-error method. We saw that in order to balance a chemical equation we must use coefficients, which we place in front of the formulas of each substance. We also discovered that there are four types of inorganic reactions: combination reactions, decomposition reactions, single replacement reactions, and double replacement reactions. We looked at examples of each type of reaction. We learned about redox reactions, and how we could tell whether a given substance is being oxidized or reduced. We also learned the meaning of oxidizing agent and reducing agent. Finally, we looked at something called the activity series, and saw how this series could help us predict the products of a single replacement reaction.

EXERCISES

1. Write a balanced chemical equation from the following word equations.
   (a) potassium + water → potassium hydroxide + hydrogen gas
   (b) magnesium + copper(II) nitrate → magnesium nitrate + copper
   (c) sulfuric acid + iron(III) hydroxide → iron(III) sulfate + water
   (d) sodium metal + bromine → sodium bromide
   (e) aluminum + mercury(II) acetate → aluminum acetate + mercury

2. Identify each of the reactions in Exercise 1 as a combination reaction, decomposition reaction, single replacement reaction, or double replacement reaction.

3. For each of the combination reactions that follow, predict the product(s) and balance the equation.

   (a) $Mg + O_2 \xrightarrow{heat}$ (b) $H_2 + I_2 \xrightarrow{heat}$

   (c) $SO_3 + H_2O \longrightarrow$ (d) $N_2 + H_2 \xrightarrow[\text{high pressure}]{\text{heat}}$

4. For each of the decomposition reactions that follow, predict the product(s) and balance the equation.

   (a) $H_2O \xrightarrow{electrolysis}$ (b) $HgO \xrightarrow{heat}$

   (c) $MgCO_3 \xrightarrow{heat}$ (d) $NaCl \xrightarrow{electrolysis}$

5. For each of the single replacement reactions that follow, predict the product(s) and balance the equation. Don't forget to check the activity series to make sure that the reaction occurs.
   (a) $Mg + HCl \rightarrow$ (b) $Cu + H_2SO_4 \rightarrow$
   (c) $Zn + Ni(NO_3)_2 \rightarrow$ (d) $Mg + H_3PO_4 \rightarrow$

6. For each of the double replacement reactions that follow, predict the product(s) and balance the equation. Don't forget to check the solubility table to make sure that the reaction occurs.
    (a) $HBr + Mg(OH)_2 \rightarrow$
    (b) sulfuric acid + iron(III) hydroxide $\rightarrow$
    (c) $NaCl(aq) + Ba(NO_3)_2(aq) \rightarrow$
    (d) barium chloride(aq) + sodium sulfate(aq) $\rightarrow$

7. For each of the redox reactions that follow, determine which substance gets oxidized and which substance gets reduced.
    (a) $Mg + H_2SO_4 \rightarrow MgSO_4 + H_2$
    (b) $2\,Na + 2\,H_2O \rightarrow 2\,NaOH + H_2$
    (c) $2\,Hg + O_2 \rightarrow 2\,HgO$
    (d) $2\,Al + Fe_2O_3 \rightarrow Al_2O_3$

8. For each of the redox reactions in Exercise 7, determine which substance is the oxidizing agent, and which substance is the reducing agent.

9. The redox reactions in Exercise 7 may also be classified as combination reactions, decomposition reactions, single replacement reactions, or double replacement reactions. Determine, for each example, the type of reaction it is.

10. For each of the reactions that follow, write a balanced chemical equation. Be sure to check the activity series or solubility table if necessary.
    (a) potassium + water $\rightarrow$ potassium hydroxide + hydrogen gas
    (b) $H_2 + Br_2 \rightarrow$
    (c) lithium iodide $\xrightarrow{\text{electrolysis}}$
    (d) $Zn + Cu(NO_3)_2 \rightarrow$
    (e) $CuCl_2(aq) + Na_3PO_4(aq) \rightarrow$
    (f) $Ag + ZnCl_2 \rightarrow$
    (g) sulfurous acid (H_2SO_3) + oxygen gas $\rightarrow$ sulfuric acid
    (h) $SrCl_2(aq) + KNO_3(aq) \rightarrow$

Chapter 6

Energy: Where Does It Come from and Where Does It Go?

Some Things You Should Know After Reading This Chapter

You should be able to:

1. Explain in your own words, the Law of Conservation of Mass and Energy.
2. State what is meant by exothermic and endothermic reactions.
3. Explain how the energy of a chemical reaction is measured in calories, kilocalories, or joules.
4. Explain the significance of the positive and negative values associated with ΔH.
5. Explain the concept of power and state a unit in which power is measured.
6. Give some examples that explain how the sun is the source of nearly all energy on our planet.
7. State what is meant by the water cycle.
8. List some early uses of energy and explain the different ways that energy was generated over the past 2000 years.
9. Explain why we'll run out of fossil fuels in the not too distant future.
10. Explain how nature produces coal and oil.
11. Explain why we can't keep recycling energy.
12. Cite some ways that energy will be produced in the near future.

INTRODUCTION

Turn on a light, take a ride in your car, cook a roast in your oven, or just turn on your portable radio. All of these devices use energy. Where does this energy come from? Usually, from chemical reactions. The entire world operates on energy. In fact, the use of energy from chemical reactions is the basis of life itself. In this chapter we discuss the energy changes that occur in chemical reactions, and examine some of their practical applications.

SCENARIO

FLIGHT INTO FANTASY · · · OR THE FUTURE*

The scene is Hartsfield Atlanta International Airport in the year 2029.

Captain Lyle Jones, a 25-year-veteran Delta Air Lines pilot, is at the controls of the Hypersonic One, the DC-2000, a sleek 500-passenger, wide cabin airliner capable of flying at six times the speed of sound (Fig. 6-1). He has long since received his computerized flight plan and clearance but must check with control before departure.

The big McDonnell Douglas jetliner begins its takeoff roll, moving slowly, considering its cruise speed of more than 4000 miles per hour. Once airborne, Capt. Jones turns to his first officer, Bill Williams, a former U.S. Air Force pilot with more than 3000 hours in very large multi-engine cargo aircraft. "I estimate Sydney in 2 hours and 50 minutes," says Capt. Jones, "based on our flight plan. The computer says we'll need to maintain a setting of .5 on our anti-gravity system so the passengers get an easier ride. Just think, it was only 50 years ago when it used to take four or five hours just to get from Los Angeles to Atlanta. Passengers must have had a lot more time in those days."

Streaking along at Mach 6, the passengers barely have time to enjoy their computer games, closeup televised view of the scenery below, cocktails and meal before arriving in Sydney. They care little that all flights have been limited—no matter how far—to less than three hours since hypersonic aircraft first entered service about 20 years ago. Before that, even the Advanced Supersonic Transport, which began operating around 1992, traveled at only a little more than twice the speed of sound, with Los Angeles to Tokyo range.

Figure 6-1 Artist's conception of a hypersonic aircraft. (McDonnell Douglas Corporation, St. Louis, Missouri)

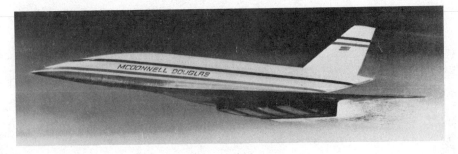

Hypersonic One, the DC-2000, is high over the Pacific Ocean above the weather systems and most of the friction caused by the Earth's atmosphere. The aircraft is performing flawlessly. "All systems check out normally," says 2nd Officer Joan Swenson.

"By the way, we've crossed the International Dateline," says Bill Williams. "It's time to contact Sydney control to confirm the prefiled descent plan."

"Sydney Control, this is Delta Hypersonic One. We've crossed the International Dateline and are now south of the Fiji Islands at Larex Intersection, flight level one zero zero zero. Request confirmation of descent plan."

"Delta this is Sydney Control. Continue approach according to flight plan. Reduce speed to Mach 3 and descend to flight level six zero zero. Maintain heading two three five degrees inbound on Sydney Omni Laser."

"Roger, Sydney Control. Reduce speed to Mach 3, descend to flight level six zero zero and maintain heading two three five degrees. Delta Hypersonic One."

(The crew begin talking about their new aircraft once again.) "NASA's rocket technology of the 1970's and 1980's helped us a lot in terms of power plants for this machine as well as new fuels. We finally beat the energy shortage, when the breeder reactor made liquid hydrogen economically viable for hypersonic transport," says Swenson.

"And wouldn't the passengers be impressed with the fancy name for the engines—'variable ducted liquid rocket propulsion system.'"

"It must have taken the power plant people a long time to figure out how to make it work, but the principle is simple—you take a rocket motor using liquid hydrogen fuel in conjunction with an onboard oxygen system for speeds from takeoff to about Mach 2. After that, we can open the inlets and use outside air instead of oxygen until we get so high that the air becomes too thin and we need our on-board oxygen again. We can use computer driven throttles to control our engine speed and to conserve fuel."

"That's how we can achieve long nonstop ranges like this Atlanta to Sydney hop of about 9200 miles. Not only that, we've also licked our emissions problems, since liquid hydrogen leaves only water as the end product of combustion."

"Actually, aviation has come pretty far in the last 50 years," says Capt. Jones.

* From Charles W. Heathco, "Flight into Fantasy or the Future," *Sky Magazine*, June 1979. Used by permission.

The foregoing scenario may seem incredible to us today. But recall what has happened in the last 50 years, when aviation moved from wooden cloth-covered biplanes to the supersonic Concorde. All these predictions may prove to be too conservative, especially when you consider that we may be able to transport people and freight by beaming their molecules from city to city in 50 years, à la "Star Trek"—just as we send radio waves today. But now let's return to the present.

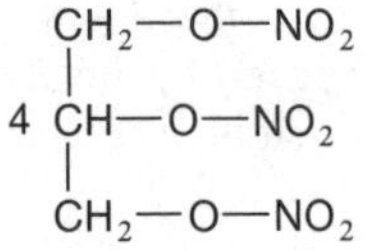

$$CH_2-O-NO_2$$
$$4 \; CH-O-NO_2 \longrightarrow 12\,CO_2 + 10\,H_2O + 6\,N_2 + O_2 + KABOOM!$$
$$CH_2-O-NO_2$$

Nitroglycerin
(major component
of dynamite)

Figure 6-2 The explosion of a stick of dynamite is a chemical reaction that releases a tremendous amount of heat energy to the environment.

MATTER AND ENERGY

Every substance has a specific amount of energy associated with it (a sort of chemical potential energy). When a substance reacts to form a new substance, it can either lose energy to the surroundings or gain energy from the surroundings. This is because in chemical reactions, old bonds between atoms are broken and new bonds are formed. Breaking a bond usually requires additional energy, while forming a bond usually releases energy. Depending on the number of bonds broken and the number of bonds formed, as well as the strengths of the bonds, energy is either released to the surroundings or absorbed from them.

Reactions that release energy to the surroundings are called *exothermic* reactions. For example, the explosion of a stick of dynamite is a chemical reaction that releases a tremendous amount of heat energy to the environment (Fig. 6-2). Reactions that absorb energy from the surroundings are called *endothermic* reactions. For example, plants live and grow by a process called *photosynthesis*. In this process the plants use carbon dioxide, water, and energy from the sun (sunlight) to make glucose and oxygen.

THE LAW OF CONSERVATION OF MASS AND ENERGY

Before we continue our discussion of energy and chemical reactions we want to discuss with you some very important scientific laws—the *Law of Conservation of Mass* and the *Law of Conservation of Energy*, or perhaps we should really say the *Law of Conservation of Mass and Energy* because these once-separate laws have been merged into one law in our century. But let us tell you the entire story.

We already have some familiarity with the Law of Conservation of Mass from Chapter 1. You should remember that it was the brilliant French chemist Antoine Lavoisier who did some experiments on matter. He heated a measured amount of tin, and found that part of it changed to a powder, and that the *product* (powder plus tin) weighed *more* than the original piece of tin. So he began heating metals in sealed glass jars. (The jars had air in them, of course.)

He measured the mass (or weight) of his starting materials (*reactants*), and when the reaction was finished (that is, when the metal and powder in the sealed jar got to the point of not changing any more), he carefully measured the mass of the resulting products. In every reaction that he carried out in these sealed jars, the mass of the reactants (the oxygen from the air in the jar, plus the original metal) equaled the mass of the products (the metal plus the powder). Today we know that after all the oxygen in the sealed jar was used up (because it combined with the metal to form the powder) the reaction just came to a standstill. But Lavoisier—although he didn't understand this—came to another correct conclusion. He maintained that *matter is neither created nor destroyed*; it just changes from one form to another.

We've been talking about matter, and the way it changes from one form to another. Okay—but what about energy? Back in the 1840's, more than half a century after Lavoisier, three scientists—the Englishman James Joule and the Germans Julius von Mayer and Hermann von Helmholtz—did various experiments which proved that, in a reaction, *energy is neither created nor destroyed*; it just changes from one form to another. This is the *Law of Conservation of Energy*.

A car motor is a good example of how one kind of energy is converted to a different kind. *Electrical* energy from the battery generates a spark that produces *heat* energy; the heat energy ignites the gasoline–air mixture in the cylinders, and this makes the pistons rise and fall. The rising and falling of the pistons is the *mechanical* energy that makes the car move. And this movement is also mechanical energy.

What's the significance of these facts? Let's think of the universe as a giant chemical reaction. At any given time there are certain amounts of matter and energy present, with the matter having a certain mass. The matter is always changing from one form to another, and so is the energy. In this century, it was also shown (by Einstein) that matter and energy are different forms of the same thing and that they also could interconvert (in other words, matter could become energy and energy could become matter). But *the total sum of matter (or mass) and energy in the universe is always the same* (constant). This is called the *Law of Conservation of Mass and Energy*.

HOW DO WE MEASURE ENERGY?

We can usually calculate the energy changes that result from reactions by measuring the heat released or absorbed during the reaction. This is because each chemical substance has a certain heat content associated with it. The heat content is called *enthalpy*, and has the symbol H. One mole of a chemical substance has a definite heat content (enthalpy). Although we can't measure the heat content of specific substances, we can measure the *change* in heat content during a chemical reaction. This is called the ΔH of the substance (read "delta-H" and indicates the change in enthalpy of the substance).

TABLE 6-1   *A comparison of energy values required or produced by some interesting phenomena*

PHENOMENON	APPROXIMATE AMOUNT OF ENERGY (kcal)
A flying housefly (30 mph)	7×10^{-8}
A pitched baseball (90 mph)	2.7×10^{-2}
A car (55 mph or 88 km/hr)	1.0×10^2
Exploding 1 kg of TNT	3.6×10^3
Burning 1 kg of wood	4.2×10^3
Burning 1 kg of anthracite coal	7.0×10^3
Burning 1 kg of gasoline	1.1×10^4
Burning 1 kg of natural gas	1.3×10^4
A 747 jumbo jet flying at 640 mph	3.2×10^6
Nuclear fission from 1 kg of ^{235}U	1.9×10^{10}
Nuclear fusion from 1 kg of ^2H	7.0×10^{10}
U.S. daily energy consumption	6.0×10^{13}
World daily energy consumption	2.0×10^{14}
Daily output of the sun	7.8×10^{27}

We usually measure the ΔH in units of calories (cal for short) or in kilocalories (kcal for short). A calorie is defined as the amount of heat needed to raise the temperature of 1 gram of water 1 degree Celsius. For example, if we were to heat 1000 g of water (that's 1 liter of water, which is slightly more than 1 quart) from room temperature (20°C) to boiling (100°C), we would require 80,000 cal of energy (which can also be reported as 80 kcal).

$$? \text{ cal} = \left(\frac{1 \text{ cal}}{g - °C}\right)(1000 \text{ g})(80 °C) = 80,000 \text{ cal}$$

Just to give you a better idea of what a calorie is, Table 6-1 shows a comparison of values of energies required or produced by some interesting phenomena and Table 6-2 shows the caloric value of some foods.

With the adoption of the SI metric system (the revised metric system) a few years ago, the joule was recognized by scientists as the basic unit of energy, replacing the calorie. However, the calorie is still used by many scientists to express the quantity of energy something contains, and it might be some time before it disappears from use. In case you find it necessary to convert from calories to joules or vice versa, remember that 1 calorie equals 4.18 joules.

1 calorie = 4.18 joules

TABLE 6-2 *Caloric value of some foods*

KINDS OF FOODS	AMOUNT	CALORIC VALUE (kcal)[a]
BEVERAGES, MILK, FATS		
Beer	354 ml (12 oz)	150
Butter	$\frac{1}{2}$ cup	810
Cola drinks	354 ml (12 oz)	145
Cream, sour	1 tbsp.	25
Wine, table	103 ml ($3\frac{1}{2}$ oz)	85
BREADS, CEREALS, SNACKS		
Bagel	1	165
Bread, white	1 slice	70
Cornflakes	1 cup	100
Puffed rice	1 cup	60
Fudge	28 g (1 oz)	115
Milk chocolate	28 g (1 oz)	145
Corn muffin	1	130
Honey	1 tbsp.	65
Ice cream	1 cup	200
Pizza, cheese	1 slice	185
Popcorn, plain	1 cup	25
Spaghetti, plain	1 cup	155
Whipped cream	1 tbsp.	10
FRUITS and FRUIT JUICES		
Apple	1	70
Apple juice	1 cup	120
Banana	1	100
Orange juice	1 cup	110
Prunes, cooked	1 cup	295
MEATS, CHEESE, EGGS, POULTRY, FISH		
Bacon (slices)	2	90
Beef, hamburger	84 g (3 oz)	245
Cheese, American	28 g (1 oz)	105

TABLE 6-2   (cont.)

KINDS OF FOODS	AMOUNT	CALORIC VALUE (kcal)[a]
Chicken, fried	84 g (3 oz)	155
Tuna, canned	84 g (3 oz)	170
Hotdog	1	170
Salami	84 g (3 oz)	390
VEGETABLES, NUTS		
Green beans	1 cup	30
Raw carrot	1	20
Corn, canned	1 cup	170
Peanut butter	1 tbsp.	95
Potato, baked	1	90
Walnuts	1 cup	790

[a] Remember that what the food industry calls the Calorie or big Calorie is actually what scientists call kilocalories or kcal. The values reported here are in kilocalories.

TABLE 6-3 *Heats of formation at 25°C and 1 atmosphere pressure*

SUBSTANCE	ΔH_f (kcal/mole)	SUBSTANCE	ΔH_f (kcal/mole)
$BaCl_2(s)$	-205.56	$CaCO_3(s)$	-288.5
$BaO_2(s)$	-150.5	$HI(g)$	$+6.2$
$CH_4(g)$	-17.9	$HCl(g)$	-22.1
$C_2H_2(g)$	$+54.2$	$H_2O(g)$	-57.8
$C_2H_6(g)$	-20.2	$H_2O(l)$	-68.3
$C_8H_{18}(l)$	-59.97	$H_2O_2(l)$	-44.8
$C_6H_{12}O_6(s)$	-304.4	$H_2S(g)$	-4.8
$CO(g)$	-26.4	$H_3PO_4(aq)$	-308.2
$CO_2(g)$	-94.1	$SO_2(g)$	-70.96
$CaO(s)$	-151.9	$SO_3(g)$	-94.45

Chemists have calculated the ΔH's of a number of chemical substances (Table 6-3). The ΔH's shown in Table 6-3 are known as the *heats of formation* of these compounds from their elements. For our purposes, we can define the heat of formation as the *heat that would be gained or lost when a compound is formed from its elements.* Notice that some of the values in Table 6-3 have a negative sign in front of them. This indicates an *exothermic* heat of formation. Other values have a positive sign in front of them. This indicates an *endothermic* heat of formation. These heats of formation are used to calculate energy changes in various chemical reactions.

Now that we are a little more familiar with energy and energy changes in chemical reactions, let's see how we got ourselves into our current energy mess. But before we do that, we should discuss the difference between energy and power.

POWER AND ENERGY

Energy is usually defined as the ability to do work, and it comes in two forms: the *potential* or stored form, and the *kinetic* or working form. The amount of energy consumed or produced in a given situation may be measured in calories. However, often we are more interested in the *rate* at which this energy is consumed or produced, not just the amount. It is here where the definition of *power* comes into use. *Power is the rate at which energy is consumed or produced.* For our purposes, a useful unit of power is the *watt*. A *watt* is equivalent to consuming or producing energy at the rate of 0.239 cal/sec. This means that

$$1 \text{ watt} = 0.239 \text{ cal/sec}$$

or
$$1 \text{ kilowatt} = 239 \text{ cal/sec}$$

We'll be using the terms "watt" and "kilowatt" in the rest of this chapter.

OUR SUN: THE SOURCE

The source of nearly all energy on our planet is the sun. The sun uses a process called *nuclear fusion* to produce fantastic amounts of heat and light. In this process, hydrogen is continuously converted into helium. (We have more to say about the fusion process in Chap. 11.) And although the earth receives only a small fraction of this solar energy (0.02 part per billion), this still amounts to an enormous amount. It comes to about 4.13×10^{16} cal/sec. That is

$$41{,}300{,}000{,}000{,}000{,}000 \text{ cal/sec} \qquad \text{or} \qquad 1.73 \times 10^{14} \text{ kilowatts}$$

(Just think of all the light bulbs you could work with that power!) The amount of energy from other sources like the tides and the earth's internal heat amount to only a small fraction of that contributed by the sun (Table 6-4).

Is all of this energy from the sun useful to us? Much of it is, in one way or another. However, almost 30% of the sun's rays are reflected back into space.

TABLE 6-4   *A comparison of solar power versus other sources for the earth*

SOURCE	POWER (kilowatts)[a]	PERCENT OF TOTAL
Solar	1.73×10^{14}	99 +
Internal heat	0.00032×10^{14}	0.02
Tides	0.00003×10^{14}	0.002

But 47% is converted to heat that warms our environment and makes our planet habitable. Another 23% powers the *water cycle*. (The water cycle is a process by which sunlight evaporates water from land and sea areas. In this manner, the radiant energy from the sun is converted to potential energy held by the water, either in the form of vapor or droplets in clouds. The potential energy of the water vapor and droplets is reconverted back to kinetic energy in the form of falling rain and snow.)

Only about 0.02% of the solar energy is used to directly power life processes on earth. This occurs in the process of photosynthesis, when green plants, in the presence of sunlight, convert carbon dioxide and water into glucose. Glucose is an energy-rich compound. Why is this process of photosynthesis so important? Because all animals, including fungi and bacteria, depend on green plants for their survival.

EARLY USES OF ENERGY

Let's journey back to when human beings first made their appearance on this planet. The chief source of energy was muscle power. This muscle power was obtained indirectly from the sun. Here's how it worked. Sunlight, as we've just explained, is responsible for the growth of green plants. The plants represent a form of stored energy. People ate these plants, or they ate the animals that ate these plants. The energy obtained from these foods allowed human beings to go about doing work that required muscle power. Of course, if more energy was needed to do a job than just mere human muscle power, animals could be used to do the work. However, this didn't really increase the availability of energy very much.

The first really big advance in energy availability was the *waterwheel*. This occurred about 2000 years ago in ancient Egypt. Waterwheels produced enough energy to grind grain and later to run small textile mills. The power produced by such wheels was between 2 to 10 kilowatts. Waterwheels remained an important source of energy up until the Renaissance.

The next advance in energy availability was the *windmill*. Windmills gained popularity around the 1400's, especially in countries like Holland. Old-style windmills produced better than 10 kilowatts of power, depending on the wind velocity. (We'll have more to say about modern windmills later in this chapter.) The beginning of the Industrial Revolution was due in part to windmills and waterwheels.

Another major advance in energy availability was the *steam engine*. This occurred about 1850 and jumped the energy availability by an enormous factor. The early steam engines were able to produce about 10,000 kilowatts. They used coal and wood to convert water into steam. It was the steam engine that really put the Industrial Revolution into high gear. Here's why. First, factories didn't have to be built by waterways. Second, the device could be used for propulsion as it was in the steam locomotive and steam ship. The invention of the *internal combustion engine*, a few decades later, further extended the use of engines to supply energy for humankind. These engines used oil for power.

It is interesting to note that the Industrial Revolution began with wood as as a major fuel to power steam engines. However, by 1900, most energy production came from the burning of coal (about 95%). By 1950, oil had replaced coal as the major source of energy.

The next major advance in energy production was the development of *nuclear fission* to generate electricity. This occurred in the late 1940's and early 1950's. These reactors produce energy in the millions of kilowatts (Fig. 6-3). For the first time in history, humankind had succeeded in obtaining an energy source that was independent of the sun. (All of the other fuels that we've discussed up to this point were produced with the help of solar energy.) We'll have more to say about nuclear fission in Chapter 11.

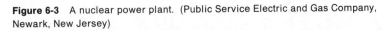

Figure 6-3 A nuclear power plant. (Public Service Electric and Gas Company, Newark, New Jersey)

Our story wouldn't be complete unless we told you about the next major step in energy availability—the fusion reactor. Once perfected, this reactor will use the process of the sun itself to produce practically limitless and pollution-free energy. We'll talk more about fusion power later in this chapter and again in Chapter 11.

THE AGE OF FOSSIL FUELS:
IT WILL SOON BE ENDING

The Industrial Revolution and much of our modern industry has been based on the use of fossil fuels—coal, oil, and natural gas. These fossil fuels were produced by decaying plant and animal organisms over 600 million years ago.

Scientists believe that coal was formed by decaying plant material that became buried under mud and water in huge swampy areas. Under conditions like this where oxygen is limited, the plants only partially decay. Plants contain large amounts of cellulose, a compound rich in carbon, hydrogen, and oxygen. It is thought that the hydrogen and oxygen turn into gaseous products, leaving behind a substance rich in carbon. This substance is what we call coal. This process takes millions of years to form what is called *hard coal*, or *anthracite coal*. Actually, the production of coal is thought to take place in a series of steps. The first step is the production of *peat*—a dark, wet, spongy material. Next, after 1 million years or more, according to some, the peat turns into *lignite*, a soft kind of coal, which has a woody appearance. Millions of years must again pass forming another type of soft coal called *bituminous coal*. This coal still has a lot of impurities in it, such as pitch and tar. And because of this, when we dig it up, bituminous coal must be cleaned before being used in our homes and factories. This is done by heating the coal to about 1400°C to burn off the impurities.

After a few million years more, the bituminous coal turns into anthracite or hard coal. This type of coal is densest and has the highest carbon content of the various types of coal. (However, even this type of coal has impurities in it, which makes coal burning a potential air pollution problem.)

All of the processes of coal production are still going on in nature today. The various forms of coal can be found in different areas of the earth. Does this mean that coal is a renewable resource? Not really, because we will have consumed in a few hundred years what has taken nature millions of years to produce.

Although *coal* appears to be derived from *plant matter*, *oil* appears to be a product of *animal decomposition*, decomposition that has taken place over millions of years. The current theory is that oil and natural gas were formed from the fats of ocean-dwelling microorganisms. That's because oil is always found in rocks of oceanic origin. Fats are compounds that contain lots of carbon, hydrogen, and oxygen in their molecules. If the oxygen is removed from such molecules with a simultaneous rearrangement of the carbon and

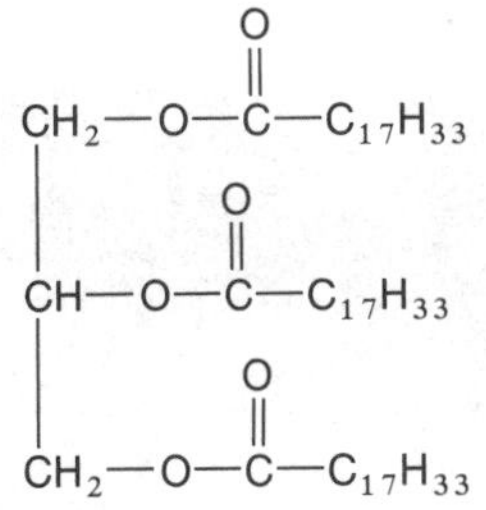

$$CH_2\!-\!O\!-\!\overset{\displaystyle O}{\overset{\|}{C}}\!-\!C_{17}H_{33}$$

$$CH\!-\!O\!-\!\overset{\displaystyle O}{\overset{\|}{C}}\!-\!C_{17}H_{33}$$

$$CH_2\!-\!O\!-\!\overset{\displaystyle O}{\overset{\|}{C}}\!-\!C_{17}H_{33}$$

Triolein, a typical fat

$$CH_3\!-\!CH_2\!-\!CH_2\!-\!CH_2\!-\!CH_2\!-\!CH_2\!-\!CH_3$$

n-heptane, a typical hydrocarbon
molecule found in petroleum

Figure 6-4 Fats are compounds that contain lots of carbon, hydrogen, and oxygen in their molecules. If the oxygen is removed from such molecules with a simultaneous rearrangement of the carbon and hydrogen atoms, we obtain a typical hydrocarbon molecule found in petroleum.

hydrogen atoms, we obtain a typical hydrocarbon molecule found in petroleum (Fig. 6-4).

Fossil fuels are still being formed in nature. For example, swamps that are a few hundred years old contain peat bogs which are the beginnings of the formation of coal (Fig. 6-5).

Figure 6-5 The formation of coal begins in swamps with the formation of peat bogs. (National Coal Association)

TABLE 6-5 *Estimates of how long various fuels will last*[a]

FUEL	NUMBER OF YEARS AT PRESENT RATE OF CONSUMPTION	NUMBER OF YEARS IF CONSUMPTION CONTINUES TO DOUBLE EVERY 14 YR. FOR COAL, 16 YR. FOR OIL, 8 YR. FOR GAS, AND 3 YR. FOR URANIUM
Coal	683 (U.S.)	60 (U.S.)
	3600 (world)	3600 (world consumption has been constant over the years)
Oil	69 (U.S. and world)	25 (U.S. and world)
Gas	276 (U.S.)	20 (U.S.)
	621 (world)	45 (world)
Uranium, ^{235}U for conventional fission reactors	—	19 (U.S.)

[a] Using 1983 as the base year and calculated on present known reserves.

It is said that we are using fossil fuels 50,000 times faster than they are being renewed. The question that we would all like to know is: How long will they last? No one can really be sure. But we can make some estimates based on the present rates of consumption and the known reserves (Table 6-5). The data in Table 6-5 indicate that our oil and natural gas will be depleted in the not too distant future. Coal, on the other hand, appears to have a brighter outlook. Coal reserves in the United States could last anywhere from 100 to 600 years, and world supplies even longer than that. The uranium needed for conventional fission reactors will be exhausted by the year 2000, unless new sources are discovered.

IF ENERGY CAN BE NEITHER CREATED NOR DESTROYED, WHY CAN'T WE KEEP RECYCLING WHAT WE HAVE?

The title of this section asks a very legitimate question. The answer to this question lies in a series of statements known as the *First and Second Laws of Thermodynamics*.

The First Law of Thermodynamics

We've already discussed the *First Law of Thermodynamics*. It is the *Law of Conservation of Energy*. This law says that "Energy can never be created or destroyed, it can just change from one form to another." This sounds very encouraging, because if we could recycle energy, we would never run out of it. Unfortunately, nature has seen fit to make this dream impossible. The reason is found in the *Second Law of Thermodynamics*.

The Second Law of Thermodynamics

To paraphrase the *Second Law*: Every time we use energy, some of it changes into heat and cannot be used to do work. For example, much of the energy used to light an ordinary light bulb is wasted as heat. Therefore, part of the energy is "lost" in a manner of speaking, as it dissipates into the environment—lost in terms of doing useful work. And because the heat dissipates, there's no way of getting it back, or concentrating it again.

So our energy storehouse continuously runs downhill. In fact, the Laws of Thermodynamics also seem to apply to the entire universe. If this is true, we can think of the universe as having started with a certain quantity of energy some 20 billion years ago. This energy was in a very concentrated form. As the universe aged its energy supply began to run down, as much of the energy dissipated into space as heat. This process continues today.

What does this mean for us on earth? Well, we don't have to worry about the universe running out of energy. That's a long way off (and we're not even sure if it's going to happen). We also don't have to worry about the sun running out of energy, although we're pretty certain it will in about 4 or 5 billion years. (Perhaps if humankind survives that long, we'll be able to move our entire civilization to another star system.) What we do have to worry about is making the most of what we have here on earth. We realize that our energy supply is not infinite, at least in terms of conventional fuels, and we have to find alternative ways to generate energy in the future.

ENERGY PRODUCTION IN THE NEAR FUTURE

In recent years our energy production has barely kept up with our energy needs. What can we do about it? There are essentially three things that we can do:

1. Use less energy.

2. Use the energy sources that we have more efficiently.

3. Find new energy sources.

Of course, the first two options are only stopgap measures. However, they are important to buy us time to implement option 3—find new energy.

What types of new energy sources are being considered today? Table 6-6 lists a number of them. You can see that some of these energy sources are not

TABLE 6-6 *Future energy sources*

ENERGY SOURCE	DESCRIPTION AND COMMENTS
Methanol	Also known as methyl alcohol or wood alcohol, it can be mixed with gasoline up to about 10% and sold as "gasohol." No engine adjustments are necessary. With proper engine adjustments, automobiles can run on 100% methanol. Methanol can be produced from wood, plant materials, and even garbage. (Ethanol or ethyl alcohol, the substance we enjoy in our favorite alcoholic beverages, can also be used as a fuel with gasoline much like methanol. Ethanol can be produced by fermenting many different types of food substances.)
Hydrogen	When hydrogen burns energy is produced: $$H_2 + \tfrac{1}{2}O_2 \longrightarrow H_2O(l) + 68 \text{ kcal}$$ The product of combustion is water—no pollution. The hydrogen can be obtained from a variety of sources, including the electrolysis of water.
Thermal energy from the oceans	Can be used where temperature gradients exist between the various layers (levels) of ocean water. Pilot plants are being constructed in the Gulf of Mexico and Hawaii. These plants will be on floating platforms (50 m high and 100 m in diameter) and will produce about 1×10^8 watts of electricity.
Geothermal energy	Heat from the depths of the earth, for example the heat from active volcanoes and geysers, can be turned into electricity. This energy is free for the taking with only one serious problem —the hot water from geysers is loaded with salts and other corrosive materials, as well as with odorous substances such as hydrogen sulfide and ammonia. Once this problem is solved, geothermal will be an excellent source of energy in certain parts of the world.
Oil shale	Sizable deposits of the rock called shale occur in Colorado, Wyoming, and Utah. Because this shale has oil "soaked into it," it is known as oil shale. The oil can be obtained by crushing and heating the rocks. This is an expensive proposition which is now becoming economically worthwhile.
Solar energy	A clean, free, and safe energy source. Solar energy can be used in many ways. On individual homes, solar panels can be placed on roofs and used to heat water. Solar panels may one day be placed in space. These panels will collect solar energy and transmit it back to earth in the form of microwaves. The microwaves will be converted to electricity for distribution around the country.
Nuclear fusion	Uses deuterium, an isotope of hydrogen, ^{2}H, to produce energy. This is the process that stars, such as our sun, use to produce temperatures of 10 million degrees. Humankind has produced uncontrolled fusion in the hydrogen bomb. But controlled fusion has yet to be accomplished. Once attained, this energy source is practically limitless since there is enough deuterium in the oceans to last for hundreds of thousands of years.

Figure 6-6 A modern windmill. (NASA)

so new. For example, windmills of modern design can generate enough power to supply electricity for homes or small stores (Fig. 6-6).

Some of the other methods of future energy generation deserve mention. For example, pilot plants are currently being developed to use the *thermal energy* in the oceans to generate electricity. This is made possible because of the temperature variation in the various layers of ocean water.

Oil shale represents an enormous storehouse of energy in the United States. However, the process of removing the oil from the shale is expensive. But as petroleum prices continue to rise, oil shale will become competitive. Once full-scale production plants are erected, oil shale may turn out to be a new source of petroleum (Fig. 6-7).

(a)

(b)

(c)

Figure 6-7   (a) An oil shale facility in Rifle, Cororado. (American Petroleum
Institute) (b) A mountain containing oil shale. (American Petroleum Institute)
(c) Part of a petroleum refinery. (Atlantic Richfield Company)

(a)   (b)

Figure 6-8   (a) Solar collectors atop a building at the Marshall Space Flight Center. (b) Solar collectors on a new office building in Huntsville, Alabama. (NASA)

Solar power represents another important pollution-free energy source for the future. In fact, the technology exists now for using solar power in many parts of the country to supply hot water and heat for homes. These systems have become commercially available in the past few years (Fig. 6-8). Perhaps as the efficiency of these devices increases and their costs come down, more people will install them. (Just imagine having a very efficient solar collector on your house, hooked up to storage batteries, that could supply all your energy needs, day and night, without monthly utility bills!)

Plans are also on the drawing boards to erect giant solar collectors in space. These solar collectors will collect the energy of the sun and then turn it into microwaves for transmission to earth (Fig. 6-9). The microwaves will be collected at a central location and turned into electricity. Of course, a few technical difficulties have to be ironed out. For example, locations for the collecting stations must be found. After all, you don't want microwaves bouncing around a populated area—the long-range effects of low-level microwave radiation on human beings is not yet known. However, after all the problems are solved, this might be an efficient way to supply solar radiation to millions of people. (The utility companies would like this system very much since they would be involved in running it. This way, they would in effect be metering the sun!)

Our best hope for practically limitless power is the process of controlled *nuclear fusion*. As we've already said, this is the process that the stars use to generate fantastic amounts of energy. In this process, hydrogen atoms are turned (or fused) into helium atoms with the liberation of tremendous quantities of heat, and no waste products. Hydrogen is the most abundant element in the universe. Once tamed, this process can be used to generate limitless quantities of electricity. The electricity could then be used to run just about everything else. (We have more to say about the process of nuclear fusion in Chap. 11.)

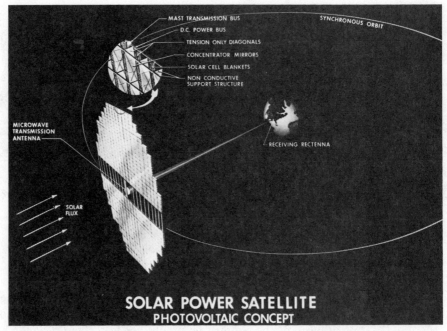

Figure 6-9 Artist's conception of how solar collectors in space will convert solar power to microwaves for transmission to earth: (a) and (b) show the collectors in space, (c) shows a receiving station on earth located offshore. (NASA)

SUMMARY

In this chapter you learned many things about energy and how it is produced. The important points to remember are:

1. The way the Law of Conservation of Mass and Energy works

2. How energy is produced in chemical reactions

3. How energy is measured in calories, kilocalories, and joules

4. The meaning of enthalpy

5. The difference between an exothermic and endothermic reaction

6. The meaning behind the negative and positive values of ΔH

7. The difference between power and energy

8. Units of power

9. How the sun is responsible for most of the energy on earth

10. What is meant by the water cycle

(b)

(c)

11. How power was produced over the course of the last 2000 years

12. How coal and oil are produced by nature

13. The importance of the First and Second Laws of Thermodynamics

14. Various methods of energy production for the near future

EXERCISES

1. Explain from a molecular point of view how energy is produced or consumed in a chemical reaction.

2. The two basic substances that compose our universe are _____ and _____ .

3. Reactions that release energy to the surroundings are called _____ reactions.

4. Reactions that absorb energy from the surroundings are called _____ reactions.

5. Explain in your own words the Law of Conservation of Mass and Energy.

6. Define a calorie. Based on your definition, how many calories are needed to raise the temperature of $5\overline{00}$ g of water from $3\overline{0}°C$ to $7\overline{0}°C$? How many kcal is this?

7. Define the terms ΔH_R and ΔH_f.

8. Explain the difference between power and energy.

9. A light bulb produces 23.9 cal/sec. How many watts of power is this?

10. An industrial machine consumes 5000 kilowatts of power. How many calories per second of energy have been consumed? How many kilocalories per second in this?

11. Give some examples that explain how the sun is the source of nearly all energy on our planet.

12. Explain how the water cycle works.

13. Match the energy-producing device with its description.
    (1) muscles
    (2) waterwheel
    (3) windmill
    (4) steam engine
    (5) fission reactor
    (6) fusion reactor

    (a) Represents the first energy-producing device, completely independent of the sun.
    (b) This device gained popularity around the 1400's in countries such as Holland.
    (c) An energy device that will use the process of the sun itself to produce practically limitless amounts of energy.
    (d) The chief source of power used by early man.
    (e) A major breakthrough in energy production that occurred in the 1850's.
    (f) A major advance in energy availability that occurred about 2000 years ago.

14. Explain why we will run out of fossil fuels in the not too distant future.

15. Explain how nature produces coal and oil.

16. What are some of the consequences of the First and Second Laws of Thermodynamics?

17. List at least four ways that energy will be produced after the year 2000. For each method you list, state the advantages and disadvantages.

18. What fuel do you use to heat your home? What are the advantages and disadvantages of this fuel?

19. Given a choice of living near a coal-burning power plant or a nuclear fission plant, which would you choose? List the reasons for your choice.

Chapter 7

The Three States of Matter: Solids, Liquids, and Gases

Some Things You Should Know After Reading This Chapter

You should be able to:

1. Name the three states of matter: solid, liquid, and gas.
2. Compare and contrast the properties of solids, liquids, and gases.
3. Describe what is meant by crystalline and amorphous solids.
4. Describe the difference between atomic, molecular, and ionic crystals.
5. Define what is meant by a crystal lattice.
6. Describe the forces that hold crystals together.
7. Describe the process of evaporation.
8. Explain what happens when a solid changes to a liquid, and when a liquid changes to a gas.
9. Discuss the major points of the kinetic theory of gases.
10. Convert °C to °K and °K to °C.
11. Convert atmospheres to torr and torr to atmospheres.
12. Use the combined gas law to solve gas law problems.
13. Explain Dalton's law of partial pressures.
14. Explain Henry's law.
15. Discuss the process of respiration in terms of how oxygen moves from the lungs to the cells of the body, and how carbon dioxide moves from the cells of the body to the lungs.
16. Explain the concept behind oxygen hyperbaric therapy and give some of its uses.

INTRODUCTION

It's true, an ice cream soda contains all three states of matter! The ice cream is solid, the water is liquid (that is, the water that makes up the soda), and the soda water has carbon dioxide gas dissolved in it. Although the various substances we are all familiar with usually appear as one of the three states of matter, many substances can exist in all three states—water, for example, can exist as solid ice, liquid water, or gaseous steam (Fig. 7-1). A comparison of the three states of matter can be seen in Table 7-1.

Each of the three states of matter is important to living beings like ourselves. Our bodies are composed of solids, liquids, and gases. Solids compose our bones, skin, and teeth. Liquids compose the fluids in our cells and blood. Gases are used in our bodies in the process of respiration. And let's not forget that the foods we eat are also composed of solids, liquids, and gases. For these reasons and more, it is important that we study the three states of matter.

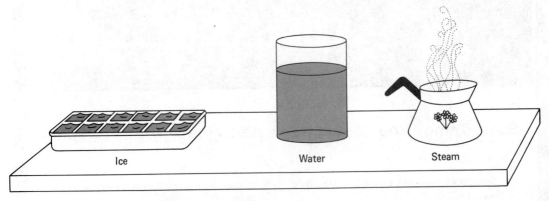

Ice   Water   Steam

Figure 7-1 The three states of H_2O: solid ice, liquid water, and gaseous steam.

TABLE 7-1 *Properties of solids, liquids, and gases*

PROPERTY	SOLIDS	LIQUIDS	GASES
Volume	Have a definite volume	Have a definite volume	Have no definite volume (gases expand to fill their containers)
Shape	Have a definite shape (most solids have a well-defined crystalline arrangement)	Have no definite shape (liquids assume the shape of their containers)	Have no definite shape
Compressibility	Cannot be compressed (they are practically incompressible)	Tend to be incompressible	Are easily compressed

THE SOLID STATE

The major characteristics of solids are their fixed shape and fixed volume. Most solids are of the crystalline variety. This means that the particles which compose the solid are arranged in a definite geometric pattern (Fig. 7-2). Some of these geometric patterns are shown in Table 7-2. Some solids, such as glass and paraffin, have no definite geometric pattern. In other words, they are not crystalline. These solids are said to be *amorphous*. Although amorphous solids are interesting to study, we'll confine our discussion to crystalline solids.

Solids can be composed of three types of particles: atoms, molecules, and ions. Solids such as carbon, silicon, copper, and gold are composed of atoms. Solids such as dry ice (CO_2), iodine, and water (ice) are composed of molecules. Solids such as sodium chloride, calcium chloride, and sodium bromide are composed of ions. These three types of solids have very different properties. Table 7-3 sums up some of the differences.

All About Crystals

A crystalline solid is obviously composed of crystals. But what is a *crystal*? The dictionary says that

> *a crystal is a solid body having a characteristic internal structure and enclosed by symmetrically arranged plane surfaces, intersecting at definite and characteristic angles.*

Figure 7-2 Crystals of sodium chloride. (American Museum of Natural History)

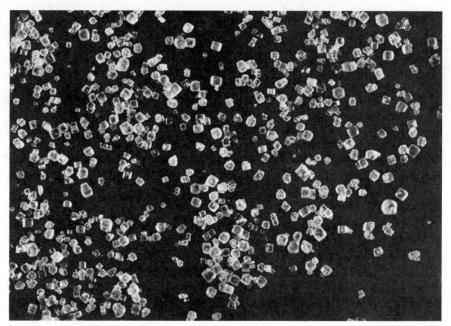

TABLE 7-2 *Some crystal structures*

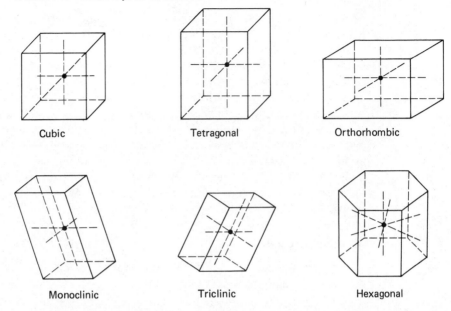

| Cubic | Tetragonal | Orthorhombic |

| Monoclinic | Triclinic | Hexagonal |

TABLE 7-3 *Some properties of crystalline solids*

TYPE	MELTING POINT	HARDNESS	CONDUCTIVITY	EXAMPLE
Ionic	High	Hard and brittle	Nonconductors of electricity	Sodium chloride, NaCl
Molecular	Low	Soft	Nonconductors of electricity	Dry ice, CO_2
Atomic				
Metallic	High	Hard	Good conductors of electricity	Copper, Cu
Nonmetallic	High	Hard and brittle	Nonconductors of electricity	Diamond, C

In simpler terms this means that a crystal is a symmetrical object. The symmetrical structure formed by the particles in a crystal is called the *crystal lattice*. The crystal lattice is the repeating unit within the crystal that gives the crystal its pattern. Actually, the crystal lattice is repeated over and over again in the crystal. One repeating unit, however, is called a *unit cell*. As we said before, there are three types of particles that compose solids: atoms, molecules, and ions. It is also these three particles which occupy the lattice positions in a crystal (Fig. 7-3).

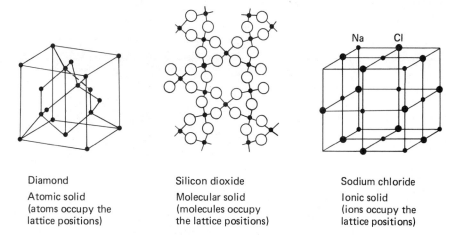

| Diamond | Silicon dioxide | Sodium chloride |
| Atomic solid (atoms occupy the lattice positions) | Molecular solid (molecules occupy the lattice positions) | Ionic solid (ions occupy the lattice positions) |

Figure 7-3 Some examples of different types of crystalline solids.

Atomic, Molecular, and Ionic Crystals: A Look at What Holds Them Together

The properties of a crystalline solid are partly the result of the forces that bind together the particles that make up the crystal. Let's look briefly into the bonding forces of the three basic types of crystals (ionic, molecular, and atomic).

In an *ionic* crystal, it is the attractive forces between oppositely charged ions that hold the crystal together. Since this kind of bonding force is very strong, ionic solids have very high melting points. For example, table salt (NaCl) has a melting point of 800°C. It takes a lot of heat energy to force those ions apart.

Molecular crystals, on the other hand, are usually held together by weak electrical forces. Only molecular crystals whose molecules are polar have binding forces that are fairly strong. The forces binding most molecular crystals together are so weak that these crystals frequently have much lower melting points than ionic crystals. For example, ice melts at 0°C.

Solids composed of *atomic* crystals have strong binding forces. Table 7-4 shows how strong. Look at the melting points of some of these substances. When the atoms are nonmetallic, they are held together by covalent bonds. An example is diamond, made up of carbon atoms that bond covalently. It takes a temperature of 3500°C to break those carbon-carbon bonds.

When atomic crystals are composed of metallic atoms, it is the attraction of opposite charges that makes the bonds so strong. Scientists think of the metallic lattice as composed of positive ions surrounded by a cloud of electrons. The electrons are given off by the metallic atoms, but are considered to be part of the entire crystal—electrons at large, you might say. Since the electrons are free to wander throughout the crystal, they are able to conduct electricity. This is the reason that metals are good conductors of electricity.

TABLE 7-4 *Melting points of some atomic crystals*

SUBSTANCE	MELTING POINT (°C)
Iron	1535
Copper	1083
Diamond	3500
Silicon	1410

THE LIQUID STATE

The major characteristics of liquids are their fixed volume and indefinite shape. This means that the particles of a liquid are not held together as rigidly as a solid. In other words, the particles of a liquid may move about each other, so that they can take the shape of their container. The fact that the particles of a liquid can move about gives rise to an interesting property of liquids called *evaporation*.

Evaporation: The Case of the Disappearing Liquid

When a liquid *evaporates*, it seems to disappear. But actually, the molecules of liquid have turned into molecules of gas. If you leave a glass of water on a table for a few days, you will notice that the water level in the glass drops due to the process of evaporation. But what causes the evaporation to take place? Surely, the temperature of the liquid as well as the exposed surface area of the liquid should affect the speed of evaporation. But there's more to the process than that and it has to do with the fact that *the molecules of a liquid are in constant motion about each other*. This means that the molecules of liquid have a specific amount of energy, and as these molecules move about and collide with each other, some liquid molecules transfer some of their energy to other liquid molecules (Fig. 7-4). If a molecule on its way up to the surface of the liquid

Figure 7-4 Liquid molecules move about and collide with one another. In these collisions some liquid molecules transfer energy to other liquid molecules.

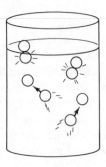

collides with a molecule already on the surface, the surface molecule may get enough energy from this collision to escape from the liquid. After this process has continued for a period of time, we notice that there is less liquid in the container. We say that the liquid has evaporated.

As a liquid evaporates, its temperature decreases. This is because the more energetic molecules (the warmer ones) are leaving. Think what this means to us! When we exercise or perform hard work, heat is brought to the skin, where it is transferred to water molecules. These molecules of water evaporate and our skin feels cooler. In this manner our body temperature can be maintained at 37°C (98.6°F). This is extremely critical for good health because at temperatures a few degrees above 37°C we can suffer from heat prostration and even death due to the malfunction of our central nervous system. This critical temperature is about 41°C (106°F). (If you don't know how to convert from °C to °F, and vice versa, see Appendix B.)

Equilibrium Vapor Pressure

If we put a cover on a container that is partly filled with water, evaporation begins, but seems to stop after a time. How can we explain this? Picture the situation shown in Figure 7-5. Water molecules begin to escape from the surface of the liquid. These molecules occupy the air space in the partly filled container. As this process continues, more and more escaping molecules occupy the same space. This increases the chance that they will collide with one another, and that some of them will be bumped back into the water. At first many molecules escape, and few return; but after a while, as the concentration of the molecules in the vapor state increases, the number of molecules returning to the liquid also increases. Eventually, we reach a point at which the number of molecules escaping from the water equals the number of molecules returning to it. The net result is that no additional liquid seems to evaporate from the container. But, in reality, a *dynamic equilibrium* is in progress. This means that the two opposing processes are proceeding at equal rates (Fig. 7-6).

The pressure of the molecules that have escaped in our partly filled container at the equilibrium point is called the *equilibrium vapor pressure* of

Figure 7-5 Liquid molecules escaping from the surface of the liquid.

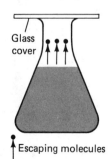

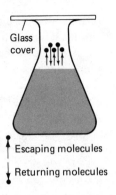

Figure 7-6   Dynamic equilibrium.

the liquid. We can define the equilibrium vapor pressure of any substance as the pressure exerted by a vapor when it is in equilibrium with its liquid at any given temperature. For every liquid at a given temperature there is a characteristic vapor pressure at equilibrium.

Let's go back to our covered container that is partly filled with water. Remember that the water and the water vapor are at equilibrium. What happens to the equilibrium vapor pressure if we raise the temperature of the water? The water molecules should gain energy as their temperature is increased. As a result, a greater number of molecules should have enough energy to escape from the surface of the liquid and become vapor. In fact, that's just what happens. The equilibrium that has existed is now upset, but it eventually reestablishes itself at the new temperature. The overall effect is a higher equilibrium vapor pressure at the higher temperature (Fig. 7-7).

Comparing equilibrium vapor pressures of different liquids at the same temperature tells us how *volatile* a liquid is (that is, how readily it vaporizes). Liquids that have high equilibrium vapor pressures tend to evaporate easily. Some examples are gasoline, ethyl alcohol, and ethyl ether. Liquids that have

Figure 7-7   Liquids have a higher equilibrium vapor pressure at a higher temperature.

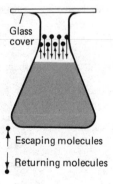

low equilibrium vapor pressures tend to evaporate less easily. Examples of these are motor oil, water, and glycerine.

The Phenomenon of Boiling

Using our information about the equilibrium vapor pressures of liquids, let's now try to explain the phenomenon of boiling.

If we heat a beaker of water over a flame, bubbles form on the bottom of the beaker. What are these bubbles? Is it steam being formed? No, these first bubbles are merely dissolved air (oxygen, nitrogen, and so forth) being driven out of the water by the heat. As the temperature rises, the molecules become more active. Soon other bubbles form and begin to rise. These bubbles are water vapor (steam). The reason they form on the bottom of the beaker is that the water is hottest there. These bubbles disappear as they rise and reach the cooler part of the water. After a while the bubbles stop disappearing and rise all the way to the surface. At this point we say that the water is *boiling*. Actually, the vapor pressure exerted by the bubbles of steam has become equal to the atmospheric pressure (Fig. 7-8).

The *boiling point* of a liquid is the temperature at which the equilibrium vapor pressure of the liquid equals the atmospheric pressure. Knowing this,

Figure 7-8 Boiling water: the vapor pressure exerted by the bubbles of steam has become equal to the atmospheric pressure.

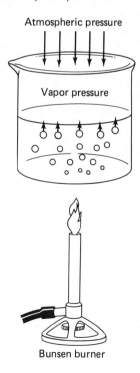

Atmospheric pressure

Vapor pressure

Bunsen burner

TABLE 7-5 *The boiling point of water at various pressures*

PRESSURE (torr)	BOILING POINT (°C)
18	20
355	80
526	90
760	100
906	105
2026	130

we can say that increasing the atmospheric pressure on the liquid should increase the boiling point. By the same token, decreasing the atmospheric pressure should decrease the boiling point. Table 7-5 shows that this is actually true. So remember that when we discuss the *normal boiling point* of a liquid, we are talking about its boiling point at what is called normal pressure—that is, 1 atmosphere.

Figure 7-9 Hot water at 1 atm pressure changes from a solid to a liquid to a gas. Below 0°C water is in the solid state—ice. In this diagram we begin with ice at −10°C. Energy (heat is supplied so that the temperature of the ice rises. When it reaches 0°C, the ice starts to melt into water. As more energy is supplied, more ice melts into water. This water remains at 0°C until all the ice melts. After all the ice has melted, the energy supplied starts to warm the water. This continues until the water temperature reaches 100°C. At this point the water starts to vaporize into steam. The temperature of the water remains constant at 100°C as the liquid is turned into gas.

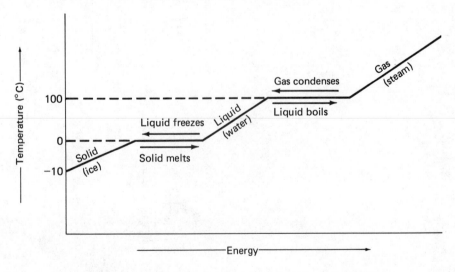

If we measure the temperature of the liquid at its boiling point, we find that it remains constant. (The boiling liquid is, of course, the same temperature as its vapor.) For example, if you heat a container of water, you find that it boils at 100°C (at 1 atmosphere pressure), and remains at this temperature *regardless of how much heat you give to the water*. The reason that the temperature of the water remains constant during boiling is that it takes energy to change the liquid water into steam. The heat energy you are supplying is used to separate the molecules of water, so that the water changes from the arrangement of a liquid to the arrangement of a gas (Fig. 7-9). (We have more to say about liquids when we examine solutions in Chap. 8.)

THE GASEOUS STATE

The major characteristics of gases are their indefinite shape and volume. This means that the particles of a gas behave as if they are independent of each other. In other words, the molecules of a gas move about a container until they fill it. Gases have been extensively studied since the 1500's and scientists have discovered a number of their properties. We can summarize these as follows:

1. Nearly all gases are composed of molecules. The so-called noble gases are composed of atoms, not molecules. Remember, atoms of noble gases don't combine readily with other atoms.)

2. Since the distances between the gas molecules are so very large, the forces of attraction between the gas molecules are very slight. Under conditions of normal atmospheric pressure and room temperature, the distances between the gas molecules are large compared to the size of the molecules.

3. Gas molecules are always in motion. They often collide with other gas molecules or with their container. After collisions occur, the molecules do not stick together; and they do not lose any energy as a result of the collisions.

4. Gas molecules move faster when the temperature rises and slower when the temperature falls.

5. All gas molecules (heavy as well as light) have the same average kinetic energy—that is, energy of motion—at the same temperature.

These five statements are known as the *kinetic theory of gases*. Keep these five statements in mind as we continue our study of gases.

The *P, V, T,* of Gases

The kinetic theory can be used to explain the properties of gases. However, a more quantitative explanation can be obtained from a series of gas law equations. The first gas law was proposed by Robert Boyle in 1660, and related the pressure of a gas with its volume (at a constant temperature, although Boyle forgot to say so). Another gas law was discovered in 1780 by the French physicist Jacques Charles, and related the temperature of a gas with its volume (at constant pressure). In mathematical terms, Boyle's law can be stated as

$$\frac{P_i}{P_f} = \frac{V_f}{V_i} \tag{7-1}$$

The P stands for pressure and the V for volume. The subscript i indicates initial conditions, and the subscript f indicates final conditions. The formula says that the volume of a gas varies inversely with pressure (at a constant temperature).

Another way of stating Boyle's law is to say that

$$P_i V_i = P_f V_f$$

This equation is obtained by cross-multiplying the terms in equation 7-1. It says that the product of the initial conditions equals the product of the final conditions. Either form of the equation may be used to solve Boyle's law problems. Use the form that's easier for you!

By the way, pressure is measured in units called *torr*, in honor of Evangelista Torricelli, the man who invented the barometer, a device used to measure air pressure.

Torricelli filled a glass tube, which was sealed on one end, with mercury. He then turned the tube upside down and placed it in a bowl, also filled with mercury (Fig. 7-10). Torricelli noticed that most of the mercury remained in the upright tube. He concluded that the surrounding air exerted pressure on the surface of the mercury in the bowl, which in turn supported the column of mercury in the tube.

Figure 7-10 A simple barometer.

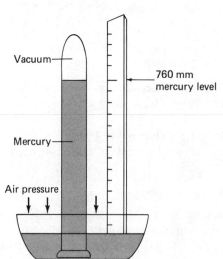

At sea level the height of mercury supported in the tube is 760 mm. This height is now expressed in pressure units of torr. In other words, normal atmospheric pressure at sea level is 760 torr. For larger pressures the unit of *atmosphere* is used.

$$1 \text{ atmosphere} = 760 \text{ torr} = 760 \text{ mm of mercury}$$

The abbreviation for atmosphere is atm. (Another unit of pressure is pounds per square inch, abbreviated psi. Normal atmospheric pressure is 14.7 pounds per square inch, 14.7 psi.)

$$1 \text{ atmosphere} = 760 \text{ torr} = 14.7 \text{ psi}$$

Boyle's law allows us to understand the process of breathing. Our lungs are elastic, which means that they can expand and contract. They do so within an airtight chamber called the thoracic cavity. A muscle called the diaphragm is part of the thoracic cavity. The contraction of the diaphragm causes an *increase in the volume* of the thoracic cavity and a corresponding *increase in the volume of the lungs*. This causes a *pressure decrease in the lungs*. The pressure of the atmosphere is now greater than the pressure within the lungs and the pressure difference causes air to rush into the lungs (Fig. 7-11). When the diaphragm returns to its normal position, the volume of the thoracic cavity decreases, and so does the volume of the lungs. This causes the air pressure in the lungs to be greater than atmospheric, and air flows out of the lungs (Fig. 7-12).

Figure 7-11 The process of breathing: inhalation.

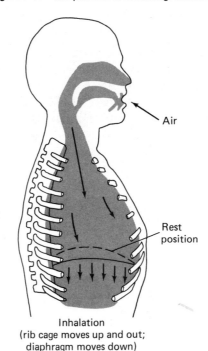

Inhalation
(rib cage moves up and out;
diaphragm moves down)

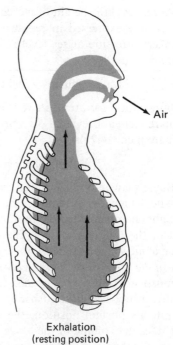

Exhalation
(resting position)

Figure 7-12 The process of breathing: exhalation.

Now that we've discussed the importance of Boyle's law, let's see if we can solve a gas law problem involving equation 7-1.

Example 7-1 An adult moves about $5\overline{00}$ ml of air in and out of his or her lungs in one inhalation–exhalation cycle. Let's say that this occurs at 760 torr pressure. If the pressure on the lungs should be increased to 2280 torr, what would be the volume of air in the lungs?

Solution Let's organize the data.

$$P_i = 76\overline{0} \text{ torr} \qquad P_f = 2280 \text{ torr}$$
$$V_i = 5\overline{00} \text{ ml} \qquad V_f = \,?$$

Now we'll write Boyle's law (equation 7-1), and solve it for V_f.

$$\frac{P_i}{P_f} = \frac{V_f}{V_i}$$

$$V_f = \frac{P_i V_i}{P_f}$$

$$V_f = \frac{(76\overline{0} \text{ torr})(5\overline{00} \text{ ml})}{2280 \text{ torr}} = 167 \text{ ml}$$

Charles' Law and Kelvin Temperature

In mathematical terms, Charles' law can be stated as

$$\frac{V_i}{V_f} = \frac{T_i}{T_f} \tag{7-2}$$

The V stands for volume, and the capital T stands for temperature in degrees Kelvin. The Kelvin temperature scale, which is also called the absolute temperature scale, was named after Lord Kelvin, the scientist who proposed it. This temperature scale is related to the activity of gas molecules. Scientists theorize that at a temperature of $0°K$ ($-273°C$), the molecular motion of gas molecules would cease. The value of $0°K$ is sometimes called *absolute zero*. The relationship between Celsius ($°C$) and Kelvin ($°K$) temperature is

$$°K = °C + 273 \tag{7-3}$$

By using the Kelvin temperature in the Charles' law equation, we can express the relationship between temperature and volume as a direct proportion. In other words, the volume of a gas varies directly with its Kelvin temperature, at constant pressure. [*Note:* In the following examples small "t" will stand for $°C$ and capital "T" for $°K$.]

Example 7-2 Figure 7-13 illustrates the situation discussed in this example. Let's say that we have $5\overline{00}$ ml of a gas in a balloon at a temperature of $27°C$. What would be the volume of the balloon if the temperature of the gas is increased to $327°C$?

Solution Let's organize the data.

$$V_i = 5\overline{00} \text{ ml} \qquad V_f = ? \text{ ml}$$
$$t_i = 27°C \qquad t_f = 327°C$$

Figure 7-13 A balloon attached to a can has a temperature of 27°C and a volume of 500 ml. The can, with the balloon attached, is placed in a hot-oil bath that is at a temperature of 327°C. The air in the can and in the balloon soon reaches this temperature and the balloon expands, increasing its volume.

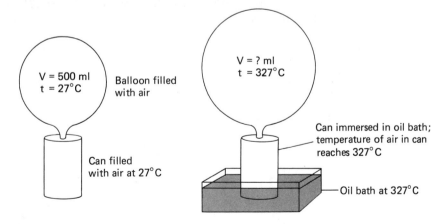

Use equation 7-3 to change °C to °K.

$$T_i = 27°C + 273 = 3\overline{00}°K$$
$$T_f = 327°C + 273 = 6\overline{00}°K$$

Now we'll write Charles' law (equation 7-2), and solve for V_f.

$$\frac{V_i}{V_f} = \frac{T_i}{T_f}$$

$$V_f = \frac{V_i T_f}{T_i}$$

$$V_f = \frac{(5\overline{00} \text{ ml})(6\overline{00}°K)}{3\overline{00}°K} = 1\overline{00}0 \text{ ml}$$

We can combine Boyle's law and Charles' law into a general law called the *combined gas law*.

$$\frac{P_i V_i}{T_i} = \frac{P_f V_f}{T_f} \tag{7-4}$$

This equation relates the initial pressure, volume, and Kelvin temperature of a gas within its final pressure, volume, and Kelvin temperature. If we are given any five quantities in equation 7-4, we can solve for the remaining quantity. Also, if one of the quantities in equation 7-4 should be a constant, the equation reduces to a *four-letter* equation. For example, if you are given a problem with T constant, equation 7-4 becomes

$$P_i V_i = P_f V_f$$

which is simply Boyle's law. The following examples will show you how to use equation 7-4 to solve all types of P, V, T problems.

Example 7-3   A gas has a volume of $1\overline{00}$ ml at a pressure of 3.00 atm. What will be the volume if the pressure is decreased to 0.500 atm? Assume that the temperature is constant.

Solution   Let's organize the data.

$$P_i = 3.00 \text{ atm} \qquad P_f = 0.500 \text{ atm}$$
$$V_i = 1\overline{00} \text{ ml} \qquad V_f = ? \text{ ml}$$

Now we'll use equation 7-4.

$$\frac{P_i V_i}{T_i} = \frac{P_f V_f}{T_f}$$

However, T is constant, so equation 7-4 becomes

$$P_i V_i = P_f V_f$$

Now we'll solve for V_f and plug in the values.

$$V_f = \frac{P_i V_i}{P_f}$$

$$V_f = \frac{(3.00 \text{ atm})(10\overline{0} \text{ ml})}{0.500 \text{ atm}} = 60\overline{0} \text{ ml}$$

Example 7-4 A gas is enclosed in a flexible container. It has a volume of $80\overline{0}$ ml at a pressure of $76\overline{0}$ torr and a temperature of 127°C. What temperature would be necessary to increase the volume to $160\overline{0}$ ml at constant pressure?

Solution Let's organize the data.

$$P_i = 76\overline{0} \text{ torr} \qquad P_f = 76\overline{0} \text{ torr}$$
$$V_i = 80\overline{0} \text{ ml} \qquad V_f = 160\overline{0} \text{ ml}$$
$$t_i = 127°\text{C} \qquad t_f = \text{?} °\text{C}$$

Now we'll use equation 7-4.

$$\frac{P_i V_i}{T_i} = \frac{P_f V_f}{T_f}$$

However, P is constant, so equation 7-4 becomes

$$\frac{V_i}{T_i} = \frac{V_f}{T_f}$$

Before we substitute the values into the equation, we must first change t_i (127°C), into °K, using equation 7-3.

$$T_i = 127°\text{C} + 273 = 40\overline{0}°\text{K}$$

$$T_f = \frac{T_i V_f}{V_i}$$

$$T_f = \frac{(40\overline{0}°\text{K})(160\overline{0} \text{ ml})}{80\overline{0} \text{ ml}} = 80\overline{0}°\text{K}$$

Now we'll change °K back to °C using equation 7-3.

$$°\text{K} = °\text{C} + 273$$

Therefore,

$$°\text{C} = °\text{K} - 273$$
$$t_f = 80\overline{0}°\text{K} - 273 = 527°\text{C}$$

Example 7-5 If your lungs can hold $50\overline{0}$ ml of air at $2\overline{0}°$C and 1.0 atm pressure, what would happen if you took a deep breath, held it, then dove into the ocean where the pressure was 1.5 atm and the temperature was $1\overline{0}°$C?

In other words, what would be the volume of air in your lungs under these conditions?

Solution   Let's organize the data.

$$P_i = 1.0 \text{ atm} \qquad P_f = 1.5 \text{ atm}$$
$$V_i = 50\overline{0} \text{ ml} \qquad V_f = ? \text{ ml}$$
$$t_i = 2\overline{0}°C \qquad t_f = 1\overline{0}°C$$

Now we'll use equation 7-4.

$$\frac{P_i V_i}{T_i} = \frac{P_f V_f}{T_f}$$

Before we solve for V_f and substitute the values into the equation, we must first change t_i and t_f into °K.

$$T_i = 2\overline{0}°C + 273 = 293°K$$
$$T_f = 1\overline{0}°C + 273 = 283°K$$
$$V_f = \frac{P_i V_i T_f}{P_f T_i}$$
$$V_f = \frac{(1.0 \text{ atm})(50\overline{0} \text{ ml})(283°K)}{(1.5 \text{ atm})(293°K)} = 320 \text{ ml}$$

Dalton's Law of Partial Pressures

It was around the year 1800 when John Dalton discovered an important property of gases, which came to be known as *Dalton's law of partial pressures*. This law states that the gases in a mixture do not affect each other unless they react chemically. In other words, the pressure exerted by each gas in a mixture remains the same and is not affected by other gases in the mixture. This means that the pressures of gases in a mixture are additive. Therefore, if you place oxygen gas in a container so that it has a pressure of 300 torr, and then add to it nitrogen gas with a pressure of 200 torr, the total pressure in the container will be 500 torr, the sum of the individual pressures. The individual pressure of each gas in the mixture is called the *partial pressure* of the gas (Fig. 7-14).

Example 7-6   Our atmosphere is a mixture of gases. There are, however, three major components in this mixture: nitrogen gas, 78.1% by volume; oxygen gas, 20.9% by volume; and argon gas, 0.9% by volume. If normal atmospheric pressure is 76\overline{0} torr, what is the partial pressure of each gas?

Solution   Dalton's law tells us that the partial pressure of each gas should be proportional to its percentage in the gas mixture.

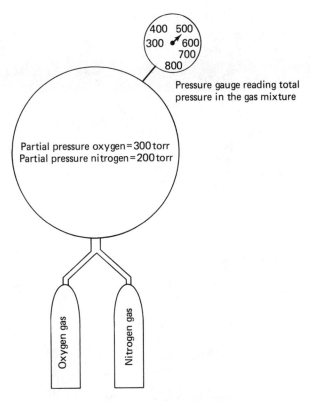

Figure 7-14 The sum of the partial pressures of each gas equals the total pressure of the gas mixture.

Partial pressure of N_2 = (0.781)(76$\overline{0}$ torr) = 594 torr

Partial pressure of O_2 = (0.209)(76$\overline{0}$ torr) = 159 torr

Partial pressure of Ar = (0.009)(76$\overline{0}$ torr) = 7 torr

Notice that the sum of the partial pressures adds up to the total pressure of the gas mixture, 76$\overline{0}$ torr.

We can use Dalton's law, and the fact that gases flow from areas of high pressure to areas of low pressure, to explain the movement of oxygen and carbon dioxide in our own bodies (Table 7-6). If you review Table 7-6, you will see that the partial pressure of oxygen in the air we inhale (*inspired air*) is about 158 torr, whereas it is only 100 torr in the *alveoli*, which are the small sacs at the end of the lungs. It is in the alveoli where the transfer of oxygen and carbon dioxide takes place (Fig. 7-15). Due to this difference in pressure, oxygen, from inspired air, flows to the alveoli. Blood, which is circulating through the capillaries of the alveoli, has a partial pressure of oxygen of about 40 torr, so oxygen moves into the blood. The blood is now called *arterial blood*, and has a built up oxygen supply of about 100 torr.

TABLE 7-6 *Partial pressures of respiratory gases (torr)*

GAS	INSPIRED AIR	ALVEOLI	ARTERIAL BLOOD AS IT PASSES THROUGH THE ALVEOLI	ARTERIAL BLOOD AFTER IT LEAVES ALVEOLI	TISSUES IN THE BODY	VENOUS BLOOD	EXPIRED AIR
Oxygen, O_2	158	100	40	100	35	40	116
Carbon dioxide, CO_2	0.3	40	40	40	50	46	28

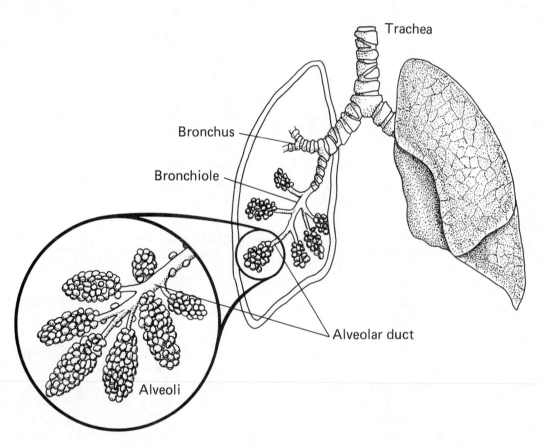

Figure 7-15 It is in the alveoli that the transfer of oxygen and carbon dioxide takes place.

Follow That Oxygen (and Carbon Dioxide)

The oxygenated blood, which has passed through the alveoli, now continues on its way to the tissues of the body. The partial pressure of oxygen in these tissues is about 30 to 35 torr. This causes the oxygen to move quickly into the cells of these tissues (Fig. 7-16). The pressure of oxygen gas in the blood now drops to about 40 torr, at which point it becomes *venous blood*, and is ready to return to the lungs to repeat the cycle.

At the same time that oxygen is moving from the lungs to the cells, the same forces are moving carbon dioxide from the cells to the lungs. The carbon dioxide pressure in an active cell is about 50 torr, whereas the carbon dioxide pressure in arterial blood is about 40 torr. This pressure difference causes the carbon dioxide to move out of the cells and into the blood. The blood (now called venous blood) has a carbon dioxide pressure of 46 torr by the time it reaches the alveoli. The carbon dioxide is now transferred to the alveoli,

Figure 7-16 Flow of oxygen gas through the body. The oxygen moves from areas of high partial pressure to low partial pressure.

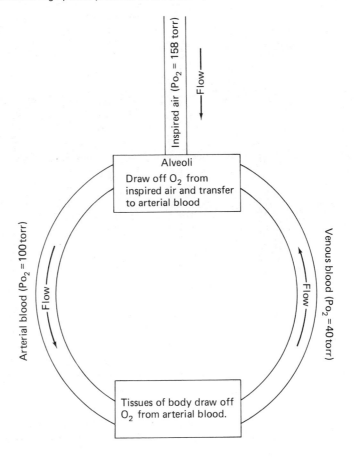

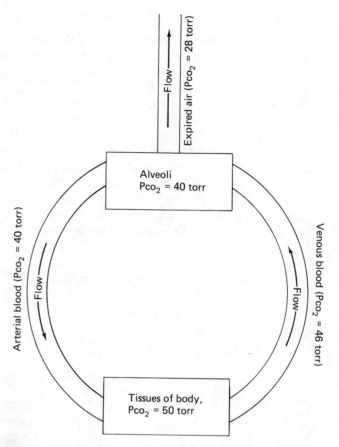

Figure 7-17 Flow of carbon dioxide gas through the body. The carbon dioxide moves from areas of high partial pressure to low partial pressure.

which only has a carbon dioxide pressure of 40 torr. From here the carbon dioxide moves out of the lungs with the *expired* (exhaled) air, which has a partial pressure of carbon dioxide of about 28 torr (Fig. 7-17).

Henry's Law: Just One More Gas Law

We've just been talking about transportation of gases through the blood. The blood, of course, is a liquid. In 1801, a doctor by the name of William Henry discovered that the solubility of a gas in a liquid is directly proportional to the pressure of the gas at the surface of the liquid. In other words, the greater the pressure of the gas above the liquid, the greater the amount of gas that can be dissolved in that liquid (Fig. 7-18). It is this principle that gave rise, in the 1960's, to *oxygen hyperbaric therapy*. This therapy involves the use of a high-pressure chamber where oxygen can be administered to an individual at a pressure of 2 or 3 atm (Fig. 7-19). This increased pressure of oxygen, over a

Figure 7-18 Henry's law says that the greater the pressure of the gas above the liquid, the greater is the amount of gas that can be dissolved in that liquid. Professor Emeritus has discovered that the reverse of this statement is also true.

Figure 7-19 A hyperbaric chamber. (Vacudyne Altair)

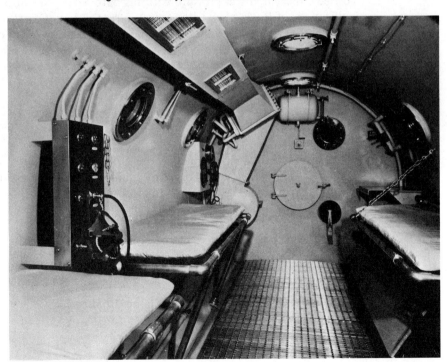

period of 5 hours, allows nearly 20 times more oxygen to be dissolved in a patient's blood. This, of course, is a great advantage to people recovering from heart attacks, strokes, gas gangrene, and various other illnesses. Babies born with hyaline membrane conditions of the lung can often be saved by treatment in a hyperbaric chamber. Although oxygen hyperbaric therapy is a great life-saving technique, it must be monitored very carefully because high concentrations of oxygen over a prolonged period of time can cause the alveoli in the lungs to collapse.

SUMMARY

In this chapter we discussed the three phases of matter: solids, liquids, and gases. We looked at crystalline solids and learned about the crystal lattice. We discovered that there are three types of crystalline solids: ionic, molecular, and atomic, and that each type has its distinct properties.

We looked at the liquid state and the process of evaporation. We learned that the process of evaporation allows us to maintain normal body temperature when we exercise. We also discussed the equilibrium vapor pressure of liquids and the phenomenon of boiling.

Our discussion of gases began with a review of the kinetic theory of gases. We then learned about Boyle's law, Charles' law, and the combined gas law. We discussed the gas variables: pressure, volume, and temperature, and how Torricelli measured gas pressure with his barometer. We also discussed the implications of Boyle's law with respect to respiration. Our study of Dalton's law of partial pressures allowed us to understand the movement of oxygen and carbon dioxide through the body. And our study of Henry's law enabled us to see how oxygen hyperbaric therapy can force more oxygen into the blood than is normally possible.

EXERCISES

1. Define the following terms.
   (a) crystal lattice
   (b) equilibrium vapor pressure
   (c) normal boiling point
   (d) torr

2. Compare and contrast the three types of crystalline solids.

3. Describe the major characteristics of solids, liquids, and gases.

4. An unknown solid has a low melting point. It is soft and does not conduct electricity. What type of solid is it (ionic, molecular, or atomic)?

5. How will the boiling point of a liquid be affected by
   (a) an increase in atmospheric pressure?
   (b) a decrease in atmospheric pressure?

6. How will the melting point of a substance be affected by
   (a) an increase in atmospheric pressure?
   (b) a decrease in atmospheric pressure?

7. Explain how the evaporation of moisture from our skin helps us to maintain our body temperature.

8. Scientists believe that the molecules of a liquid are constantly in motion. What simple experiment could you perform to show that this is true?

9. Explain why oxygen flows from the alveoli into the blood, and not the reverse.

10. If the pressure of a gas is increased at constant temperature, what happens to its volume? Does it increase or decrease?

11. If the temperature of a gas is increased at constant pressure, what happens to its volume? Does it increase or decrease?

12. When you open a bottle of soda pop, carbon dioxide seems to escape from the liquid. This is an example of what gas law?

13. A device called an *iron lung* was used frequently at one time to help children who were afflicted with polio. The iron lung was used to help the patient breathe. What gas law is the principle behind the iron lung?

14. Change the following pressures in torr to pressures in atmosphere.
   (a) $38\overline{0}$ torr   (b) 152 torr
   (c) $190\overline{0}$ torr   (d) 4560 torr

15. Change the following pressure in atmosphere to pressures in torr.
   (a) 1.50 atm   (b) 0.10 atm
   (c) 0.75 atm   (d) 3.50 atm

16. Change the following temperatures in °C to temperatures in °K.
   (a) 25°C   (b) 37°C
   (c) −25°C   (d) −273°C

17. Change the following temperatures in °K to temperatures in °C.
   (a) 373°K   (b) 300°K
   (c) 0°K   (d) 100°K

18. In a hospital, oxygen is sometimes brought into a patient's room in a large cylinder. This cylinder may contain about 423 liters of oxygen at a pressure of 136 atm. What would be the volume of this oxygen at a pressure of 1 atm? Assume that the temperature of the gas remains constant.

19. A balloon is filled with $20\overline{0}$ ml of gas at a pressure of $76\overline{0}$ torr and a temperature of 25°C. What would the volume be if the temperature is increased to $5\overline{0}$°C and the pressure remains at $76\overline{0}$ torr?

20. A child's lungs can hold about $2\overline{00}$ ml of air. Suppose that a child took a deep breath ($V = 2\overline{00}$ ml, $t = 2\overline{0}°C$, $P = 76\overline{0}$ torr), then dove into the ocean, where the temperature was 5°C and the pressure was $10\overline{00}$ torr. What would be the volume of air in his lungs?

21. A mixture of carbon dioxide (5% by volume) and oxygen (95% by volume) is sometimes used as a respiratory stimulant. If a container of this gas mixture has a total pressure of $76\overline{0}$ torr, what is the partial pressure of each gas in the mixture?

Chapter 8

Solutions of Acids, Bases, and Salts

Some Things You Should Know After Reading This Chapter

You should be able to:

1. Define the terms *solution*, *solute*, and *solvent*.
2. Discuss the differences between a true solution, a colloidal dispersion, and a suspension.
3. Calculate the percent by weight-volume of a solution given the necessary information.
4. Calculate the molarity of a solution given the necessary information.
5. State the four classes of inorganic compounds, as acids, bases, salts, and oxides.
6. Define the terms acid, base, and salt.
7. State some of the important characteristics of acids and bases.
8. Discuss the concept of ionization.
9. State what is meant by a strong electrolyte and a weak electrolyte.
10. Write an equation to show the ionization of water.
11. State that pure water has a hydrogen ion concentration of 10^{-7} mole per liter and a hydroxide ion concentration of 10^{-7} mole per liter.
12. Define pH as the negative logarithm of the hydrogen ion concentration in moles per liter.
13. Calculate the pH of a solution given its hydrogen ion concentration.
14. State the relationship between hydrogen ion concentration and hydroxide ion concentration in aqueous solutions.
15. Calculate the pH of a solution given its hydroxide ion concentration.

INTRODUCTION:
THE STRANGE CASE OF UNCLE CHARLEY

Uncle Charley is a patient whom any doctor would love. After all, in his 65 years he's had just about every ailment one human being could have. You might think that Uncle Charley would be a bitter, angry man, but he is not. Throughout it all he has kept his sense of humor. In fact, Uncle Charley is a 24-hour comedy show.

Uncle Charley's problems began in 1932, when he was stricken with diabetes mellitus at the age of 20. Luckily, just 10 years before in 1922, three researchers, Banting, Best, and Macleod, discovered that insulin from pigs could be injected into human beings to control some forms of diabetes. Uncle Charley was one of the lucky people whose diabetes was controlled (at least to some degree). In general, life went well for Uncle Charley until about 1960. It was then when his medical problems really started to surface, perhaps in part due to his diabetes. In rapid succession he suffered a major heart attack (a myocardial infarction), gallbladder stones (requiring an operation for their removal), glaucoma (which almost cost him his sight), and an enlarged prostate (which required surgery). But the most serious thing that happened to Uncle Charley was the one day he tried to do two things at the same time; take an aspirin for his headache and test his urine for sugar with Clinitest tablets. These tablets, we should point out, are highly caustic and can cause severe skin burns

Figure 8-1   Uncle Charley's stomach, or is it the map of Florida?

and death if taken internally. Now if things were to be done properly, Uncle Charley should have swallowed the two aspirin tablets and placed the Clinitest tablets into a cup that contained a sample of his urine. Unfortunately, that's not what happened. Uncle Charley swallowed the Clinitest tablets and placed the aspirin in the cup of urine. Realizing what he had done, Uncle Charley ran to the refrigerator and drank a 16-ounce bottle of lemon juice. This action saved his life, but unfortunately damage had already been done to his esophagus. An emergency six-hour operation and six months in the hospital were required to repair the damage. Today Uncle Charley is well and living in Miami Beach. And he's still telling jokes, like the one about his stomach. "You know", he says, "I have the only stomach that looks like the map of Florida, and I owe it all to my operations" (Fig. 8-1).

Why did we tell you about Uncle Charley? Because all of his problems had something to do with solution chemistry. And the last problem we discussed had to do with the solutions of acids and bases. Many of the reactions that take place in our bodies occur in solution. Some of these solutions contain acids, bases, and salts. In this chapter we see just how important solutions are to all of us.

SOLUTIONS DEFINED

In Chapter 2 we defined a solution as a homogeneous mixture. A solution can contain two substances, or three, or more. The most common types of solutions are made by dissolving a solid in a liquid: for example, salt in water. However, you can make solutions by mixing any of the three states of matter. Table 8-1 shows six possible types of solutions, and gives an example of each.

Since solutions are always composed of at least two substances, we need to be able to identify the role that each substance plays. The *solute* is the substance that is being dissolved. The *solvent* is the substance that is doing the dissolving. For example, in a salt-and-water solution, salt is the solute and water is the solvent. When we deal with solutions that are composed of the

TABLE 8-1   *Various types of solutions*

	SOLID	LIQUID	GAS
Solid	Copper metal dissolved in silver metal (for example, coins)	—	—
Liquid	Salt dissolved in water	Ethyl alcohol dissolved in water	—
Gas	Hydrogen dissolved in platinum metal	Carbon dioxide dissolved in water (soda water)	Oxygen gas dissolved in nitrogen gas

same states of matter—such as liquid–liquid solutions—it's hard to establish which substance is the solute and which is the solvent. A rule to go by is that the substance present in the larger amount is called the solvent. If we have a solution containing 10 ml of ethyl alcohol and 90 ml of water, the water is the solvent and the ethyl alcohol is the solute.

Example 8-1   Which is the solute and which is the solvent in the following solutions?

(a) sugar and water

(b) hydrogen chloride and water

(c) 70 ml of ethyl alcohol and 30 ml of water

(d) 80 ml of nitrogen and 20 ml of hydrogen

(e) soda water (which contains carbon dioxide gas)

Solution   (a) Sugar is the solute; water is the solvent.

(b) Hydrogen chloride is the solute; water is the solvent.

(c) Water is the solute; ethyl alcohol is the solvent.

(d) Hydrogen is the solute; nitrogen is the solvent.

(e) Carbon dioxide is the solute; water is the solvent.

TRUE SOLUTIONS VERSUS COLLOIDAL DISPERSIONS AND SUSPENSIONS

A *true solution* is one where the solute particles have dissolved to the point of ions, atoms, or molecules into the solvent. In a true solution you can't visually distinguish the solute and solvent. Also, the solute particles will not settle out of the solvent after a period of time.

In a *colloidal dispersion*, the solute particles don't dissolve to the point that they do in a true solution. Instead, the solute particles form groups of ions, atoms, or molecules. Of course, the solute particles in a colloidal dispersion are evenly dispersed through the solvent. Because of this we can think of a colloidal dispersion as a solution of sorts. But unlike a true solution, a colloidal dispersion will appear cloudy when a beam of light shines through it (Fig. 8-2). This is called the *Tyndall effect*, named after a famous nineteenth-century British physicist.

Some examples of colloidal dispersions are fog, airborne dust, egg white, and homogenized milk (the colloidal dispersion is between the cream and the skim milk fraction). The solute particles in a colloidal dispersion will, like the solute particles in a true solution, pass through most paper filters. But colloidal particles will not pass through semipermeable membranes, such as cellophane or cell walls (Fig. 8-3). This is important in the process of digestion. The membranes that line the small and large intestine allow the particles of a true solution

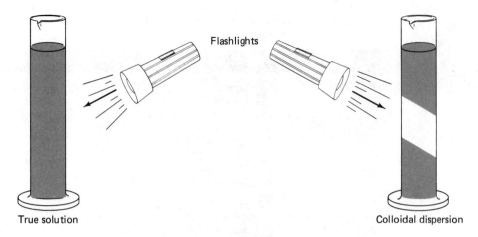

Flashlights

True solution

Colloidal dispersion

Figure 8-2 A colloidal dispersion will appear cloudy when a beam of light shines through it. This is called the Tyndall effect. The light beam appears to become visible when it passes through the colloidal dispersion. A true solution allows the light to pass through.

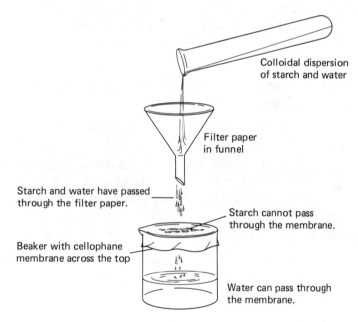

Colloidal dispersion of starch and water

Filter paper in funnel

Starch and water have passed through the filter paper.

Starch cannot pass through the membrane.

Beaker with cellophane membrane across the top

Water can pass through the membrane.

Figure 8-3 Colloidal particles will pass through most filter papers, but not through semipermeable membranes.

to pass into the blood and lymphatic systems while keeping out colloid-size particles.

Suspensions are not solutions in any sense of the word; they're really just mixtures. In a suspension the suspended particles settle out and can be easily separated from the solvent by filtering. Many common pharmaceutical agents

TABLE 8-2 *Some major characteristics of true solutions, colloids, and suspensions*

	TRUE SOLUTIONS	COLLOIDS	SUSPENSIONS
Particles that compose them	Ions, atoms, or molecules	Groups of ions, atoms, or molecules	Large groups of insoluble particles
Size of particles	Less than 1 nm	1–100 nm	Greater than 100 nm
Separation of solute and solvent by filtering	Will not effect separation	Will not effect separation	Will effect separation
Separation of solute and solvent by a semipermeable membrane	Will not effect separation	Will effect separation	Will effect separation

are suspensions, and must be shaken well before they are used. Milk of magnesia is an excellent example. Table 8-2 summarizes the major characteristics of true solutions, colloidal dispersions, and suspensions.

Example 8-2   Determine whether each of the following are true solutions, colloidal dispersions, or suspensions.

(a) The solute and solvent are shaken well, but the solute settles upon standing.

(b) The solute and solvent are shaken well, then filtered. Nothing remains on the filter paper. The solution is then poured through a semipermeable membrane. The solute and solvent separate.

Solution   (a) This must be a suspension, because the solute settles upon standing.

(b) This must be a colloidal dispersion, because the solute and solvent are separated by a semipermeable membrane.

CONCENTRATIONS OF SOLUTIONS

A particular solution can be made in various concentrations simply by adding more solute to a given amount of solvent. However, we should also mention that you can only vary the concentration of a solution up to a point, and that is the *point of saturation*. The point of saturation occurs when no more solute dissolves in the solution. In this chapter we look at two ways in which chemists measure the concentrations of solutions: *percent by weight-volume*, and *molarity*.

Percent by Weight-Volume

Every solution has a solute and a solvent. In a percent by weight-volume solution, the solute is usually measured in grams, and the solvent in milliliters. The formula for calculating the percent by weight-volume of a solution is

$$\% \text{ by weight-volume} = \frac{\text{grams of solute}}{\text{ml of solution}} \times 100 \tag{8-1}$$

For example, if we dissolve 1.8 grams of NaCl in enough water to make exactly 200 ml of solution we would have a 0.9% NaCl solution.

$$\% \text{ NaCl by weight-volume} = \frac{1.8 \text{ g NaCl}}{200 \text{ ml solution}} \times 100 = 0.9\%$$

A 0.9% weight-volume sodium chloride solution is sometimes called physiological saline solution. This is the salt solution we sometimes receive in a hospital before or after an operation. This solution is administered intravenously to maintain the normal salt concentration in the body.

Example 8-3 A solution is prepared by dissolving 5 g of glucose in enough water to make 250 ml of solution. What is the percent weight-volume of the solution?

Solution We use equation 8-1.

$$\% \text{ glucose by weight-volume} = \frac{5 \text{ g glucose}}{250 \text{ ml solution}} \times 100 = 2\%$$

Example 8-4 How many grams of sucrose must you mix with water to prepare 200 ml of 5% by weight-volume sucrose solution?

Solution A 5% weight-volume sucrose solution means that there are 5 g of sucrose per 100 ml of solution, or

$$\frac{5 \text{ g sucrose}}{100 \text{ ml solution}}$$

Therefore, we can use the factor-unit method to solve for the number of grams of sucrose in 200 ml of water.

$$? \text{ g sucrose} = (200 \text{ ml solution})\left(\frac{5 \text{ g sucrose}}{100 \text{ ml solution}}\right)$$

$$= 10 \text{ g sucrose}$$

Molarity

Another frequently used unit of concentration is *molarity*. We may define molarity as *the number of moles of solute per liter of solution*, or in terms of a formula,

$$\text{Molarity} = \frac{\text{moles of solute}}{\text{liter of solution}}$$

or in an abbreviated form as

$$M = \frac{\text{moles}}{\text{liter}} \qquad\qquad (8\text{-}2)$$

Molarity is an extremely useful concentration unit because it is based on the mole. This is important information for a chemist who is carrying out chemical reactions in solutions. (By the way, if you've forgotten what a mole is, go back and reread (Chap. 2.)

Figure 8-4 How to prepare 1 liter of 2 *M* NaCl.

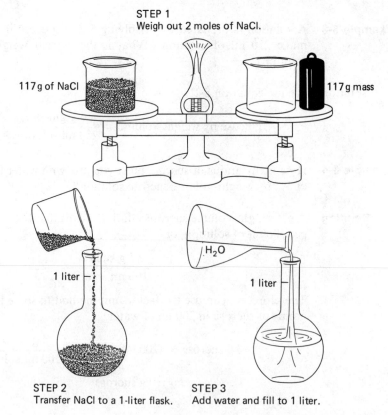

STEP 1
Weigh out 2 moles of NaCl.

117g of NaCl

117g mass

1 liter

H_2O

1 liter

STEP 2
Transfer NaCl to a 1-liter flask.

STEP 3
Add water and fill to 1 liter.

If you wanted to prepare a 2 molar sodium chloride solution (which you would label 2 M NaCl), you would need to obtain 2 moles of NaCl for every liter of solution you prepared. A mole of NaCl weighs 58.5 g; therefore, you would need 117 g of NaCl for each liter of solution. Figure 8-4 shows how you would prepare 1 liter of 2 M NaCl solution. Sometimes the concentrations of certain pharmaceutical agents are given in units of molarity. See if you can use this concentration unit in the examples that follow.

Example 8-5 The anticancer agent 5-fluorouracil (abbreviated 5-FU) has a molecular weight of 130 and comes in a strength of 50.0 mg/ml. This means that there are 50.0 mg of 5-FU for every milliliter of solution, or 50.0 g of 5-FU for every liter of solution. What is the concentration of this material in terms of molarity?

Solution Molarity is defined as moles of solute per liter of solution. We know that there are 50.0 g of solute per liter of solution. So to use equation 8-2 we must change the 50.0 g of 5-FU into moles.

$$? \text{ moles 5-FU} = (50.0 \text{ g})\left(\frac{1 \text{ mole}}{130 \text{ g}}\right) = 0.38 \text{ mole}$$

$$? M = \frac{\text{moles}}{\text{liter}} = \frac{0.38 \text{ mole}}{1.0 \text{ liter}} = 0.38 \text{ } M$$

Example 8-6 Lactated Ringers solution is used intravenously to maintain the electrolytic balance in patients who otherwise might go into shock. One of the ingredients in lactated Ringers is sodium chloride, whose concentration is about 0.13 M. How many grams of sodium chloride (NaCl) would be present in $5\overline{00}$ ml of this solution?

Solution We will use equation 8-2 to find the moles of NaCl in $5\overline{00}$ ml of 0.13 M solution. Then we will convert the moles of NaCl to grams of NaCl. (*Note*: $5\overline{00}$ ml = $0.5\overline{00}$ liter.)

$$M = \frac{\text{moles}}{\text{liter}} \qquad \text{therefore,} \qquad \text{moles} = (M)(\text{liter})$$

$$? \text{ moles} = (0.13 \text{ } M)(0.500 \text{ liter}) = 0.065 \text{ mole}$$

$$? \text{ grams of NaCl} = (0.065 \text{ mole})\left(\frac{58.5 \text{ g}}{1 \text{ mole}}\right) = 3.80 \text{ g}$$

Example 8-7 A popular cough syrup and expectorant is elixir terpin hydrate with codeine. The concentration of codeine in this medicine is 0.0070 M. How many milliliters of cough syrup would you have to consume in order to ingest $1\overline{0}$ mg of codeine? The molecular weight of codeine is $3\overline{00}$.

Solution First, $1\overline{0}$ mg of codeine is 0.010 g of codeine. Now let's change this to moles of codeine.

$$? \text{ moles} = (0.010 \text{ g})\left(\frac{1 \text{ mole}}{30\overline{0} \text{ g}}\right) = 3.3 \times 10^{-5} \text{ mole}$$

Now we can use equation 8-2 to solve for the liters of cough syrup that contain 3.3×10^{-5} mole of codeine. Note that the cough syrup is 0.0070 M with respect to the codeine. This is the same as $7.0 \times 10^{-3} M$.

$$M = \frac{\text{moles}}{\text{liter}} \qquad \text{therefore,} \qquad \text{liter} = \frac{\text{moles}}{M}$$

$$? \text{ liter} = \frac{3.3 \times 10^{-5} \text{ mole}}{7.0 \times 10^{-3} M}$$

$$= 4.8 \times 10^{-3} \text{ liter} \quad \text{or} \quad 4.8 \text{ ml}$$

In other words, you would have to drink 4.8 ml of cough syrup to ingest $1\overline{0}$ mg of codeine.

CLASSES OF INORGANIC COMPOUNDS: ACIDS, BASES, SALTS, AND OXIDES

If we attempted to classify the thousands and thousands of inorganic compounds, we would find that they fall into four broad categories:

1. *Acids* Substances that release hydrogen ions in solution and counteract bases. (We should point out that our definition of an acid is the simplest of the possible concepts and is only one of the various aspects of acidity. The same is true for our definition of a base.)

2. *Bases* Substances that release hydroxide ions in solution and counteract acids.

3. *Salts* Substances composed of the positive ions of a base and the negative ions of an acid.

4. *Oxides* Substances composed of any element combined with oxygen.

Although all four classes of compounds are important, we'll concentrate our efforts on acids, bases, and salts.

Acids: What Have They Got in Common?

The acids listed in Table 8-3 have many common characteristics.

1. *Each has a sour taste* (The word "acid" comes from the Latin *acidus*, meaning sour.)

TABLE 8-3 *Some common acids*

NAME OF ACID	FORMULA
Hydrochloric acid	HCl
Nitric acid	HNO_3
Sulfuric acid	H_2SO_4
Phosphoric acid	H_3PO_4
Acetic acid	$HC_2H_3O_2$
Boric acid	H_3BO_3

2. *Each can change the color of certain dyes* For example, blue litmus dye or litmus paper turns red in the presence of an acid. These dyes are called *indicators* because they indicate whether a substance is an acid or a base (Fig. 8-5).

3. *Each can react with certain metals, such as zinc and magnesium, to produce hydrogen gas*

$$Zn + 2\,HCl \longrightarrow ZnCl_2 + H_2(g)$$

$$Zn + H_2SO_4 \longrightarrow ZnSO_4 + H_2(g)$$

4. *Each can neutralize bases* (We'll hold our discussion of this until later.)

Figure 8-5 Hydrochloric acid changes blue litmus dye to red.

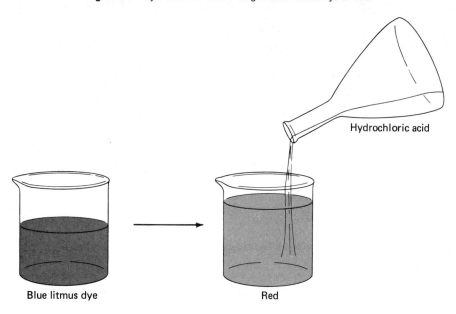

Hydrochloric acid

Blue litmus dye   Red

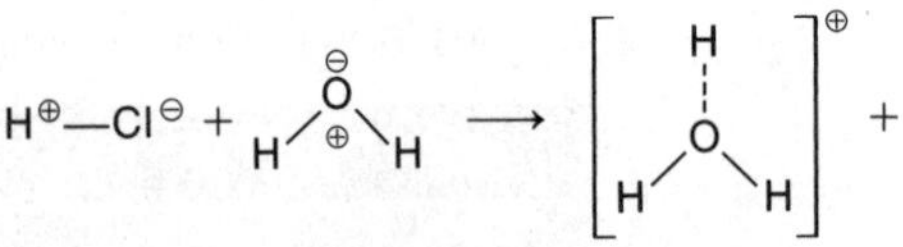

$$H^{\oplus}\!-\!Cl^{\ominus} + \underset{H}{\overset{\overset{\ominus}{O}}{\underset{\oplus}{\diagup}}}H \longrightarrow \left[\; \underset{H}{\overset{\overset{H}{\underset{\vdots}{O}}}{\diagup}}\overset{}{\diagdown}H \;\right]^{\oplus} + Cl^{\ominus}$$

Figure 8-6 Hydrogen chloride gas reacts with water. The covalently bonded hydrogen chloride molecule is ionized by the water into a hydrogen ion and a chloride ion. The positive hydrogen ion is attracted to the negative end of the water molecule forming a hydronium ion.

But the most important characteristic of acids is the one we used in the definitions: *They dissolve in water to produce hydrogen ions* (H^{+1}). These ions are responsible for the other common properties of acids.

Let's take a closer look at how an acid dissolves in water. In Chapter 4 we said that water is a polar molecule, with the oxygen end negatively charged and the hydrogen end positively charged. Suppose that we take hydrogen chloride gas (HCl) and bubble it into a beaker of water. What happens? The gas seems to dissolve in the water. But actually much more is happening. The HCl gas is reacting with the water (Fig. 8-6). The covalently bonded hydrogen chloride molecules are being broken apart (ionized) by the water into hydrogen ions (H^{+1}) and chloride ions (Cl^{-1}). The hydrogen ions are attracted to the negative ends of the water molecules, forming ions of a new type. The new kind of ion, $(H_3O)^{+1}$, is called a *hydronium ion*. The reaction can be written as

$$HCl + H_2O \longrightarrow (H_3O)^{+1} + Cl^{-1}$$

If we examine the other acids in Table 8-3, we find that similar reactions occur.

$$H_2SO_4 + H_2O \rightleftharpoons (H_3O)^{+1} + (HSO_4)^{-1}$$

$$H_3PO_4 + H_2O \rightleftharpoons (H_3O)^{+1} + (H_2PO_4)^{-1}$$

Some Acids: Their Discovery and Uses

The ancient Greeks knew how to ferment grapes to make wine. They also knew that if they let the fermentation process continue for too long, they got vinegar (which means sour wine). Vinegar was the strongest acid known to the Greeks. Centuries later an Arab alchemist named Geber distilled vinegar and got the substance responsible for vinegar's acidic properties. Today we call this substance acetic acid, and we use it as a solvent for manufacturing rubber, plastics, acetate fibers, drugs, and photographic chemicals.

We get acids such as acetic acid and citric acid from living things, and therefore we call them *organic acids*. For many centuries, organic acids were the only acids alchemists knew. However, in the 1200's, another alchemist, also called Geber, found a method of preparing acids from minerals. Two of the *mineral acids* he prepared were *sulfuric acid* and *nitric acid*. Since these were much stronger than the organic acids, alchemists could dissolve many substances that they thought were inert. This meant that new chemical reactions were possible. Mineral acids were essential to the great advances of modern chemistry and medicine. Let's look at how the acids in Table 8-3 are used.

Hydrochloric acid (HCl), also known commercially as muriatic acid, is used, in its more concentrated forms, to clean brick, cement, and metals. Although hydrochloric acid is considered to be a mineral acid, it is also produced in the stomach, where it is used for the digestion of proteins. In this capacity, hydrochloric acid has also been called *stomach acid*.

Nitric acid (HNO_3), another mineral acid, is very reactive. If nitric acid gets on your skin, it will turn it yellow. This is because of a reaction between the protein in your skin and the nitric acid. Because it has the ability to coagulate protein, nitric acid is also used to test for the presence of albumin in urine. However, the major uses of nitric acid are in the production of fertilizers, dyes, plastics, and explosives.

Sulfuric acid (H_2SO_4) is used primarily in the production of fertilizers, explosives, dyestuffs, other acids, paper, and glue. In its concentrated form, sulfuric acid is highly corrosive to our skin, and inhalation of its vapors can cause serious damage to our lungs. In its diluted form, sulfuric acid has been used as an appetite stimulant and in the control of serious diarrhea, but please don't try it as a home-brew remedy. Sulfuric acid is also the acid found in lead storage batteries. It is the acid that is commonly called "battery acid."

Phosphoric acid (H_3PO_4), is a weak mineral acid. Its uses range from flavoring soft drinks to producing fertilizers. The false teeth of millions of people are held in place by dental cements containing phosphoric acid.

Acetic acid is found in vinegar, as we have seen. According to the pure-food laws of the United States, vinegar must contain no less than 4% acetic acid. Concentrated acetic acid is strong enough to be used in making synthetic fibers called *acetates*. It is also used in manufacturing various plastics. Remember, though, that acetic acid, unlike the other acids we have just discussed, is an organic acid.

Boric acid (H_3BO_3) is a weak acid that has been used frequently and safely by people as an antiseptic, germicide, and eyewash. However, misuse of this acid has led to boric acid poisoning in certain individuals, and even to death. This has occurred from boric acid absorption through the skin when it was applied to certain types of wounds. Ingestion of less than 5 g of boric acid in infants and from 5 to 20 g in adults has also caused death.

Bases: What Have They Got in Common?

The bases listed in Table 8-4 have many common characteristics.

1. *Each has a bitter taste*

2. *Each can change the color of certain dyes* For example, red litmus turns blue in the presence of base (Fig. 8-7).

TABLE 8-4 *Some common bases*

NAME OF BASE	FORMULA
Lithium hydroxide	LiOH
Sodium hydroxide	NaOH
Potassium hydroxide	KOH
Calcium hydroxide	$Ca(OH)_2$
Magnesium hydroxide	$Mg(OH)_2$
Iron(III) hydroxide	$Fe(OH)_3$
Ammonium hydroxide	NH_4OH

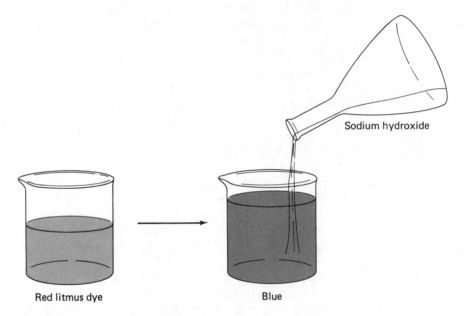

Sodium hydroxide

Red litmus dye

Blue

Figure 8-7 Sodium hydroxide changes red litmus dye to blue.

3. *Each has a slimy or soapy feeling*

4. *Each can neutralize acids*

But the most important characteristic of bases is the one we used in the definition: *They dissolve in water to produce hydroxide ions* or $(OH)^{-1}$. It is these hydroxide ions that are responsible for the other common properties of bases.

How do bases dissolve in water to produce hydroxide ions? Take for example, solid sodium hydroxide (NaOH). Sodium hydroxide is an ionic

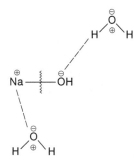

Figure 8-8 When you add sodium hydroxide to water, the negative end of the water molecule attracts the sodium ions, and the positive end of the water molecule attracts the hydroxide ions.

compound made up of sodium ions and hydroxide ions, or Na^{+1} and $(OH)^{-1}$. The ions are held together by the attraction of a plus charge to a minus charge. When you add sodium hydroxide to water, the negative end of the water molecule (the oxygen) attracts the sodium ions, and the positive end (the hydrogen) attracts the hydroxide ions. This attraction breaks the bonds between the sodium ions and the hydroxide ions (Fig. 8-8). There are now free sodium ions and free hydroxide ions in solution. The reaction can be represented by the following equation:

$$NaOH \xrightarrow{\text{water}} Na^{+1} + (OH)^{-1}$$

Some Bases and Their Uses

Like acids, bases play an important role in our daily lives, and are just as dangerous to our skin and clothes. They are used in our homes, in industrial processes, and in manufacturing various pharmaceutical products.

For example, sodium hydroxide, NaOH, is a base used in drain cleaners, or in what we call *lye*. It is also used to make soaps and cellophane.

Calcium hydroxide, $Ca(OH)_2$, which in solution is also known as *lime-water*, is sometimes used as an antacid. However, its major uses are in the construction industry, where it is used in mortar, plaster, cement, and paving materials.

Ammonium hydroxide, NH_4OH, is the household cleaner we know as ammonia water, and it has an intense, almost suffocating odor. Ammonium hydroxide is an ingredient of *smelling salts*. It is excellent for removing stains, and it is also used in manufacturing textiles such as rayon, and in making plastics and fertilizers.

Magnesium hydroxide, $Mg(OH)_2$, commonly called milk of magnesia, has many uses as a medicine. In small doses of a few hundred milligrams, the magnesium hydroxide acts as an antacid. In large doses of 2 to 4 g, it acts as a laxative.

Salts: What Have They Got in Common?

We have learned that acids and bases react with each other in what are called neutralization reactions (Chap. 5). The products of these reactions are salts and water.

$$HCl + NaOH \longrightarrow NaCl + H_2O$$

Sodium
chloride

$$HCl + NH_4OH \longrightarrow NH_4Cl + H_2O$$

Ammonium
chloride

$$H_2SO_4 + 2NH_4OH \longrightarrow (NH_4)_2SO_4 + 2H_2O$$

Ammonium
sulfate

If we examine these reactions closely, we can see what takes place. The hydrogen ions from the acid combine with the hydroxide ions from the base to produce water. At the same time, the positive ions of the base combine with the negative ions of the acid to produce the compound that we call a salt. Most salts are

TABLE 8-5 *Some salts and their uses*

FORMULA	CHEMICAL NAME	MEDICAL USE
$AlCl_3$	Aluminum chloride	Deodorant
$AlPO_4$	Aluminum phosphate	Antacid
NH_4Cl	Ammonium chloride	Expectorant
$BaSO_4$	Barium sulfate	X-ray work
$CaCO_3$	Calcium carbonate	Antacid
$CaCl_2$	Calcium chloride	Electrolyte replacement
$FeSO_4$	Iron(II) sulfate	As an iron supplement
$MgCl_2$	Magnesium chloride	Laxative
$MgSO_4$	Magnesium sulfate (as the heptahydrate, $MgSO_4 \cdot 7H_2O$, this substance is known as epsom salts)	Laxative
$HgCl_2$	Mercury(II) chloride	Topical antiseptic
$NaCl$	Sodium chloride	Used in physiological saline solution
NaF	Sodium fluoride	Prevents dental decay
SnF_2	Tin(II) fluoride	Prevents dental decay

composed of a metal ion combined with a nonmetal (or polyatomic) ion that has a negative oxidation number. Examples of these salts are sodium chloride (NaCl), which is everyday table salt, silver bromide (AgBr), potassium sulfate (K_2SO_4), and iron(III) phosphate ($FePO_4$). Some salts are composed of a polyatomic ion with a positive oxidation number combined with a nonmetal (or polyatomic) ion that has a negative oxidation number. Examples of these salts are ammonium chloride (NH_4Cl) and ammonium nitrate (NH_4NO_3).

Some Salts and Their Uses

Salts have a tremendous number of uses in the body. For example, calcium phosphate is the main constituent of our bones and teeth. Iron salts are necessary for the production of hemoglobin. Sodium and potassium salts help maintain the acid–base balance in our bodies. And the salt sodium iodide is necessary for the proper functioning of our thyroid gland. Salts are also responsible for the proper functioning of our muscles, including the heart, and they also maintain the fluid balance in our cells. A summary of salts and their uses can be found in Table 8-5.

IONIZATION AND THE CONCEPT OF ELECTROLYTES

What happens when things dissolve? For years chemists have tried to find out what happens to one substance as it dissolves in another. Chemists of the 1800's were also puzzled because some solutions would conduct electric current, whereas others would not.

One of the scientists who tried to explain this phenomenon was Michael Faraday. Faraday classified two kinds of solutions in the following way.

1. *Electrolytic solutions* are ones that conduct electric current.

2. *Nonelectrolytic solutions* are ones that do not conduct electric current.

Faraday said that electrolytic solutions—such as sodium chloride (NaCl) in water—have charged particles in them. He called these particles *ions* (from a Greek word that could be translated as "wanderer"), and said that ions wandered through the solution carrying the electric current (Fig. 8-9). In Faraday's time, questions arose about the nature of these ions. For example, why are there ions in a salt-and-water solution, and no ions in a sugar-and-water solution? It was left to Arrhenius to come up with the answer.

In 1884, in his Ph.D. thesis, Arrhenius advanced his ideas about ions. He suggested that Faraday's ions were really simple atoms (or groups of atoms) carrying a positive or negative charge. He said that when some substances dissolve in solution, they break up into ions. This process is called *ionization*.

$$NaCl(s) + H_2O \longrightarrow Na^{+1}(aq) + Cl^{-1}(aq)$$

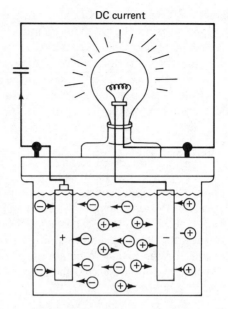

Figure 8-9 An electrolytic solution conducts electric current.

Solutions that contain ions are electrolytic, since they have charged particles that carry electric current (Fig. 8-10).

On the other hand, there are some substances that dissolve in solution and do *not ionize*; they simply break up into their neutral molecules and become surrounded by the molecules of solvent.

$$C_6H_{12}O_6(s) + H_2O \longrightarrow C_6H_{12}O_6(aq)$$
Glucose

Figure 8-10 An electrolytic solution contains ions.

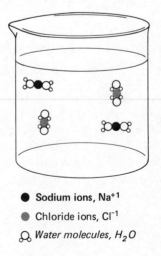

● Sodium ions, Na^{+1}

● Chloride ions, Cl^{-1}

⚬ *Water molecules, H_2O*

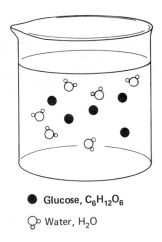

● Glucose, $C_6H_{12}O_6$

�Ⓞ Water, H_2O

Figure 8-11 Nonelectrolytic solutions have no charged particles or ions.

These solutions are nonelectrolytic, since they have *no* charged particles to carry electric current (Fig. 8-11).

Strong and Weak Electrolytes

If we study solutions of electrolytes very carefully we will discover that we can further subdivide them into two categories: strong electrolytes and weak electrolytes. The *strong electrolytes* are those which essentially ionize completely. For example, one mole of sodium chloride in a liter of solution contains essentially one mole of sodium ions (Na^{+1}) and one mole of chloride ions (Cl^{-1}). There are no sodium chloride molecules! Strong electrolytes conduct current well and light the bulb in Figure 8-9 very brightly. Strong electrolytes can be compounds of acids, bases, or salts. Therefore, an acid that ionizes completely in solution is called a *strong acid*, and a base that ionizes completely in solution is called a *strong base*. A list of some strong electrolytes is given in Table 8-6.

 Weak electrolytes are those which don't ionize completely. For example, if you were to analyze a solution containing one mole of $HC_2H_3O_2$ in 1 liter of solution (this would be a 1 molar acetic acid solution), you would find that most of the acetic acid stays in the form of molecules and only about 0.4% ionizes to form hydrogen ions and acetate ions. We can represent the dissociation

TABLE 8-6   *Some strong electrolytes*

ACIDS	BASES	SALTS
Hydrochloric acid, HCl	Sodium hydroxide, NaOH	Sodium chloride, NaCl
Nitric acid, HNO_3	Potassium hydroxide, KOH	Potassium chloride, KCl

TABLE 8-7   *Some weak electrolytes*

ACIDS	BASES	SALTS
Acetic acid, $HC_2H_3O_2$	Ammonium hydroxide, NH_4OH	Barium fluoride, BaF_2
Carbonic acid, H_2CO_3	Dimethylamine, $(CH_3)_2NH$	Silver sulfate, Ag_2SO_4

of a weak electrolyte in the following manner:

$$HC_2H_3O_2 \underset{\text{water}}{\rightleftharpoons} H^{+1} + (C_2H_3O_2)^{-1}$$

The *double arrow* indicates that in a given solution of acetic acid an equilibrium exists between the acetic acid molecules, the hydrogen ions, and the acetate ions. Weak electrolytes can also be compounds of acids, bases, or salts (Table 8-7). An acid that does not ionize completely is called a *weak acid*, and a base that does not ionize completely is called a *weak base* (Fig. 8-12).

The Ionization of Water

We've just finished discussing the ionization of weak acids and weak bases, but did you know that water also has a tendency to ionize? We can represent the ionization of water as follows:

$$H\!-\!OH + H_2O \rightleftharpoons \underset{\substack{\text{Hydronium}\\\text{ion}}}{(H_3O)^{+1}} + \underset{\substack{\text{Hydroxide}\\\text{ion}}}{(OH)^{-1}} \qquad (8\text{-}3)$$

Figure 8-12   An acid that does not ionize completely is called a weak acid, and a base that does not ionize completely is called a weak base. Acetic acid and ammonia water (also known as ammonium hydroxide) are examples.

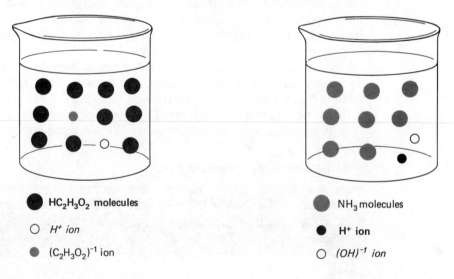

● $HC_2H_3O_2$ molecules

○ *H^+ ion*

● *$(C_2H_3O_2)^{-1}$ ion*

● NH_3 molecules

● H^+ ion

○ *$(OH)^{-1}$ ion*

We purposely wrote the formula of the first water molecule as H—OH so that you could see how the ionization occurs. A hydronium ion and hydroxide ion are formed. However, we can represent this ionization in a simpler way as follows:

$$H_2O \rightleftharpoons H^+ + (OH)^{-1} \tag{8-4}$$

This shows that a hydrogen ion and hydroxide ion are formed. Even though equation 8-3 is more technically correct, equation 8-4 is easier to use and gets the point across. However, we must point out that the ionization of water occurs only to a slight extent. In fact, if you had exactly 1 liter of water, at 25°C, which is 55.6 moles of water, only 0.0000001 mole (which can be written as 10^{-7} mole) of water would actually dissociate in hydrogen and hydroxide ions. In other words, there would be 10^{-7} mole of hydrogen ions and 10^{-7} mole of hydroxide ions present per liter of water. Because there are equal amounts of hydrogen ions and hydroxide ions, pure water is said to be neutral.

But what happens if you add hydrochloric acid to pure water? Adding hydrochloric acid to pure water increases the hydrogen ion content. (Remember, we've just learned that hydrochloric acid is a strong acid, which means that it dissociates into H^{+1} ions and Cl^{-1} ions.) This of course makes the solution of water and hydrochloric acid acidic. Just how acidic depends on the concentration of the hydrogen ions. It is just this idea that gives rise to the concept of pH.

THE pH SCALE

The concept of pH is a means of measuring the acid–base strength of solutions. The formal definition of pH is that pH is equal to the negative logarithm of the hydrogen ion concentration (in moles per liter). Symbolically, this is

$$pH = -\log [H^{+1}]$$

The brackets, $[\cdot]$, indicate moles per liter. The pH scale runs from zero to 14 (Fig. 8-13). A solution whose pH is equal to 7 is neutral. Why? Remember that in our discussion of the ionization of water, the concentration of hydrogen

Figure 8-13 The pH scale.

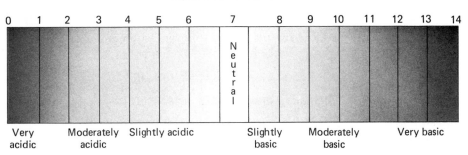

| Very acidic | Moderately acidic | Slightly acidic | | Slightly basic | Moderately basic | Very basic |

TABLE 8-8 *The pH value of some common substances*

SUBSTANCE	pH
Lemons	2.3
Vinegar	2.8
Soft drinks	3.0
Oranges	3.5
Tomatoes	4.2
Bananas	4.5
Rainwater	6.2
Milk	6.5
Pure water	7.0
Seawater	8.5
Blood	7.4
Urine	5.5–7.0
Saliva	6.5–7.5
Gastric juices	1.0–3.0

ion was 10^{-7} mole/liter. Therefore, the pH of water, which is neutral because it has equal amounts of hydrogen ion and hydroxide ion, is calculated as follows:

$$pH = -\log[H^{+1}] = -\log[10^{-7}] = 7 \quad (\text{at } 25°C)$$

(The logarithm of an exponential number is simply that exponent.)

A solution that has a pH between zero and 2 is considered to be very acidic. A solution that has a pH between 2 and 5 is moderately acidic, and one with a pH between 5 and 7 is slightly acidic. A solution whose pH is 7 is neutral. A solution whose pH is between 7 and 9 is slightly basic, one with a pH between 9 and 11 is moderately basic, and one with a pH between 11 and 14 is very basic. Table 8-8 lists the pH values of some common substances.

You can determine the pH of a solution by many methods. A simple method is to place a few drops of a chemical dye or a piece of paper tape that contains the dye in the solution. The dye or paper tape turns color, according to the pH of the solution. There is also a device called a pH meter, which directly reads the pH of a solution.

Example 8-8 Determine the pH of the following solutions, then use Figure 8-13 to state how acidic or basic the solution is.

(a) A solution whose $[H^{+1}] = 10^{-3} M$

(b) A solution whose $[H^{+1}] = 10^{-12} M$

(c) A solution whose $[H^{+1}] = 10^0 M$

Solution pH $= -\log[H^{+1}]$; therefore,

(a) pH $= -\log[10^{-3}] = 3$, which is moderately acidic.

(b) pH $= -\log[10^{-12}] = 12$, which is very basic.

(c) pH $= -\log[10^0] = 0$, which is very acidic.

THE RELATIONSHIP BETWEEN HYDROGEN IONS AND HYDROXIDE IONS

There is an interesting relationship between the hydrogen ion concentration and the hydroxide ion concentration in aqueous (water) solutions. The product of the $[H^{+1}]$ and $[OH^{-1}]$ is always equal to 10^{-14}. In other words,

$$[H^{+1}][OH^{-1}] = 10^{-14}$$

Remember that neutral water had an H^{+1} concentration of 10^{-7} M and an $(OH)^{-1}$ concentration of 10^{-7} M. The product of these two numbers would be 10^{-14}.

$$[H^{+1}][OH^{-1}] = [10^{-7}][10^{-7}] = 10^{-14}$$

The value of 10^{-14} is called the dissociation constant of water. It is usually given the symbol K_w. Therefore,

$$[H^{+1}][OH^{-1}] = K_w = 10^{-14}$$

If a solution has a hydrogen ion concentration of 10^{-5} M, its hydroxide ion concentration has to be 10^{-9} M, because

$$[10^{-5}][OH^{-1}] = 10^{-14} \quad \text{therefore,} \quad [OH^{-1}] = \frac{10^{-14}}{[10^{-5}]} = [10^{-9}]$$

A summary of the relationships between hydrogen ion concentration, hydroxide ion concentration, and pH is given in Table 8-9.

Example 8-9 Determine the pH of the following solutions.

(a) a 0.001 M HCl solution

(b) a 0.01 M HNO$_3$ solution

(c) a solution whose $[OH^{-1}] = 10^{-3}$ M

(d) a solution whose $[OH^{-1}] = 10^{-11}$ M

(e) a 0.0001 M NaOH solution

Solution (a) A 0.001 M HCl solution releases 0.001 M hydrogen ions, which is 10^{-3} M hydrogen ions. Therefore, the pH $= -\log[H^+] = -\log[10^{-3}] = 3$.

TABLE 8-9 *The relationship between hydrogen ion concentration and hydroxide ion concentration*

$[H^{+1}]$	pH	$[(OH)^{-1}]$
$10^0 = 1$	0	10^{-14}
$10^{-1} = 0.1$	1	10^{-13}
$10^{-2} = 0.01$	2	10^{-12}
$10^{-3} = 0.001$	3	10^{-11}
$10^{-4} = 0.0001$	4	10^{-10}
$10^{-5} = 0.00001$	5	10^{-9}
$10^{-6} = 0.000001$	6	10^{-8}
$10^{-7} = 0.0000001$	7	10^{-7}
$10^{-8} = 0.00000001$	8	10^{-6}
$10^{-9} = 0.000000001$	9	10^{-5}
$10^{-10} = 0.0000000001$	10	10^{-4}
$10^{-11} = 0.00000000001$	11	10^{-3}
$10^{-12} = 0.000000000001$	12	10^{-2}
$10^{-13} = 0.0000000000001$	13	10^{-1}
$10^{-14} = 0.00000000000001$	14	10^0

(b) A 0.01 M HNO_3 solution releases 0.01 M hydrogen ions, which is 10^{-2} M hydrogen ions. Therefore, the pH $= -\log[H^{+1}] = -\log[10^{-2}] = 2$.

(c) A solution whose $[OH^{-1}]$ is 10^{-3} M has a $[H^{+1}] = 10^{-11}$ M (see Table 8-9). Therefore, the pH $= -\log[H^{+1}] = -\log[10^{-11}] = 11$.

(d) A solution whose $[OH^{-1}]$ is 10^{-11} M has a $[H^{+1}] = 10^{-3}$ M (see Table 8-9). Therefore, the pH $= -\log[H^{+1}] = -\log[10^{-3}] = 3$.

(e) A 0.0001 M NaOH solution releases 0.0001 M hydroxide ions, which is 10^{-4} M hydroxide ions. A solution whose $[OH^{-1}]$ is 10^{-4} M has a $[H^{+1}] = 10^{-10}$ M (see Table 8-9). Therefore, the pH $= -\log[H^{+1}] = -\log[10^{-10}] = 10$.

SUMMARY

In this chapter we learned about solutions of acids, bases, and salts. We also talked about colloidal dispersions and suspensions. We discussed how concentrations of solutions are measured using percent weight-volume and molarity, and we talked about ionization and the concept of strong and weak electrolytes. We also learned about pH and how it is used to measure the acid–base strength of a solution.

EXERCISES

1. Which is the solute and which is the solvent in the following solutions?
   (a) potassium chloride in water
   (b) 7% weight-volume sodium chloride solution
   (c) 5 liters of oxygen gas and 3 liters of nitrogen gas
   (d) 3 M HCl solution

2. Determine whether each of the following are true solutions, colloidal dispersions, or suspensions.
   (a) A bottle of milk of magnesia from the pharmacy. The directions on the bottle say, "Shake well before using."
   (b) A solute and solvent are combined and shaken well, then filtered. Nothing remains on the filter paper. The solution is then poured through a semipermeable membrane. The solute and solvent do not separate.

3. What is the percent weight-volume of each of the following solutions?
   (a) 25.0 g of sodium chloride dissolved in $20\overline{0}$ ml of water
   (b) 9 g of alcohol dissolved in $3\overline{0}$ ml of water
   (c) 0.8 g of glucose in $4\overline{0}$ ml of water

4. How many grams of potassium chloride must you mix with water in order to prepare a 15% by weight-volume potassium chloride solution? (The final volume of the solution will be 500 ml.)

5. Determine the molarity of the following solutions.
   (a) $18\overline{0}$ g of glucose ($C_6H_{12}O_6$) in $50\overline{0}$ ml of solution
   (b) 5.85 g of NaCl in $25\overline{0}$ ml of solution
   (c) 72 g of HCl in 2.0 liters of solution

6. How many grams of calcium hydroxide, $Ca(OH)_2$, are necessary to prepare $60\overline{0}$ ml of a 2.0 M solution?

7. How many milliliters of a 0.10 M glucose solution are necessary to obtain 36 g of glucose?

8. A popular pharmaceutical agent used in the control of seizures is phenobarbital. The concentration of phenobarbital in its elixir is 0.017 M. The typical dosage of this elixir is 5.0 ml. How many milligrams of phenobarbital are there in one dose? (*Hint*: The molecular weight of phenobarbital is 232.)

9. Define the following terms.
   (a) acid   (b) base   (c) salt   (d) oxide

10. List three properties of all acids.

11. List three properties of all bases.

12. Complete the equation: $HCl + H_2O \rightarrow$

13. Complete the equations and balance the following neutralization reactions.
    (a) $HCl + NaOH \rightarrow$ (b) $H_2SO_4 + KOH \rightarrow$
    (c) $HC_2H_3O_2 + Ca(OH)_2 \rightarrow$ (d) $H_3PO_4 + Mg(OH)_2 \rightarrow$

14. Define electrolytic and nonelectrolytic solutions. Explain what makes them different from each other in terms of the particles that compose them.

15. What is the difference between a strong and weak electrolyte? Give some examples of each.

16. Write the reaction for the ionization of water.

17. Given the hydrogen ion concentration of each of the following aqueous solutions, determine the hydroxide ion concentration.
    (a) $[H^{+1}] = [10^{-6}]$ (b) $[H^{+1}] = [10^{-9}]$ (c) $[H^{+1}] = [10^{-1}]$

18. Determine the pH of each solution in Exercise 17.

19. Determine the pH of each solution.
    (a) $1\ M\ HCl$ (b) $0.01\ M\ HCl$
    (c) $0.01\ M\ NaOH$ (d) $0.0001\ M\ LiOH$

Chapter 9

Water and Water Pollution

Some Things You Should Know After Reading This Chapter

You should be able to:

1. State the importance of water to the survival of humankind.
2. Explain how pollutants in water can turn up in the foods that we eat.
3. Define the term potable and list sources of potable water.
4. List some common water pollutants.
5. List the physical properties of water.
6. Explain the term hydrogen bond.
7. Write equations for the reaction of water with oxides of metals and nonmetals.
8. Define the term hydrate and give an example.
9. Trace the path of water from falling rain to the ocean.

10. Explain the difference between aerobic and anaerobic organisms.
11. Explain the process of potable water treatment.
12. Explain the processes of primary, secondary, and advanced wastewater treatment.
13. Explain the meaning of eutrophication and BOD.
14. Explain the importance of chlorination in the processes of potable water treatment and wastewater treatment.

INTRODUCTION

In the many thousands of years that human beings have been on this earth, they have done a great deal to shape their environment. Unfortunately, it seems that they have shaped it into the form of a noose ready to hang them before many more years have passed.

Water—one of the most necessary materials for human survival—must be free from contamination of deadly chemicals and bacteria in order to be potable (drinkable). In the United States, which uses more than 35 billion gallons (132 million cubic meters) of water every day, pure water is becoming scarce. Most of the earth's water is in the sea, and is therefore unfit for drinking. Of our sources of potable water—lakes, rivers, springs, and wells—the last two are rapidly being drained. And we're using more water per person with each year that goes by. For that reason most big cities and towns are turning to their rivers and lakes.

To be considered potable, water must meet standards set by the U.S. Public Health Service (Table 9-1). But we run into problems when we apply these standards to our rivers and lakes.

In the past a river was usually the discharge point for used water. Raw domestic sewage by the millions of gallons is still being dumped into our rivers. In the United States, this waste amounts to more than 260 million gallons per day. The pollutants in this sewage come from many sources. Industries release poisonous chemicals containing mercury, arsenic, lead, and other metals. Farms use chemical insecticides such as DDT, which "leach" down through the soil and pollute the groundwater, which in turn pollutes lakes and rivers. For example, in 1969 the commercial salmon fishermen who fished in Lake Michigan had to suspend operations because DDT was found in the canned salmon from the lake. DDT accumulates inside fish, in their tissues, in alarming concentrations, because it isn't excreted. Then people eat the fish and the DDT begins to accumulate in *their* tissues. Birds eat the insects that are killed by the DDT, and the substance builds up in the birds to such an extent that their eggshells become thin. The thin shells break, and the young birds die before they hatch. So the population of birds begins to dwindle, which in turn means that there aren't enough birds to eat the insects. The insects then multiply and destroy the crops, so that farmers need still more pesticides, which leach down into the water supply, and so on and so on.

Farms also are responsible for tremendous amounts of phosphates that have filtered into the Great Lakes, making them anything but "great." These phosphates (from fertilizers) have destroyed the natural balance of life in the Great Lakes, causing huge growths of algae, especially in Lake Erie. The algae consume large amounts of oxygen, so much that fish and other forms of life die for lack of oxygen.

Any chemical that gets into the water supply is taken up by these algae, or by bacteria or plankton, all of which are then eaten by other forms of life,

TABLE 9-1 *U.S. Public Health Service standards for potable water*

CONTAMINATING ION(S)	MAXIMUM CONCENTRATION (mg/liter)
Arsenic	0.05
Barium	1.00
Cadmium	0.01
Chloride	250
Chromium	0.05
Copper	1.00
Cyanide	0.20
Fluoride	2.00
Iron	0.30
Lead	0.05
Manganese	0.05
Nitrate	45
Organics	0.20
Selenium	0.01
Silver	0.05
Sulfate	250
Zinc	5.00
Total dissolved solids	500

including fish—and including, finally, people—so that we, as the final link in the chain, become the victims of our own pollution (Fig. 9-1).

In better days, lakes and rivers cleaned themselves. But pollution begets more pollution. Millions of dead and decaying organisms in our rivers and lakes form toxic products that in turn kill more organisms. Once the natural, healthy balance is upset, the normal cleansing action of the river or lake stops. This is the water supply that we must now turn to for potable water.

If we ourselves are going to survive, it will take massive action on the part of all of us to stop pollution. The technology to do this exists today; it is the *cost* that is holding us back. We must learn that the cost has to be faced if we are to have water to drink.

In this chapter we are going to take an in-depth look at water and water pollution.

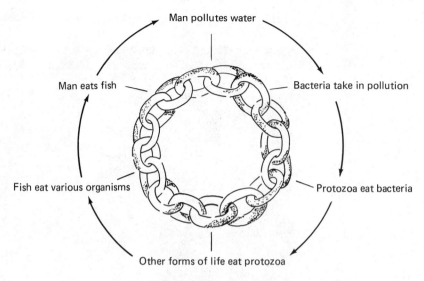

Figure 9-1 The food chain.

SCENARIO

THE PROBE

On the planet Earth, third planet from the sun (called Sol), it is the year 4000. Meanwhile, in another solar system, a historical event is taking place in the control room of a space flight center.

The landing probe is entering the planet's atmosphere. There is excitement in the control room as the technicians and scientists make their final adjustments. This is it, the first interstellar probe to land on a planet in *another* solar system. The rocket had traveled 4 light years at about one-half the speed of light, so it was eight years since it left the planet Terra, fourth planet from the sun in the star system of Alpha Centauri. Its target is Earth, the third planet from the sun, Sol, in the adjacent solar system. Previous fly-by missions had shown that Earth might be capable of supporting intelligent life much like that which existed on Terra. The necessary elements were there. So, too, was an atmosphere similar to that on Terra. There also seemed to be abundant water on the planet. In fact, four-fifths of the planet's surface seemed to be covered by water, forming great oceans. The planet was at the proper distance from its sun so that its temperatures could support life. The planet's age of about 4 billion years allowed sufficient time for intelligent life to have evolved. In fact, all the conditions that allowed life to evolve on Terra seemed to be present on this planet. The unmanned probe would have to provide the final proof. It was decided to land the probe in what appeared to be an ocean. This was because no one on Terra was sure how far life might have evolved on Sol's third planet, Earth. However, since life on Terra evolved from the oceans, it was thought that the same mechanism might exist on this planet as well.

Mission controller Robert Scott was now in command.

"Our probe is 50 miles above the planet's surface. All lights on the master console are green," he announced.

Chief mission engineer Michael Gregory turned to Robert Scott: "The warp radio transmission units are working perfectly. Imagine, instant communications with the probe even when it's 4 light years away! Everything on our end is *go* for landing."

"Remember," said Scott, "We have to be at a descent rate of 2 feet per second when we impact on the ocean."

"No problem," replied Gregory. "We'll be in that range when we reach an altitude of 500 feet."

The probe continued its descent. There were 300 people in the control room, but not a sound to be heard; then the voice of Scott monitoring the last leg of the flight.

"The probe is at 500 feet and descending at 2 feet per second," announced Scott. "Everything is operating properly. I sure hope that what we think is an ocean really contains H_2O. We're descending through the last 100 feet. Be sure to turn off all rocket motors when the probe impacts on the ocean. Ten feet to go, 6, 2, IMPACT—All motors switched to off. All cameras activated. Status report from all consoles."

The status reports appeared on the screen. The probe had landed in an ocean composed of liquid H_2O, and had sunk to a depth of 100 feet, about 20 miles off the coast of one of the continents. A land rover was released from the probe which would move along the bottom of the ocean and then on to the land to search for additional life forms.

Gregory turned to Scott and began to speak, "The televised pictures of the oceans are clear but I don't see anything that appears to be alive. There are no fish or plants similar to those in the oceans on Terra."

"I know," said Scott. "All of the preliminary data seemed so promising. I thought that perhaps we would come eye to eye with human beings such as ourselves, but at this point our instrument packages haven't even picked up any simple forms of life."

The hours passed as the analyses proceeded. The televised pictures from the land rover showed a desolate beach and a blue clouded sky. It was all so much like Terra—but without people or other life forms. Just then the data screens in the control room began flashing. A complete chemical analysis of the ocean water began to appear on the screen. The scientists watched in amazement. Of course, the ocean water had the same basic ingredients as the ocean water on Terra. But there were some other ingredients as well. Among these "other" ingredients were enormous concentrations of organic chemicals. These organic chemicals were similar to those compounds used as pesticides on Terra. There were also a number of other organic chemicals that resembled those found in industrial wastes on Terra. The concentrations of these substances in the ocean waters would certainly be toxic to any evolving life forms.

"That explains it," exclaimed Scott, "life couldn't evolve on this planet because of the formation of these chemicals. All of the other conditions were favorable for life, but these chemicals in the water prevented life from evolving. Too bad! These chemicals must have formed in the natural evolution of the planet."

"Wait a minute," interrupted Gregory, "there is another explanation."

"What's that?" asked Scott.

"It's just possible that intelligent life did evolve on the planet, and that in the course of their civilization they created these chemicals that polluted the water on the planet and which eventually destroyed all living organisms. After all, our chemists on Terra created these chemicals for our civilization."

"Yes, yes," exclaimed Scott indignantly, "But do you think that intelligent life forms like ourselves would ever do something as stupid as that—pollute the very substance that gives life to all living things on the planet? I think we can conclude that my explanation is correct—intelligent life never evolved on the third planet from Sol."

And so it was recorded in the history books on the planet Terra: life never evolved on planet Earth.

The purpose of this scenario was to show you the importance of water to living organisms. Let's now take a more in-depth look at water and water pollution before our scenario becomes our legacy.

THE CHEMISTRY OF WATER

Physical Properties of Water

Pure water is a tasteless, odorless, colorless substance that has a melting point of 0°C and a boiling point of 100°C at 1 atm pressure. These two temperatures serve as reference points on the Celsius temperature scale.

Figure 9-2 Establishing the density of water (at 4°C). The experiment shows that 1 ml of water has a mass of 1 g (in other words, the density of water is 1 g/ml).

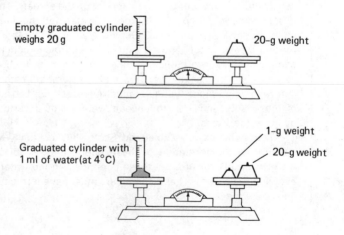

Empty graduated cylinder weighs 20 g

20-g weight

Graduated cylinder with 1 ml of water (at 4°C)

1-g weight

20-g weight

The density of water is another reference standard. It relates metric units of volume (milliliters) to metric units of mass (grams). At 4°C, water has a density of 1 gram per milliliter (1 g/ml). This means that 1 gram of water at 4°C will occupy a volume of 1 milliliter. In other words, if we measure 1 ml of water in a graduated cylinder, we find that it has a mass of 1 g (Fig. 9-2).

Water has intrigued humankind since the days of the ancient Greeks. One of its most interesting properties is its change in volume when its temperature is changed. If we take a certain amount of water at room temperature and warm it, the volume of the water increases. If we cool the water below room temperature, its volume decreases. This is the type of behavior we expect from a liquid. However, as we continue to cool the water below 4°C, the volume of the water *increases*. This behavior is very unusual, since the volumes of most liquids decrease continuously as the liquids are cooled. One result of water's strange behavior is that the density of ice is *less* than the density of water. Why? Remember that D = mass/volume (Appendix A). For any mass of water, the volume increases as the water is cooled. At 4°C, the volume of 10 g of water is 10 ml, but at 0°C the volume of the water (ice) increases to 10.9 ml. Therefore,

$$D \text{ (at } 4°C) = \frac{10 \text{ g}}{10 \text{ ml}} = 1 \text{ g/ml}$$

$$D \text{ (at } 0°C) = \frac{10 \text{ g}}{10.9 \text{ ml}} = 0.92 \text{ g/ml}$$

This calculation explains why ice floats on water. For most other substances the solid form does *not* float on the liquid form. For instance, if you freeze pure alcohol and drop it into liquid alcohol, it will sink. This is because the alcohol, like most other substances, reaches its maximum density at its freezing point.

It's fortunate for us that ice floats on water. Think of what would happen if it didn't! In the winter when the lakes, rivers, and seas froze, and water turned into ice, the ice would sink to the bottom. As time passed, more and more ice would form and sink to the bottom. The entire body of water would eventually be frozen and all life forms in it would be killed. Since scientists believe that life on earth originated in the sea, human beings would never have evolved. That is why we are lucky that ice floats on water. The layer of floating ice insulates the rest of the water and prevents it from freezing. This explains why an entire body of water does not freeze, even though the air temperature above it stays below freezing for many months (Fig. 9-3).

But there are problems caused by the unusual behavior of water. Because water expands when it freezes, if we put a bottle filled with water in a freezer, the bottle will crack when the water freezes. In cold weather, unless you put antifreeze in your car, the water in the radiator will freeze and crack the engine block. This behavior of water also accounts for the cracks and potholes in paved streets during the months of cold weather. Water exerts a tremendous force when it expands while freezing.

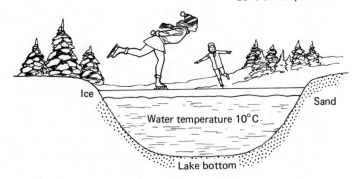

−20°C air temperature

Ice

Sand

Water temperature 10°C

Lake bottom

Figure 9-3 Floating ice insulates the lake water below.

The unusual behavior of water is the result of its bonding and the three-dimensional shape of the water molecule. If we examine a single water molecule, we find that:

1. The bonds between the hydrogen and oxygen atoms are polar. Therefore, each hydrogen atom seems to have a slight positive charge, and the oxygen atom seems to have a slight negative charge. (Remember our discussion on bonding in Chap. 4?)

2. The bond angle between the hydrogen atoms is about 105° (Fig. 9-4).

Because of these two conditions, the positively charged hydrogen atom of one water molecule can attract the negatively charged oxygen atom of another water molecule. The two molecules then are held together by a force called a *hydrogen bond* (Fig. 9-5). Several water molecules may bond together to form long chains (Fig. 9-6). It is the hydrogen bonds, which hold the water molecules together, that give water its unusually high boiling point.

But how do we explain the fact that water expands when it freezes? If we look at a three-dimensional structure of a crystal of ice, it would seem to be hexagonal, or six-sided (Fig. 9-7). This type of geometry means that ice mole-

Figure 9-4 The structure of a water molecule. The bond angle between hydrogen atoms is about 105°.

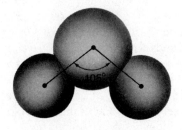

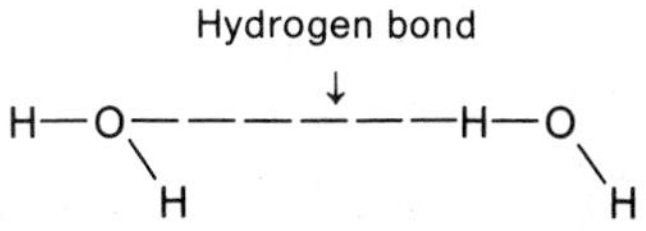

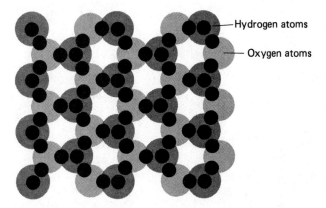

Figure 9-5 Hydrogen bonding between two water molecules.

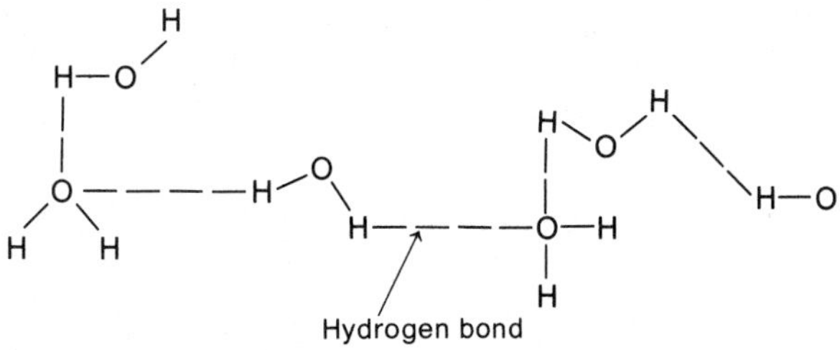

Figure 9-6 Hydrogen bonding between many water molecules.

Hydrogen atoms

Oxygen atoms

Figure 9-7 The crystal structure of ice. (Note: All the hydrogen atoms are not shown because some of them would be behind the plane of the paper.)

cules have to be farther apart from each other than liquid water molecules. Therefore, water in the form of ice has a greater volume than water in the form of a liquid.

Chemical Properties of Water

Water is one of the most stable compounds—so stable that for many years it was considered one of the basic elements of matter. Even when heated as high as 2000°C, water shows little decomposition. However, if an electric current is passed through it, water breaks down.

$$2\,H_2O \xrightarrow{\text{electricity}} 2\,H_2 + O_2$$

Water reacts with oxides of *non*metals to produce acids. For example, in the case of sulfur trioxide,

$$SO_3 + H_2O \longrightarrow H_2SO_4$$

Sulfuric
acid

and in the case of sulfur dioxide,

$$SO_2 + H_2O \longrightarrow H_2SO_3$$

Sulfurous
acid

Water reacts with oxides of metals to produce bases. For example,

$$CaO + H_2O \longrightarrow Ca(OH)_2$$
$$Na_2O + H_2O \longrightarrow 2\,NaOH$$

Many salts combine with water to form crystalline compounds called *hydrates*. Some common hydrates are:

$Na_2B_4O_7 \cdot 10\,H_2O$	Borax
$CaSO_4 \cdot 2\,H_2O$	Gypsum
$KAl(SO_4)_2 \cdot 12\,H_2O$	Alum
$MgSO_4 \cdot 7\,H_2O$	Epsom salts
$Na_2CO_3 \cdot 10\,H_2O$	Washing soda
$CuSO_4 \cdot 5\,H_2O$	Blue vitriol

The centered dot between the formula for the salt and the formula for the water is not a decimal point. It is a way of showing that there is a bond between the salt and the water. When this kind of salt is heated, the bond is broken and the water is driven off. For example,

$$CuSO_4 \cdot 5H_2O \xrightarrow{\text{heat}} CuSO_4 + 5H_2O$$

A blue   A white
compound   compound

Now that we've examined some of the chemical and physical properties of water, let's turn our attention to the problem of water quality and water pollution.

A WATER CYCLE

How does water become polluted, and how does it get pure again? Let's follow the path of some water as it begins its journey from falling rain to the ocean.

Rain falling over the mountains forms a clear mountain stream. The water is pure and refreshing. (Actually, the water isn't pure in a chemical sense. In other words, it contains more than just H_2O. There are usually various dissolved salts in the water that give it a unique character and taste.) As our small stream moves down the mountainside it merges with other streams and eventually forms a river (Fig. 9-8).

Along the river bank is a town. The town uses the river as its source of *potable* water. The water is withdrawn from the river by the town's potable water treatment plant. Even though this is the first town along the river, the water must be tested and treated to be sure that it is safe to drink. We'll have more to say about potable water treatment later in this chapter.

Individuals and industries in the town use the water in many ways. Also, some river water is taken directly from the river by certain industries and used for industrial processes. However, once the water has been used, it is considered to be wastewater. This wastewater is flushed down toilets, sink drains, and industrial sewers, and flows through a series of sewer pipes to a sewage treatment plant. At the sewage treatment plant some of the wastes are removed from the wastewater before it is allowed to return to the river. The amount of wastes removed depends on the sophistication of the plant. We'll have more to say about wastewater treatment later in this chapter. (In some towns in the country the wastewater runs directly back into the river because there is no sewage treatment plant.)

What types of waste are in the wastewater? There are the biological wastes from the people in the town, food wastes from kitchen sinks, industrial

Figure 9-8 A clear mountain stream. (EPA-DOCUMERICA; photographer, Gene Daniels)

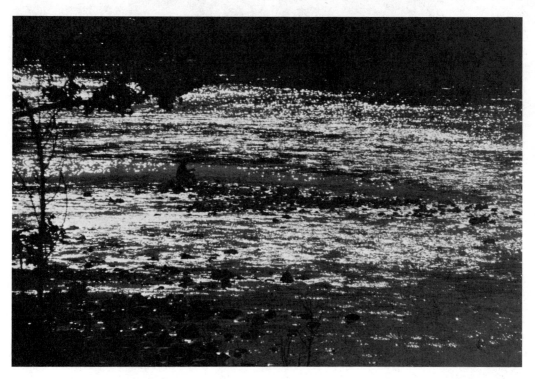

wastes from the town's industries and agricultural wastes, including fertilizers from the town's farms. Even with sewage treatment many of these waste materials end up in the river. Our once clean mountain stream now contains various amounts of human waste, food wastes, industrial chemicals, pesticides, and agricultural fertilizers. If the amounts of these materials are small and if they can be broken down biologically by the microorganisms in the river, the river may be able to cleanse itself. But what happens if large amounts of sewage are continuously dumped into the river?

Once in the river the organic materials act as food for the microorganisms in the water. With an abundant supply of food the microorganisms grow at a rate above normal. In consuming the waste materials the microorganisms also consume the oxygen dissolved in the river water. There is only a given amount of dissolved oxygen that can exist in a quantity of water at any one time. If this dissolved oxygen is consumed too quickly, a serious situation can develop in the river, because it takes time for more oxygen to dissolve back into the water from the air. In the meantime all of the organisms that need oxygen to survive can die; this includes fish as well as the microorganisms that break down the sewage (Fig. 9-9).

Figure 9-9 A biologist for the state of Illinois takes measurements of a carp and other fish that were the victims of a massive fish kill in the state in 1967. (EPA-DOCUMERICA)

Microorganisms that use dissolved oxygen to break down sewage into harmless materials are called *aerobic microorganisms*. They must survive if the river is to remain clean and healthy. However, if they die due to lack of oxygen, other microorganisms, called *anaerobic microorganisms*, begin to take over and break down the remaining wastes into foul-smelling and sometimes poisonous products. It is in this way that pollution begets more pollution.

Besides the organic wastes that can pollute a river, there are the agricultural fertilizers, which can have an even more serious effect. These fertilizers contain nitrates and phosphates which also act as food for algae. Large growths of algae can choke a river or lake, causing it to fill in, which is a process known as *eutrophication*. Also, dead algae exert a tremendous oxygen demand on the river or lake as microorganisms attempt to consume them (Fig. 9-10).

Industrial wastes can have two effects on a river. They can act as poisons to all life forms in the water or they can require large amounts of dissolved oxygen on the part of microorganisms in order to break them down.

The river moves on. Farther downstream is another town. This town also uses the river as its source of potable water. Again, water is withdrawn from the river by the town's potable water treatment plant. The town must be extremely careful in purifying this water because small amounts of pollution from upstream may be present. This town will also send its wastewater back to the river. And so it goes. As the river passes by each town it picks up more and more pollution. If the pollution is too great, the river and all in it can die.

Figure 9-10 Algae growth in Lake Tahoe, Nevada, photographed in May, 1972. (EPA-DOCUMERICA; photographer, Belinda Rain)

If this occurs, a town farther downstream may not be able to use the river for its drinking water or for fishing or recreation.

Eventually, the river flows into the ocean. At one time it was thought that the ocean could not be polluted because of its vastness, but today we know better. There are areas in some oceans which have been polluted to such a degree that no life exists there. However, large amounts of water evaporate from the ocean and form clouds. The clouds are carried aloft and eventually drop their moisture as rain. Pure water, again, finds its way to a clear mountain stream.

So the *water cycle* gives us back fresh, clean, and pure water, but it takes time. However, pollution still remains in the ocean. Even more important is the fact that we can't afford to lose our rivers, lakes, and streams to pollution when our demands for clean water are increasing. Also, we get a lot of food from the world's oceans in the form of fish. With food supplies in the world critical, we can hardly afford to contaminate this source of food. Our only hope for protecting our sources of water is to have thorough and efficient wastewater treatment.

In the next part of this chapter we're going to look at how a potable water treatment plant works and how a wastewater treatment plant works.

POTABLE WATER TREATMENT: A JOB WELL DONE

Why do we have to treat the water that we drink? Because it has been shown that many diseases are transmitted by water. There are many stories in the history books about plagues that killed large numbers of people. During the fourteenth century, a plague known as the Black Death swept over Europe and killed about 25% of the population. In the mid-1600's an epidemic killed 70,000 Londoners in one year. In 1854 a cholera epidemic broke out in London. Two investigators, John Snow and John York, showed that the disease was being transmitted by the water from the Broad Street Pump. They also showed that this well was being contaminated with human wastes from a broken sewer and that this sewer, in turn, was connected to the home of an individual stricken with the disease.

With the birth of the science of bacteriology in 1870, it was possible to identify the causative agents of disease. As epidemics occurred in the late 1800's bacteriologists were able to identify the causative agents of specific diseases as being present in the drinking water at the time of the epidemic. Such was the case in the Lausen, Switzerland, typhoid epidemic in 1872, and in the cholera epidemic in Hamburg in 1892. Episodes such as these convinced public health officials that something had to be done to protect potable water supplies.

Chlorine has the ability to kill pathogenic organisms in water, and if the concentration is carefully controlled, the chlorine will not harm human beings. It was for this reason that chlorinating public water supplies seemed to be the answer to the epidemic problem. Emergency chlorination of water began in

1850 whenever an epidemic occurred. However, in 1904 continuous chlorination of a public water supply was attempted in England. In 1908 it was begun at the Union Stockyards in Chicago. Then in 1909, Jersey City, New Jersey, began chlorinating its water and became the first city in the United States to chlorinate a public water supply continuously. As chlorination of public water supplies grew, death rates due to typhoid and other waterborne diseases fell (Table 9-2). However, it has recently been found that chlorination is not a panacea. The chlorine in the water can combine with other substances and form a class of compounds known as chlorinated hydrocarbons. These compounds have been shown to be cancer-causing in high concentrations. Fortunately, these chlorinated hydrocarbons have been found only in minute amounts in public water supplies. But they will have to be monitored carefully in the future.

Besides chlorination, a potable water treatment plant may treat the water with coagulants to remove a number of problem substances: for example,

1. Turbidity, inorganic and organic

2. Color

3. Harmful bacteria and other pathogens

4. Algae and other plant organisms

5. Taste- and odor-producing substances

6. Phosphates (which serve as nutrients for the growth of algae)

TABLE 9-2 *Typhoid and paratyphoid death rates in the United States, 1900–1935*

YEAR	DEATHS PER 100,000 POPULATION
1900	35
1905	28
1910	24
1915	12[a]
1920	8
1925	8
1930	3
1935	2

[a] Notice the remarkable decrease between 1910 and 1915. This represents the advent of chlorination of public water supplies in the United States.

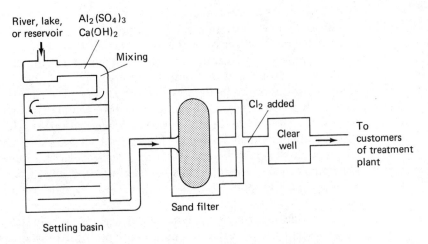

Figure 9-11 A diagram of a potable water treatment plant.

In summary, we can say that a potable water treatment plant must produce a safe and aesthetically pleasing product (Fig. 9-11).

We should also mention that in some regions of the country some consumers add additional treatment of their own to their water. In areas where the water is "hard," many people purchase *water softeners*. Hard water contains calcium and magnesium salts which inhibit the cleansing action of soaps and cause a "scum" to form on bathtubs and sinks. Water softeners chemically remove these hardness-causing substances and make the water "soft." Some water softeners work by exchanging sodium ions for calcium and magnesium ions. Sodium salts are more soluble than calcium and magnesium salts and they don't interfere with the cleansing action of soaps.

WASTEWATER TREATMENT: SAVING OUR LAKES, RIVERS, AND STREAMS

The only way that we're going to save our lakes, rivers, and streams is to have adequate wastewater treatment. Pollutants must be removed from the water once the water has been used, before it is returned to the river.

Scientists and engineers who work in wastewater treatment plants categorize the pollutants in wastewater as *dissolved* or *undissolved solids*. The undissolved solids are further classified into *settleable* and *suspended solids*. Various treatment methods are used to remove these solids (pollutants) from wastewater. The three methods are known as *primary*, *secondary*, and *advanced wastewater treatment*.

Primary Treatment

The most common type of wastewater treatment plant is the primary plant (Fig. 9-12). In primary treatment, water from the sanitary sewer system passes

Figure 9-12 A primary settling tank at a wastewater treatment plant. (EPA-DOCUMERICA; photographer, Belinda Rain)

through a *bar screen* at the head of the plant. This bar screen removes large particles from the wastewater. After passing through the bar screen, the water is channeled into a large settling tank. In the settling tank the flow of the water is slowed and the settleable solids have a chance to settle out of the water. Also, a skimmer removes some suspended solids on the surface of the water. The final step in primary treatment is the chlorination of the water as it leaves the plant.

Primary treatment removes about 30% of the pollutants from the water. Actually, the efficiency of a wastewater treatment plant is measured in terms of biological oxygen demand (abbreviated BOD). The BOD test measures the amount of oxygen required by the microorganisms in the sewage to break down the wastes under aerobic conditions. The test is performed under controlled conditions on a sample of the wastewater.* Typical domestic sewage has a BOD value of 200 mg/liter. In other words, for untreated domestic sewage 200 mg of oxygen is needed by the microorganisms, per liter of sewage water, to break down the wastes aerobically.

If the primary treatment plant is running efficiently, the BOD of the processed sewage (known as effluent) should be about 140 mg/liter. As you can see, there's still lots of wastes left in the water after primary treatment. How come? Remember, primary treatment removes only the settleable solids and some suspended solids. It does not remove dissolved solids. It is for this reason

* The BOD test measures the amount of oxygen consumed by the microorganisms over a 5-day period when the water is kept at 20°C.

that many primary treatment plants in the country are currently being upgraded to secondary treatment plants.

Secondary Treatment

Secondary treatment involves aerating the wastewater in the presence of large numbers of microorganisms. In this way the microorganisms convert the dissolved solids into more stable inorganic substances.

There are two forms of secondary treatment. In the *activated sludge process*, water from the treatment plant's primary system enters a huge aeration tank (Fig. 9-13). The water is aerated with a material called *sludge*. The sludge is composed of microorganisms and other inorganic materials from previously treated sewage. During the aeration process more sludge is formed as microorganisms consume the organic wastes in the sewage. The sludge is allowed to settle out in a *settling tank* before the water is allowed to move on.

The other method of secondary treatment is the *trickling filter process*. In this process the wastewater from the treatment plant's primary system is pumped into a long rotating device which distributes the water over a series of rocks (Fig. 9-14). The rocks are coated with a gelatinous film of microorganisms. As the water passes over the rocks the microorganisms consume the waste material. Sometimes the gelatinous material sloughs off of the rocks and is carried away with the water. For this reason the water passing through the trickling filter is directed to a settling tank before being chlorinated and allowed to leave the plant.

The addition of secondary treatment increases the overall removal of wastes from sewage water to 80 to 90% of the BOD. For typical domestic sewage this means that the effluent would have a BOD of only 20 to 40 mg/liter. This is quite an improvement over just primary treatment.

Figure 9-13   A diagram for the activated sludge method of secondary wastewater treatment.

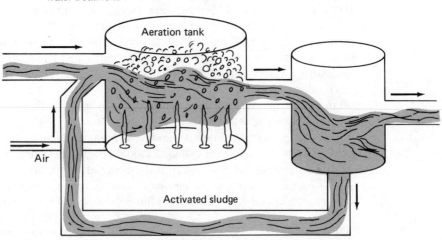

Figure 9-14 A close-up of a trickling filter. (U.S. Environmental Protection Agency)

Advanced Treatment

There are very few advanced wastewater treatment plants operating in the United States, at present. The costs involved are high and the return is small in the sense that only 10 to 20% of the BOD remains to be removed. However, in some areas of the country removing a good part of that remaining BOD is critical. For example, in Lake Tahoe, California, an advanced treatment system was built for the chemical removal of nitrates and phosphates from the wastewater to prevent eutrophication of the lake. The lake (Lake Tahoe) is an important recreational resource, which residents and businesspeople felt had to be protected, regardless of cost.

WASTEWATER TREATMENT: A SHORT SUMMARY

In this section of the chapter we examined the three types of wastewater treatment. A good way to summarize these three forms of treatment is to say that:

1. *Primary treatment* is a physical process, whereas

2. *Secondary treatment* is a biological process, and

3. *Advanced treatment* is a chemical process.

Although primary treatment is the most common type of treatment used by cities and towns today, it seems almost inevitable that secondary and advanced treatment plants will be needed as populations grow in particular areas. This additional treatment will be necessary just to maintain current water quality standards that are applied to our lakes, rivers, and streams that receive the treated sewage.

SUMMARY

In this chapter we've reviewed various aspects of water and water pollution. We began by looking at the importance of water to human survival on our planet. Next, we examined the chemical and physical properties of water. We discovered that water does indeed have some very unusual properties. Our attention then turned to the study of water treatment and water pollution. We traced the path of a river and saw how clean, fresh water is turned into used and polluted water. Then we looked at the process of potable water treatment. Finally, we examined the various processes of wastewater treatment.

All of us use water every day of our lives. Perhaps now you have a better understanding of this unique substance.

EXERCISES

1. State the importance of water to the survival of humankind.

2. Trace the path of pollutants from water into the food we eat.

3. Discuss the possibility of the scenario "The Probe" becoming our legacy.

4. Name four sources of potable water.

5. What is unique about the density of water and its relationship with temperature?

6. Draw a sketch showing the hydrogen bonding between water molecules.

7. Complete the following equations; then balance them.

   (a) $H_2O \xrightarrow{\text{electricity}}$　　　　(b) $SO_2 + H_2O \rightarrow$

   (c) $Na_2O + H_2O$　　　　(d) $CaSO_4 \cdot 2\,H_2O \xrightarrow{\text{heat}}$

8. List some common water pollutants.

9. Trace the path of water from falling rain over the mountains back to the ocean.

10. Explain the difference between aerobic and anaerobic microorganisms, and relate their significance to the health of a stream.

11. Explain the process of eutrophication and how it occurs.

12. State some of the processes involved in potable water treatment. What process is the most important one? Why?

13. How are pollutants in wastewater classified?

14. Discuss the process of primary sewage treatment.

15. Discuss the two processes used in secondary sewage treatment.

16. What is the purpose of advanced wastewater treatment?

17. Match the words on the left with the descriptions on the right.

    (1) primary treatment   (a) biological process
    (2) secondary treatment   (b) chemical process
    (3) advanced treatment   (c) physical process

18. What is BOD, and how is it used in the treatment of wastewater?

19. Why is it important to chlorinate wastewater before it leaves the plant?

20. Can you think of any reasons why you wouldn't want your drinking water chlorinated?

Chapter 10

Air Pollution

Some Things You Should Know After Reading This Chapter

You should be able to:

1. Explain what happened in Donora, Pennsylvania, in late October 1948.
2. Calculate the percent by volume of a gas in a mixture given sufficient data.
3. Calculate the parts per million of a pollutant in the air given sufficient data.
4. Calculate the micrograms per cubic meter of a pollutant in the air given the necessary information.
5. Name the major constituents of air.
6. List the major air pollutants: oxides of sulfur, carbon monoxide, photochemical oxidants, hydrocarbons, oxides of nitrogen, particulate matter.
7. State some of the problems associated with each of the pollutants listed in objective 6.
8. Name the three hazardous pollutants: asbestos, beryllium, and mercury.
9. List some of the problems associated with each of the hazardous pollutants.
10. Write some of the chemical reactions associated with the burning of coal.
11. Write some of the chemical reactions associated with automotive pollution.
12. List some pollution control devices and tell how they are used.

INTRODUCTION

Never before has our world been faced with so many environmental problems, because never before have we had so many people around to contaminate the place. The air we breathe, the water we drink, and the land we live on are rapidly being polluted. We ask ourselves: How can we stop this pollution? To stop it we must first understand how it is caused. In this section we look briefly at the mechanisms of air pollution from the chemist's point of view.

We can define air pollution as the presence of a contaminant in the outdoor atmosphere in a concentration large enough to injure human, plant, or animal life, or interfere with the enjoyment of life or property.

SCENARIO

> ### THE HALLOWEEN THAT ALMOST WASN'T (A TRUE STORY)
>
> It is late in October 1948. The place is Donora, Pennsylvania, a small industrial town located at the bottom of a river valley. An unusually thick smog has settled over the town, and people look up at the leaden sky and sniff uneasily.
>
> **Figure 10-1** The effect of an atmospheric inversion over St. Louis. (Courtesy of the St. Louis Post Dispatch. Photo by Robert C. Holt, Jr. Used by permission.)
>
>

The smog is so thick that it makes people's eyes sting and their throats become raw. Sulfur dioxide, dust, and waste products from nearby zinc, iron, and steel mills contribute to the smog. The weather conditions have created an atmospheric inversion, which means that the polluted air is kept near the ground (Fig. 10-1). By the time the smog lifts some five days later, 17 people have died as a direct result of it, and almost half the population of Donora has been affected to some degree. The nightmare of environmental pollution has made its first widely recognized appearance in the United States.

MEASURING POLLUTION CONCENTRATION IN THE AIR

Before we continue our discussion of air pollution, we must first tell you how the concentration of pollutants and other substances are measured in the air.

The major components of air are usually measured in units of percent by volume. For example, imagine if you mixed 78 liters of nitrogen with 21 liters of oxygen and 1 liter of some "other gases" (Fig. 10-2). (This mixture would be pretty close to the composition of our atmosphere.) The percent by volume for each component in the mixture would be calculated as follows:

$$\text{Percent nitrogen} = \frac{78 \text{ liters nitrogen}}{100 \text{ liters total gas}} \times 100 = 78\%$$

$$\text{Percent oxygen} = \frac{21 \text{ liters oxygen}}{100 \text{ liters total gas}} \times 100 = 21\%$$

$$\text{Percent "other gases"} = \frac{1 \text{ liter "other gases"}}{100 \text{ liters total gas}} \times 100 = 1\%$$

Example 10-1   Determine the percent by volume of each gas in the following mixture. The mixture contains 20 liters of oxygen gas, 30 liters of nitrogen gas, and 10 liters of hydrogen gas.

Solution   Total volume of gas is 20 liters + 30 liters + 10 liters = 60 liters.

$$\text{Percent oxygen} = \frac{20 \text{ liters oxygen}}{60 \text{ liters total gas}} \times 100 = 33\%$$

Figure 10-2   A gas mixture similar to our atmosphere.

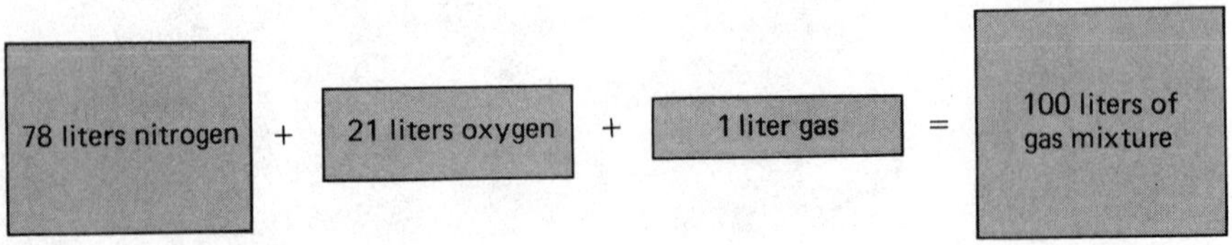

| 78 liters nitrogen | + | 21 liters oxygen | + | 1 liter gas | = | 100 liters of gas mixture |

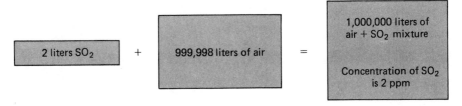

Figure 10-3 A mixture of 2 parts per million sulfur dioxide in air.

$$\text{Percent nitrogen} = \frac{30 \text{ liters nitrogen}}{60 \text{ liters total gas}} \times 100 = 50\%$$

$$\text{Percent hydrogen} = \frac{10 \text{ liters hydrogen}}{60 \text{ liters total gas}} \times 100 = 17\%$$

The minor components in the air (which include most pollutants) are usually measured in *parts per million* (abbreviated *ppm*). For example, imagine that we mix 2 liters of sulfur dioxide (SO_2) with 999,998 liters of air (Fig. 10-3). The concentration of the sulfur dioxide would be 2 ppm. In other words, there would be 2 liters of sulfur dioxide out of 1 million liters of gas mixture. (At this point you might be asking yourself: "Why don't they measure the concentration of pollutants in percent, also?" To convert the 2 ppm to percent would yield a very small number, so it's easier to use the ppm notation. Let's show you what we mean. To convert 2 ppm to percent, we ask the question: "2 out of 1 million is to how many out of 100?" Mathematically stated:

$$\frac{2 \text{ liters sulfur dioxide}}{1,000,000 \text{ liters gas mixture}} = \frac{x \text{ liters sulfur dioxide}}{100 \text{ liters gas mixture}}$$

$$x = 0.0002 \text{ liter sulfur dioxide}$$

It's much easier to express whole numbers than decimal numbers; that's why we use the ppm notation.)

Because the volumes of gases are directly proportional to their number of molecules, we can also say that a 2 ppm sulfur dioxide concentration means that 2 molecules out of every million molecules in our gas mixture is sulfur dioxide.

Example 10-2 If you had 500,000 pennies and all of them were copper pennies except for *one*, which was a steel (World War II) penny, what is the *concentration* of steel pennies in parts per million?

Solution One penny out of 500,000 is steel. So the question is: How many out of 1,000,000 would be steel? Mathematically stated:

$$\frac{1 \text{ steel penny}}{500,000 \text{ total pennies}} = \frac{x \text{ steel pennies}}{1,000,000 \text{ total pennies}}$$

$$x = 2 \text{ steel pennies}$$

Example 10-3   Consider a sample of air that contains 2 molecules of carbon monoxide (CO) in 200,000 molecules of air. What is the concentration of CO in ppm?

Solution   Two molecules out of 200,000 molecules are CO. So the question is: How many out of 1,000,000 molecules would be CO? Mathematically stated:

$$\frac{2 \text{ molecules CO}}{200,000 \text{ molecules air}} = \frac{x \text{ molecules CO}}{1,000,000 \text{ molecules air}}$$

$$x = 10 \text{ molecules CO}$$

When you think of the concentration of a substance in parts per million it sounds small, whereas a substance whose concentration is in percent sounds large. Don't be misled. Concentration doesn't tell you anything about the effect of a pollutant on people, animals, or plants. A concentration of 0.5 ppm SO_2 is serious enough to cause long-term lung-damaging effects on people inhaling that air, whereas a concentration of 1% carbon dioxide (CO_2) is not harmful at all. It is important therefore to concern ourselves not only with the concentrations of pollutants, but with their effects on living organisms, as well.

Also, we should point out that certain types of pollutants known as particulates are measured in units of *micrograms per cubic meter of air* ($\mu g/m^3$). Particulate matter is composed of solid materials such as dust and smoke or liquid particles such as mists and sprays. The concentrations of these materials are determined by separating out the particulate matter from a sample of air. The weight of the particulate matter is obtained and the weight in micrograms per cubic meter of air is calculated. (By the way, a microgram is 0.000001 g or 1×10^{-6} g. If you're not familiar with these metric terms, see Tables 1 and 2 in Appendix C).

Example 10-4   A sample of air is analyzed for particulate matter. It is found that 0.00375 g of particulate matter is in 50 m^3 of air. What is the concentration of this particulate matter in $\mu g/m^3$?

Solution   First we convert grams to micrograms. The conversion factor is 1 $\mu g = 1 \times 10^{-6}$ g. Therefore,

$$? \mu g = (0.00375 \text{ g}) \left(\frac{1 \mu g}{1 \times 10^{-6} \text{ g}} \right) = 3.75 \times 10^{+3} \mu g$$

To get the concentration, we simply divide the number of micrograms by the cubic meters of air:

$$\text{concentration in } \mu g/m^3 = \frac{3.75 \times 10^{+3} \mu g}{50 \text{ m}^3} = 75 \mu g/m^3$$

THE AIR WE BREATHE: WHAT'S IN IT?

Open any chemistry text to the section on the composition of air. You'll discover that air consists of 78% nitrogen, 21% oxygen, and 1% other gases (such as argon, helium, hydrogen, carbon dioxide, water vapor, and inert gases). These percentages are for the most part correct, but unfortunately they don't tell the whole story. There are a number of substances missing from the list. And although their concentrations may be extremely small, they have a devastating effect on human health. These unnamed ingredients include carbon monoxide, hydrocarbons, sulfur dioxide, sulfuric acid, nitrogen oxides, and—most important—particulate matter (which are the tiny particles of smoke, dust, fumes, and aerosols in the atmosphere).

But before we look at these various pollutants, let's examine the composition of clean air, that is, air not contaminated by humankind's endeavors. Table 10-1 lists the components of "ordinary air." We chose dry air because the moisture content in air can vary around the world from practically zero in the desert to about 5% in the jungle. Also, naturally occurring particulate

TABLE 10-1 *The gaseous components of "ordinary" dry air*

PURE AIR COMPONENTS	CONCENTRATION (BY VOLUME)	
	ppm	PERCENT
Nitrogen (N_2)	780,900	78.09
Oxygen (O_2)	209,400	20.94
Inert gases		
Argon (Ar)	9,300	0.93
Neon (Ne)	18	
Helium (He)	5	
Krypton (Kr)	1	
Xenon (Xe)	1	
Carbon dioxide (CO_2)	315	
Methane (CH_4) (although a hydrocarbon, it occurs naturally in the biosphere and is not included with other hydrocarbons when calculating total hydrocarbon pollution)	1	
Hydrogen (H_2)	0.5	
NATURAL POLLUTANTS		
Oxides of nitrogen (produced by solar radiation and lightning)	0.52	
Ozone (O_3)	0.02	

matter is eliminated from our list because that, too, can vary from place to place.

Notice the concentration of the various substances in Table 10-1. The concentrations are important because a substance that is not ordinarily considered a pollutant at one concentration can become one at a different concentration. For example, a 10% methane (CH_4) level in air is explosive, and so at such a concentration, methane would be considered a pollutant.

Let's now turn our attention to the major air pollutants.

MAJOR AIR POLLUTANTS: A DEEP DEADLY BREATH

In this part of the chapter we examine the major air pollutants (Table 10-2). We examine various aspects of each pollutant: for example, its major sources and effects. We also look at the conclusions found in the Environmental Protection Agency's (EPA's) criteria document for each pollutant.

What We Know About Oxides of Sulfur

The major source of sulfur oxides is fuel combustion. Some minor sources are chemical plants, metal processing plants, and trash burning (Fig. 10-4). How do sulfur oxides form from the burning of hydrocarbon fuels? Sulfur is a nonmetallic element found in coal and fuel oil. When these fuels are burned, sulfur joins with oxygen in the air to form gaseous oxides of sulfur, including sulfur dioxide (SO_2) and sulfur trioxide (SO_3).

TABLE 10-2 *The major air pollutants*

POLLUTANT	NATIONAL AMBIENT AIR QUALITY STANDARD ($\mu g/m^3$) NOT TO BE EXCEEDED
Particulate matter	
Annual average	75
Maximum 24-hour concentration	260
Sulfur oxides	
Annual average	80 (0.03 ppm)
Maximum 24-hour concentration	365 (0.14 ppm)
Carbon monoxide	
Maximum 8-hour concentration	10 (9 ppm)
Maximum 1-hour concentration	40 (35 ppm)
Photochemical oxidants, maximum 1-hour concentration	160 (0.08 ppm)
Hydrocarbons, maximum 3-hour (6-9 A.M.) concentration	160 (0.24 ppm)
Nitrogen oxides, annual average	100 (0.05 ppm)

(a)

(b)

Figure 10-4 The major source of sulfur oxides is fuel combustion. The photograph shows the High Bridge Power Plant before pollution control devices were installed (a) and after installation of pollution control equipment (b). (Northern States Power Company, St. Paul, Minnesota)

Sulfur oxides, in combination with moisture and oxygen, can yellow leaves of plants, dissolve marble, and eat away iron and steel. They can limit visibility and cut down the light from the sun. They can affect a person's breathing: at sufficiently high concentrations, sulfur dioxide irritates the upper respiratory tract; at even lower concentrations, when carried on particulates, it appears able to do still greater harm by injuring lung tissues.

The conclusions found in the EPA's criteria document show that increased mortality (death) occurred when the annual geometric mean was as high as 115 $\mu g/m^3$. It also showed that adverse effects can be detected when sulfur oxide pollution exceeds certain levels for short periods of time. These effects are especially evident in the case of sulfur dioxide. Levels of 300 $\mu g/m^3$ for sulfur dioxide over a 3- or 4-day period have been associated with a variety of adverse health effects.

What We Know About Carbon Monoxide

The major source of carbon monoxide (CO) is the internal combustion engine in motor vehicles, primarily the automobile (Fig. 10-5). Some minor sources are various industrial processes and solid waste disposal.

Figure 10-5 The primary source of carbon monoxide is the internal combustion engine in motor vehicles. (EPA-DOCUMERICA)

Carbon monoxide, an invisible, odorless, and tasteless gas, is formed when any carbon-containing fuel (gasoline, coal, etc.) is not completely burned to carbon dioxide (CO_2). Because of its characteristics, the internal combustion engine, especially in cars, is responsible by far for the largest fraction of human-made emissions of carbon monoxide.

Compared to other common air pollutants, carbon monoxide has a unique mechanism of action. It does not irritate the respiratory tract but rather passes through the lungs directly into the bloodstream. There it combines with the red blood cell's hemoglobin, the substance that normally carries oxygen to all the tissues of the body. Because hemoglobin binds carbon monoxide over 200 times as strongly as oxygen, a low concentration of carbon monoxide in the ambient air has a greatly magnified effect on the body (Fig. 10-6). Since the heart and brain are the two tissues most sensitive to oxygen deprivation, they show the most serious effects from carbon monoxide exposure. Thus at high concentration (1000 ppm and more), carbon monoxide kills by paralyzing normal brain action. At much lower levels, effects on these two tissues are also the predominate ones.

Because of its unique mode of action, carbon monoxide is not known to have adverse effects on vegetation, visibility, or material objects.

The conclusions found in the EPA's criteria document show that exposure to 30 ppm (35 mg/m^3) will, after a few hours, inactivate about 5% of the blood's hemoglobin, thus lowering its oxygen content. This loss can impair perfor-

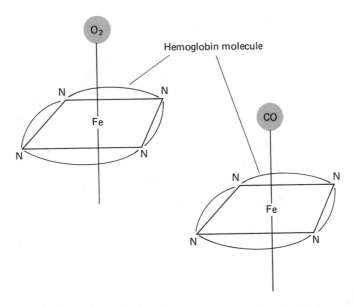

Figure 10-6 Because hemoglobin binds carbon monoxide over 200 times as strongly as oxygen, a low concentration of carbon monoxide in the ambient air has a greatly magnified effect on the body.

mance on certain psychomotor tests, indicating a significant effect on brain function. At higher exposures, excess strain is put on patients with heart disease. Exposure to 10 to 15 ppm (12 to 17 mg/m³) for several hours also affects the brain by altering time interval discrimination. In addition, there is some preliminary evidence that at even lower weekly average levels of carbon monoxide (8 to 14 ppm or 9 to 16 mg/m³), people hospitalized for heart attacks have increased death rates.

What We Know About Photochemical Oxidants

Photochemical oxidants are several different pollutants (notably ozone and a group of chemicals called peroxyacetylnitrates, or PAN, for short) which can come from several sources. All of these pollutants share three properties:

1. They are all formed by the *chemical* reaction of other pollutants.

2. The reactions forming them proceed much more rapidly in areas with intense *sunlight*.

3. They are extremely reactive chemical substances, acting as *oxidizing agents*.

Among the most effective combinations for producing this class of pollutants are the oxides of nitrogen and reactive hydrocarbon vapors. (Hydrocarbons are organic compounds composed of carbon and hydrogen. We have a lot more to say about hydrocarbons in Chap. 12.) Los Angeles, with its

(a)

(b)

(c)

Figure 10-7 Downtown Los Angeles on a clear day (a) a smoggy day (b) and a day with ground-level smog (c). (South Coast Air Quality Management District)

sunny climate and high number of cars, offers extremely good conditions for the production of photochemical oxidants, and in fact this pollution comprises the main part of that city's infamous smog (Fig. 10-7). It is not confined to Los Angeles, however. The constituents of photochemical smog can now be readily detected in many metropolitan areas.

The various components of photochemical oxidants can have several adverse effects. First, they can directly affect the lungs and eyes of people, causing respiratory irritation and possibly changes in lung function. They are extremely toxic to many kinds of plants, affecting primarily the leaves. In addition, they can physically weaken such materials as rubber and fabrics.

The conclusions found in the EPA's criteria document show that the impairment of the performance of student athletes occurred over a range of hourly oxidant levels from 0.03 to 0.3 ppm (60 to 590 $\mu g/m^3$). An increased frequency of attacks in some people with asthma has been observed when hourly averages, as determined by peak measurements, were 0.05 to 0.06 ppm (100 to 120 $\mu g/m^3$). Eye irritation occurs in people at once upon exposure to about 0.1 ppm (200 $\mu g/m^3$); this is roughly equivalent to an hourly average of 0.03 to 0.05 ppm (60 to 100 $\mu g/m^3$).

Adverse effects on vegetation have been noted at levels of about 0.05 ppm (100$\mu g/m^3$) maintained for four hours. Damage to materials, although clearly observed at levels present in many cities, has not been accurately quantitated.

What We Know About Hydrocarbons

The major source of hydrocarbons is the internal combustion engine in motor vehicles, primarily the automobile. Some minor sources are the evaporation of organic solvents (from painting, dry cleaning, etc.), agricultural burning, and gasoline marketing.

At levels of hydrocarbons currently measured in urban areas, no adverse human effects are known to be caused by the hydrocarbons in isolation. However, as discussed in the section on photochemical oxidants, hydrocarbons are an extremely important component of those materials whose effects have been observed.

Certain specific hydrocarbons do have effects. Ethylene, for example, damages plants; it can inhibit growth and cause the leaves and flowers to fall.

The EPA's criteria document states that damaging levels of photochemical oxidants are directly related to concentrations of hydrocarbons in the air, which are, if alone, without effect. However, hydrocarbon concentrations (excluding methane) of 0.3 ppm (200 $\mu g/m^3$) for three hours may produce photochemical oxidant levels of up to 0.1 ppm (200 $\mu g/m^3$) a few hours later.

What We Know About Oxides of Nitrogen

The major source of nitrogen oxides is fuel combustion. A minor source is chemical plants. Nitrogen gas is normally a relatively unreactive substance which comprises about 80% of the air around us. However, at high temperatures (and also under certain other conditions), it can combine with oxygen in

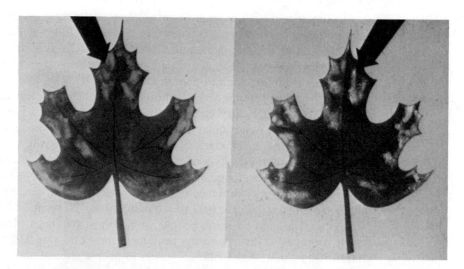

Figure 10-8 The effect of nitrogen oxide pollution on plants. Necrosis (left) is a burning of the leaf and chlorosis (right) is a bleaching effect on the leaf. U.S. Environmental Protection Agency)

the air to form several different gaseous compounds, collectively called the oxides of nitrogen (NO_x). Nitric oxide (NO) and nitrogen dioxide (NO_2) are the two most important.

The oxides of nitrogen can, at certain concentrations, cause serious injury to vegetation, including the bleaching or death of plant tissue, the loss of leaves, and a reduced growth rate (Fig. 10-8). Oxides of nitrogen can also cause fabric dyes to fade and fabrics themselves to deteriorate. Nitrate salts, formed from the oxides of nitrogen, have been associated with the corrosion of metals. Finally, nitrogen oxides can reduce visibility.

Certain members of this group of pollutants are known to be highly toxic to various animals, as well as to humans. High levels can kill; lower levels affect the delicate structure of lung tissue. This leads, in experimental animals, to a lung disease that resembles emphysema in human beings (Fig. 10-9). Exposure to NO_x lowers the resistance of animals to such diseases as pneumonia and influenza; the same may possibly occur in human beings. Exposure to high levels causes people to suffer lung irritations and potential damage. Exposure to lower levels has been associated with increased respiratory disease.

In addition, oxides of nitrogen, in the presence of sunlight, can react with hydrocarbons to form photochemical oxidants.

The EPA's criteria document states that a high incidence of chronic bronchitis has been found in children living in areas where daily averages of NO_2 varied from 0.062 to 0.083 ppm (118 to 156 $\mu g/m^3$) and where nitrate salts in the air were also at elevated levels. Adverse effects on plants have been observed when NO_2 levels exceed 0.25 ppm (470 $\mu g/m^3$) for several months. Limited evidence suggests that somewhat higher levels of NO_x (roughly

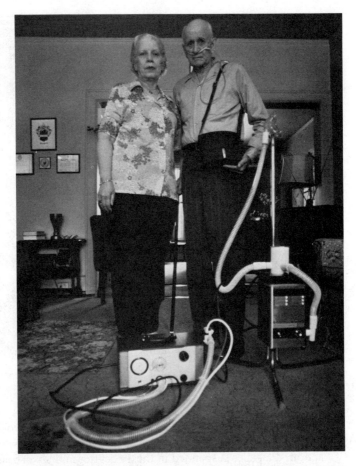

Figure 10-9 Before his death, Robert B. Jones was totally dependent on pure oxygen from equipment carried wherever he went. He had lived in Birmingham while working for the railroad for fifty years, and was suffering from emphysema. On April 21, 1973, at the age of 78, he died of pneumonia. (EPA-DOCUMERICA; photographer, LeRoy Woodson)

0.11 ppm or 214 $\mu g/m^3$) in the morning hours may be associated, under certain conditions, with the production later in the day of photochemical oxidant levels harmful to human health.

What We Know About Particulate Matter

Pollutants can exist as solid matter, liquid droplets, or gas. Both the solid and liquid matter are called particulates (which simply means particles in the atmosphere). Solid particulates consist of dust, smoke, or fumes; liquid particulates are mists and sprays. Particulate pollution results from many kinds of industrial and agricultural operations and from combustion products, including automobile exhausts.

Figure 10-10 Boston sootfall, 1960. (EPA-DOCUMERICA: photographer, Ken Metal)

Particulate matter in the respiratory tract may produce injury by itself, or it may act in conjunction with gases, altering their sites or their modes of action. Particles suspended in the air scatter and absorb sunlight, reducing the amount of solar energy reaching the earth, producing hazes and reducing visibility. Particulate air pollution causes a wide range of damage to materials. It may chemically attack materials through its own intrinsic corrosivity or through the corrosivity of substances absorbed or adsorbed by it. Merely by soiling materials, and thereby necessitating more frequent cleaning, particulates can accelerate deterioration (Fig. 10-10).

The EPA's criteria document reports that adverse health effects were noted when the annual geometric mean for particulate matter reached 80 $\mu g/m^3$.

HAZARDOUS AIR POLLUTANTS

At sufficiently high levels, almost any pollutant will cause illness or death. But some substances pose a much more immediate threat. Generally, these are substances that build up in the body over periods of time. Thus, even small amounts can be dangerous. Asbestos, beryllium, and mercury are among these dangerous pollutants. It was for this reason that EPA placed these three pollutants on their first hazardous air pollutant list.

What We Know About Asbestos

Asbestos is basically a fibrous mineral substance. Because the fibers are extremely small, they may be inhaled deep into the lungs if they become airborne. It has been known for some time that exposure to high levels of asbestos fibers over a period of years causes a serious chronic lung condition called *asbestosis*. This disease, which is common among asbestos miners, is similar to the "black lung" disease found among coal miners. It is caused by the accumulation of relatively large amounts of fibrous asbestos which physically obstruct the lung's air passages and in other ways damages its ability to function (Fig. 10-11).

It has also been found that individuals exposed for perhaps two decades to levels of asbestos that are not high enough to cause asbestosis may develop other medical problems. Specifically, the incidence of cancer—including an otherwise rare form of lung cancer called *mesothelioma*—is markedly increased in individuals so exposed. Effects of exposure to asbestos have been closely studied among workers who spray asbestos-containing insulation onto the girders of skyscrapers as a fireproofing technique.

There are some clear indications that individuals more indirectly exposed to airborne asbestos fibers also may be endangered. For example, construction

Figure 10-11 Magnified views of asbestos fibers. (EPA-DOCUMERICA)

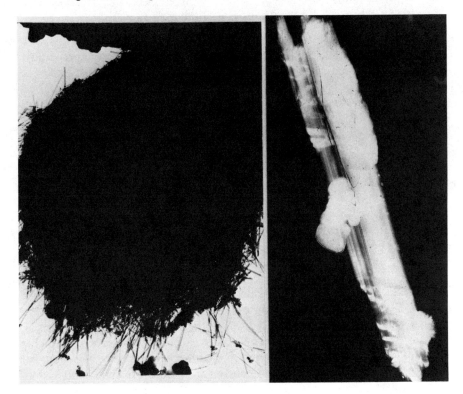

workers who do not directly use asbestos (plumbers and electricians, for example) may nonetheless be exposed to elevated levels of asbestos fibers if they are working near persons spraying asbestos-containing insulation. Similarly, those people who have an asbestos miner in their family or who live in a town with an asbestos mine or processing facility are exposed to elevated levels of asbestos dust. Some individuals in these areas have developed mesothelioma.

One of the most interesting aspects of asbestos-caused disease is that the final disease state does not develop until years or even decades of exposure. Because of this, it is not known whether there is clearly a "safe" level of exposure; indeed, it would be extraordinarily difficult to determine whether such a level exists. This uncertainty has influenced EPA's decision to classify asbestos as a "hazardous" air pollutant. The agency's concern has undoubtedly been reinforced by the discovery that from one-fourth to one-half of the lungs of urban Americans contain detectable "asbestos bodies"—these are asbestos fibers coated by the body in an attempt to isolate them from surrounding cells.

Of course, this large a fraction of the American population has not been exposed to asbestos due to their occupations. Instead, it is due to the presence in the ambient air of asbestos fibers from the construction and demolition of buildings with asbestos-containing insulation, the erosion of asbestos-containing materials (such as brake linings and even certain fabrics), and other such sources.

What We Know About Beryllium

Beryllium—as well as other substances made from it—is very dangerous to human beings. In fact, a report on this pollutant prepared for the federal government stated that "beryllium is among the most toxic and hazardous of the nonradioactive substances being used in industry." Beryllium and its derivatives are not a major item of commerce, although they are very important components in certain metal alloys and rocket fuels.

As is often the case with air pollutants, people employed in industrial facilities using beryllium were the first to develop diseases traceable to this substance. Therefore, much of the available information on the effects of beryllium on human beings comes from the field of industrial hygiene.

Two types of diseases have been traced to exposure to airborne beryllium: an acute (or short-term) disease lasting up to a few weeks, and a chronic (or permanent) condition, often called *berylliosis*. Both conditions may cause death.

If the acute disease is caused by extremely large exposures, death may occur rapidly due to serious, massive lung damage. A less serious acute condition may cause lung damage, generalized weakness, and loss of weight. Skin conditions such as rashes, and ulcers, may also occur.

The chronic disease berylliosis may not develop until after months or years of exposure. This is a long-lasting disease with a high mortality rate. It involves the lungs and many other tissues, since beryllium has the ability to interfere with basic biochemical processes in many cells of the body. Lung damage is usually serious and permanent. It often leads to death. Animal experiments suggest that continued exposure to beryllium may result in cancer.

Workers are not the only people who have developed beryllium-caused diseases. Residents of communities where facilities that use beryllium are located have also developed diseases traceable to beryllium exposure. In some instances, people developing berylliosis had no connection with the plant except that they lived within a few miles of it. Yet airborne beryllium apparently was present at high enough levels to make them seriously ill. It is for this reason that the EPA has declared beryllium a hazardous air pollutant.

What We Know About Mercury

Mercury, in any of its chemical forms, is a very toxic metal. Exposure to high levels of this pollutant can result in very serious damage to many organs of the body, particularly to the brain and kidneys. Exposure to lower levels of mercury can also have serious effects, especially on the brain. The expression "mad as a hatter" was coined to describe the aberrations found in people who worked in the hatmaking industry when mercury was used to treat felt.

Several outbreaks of mercury poisoning have been well documented over the past few years. These have involved industrially exposed workers as well as members of the general public. Two general conclusions may be drawn from these events. First, no matter what the specific source of mercury is, the effects on human beings (and even on unborn babies when the mother is exposed) are very much the same if the exposure is high enough. Second, there is no general agreement as to the "safe" level of mercury in food or air.

Some findings indicate that airborne mercury may be a significant source of mercury contamination of other parts of the environment, such as water. Mercury is mobile in the environment and once released may cycle among air, land, and water for long periods of time. Many activities release mercury into the air. Two common fuels—coal and oil—contain small amounts of naturally occurring mercury. When these fuels are burned, much of the mercury is vented into the air. Some kinds of paper are often treated with mercury during manufacture. When the paper is burned after its use (in a municipal incinerator, for example), mercury is released. These sources supplement the major industrial sources of mercury, such as the processing of mercury-containing ore and chlor-alkali plants producing mercury cells. Because of the serious detrimental effects of mercury on health, the EPA has declared mercury to be a hazardous pollutant. (By the way, mercury can also be absorbed through the skin. Individuals who play with mercury, from a broken thermometer or to make coins shiny, for example, run the risk of absorbing elemental mercury through the skin.)

REACTIONS IN THE AIR

Pollution of the air has been known since ancient times. But the causes were natural. Active volcanoes spewed out huge amounts of ash, and swamps emitted foul-smelling gases. However, as population increased and people advanced technologically, the problem arose.

In England, for example, where the Industrial Revolution had its start, coal was the main source of fuel. Now coal is basically carbon, together with some sulfur and other impurities. When the English burned their coal to protect themselves from the cold and damp air, the smoke from their millions of fires threw out huge amounts of particulates and gases. It was only in 1956 (after a disastrous 4-day smog in 1952 that killed 4000 Londoners) that Britain passed the Clean Air Act and made it illegal to burn coal in the cities; the country even switched its railroads from coal to electricity and diesel fuel. Some of the gases that burning coal forms are caused by the following reactions.

$$C + O_2 \longrightarrow CO_2$$
Carbon
dioxide

$$S + O_2 \longrightarrow SO_2$$
Sulfur
dioxide

$$2\,SO_2 + O_2 \longrightarrow 2\,SO_3$$
Sulfur
trioxide

$$SO_3 + H_2O \longrightarrow H_2SO_4$$
Sulfuric
acid

As you can see, burning coal gives you a number of side products that you'd just as soon do without. The sulfur gives rise to sulfur dioxide, which eventually is oxidized to sulfuric acid. The sulfuric acid then remains as a mist in the atmosphere, where it can be inhaled by people or can corrode materials. This sulfuric acid mist gives rise to what some people call "acid rain."

Ever since the turn of the century, cars have been on the increase, until today they are the major source of air pollution. The trouble with the car stems from flaws in the internal combustion engine. The two major problems are:

1. Not all the gasoline reacts. This leads to emissions of hydrocarbons. (Gasoline is actually a mixture of hydrocarbons.)

2. The high operating temperatures inside the engine make the nitrogen and oxygen that are present in the air react. This leads to emissions of oxides of nitrogen.

Since not all the gasoline reacts, the unburned part travels out of the exhaust pipe and into the air. Once in the air, the hydrocarbons can react with nitrogen oxides to produce *photochemical smog.*

The other major by-products of the combustion of gasoline—oxides of nitrogen—are produced in a series of steps. Nitrogen and oxygen, both present when the gasoline burns, react with each other because of the high temperatures inside the engine. The major product of their reaction is nitric oxide.

$$N_2 + O_2 \longrightarrow 2NO$$

Nitric
oxide

The nitric oxide of course gets into the atmosphere, where some of it then reacts with oxygen in the air to produce nitrogen dioxide.

$$2NO + O_2 \longrightarrow 2NO_2$$

Nitrogen
dioxide

Nitrogen dioxide, which is unstable in the presence of sunlight, can decompose into nitric oxide and atomic oxygen.

$$NO_2 + \text{sunlight} \longrightarrow NO + O$$

Atomic
oxygen

The atomic oxygen, in the presence of sunlight, can react with unburned hydrocarbons to produce other harmful chemicals. But the principal reaction is that of the atomic oxygen (O) with molecular oxygen (O_2) to produce ozone (O_3). High concentrations of ozone near ground level pose a severe health problem because ozone irritates the eyes and respiratory system. It can also react with many different substances. For example, ozone can destroy rubber by cracking it (Fig. 10-12).

Figure 10-12 The effect of ozone on samples of various rubber components. (Photo by F. H. Winslow. Copyright 1951 American Telephone and Telegraph Company. Reprinted by permission from *The Bell System Technical Journal*.)

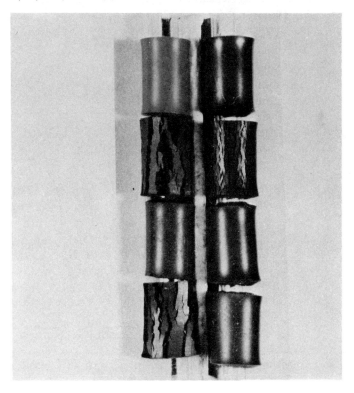

Now that we have examined some of the major air pollutants, let's investigate some methods used to put the lid on them.

The major drawbacks of both coal and oil as fuels are emissions of particulate matter and sulfur oxides. We can reduce the emissions of sulfur oxides by using coal or oil that has a low sulfur content. We can also reduce emissions by using a *scrubber*, a device that absorbs gaseous pollutants. A scrubber for removing the sulfur dioxide from a smokestack usually consists of a fine spray of water. Gas rising from the stack is passed through the scrubber, where the water absorbs the sulfur dioxide (Fig. 10-13).

We can also control particulate emissions by using mechanical devices. One such device is the *electrostatic precipitator* (Fig. 10-14), in which dust particles that have been electrically charged are collected between highly charged plates. This eliminates the dust particles from the gases spewing from the smokestack. Another such device is a baghouse (Fig. 10-15), in which the emissions from a smokestack are passed through porous bags. These bags allow gases to pass through, but retain the particulate matter. The bags are cleaned periodically to remove the collected particulate matter.

Since cars and trucks contribute half of all air pollution, eliminating emissions from them is a top-priority problem for all automobile manufacturers. One solution would be to switch from the internal combustion engine to electrically powered cars. The decade of the 1980's may be the beginning of this trend. Although this would transfer the source of the pollution to the power companies, it might be an easier source of pollution to control.

Figure 10-13 A scrubber.

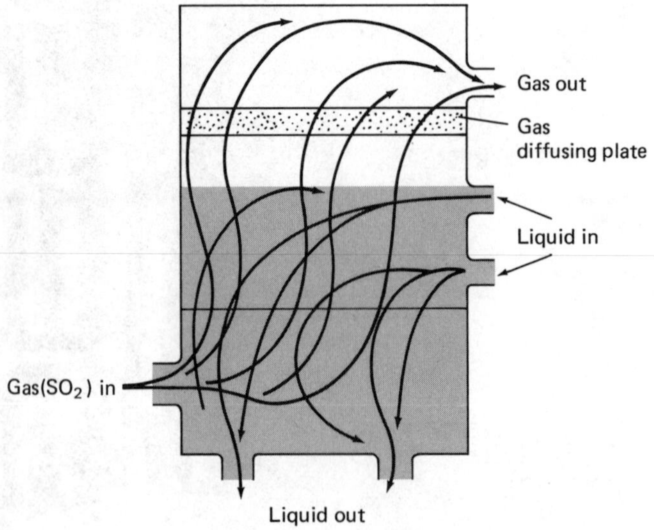

Gas out

Gas diffusing plate

Liquid in

Gas(SO_2) in

Liquid out

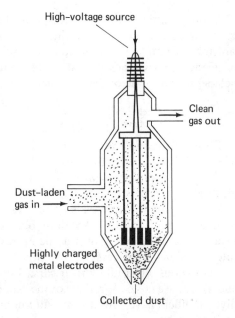

High-voltage source

Clean gas out

Dust-laden gas in →

Highly charged metal electrodes

Collected dust

Figure 10-14 An electrostatic precipitator.

Figure 10-15 A baghouse is used to control particulate matter from industrial smokestacks. (U.S. Environmental Protection Agency)

Many pollution control devices have already been installed in late-model cars. The major aim of some of these devices has been to burn gasoline more completely. Complete combustion of gasoline would produce only carbon dioxide and water vapor. We can write the reaction for complete combustion of octane, which is one of the ingredients of gasoline.

$$2\,C_8H_{18} + 25\,O_2 \longrightarrow 16\,CO_2 + 18\,H_2O$$
Octane

Efforts to reduce nitrogen oxides have focused on the use of catalytic mufflers. These mufflers either absorb the nitrogen oxides as they pass through the exhaust, or convert them to less troublesome substances. Beginning in 1975, catalytic mufflers became standard equipment on most new cars. (However, catalytic mufflers have also had their problems over the years. First, these mufflers get very hot. An automobile equipped with a catalytic muffler can start a fire if it is idling over a pile of leaves. Also, these mufflers increase sulfate emissions from cars. Many drivers have noticed the smell of these sulfate emissions when idling their cars. Finally, cars equipped with catalytic mufflers must use no-lead gasoline or the catalyst will be poisoned. So although catalytic mufflers have solved some automobile pollution problems, they have created others.)

AIR POLLUTION: A SOCIAL COMMENTARY

The control of air pollution involves much more than the application of the right technology. There are aspects of the problem that involve social and economic factors as well. For example, many people feel that we can do much to improve our air quality if we can clean up pollution from transportation sources. After all, transportation causes more than 50% of all air pollution in this country. But the question is: How do we attack such a problem without destroying our economy and way of life?

Some groups have suggested that we limit the number of private cars that may be owned by a family. This might force more people to use public transportation. Other groups have suggested that we have a national program to switch from gasoline-powered cars to electric-powered vehicles over the next few years. Still others have suggested that we limit the amount of gasoline that each car may use during the course of a year. Each of these suggestions could have far-ranging effects if implemented. For example, limiting the number of automobiles that a family could own would necessitate the development of mass transit systems in many places where they do not exist presently. The switch to electric cars is already occurring on a small scale. But a massive program to switch from gasoline-powered vehicles to electric-powered vehicles could have a number of important consequences. First, the power companies would have to handle a greater demand for electricity. This could mean the building of many new power plants. Also, there would have to be assurances from the power companies that we would not be trading one form of air pollution for another. Emissions from power company smokestacks would have to be carefully controlled. But even with that, what about all those gasoline service stations? What happens to them in a quick switch over? The economic consequences could be disastrous.

The important point that we must keep in mind when looking for a solution to our environmental problems is that our civilization is very complex. Our society is a very delicately balanced system. Changes can be made to improve the overall system, but they must be made slowly and with great care in order to gauge the impact on the other parts of the system.

SUMMARY

In this chapter we discussed the various aspects of air pollution. We began by telling you about the true story of Donora, Pennsylvania. We discussed how the concentrations of pollutants in the air are measured. Then we looked at the substances found in the air that we breathe.

We also studied about the major air pollutants, which included substances such as the oxides of sulfur, carbon monoxide, photochemical oxidants, hydrocarbons, oxides of nitrogen, particulate matter, asbestos, beryllium, and mercury. We learned about some of the chemical reactions that take place in the air. We also looked at some of the pollution control devices used to control emissions from smokestacks and vehicles. Finally, we examined the problem of air pollution from a social point of view.

EXERCISES

1. Discuss the episode that occurred in Donora, Pennsylvania, in late October 1948. Could the situation have been prevented? Explain.

2. Determine the percent by volume of each gas in the following mixture. The mixture contains 10 liters of oxygen gas, 20 liters of nitrogen gas, 30 liters of carbon dioxide, and 15 liters of hydrogen gas.

3. If you had 5 molecules of methane gas in 250,000 molecules of air, what is the concentration of methane in parts per million?

4. The pollutant concentration of sulfur dioxide in a sample of air is 0.0015%. What is this concentration in parts per million?

5. A sample of air is analyzed for particulate matter. It is found that 0.006 g of particulate matter is in 30 m^3 of air. What is the concentration of this particulate matter in $\mu g/m^3$?

6. Name the two major constituents of air.

7. How can a substance that is ordinarily not considered a pollutant become a pollutant?

8. How do sulfur oxides form from the burning of hydrocarbon fuels?

9. Carbon monoxide, unlike other air pollutants, does not irritate the respiratory tract. How, then, does it cause us problems?

10. List the three properties shared by all photochemical oxidants.

11. Photochemical oxidants are known as *secondary pollutants* because they form from the reaction of *primary pollutants* in the atmosphere. What two primary pollutants form photochemical oxidants?

12. What is the primary source of hydrocarbon pollution?

13. What is the major source of nitrogen oxide pollution?

14. What are the molecular formulas of the two most important nitrogen oxides?

15. In terms of human health, what do you consider to be the most important aspect of nitrogen oxide pollution?

16. Describe particulate matter in your own words.

17. List some of the detrimental effects of particulate matter.

18. Name the three major hazardous air pollutants.

19. What is asbestosis, and how does an individual get it?

20. Is pollution from asbestos a problem only for workers who manufacture it? Explain your answer.

21. List some of the reasons the EPA has designated beryllium as a hazardous pollutant.

22. List some of the health problems associated with mercury pollution.

23. Complete and balance the following equations.
    (a) $C + O_2 \rightarrow$ (b) $S + O_2 \rightarrow$
    (c) $SO_2 + O_2 \rightarrow$ (d) $SO_3 + H_2O \rightarrow$

24. List the two major problems associated with the combustion engine as regards air pollution.

25. Of the gaseous and particulate air pollutants that we've discussed, which do you think would be most harmful to people suffering from respiratory and lung diseases?

26. What type of a control device would you place on a smokestack of a power company that is using oil as a fuel for generating electricity?

27. What type of control device would you place on a smokestack of a city incinerator that is spewing out black smoke?

28. Complete the following equations.

    (a) $NO + O_2 \rightarrow$ (b) $NO_2 + sunlight \rightarrow$

29. "Dilution is the solution to pollution." Discuss the logic of this erroneous statement.

30. To stop the pollution of our air, discuss the steps that must be taken by

    (a) government   (b) industry   (c) the average citizen

Chapter 11

Nuclear Power

Some Things You Should Know After Reading This Chapter

You should be able to:

1. Explain what is meant by radioactivity.
2. Name the three types of nuclear radiation: alpha rays, beta rays, and gamma rays.
3. State the charge and mass of an alpha particle.
4. State the charge and mass of a beta particle.
5. Complete a nuclear equation, given the starting isotope and the particle emitted.
6. Explain the term half-life.
7. Calculate the number of grams of isotope remaining after a given time period, when told its half-life.
8. Explain the terms chain reaction, nuclear fission, and nuclear fusion.
9. Discuss the advantages and disadvantages of nuclear power plants as a source of electricity.
10. Explain how a breeder reactor works.
11. Explain the two ways in which radiation can damage or destroy cells.
12. Explain what may happen when radiation reacts with water in cells.
13. Explain and give an example of a stage 1 radiation effect.
14. Explain and give an example of a stage 2 radiation effect.
15. Name the types of cells most sensitive to radiation.
16. Name four devices used to detect radiation and explain what each one is used for.
17. Define each of the following units of radiation: curie, roentgen, rad, and rem.
18. Explain what is meant by LD_{50}^{30}.
19. Explain the relationship between rem and rad.
20. Explain how the following isotopes are used in medical diagnosis and therapy: iodine-131, chromium-151, and cobalt-50 or cobalt-60.
21. Explain the three methods by which radiation can be delivered to a malignant area: teletherapy, brachytherapy, and radiopharmaceutical therapy.

INTRODUCTION

Nuclear radiation has, over the past few decades, become a major weapon of the physician in the fight to treat diseases that were untreatable at one time. Like invisible bullets, radiation can penetrate the skin and shrink tumors lodged deeply in the brain. And radioactively tagged isotopes can pinpoint abnormalities deep within a vital organ, allowing the abnormalities to make themselves known. "A panacea," you say. Unfortunately no, because radiation can kill as well as cure. Normal cells in the body can be destroyed or changed into cancerous cells by radiation. And the increased use of radioactive materials in medicine has been paralleled by the increased use of radioactive materials in general, for example, in defense, power generation, and industrial applications. This has added additional sources of nuclear radiation to our environment, thus increasing the chance of exposure for all of us. We can only hope that the benefits derived from the use of nuclear radiation far outweigh the hazards (Fig. 11-1).

Figure 11-1 A photograph of the mushroom-shaped cloud following the Hiroshima detonation. (Los Alamos Scientific Laboratory).

In this chapter we are going to learn about nuclear radiation and look at some of its applications.

SCENARIO

THE BEGINNING OF AN ERA (A TRUE STORY)

It is August 6, 1945. A lone aircraft flies over Hiroshima, Japan, and releases a single bomb. Seconds later, the bomb explodes. A huge mushroom-shaped cloud forms and grows until it fills the sky. Within hours, approximately 75,000 persons are dead. During the next 30 years, an additional 800,000 people are to die of various diseases caused by the radiation from the bomb. The world has been catapulted into a new era, the atomic age.

HISTORY

Although the explosion of the atomic bomb over Hiroshima in 1945 is recognized as the beginning of the nuclear age, the study of nuclear chemistry actually began much earlier, in 1895, when the German physicist Wilhelm Konrad Röntgen accidentally discovered x rays. He found that x rays have a great penetrating power, so great that they can pass through walls.

In 1896, a French physicist by the name of Antoine Becquerel placed some salt that contained uranium on a photographic plate that was wrapped in black paper. When he developed the photographic plate, he found to his surprise that he had the image of the pile of salt. The uranium salt had taken a picture of itself! Becquerel concluded that the uranium was giving off some type of penetrating rays, and that these rays must be very strong to be able to pass through the black paper and expose the photographic plate (Fig. 11-2). However, when he placed a thick barrier of lead between the salt and the

Figure 11-2 Becquerel's discovery.

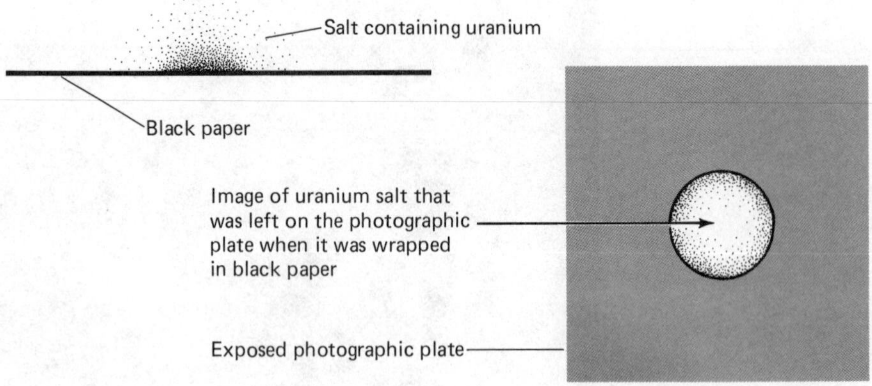

Salt containing uranium

Black paper

Image of uranium salt that was left on the photographic plate when it was wrapped in black paper

Exposed photographic plate

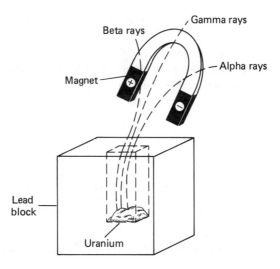

Figure 11-3 When the radiation produced by uranium passes through a magnetic field, it breaks up into three types.

photographic plate, the lead absorbed the rays. Becquerel realized that these penetrating rays were probably Röntgen's x rays.

At the same time there was in Paris a young Polish chemist, Marie Curie, who, with her husband Pierre, was working in the laboratories at the Sorbonne. The Curies became interested in Becquerel's problem. It was, in fact, Marie Curie who defined the ability of a substance to produce penetrating rays as *radioactivity*. The Curies found that a radioactive substance seems to keep on and on, year after year, emitting these powerful penetrating rays.

In 1898, they discovered that the element thorium is radioactive, and so is polonium. And so, most of all, is radium.

Although early experimenters first thought that radioactive materials produced only x rays, they soon discovered that the situation was more complex. For example, when they allowed the radiation produced by uranium to pass through a magnetic field, they could detect *three* types of radiation.

The English physicist Lord Rutherford called these three types of radiation *alpha rays*, *beta rays*, and *gamma rays*, from the first three letters of the Greek alphabet, α, β, and γ (Fig. 11-3).

GAMMA RAYS (γ RAYS)

Many experiments in passing radioactive materials through a magnetic field led to the conclusion that—although alpha and beta rays are deflected to one side or the other—gamma rays travel straight through and are not deflected. For this reason the early experimenters assumed correctly that gamma rays carried no charge. Gamma rays seemed to have properties like those of x rays, but they were more energetic. In fact, gamma rays have a very high energy content and strong penetrating ability. Today we know that because of this penetrating ability, a high concentration of gamma rays can harm human genes. The *nucleus* of an atom gives off gamma rays only under certain conditions, usually when the nucleus loses energy. And the atom, of course, has to be of the radioactivity variety.

BETA RAYS (β RAYS)

When Becquerel allowed beta rays to pass through a magnetic field, he found that they behaved like cathode rays—that is, like rays that come from a negative electrode. Therefore, Becquerel concluded that these rays were composed of speeding electrons. However, beta particles, like gamma and alpha particles, come from the *nucleus* of the atom, and electrons don't exist in the nucleus.

How, then, are beta particles formed? Scientists believe that they are formed when neutrons, during a nuclear reaction, are transformed into protons:

$$\text{neutron} \longrightarrow \text{proton} + \beta \text{ particle}$$

A beta particle, like an electron, has a zero mass number and a charge of -1. We can represent this in the following way.

$$_{-1}^{0}e = \beta \text{ particle}$$

Beta particles have only slight penetrating power compared to gamma rays. They can be stopped by a thin sheet (a few centimeters thick) of almost any metal.

ALPHA RAYS (α RAYS)

When scientists of the early 1900's (Becquerel, James Chadwick, and Werner Heisenberg, among others) investigated alpha particles in a magnetic field, they found that these particles were deflected in the opposite direction from beta rays. They therefore assumed that alpha particles must be positively charged. Further experiments showed that these particles had a mass number of 4 and a charge of $+2$. Heisenberg suggested that we picture an alpha particle as a helium-4 atom with its two electrons removed; in other words, we can imagine it as composed of two protons and two neutrons. We can represent an alpha particle in this way:

$$_{2}^{4}\text{He} = \alpha \text{ particle}$$

Alpha particles have virtually no penetrating ability. They can be stopped by a thin sheet of paper.

NUCLEAR TRANSFORMATION

Now that we are familiar with the basic particles emitted by radioactive materials, let's examine some natural nuclear reactions—that is, nuclear reactions that occur spontaneously in nature, as opposed to nuclear reactions that people set up in laboratories. Table 11-1 will help us get the picture. We can use this table to help us determine what happens to the element that is doing the emitting, once the radioactive particle has been released. An element that is emitting such particles is said to be undergoing *radioactive decay*, or *nuclear decay*.

Isotopes of some elements produce radiation naturally. An example is radium-226, $^{226}_{88}$Ra. When it is undergoing nuclear decay, radium emits alpha particles and actually becomes another element! We call this *nuclear transformation*, and we can represent it by the following equation:

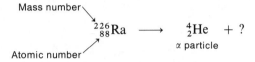

What is the unknown element? Subtraction tells us that it must be the element whose atomic number is 86 and whose mass number is 222 ($226 - 4 = 222$ and $88 - 2 = 86$). We look at the periodic table and find that element 86 is radon (Rn). So we can write the reaction in this way:

$$^{226}_{88}\text{Ra} \longrightarrow \, ^4_2\text{He} + \, ^{222}_{86}\text{Rn}$$

In the balanced nuclear equation, the sum of the superscripts, and also the sum of the subscripts, is the same on both sides of the arrow.

Take another example. The isotope thorium-234 produces radiation by emitting beta particles. We can write the nuclear reaction as

$$^{234}_{90}\text{Th} \longrightarrow \, \underset{\beta \text{ particle}}{^{\;\;0}_{-1}e} + \; ?$$

What is the other element that is being formed? Well, Table 11-1 tells us that the formation of a beta particle increases an element's atomic number by 1. Therefore, the unknown element must be the element whose atomic number

TABLE 11-1 *Summary of nuclear particles*

TYPE OF PARTICLE EMITTED	MASS NUMBER OF PARTICLE EMITTED	CHARGE OF PARTICLE EMITTED	WHAT HAPPENS TO ELEMENT'S MASS NUMBER?	ATOMIC NUMBER?
Alpha	4	+2	Decreases by 4	Decreases by 2
Beta	0	−1	No change	Increases by 1
Gamma	0	0	No change	No change

is 91. The periodic table tells us that element 91 is protactinium. So we can write our equation as

$$^{234}_{90}\text{Th} \longrightarrow ^{\ 0}_{-1}e + ^{234}_{91}\text{Pa}$$

Once again the sums of both superscripts and subscripts are the same on both sides of the arrow.

HALF-LIFE

We have been discussing some nuclear reactions, and we looked at what actually happens to a single atom of an element when it undergoes decay. We might now ask two questions:

1. Why are some elements radioactive? In other words, why do they undergo nuclear decay?

2. Can we predict when a particular atom of a radioactive substance will decay?

Although nobody really knows the exact answer to our first question, many scientists do feel that certain combinations of protons and neutrons make the nucleus of an atom unstable, and therefore make it tend to be radioactive.

The answer to our second question is no. No one can tell when a particular atom of a radioactive substance will decay. However, when we are dealing with large numbers of radioactive atoms (which is usually the case), we can predict how long it will take for a certain number or percentage (say half) of the atoms to decay. For each radioactive isotope, we can predict a certain *half-life*.

The half-life is the time it takes for one-half of the atoms originally present in a sample to decay.

The half-life of a radioactive isotope may be short or long. For example, the half-life of $^{238}_{92}\text{U}$ is 4.5 billion years, whereas the half-life of $^{257}_{103}\text{Lr}$ is 8 seconds.

Example 11-1   The half-life of strontium-90 is 28 years. If we have 100 grams of strontium-90 today, how many grams of strontium-90 will be left in our sample in 84 years?

Solution   Since the half-life of strontium-90 is 28 years, this means that half of our strontium-90 will decay every 28 years. After 84 years, our sample will contain 12.5 g of strontium-90.

TIME (years)	AMOUNT OF STRONTIUM-90 REMAINING (g)
Now	100
28	50
56	25
84	12.5

While we're on the subject of strontium-90, we should say that it has a chemical similarity to calcium (notice that Ca and Sr are in the same chemical family), which causes it to replace the calcium in bone, and stay there, undergoing its radioactive decay. When nations test atomic bombs, strontium-90 gets into the atmosphere and begins to float around the world. Gradually it falls down—on fields, for example. Since cows eat the grass from the fields, the strontium-90 gets into milk and other dairy products; it also gets into vegetables. People consume these farm products—we have, after all, no choice!—and this radioactive isotope, strontium-90, accumulates in our bones, side by side with the calcium. It then bombards the nearby tissues and organs with beta rays. That 28-year half life means that it stays around for a long time. And if a person gets a really large dose, it can cause bone cancer. That's why thoughtful people were happy in the summer of 1963 when Russia, the United States, and Great Britain (three of the five nations with nuclear programs) signed the treaty banning atmospheric nuclear tests. Unfortunately, China and France didn't sign, and they've been testing. In late spring of 1974, India exploded a nuclear device underground, and became the sixth nation to join the nuclear club. They, too, have not signed a test-ban treaty.

HUMANLY PRODUCED RADIATION

The year is 1919. The British physicist Lord Rutherford, who first theorized the existence of the atomic nucleus, bombards nitrogen-14 with alpha rays. The results are astounding. Rutherford finds that he has produced oxygen-17, a nonradioactive isotope of oxygen.

$$\underset{\alpha \text{ particle}}{^4_2\text{He}} + {}^{14}_{7}\text{N} \longrightarrow {}^{17}_{8}\text{O} + \underset{\text{A proton}}{^1_1\text{H}}$$

The dreams of the alchemists have come true! Rutherford, by having triggered the first humanly produced nuclear reaction, has "transmuted" an element (Fig. 11-4).

The year is 1934. The French physicists Irène Joliot-Curie and Frédérick Joliot-Curie (daughter and son-in-law of Marie and Pierre Curie) bombard

$$\text{}^4_2\text{He} \quad + \quad \text{}^{14}_7\text{N} \quad \longrightarrow \quad \text{}^{17}_8\text{O} \quad + \quad \text{}^1_1\text{H}$$

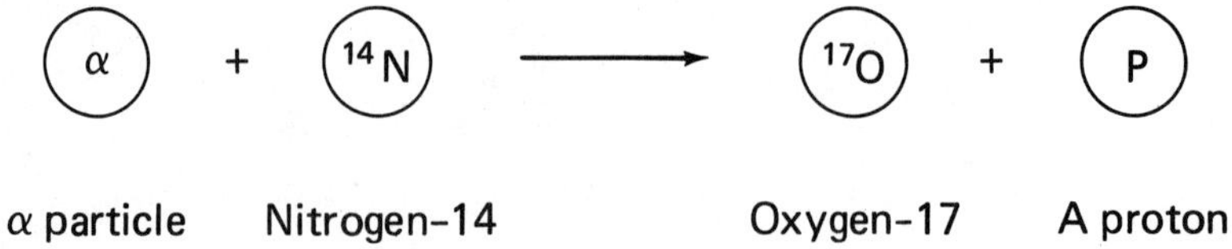

$$\boxed{\alpha} \quad + \quad \boxed{\text{}^{14}\text{N}} \quad \longrightarrow \quad \boxed{\text{}^{17}\text{O}} \quad + \quad \boxed{\text{P}}$$

α particle   Nitrogen–14   Oxygen–17   A proton

Figure 11-4   When bombarded with alpha particles, nitrogen-14 becomes oxygen-17.

boron-10 with alpha rays and obtain nitrogen-13, the first artificially produced radioactive isotope.

$$\text{}^{10}_5\text{B} + \quad \text{}^4_2\text{He} \quad \longrightarrow \quad \text{}^{13}_7\text{N} + \quad \text{}^1_0 n$$
α particle   A neutron

The Chain Reaction and Nuclear Fission

You can see that by the 1930's scientists were unraveling the atom's deepest secrets, and were about to discover the power of the atom. They were, in fact, on the verge of producing a nuclear chain reaction, which would turn very small quantities of mass into huge quantities of energy.

In 1934, the Italian physicist Enrico Fermi (who won the Nobel Prize in 1938) bombarded uranium (element 92) with neutrons in an attempt to produce element 93 (neptunium). But, instead, Fermi found himself with an isotope of barium (element 56), a mystifying outcome.

In 1938, the German physical chemist Otto Hahn, who was also working on atomic power (many German scientists were, in this pre-World War II era), proposed an explanation. He said that uranium atoms split into different atoms when they are bombarded by neutrons. (This splitting process is called nuclear *fission*.) Furthermore, when these uranium atoms split, they produce *more* neutrons, which in turn are able to split still more uranium atoms. This is called a *branching chain reaction*. You can liken it to placing dominoes in a triangular pattern. When you tip the first domino, it causes two others to tip over, and so on down the line (Fig. 11-5).

In September 1939, Europe exploded into World War II, and many nations looked to their scientists for new weapons. Scientists all over the world who were working in the field of radioactivity knew that nuclear fission could be turned into a decisive weapon of war. It was only a matter of time before someone perfected the technique. Would the Germans be first? Were they racing toward that goal? (Fortunately for the free world, Hitler decided to turn the work of his nuclear scientists in other directions. This prevented the Germans from developing the atomic bomb prior to our doing so.) Everyone knew that their scientists were second to none.

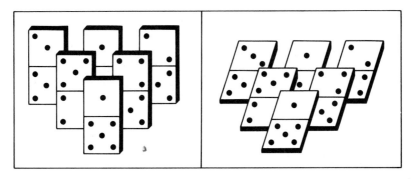

Figure 11-5 The domino chain reaction.

The United States, although not yet at war, recognized the danger and launched a research program to produce an atomic (fission) bomb. Under Enrico Fermi's leadership, the Manhattan Project (a code name) was started; it operated out of a windy, makeshift laboratory underneath the bleachers of Stagg Field at the University of Chicago. On December 2, 1942, Enrico Fermi and his team achieved the first sustained nuclear chain reaction. By July 1945, by incredible hard work and perseverance, the Manhattan Project scientists had managed to scrape together enough uranium-235 to make a fission bomb. To test it the United States had to fire it, and the first explosion of an A-bomb took place in a deserted area near Alamogordo, New Mexico. It created sickening devastation, and so, with feelings of forboding that may be imagined, the scientists set about making two more bombs at once. The major reaction produced by the exploding A-bomb may be presented as

$$^{235}_{92}U + {}^{1}_{0}n \longrightarrow {}^{141}_{56}Ba + {}^{92}_{36}Kr + 3\,{}^{1}_{0}n + \text{energy}$$

By August 1945 the other two bombs were ready. Yet such was the crisis—with Americans believing that they would lose a million American soldiers if they attempted to storm the Japanese mainland—that the United States went ahead and exploded them: the first at Hiroshima, and the second, four days later, at Nagasaki, Japan. Many Japanese people who lived in or near those two cities developed cancer as a result of their exposure to the radiation. Effects of this radiation are just now being seen in the second generation of Japanese from these two areas of Japan.

The Atom and Electric Power

Even though the first use of energy from nuclear fission was for making war, scientists could readily see that the atom was a tremendous storehouse of energy that could be used for peaceful purposes. Starting in the 1950's, several nations began to use atomic power for supplying electricity, and the number of nuclear-powered electric plants throughout the world has grown to more than 150. This enormous increase was a result of the development of nuclear reactors, which use the heat energy produced by controlled fission to generate steam. The steam in turn powers a generator, which produces electricity (Fig. 11-6).

A major advantage of nuclear power plants over conventional power plants (which use coal or oil) is that they produce virtually no air pollution. There is no particulate matter and no oxides of sulfur produced. Another advantage is cost. With oil prices continuing to increase, it is cheaper to produce electricity using nuclear power. (And let's not forget about the fact that by using nuclear power, we don't have to depend as much on foreign nations for our oil.)

However, there are certain disadvantages to nuclear power plants. For example, there's always the possibility of a major accident, which could release large amounts of radioactivity into the atmosphere. The accident at the Three Mile Island nuclear power plant in March 1979 was a near disaster for

Figure 11-6 A nuclear power plant. (Public Service Electric and Gas Company, Newark, New Jersey)

the Harrisburg, Pennsylvania, area. A release of radioactive material, including radioactive iodine, into the atmosphere *could have* resulted in numerous cancer deaths (especially thyroid cancer, because iodine is taken up by the thyroid gland). Luckily, most of the radiation was contained within the reactor building and no serious medical consequences are expected to occur to the residents of the Three Mile Island area.

Another problem with nuclear power plants is that they generate tons of radioactive waste material each year. This material must be removed from each plant and stored in a safe place. Moving large amounts of radioactive waste on our interstate highways and through our towns and cities is seen by some as an open invitation to disaster. What happens if a serious traffic accident occurs with a truck carrying this radioactive waste? Also, the question of *safe storage* is an ongoing controversy. Putting these wastes in salt mines or deep wells is seen by some as simply sweeping the dirt under the rug. Some scientists fear that over the years an earthquake could occur near a nuclear waste storage facility and release deadly radioactive material into the environment. (Remember, some of these radioactive wastes remain deadly for thousands of years. This means that future generations of human beings will have to contend with the nuclear waste problems of today.)

Finally, based on known reserves, it appears that the supply of high-grade uranium ore will run out by the year 2000. This means that the cost of refueling nuclear reactors will at best be very expensive, and at worst will be impossible. In fact, as high-grade ore is depleted, many reactors may have to be shut down.

The Breeder Reactor

Conventional nuclear reactors use uranium-235 as their fuel. However, uranium-235 makes up less than 1% of naturally occurring uranium. Most uranium occurs as the isotope uranium-238. The only problem is that uranium-238 can't be used in conventional nuclear reactors. It doesn't undergo fission like uranium-235. However, if uranium-238 could be used as a nuclear fuel, there would be sufficient uranium to run nuclear reactors for hundreds of years.

The *breeder reactor* was developed to use uranium-238. Here's how it works. A reactor is built with a core of fissionable plutonium, ^{239}Pu. The plutonium-239 core is surrounded by a layer of uranium-238. As the plutonium-239 undergoes spontaneous fission, it releases neutrons. These neutrons convert the uranium-238 to plutonium-239. In other words, this reactor *breeds* fuel (^{239}Pu) as it operates. After all the uranium-238 has been changed to plutonium-239, the reactor is refueled.

However, there are some major problems with the breeder reactor. To begin with, plutonium-239 is extremely toxic. If an individual inhales a small amount, he or she will contract lung cancer. Also, the half-life of the material is extremely long, about 24,000 years. This could create an almost impossible disposal problem if large amounts of this material are generated.

Also, because of the nature of the reactor core, water can't be used as the coolant. Instead, liquid sodium must be used. In the event of an accident a catastrophe could develop because liquid sodium reacts violently with water and air.

Although the breeder reactor could solve the uranium fuel problem, there are still a number of other problems that will have to be worked out.

Nuclear Fusion

Hydrogen is the most abundant element in the universe. It is hydrogen that is responsible for the tremendous quantities of energy produced by the sun and other stars. However, the sun and stars get their energy through the process of nuclear fusion: the fusing or combining of two hydrogen atoms to form a helium atom. This fusing process creates huge amounts of heat—on the order of 2 million degrees Celsius—and there are no waste products left over. However, there are some catches. How do you contain such fantastic heat? What vessel or tools can handle it? And also, in order to duplicate the sun's fusing process on earth, you would need a great amount of energy to start with. So far, the only device that scientists have developed from their knowledge of atomic fusion is the *hydrogen bomb* (Fig. 11-7). This type of bomb is now possessed by the United States, Russia, and China. Because it uses atomic fusion rather than fission, the hydrogen bomb is a lot more powerful than the old-fashioned atomic bomb (in fact, in the hydrogen bomb, an atomic bomb acts as the initiator or detonator).

Figure 11-7   The explosion of an H-bomb, an example of nuclear fusion. (U.S. Department of Energy)

Controlled fusion could practically end our energy worries. One gram of hydrogen upon fusing would release the energy equivalent of about twenty tons of coal. The reactions involved are:

$$\underset{\text{Deuterium}}{^2_1\text{H}} \quad + \quad \underset{\text{Tritium}}{^3_1\text{H}} \quad \longrightarrow \quad \underset{\text{Helium}}{^4_2\text{He}} \quad + \quad \underset{\text{A neutron}}{^1_0n} \quad + \text{ energy} \qquad (11\text{-}1)$$

$$\underset{\text{Lithium}}{^6_3\text{Li}} \quad + \quad ^1_0n \quad \longrightarrow \quad ^4_2\text{He} + {^3_1\text{He}} \qquad (11\text{-}2)$$

Reaction (11-1) involves the fusing of deuterium and tritium (two isotopes of hydrogen) to produce inert helium gas, a neutron and lots of energy (in the form of heat). Reaction (11-2) uses the neutrons produced in reaction (11-1) to react with lithium to produce helium and more tritium. The tritium can then be used to react with more deuterium to continue the fusion process.

The advantage of *fusion* over other forms of energy production are:

1. The principal fuel, deuterium, is readily obtained from the electrolysis of water. Fractional electrolysis is used to separate the deuterium from the other isotopes of hydrogen. Just think of all the water in the world. Even though only one hydrogen atom in 5000 is a deuterium atom, the supply of deuterium appears practically inexhaustible.

2. There are no radioactive waste products with fusion and there are no air pollutants. In fact, the only type of pollution that we have to worry about with fusion is *thermal* pollution. But we should remember that all forms of energy production cause thermal pollution.

Based on current research it is expected that a demonstration model controlled fusion reactor will be a reality by the 1990's, and that a commercial-size controlled fusion power plant will be operating early in the twenty-first century.

So it's clear that atomic fusion can lead to our doom if used unwisely. However, once we learn how to control the process, and tame it for peaceful uses, fusion may give us an unlimited power supply, with *no radioactive waste products*. In this way atomic fusion can increase the human stay on earth, by providing enough energy for all human needs, millions of years after conventional fuels have been used up. The decades of the 1980's and 1990's may see us realize this dream.

RADIATION AND ITS MEDICAL USES

Radiation and Cellular Damage

When alpha, beta, and gamma rays strike the complex molecules that compose the cells of our bodies, they can actually cause parts of these complex molecules to break away and ionize. Or the radiation can interact with these complex molecules, causing the formation of free radicals. [*Free radicals* are highly reactive *uncharged* species. They can be atoms or groups of atoms having an odd (unpaired) number of electrons. For example, a single chlorine atom, :C̤l·, would be a free radical.] Free radicals and ions can interact with other complex molecules in the body, causing their destruction (Fig. 11-8). If the molecules that are destroyed are necessary for important biological processes, the whole organism can suffer and even die. For example, if the molecule attacked by the radiation is DNA, the cells involved will not be able to reproduce themselves.

Of course, the radiation could simply damage the cells, not destroy them. But if this should happen, there is a possibility that the damaged cells will not reproduce themselves in a normal fashion, which means that they could become cancerous.

Radiation can also react with one of the major compounds in our bodies— water! Cells contain about 80% water, and ionizing radiation can cause the water to go through a series of free-radical reactions, forming a toxic substance, hydrogen peroxide. This could be fatal to the cells. The reactions would look

Figure 11-8 The effect of radiation on living cells. (a) A normal cell contains 23 pairs of chromosomes. (b) Ionizing radiation causes the chromosomes in this cell to multiply abnormally. (c) Radiation prevents this cell from dividing normally. The cell contains more than 700 chromosomes, and has grown to more than 10 times normal size. (Courtesy of Dr. C. K. Yu, National Tsing Hua University, Taiwan, China).

(a) (b) (c)

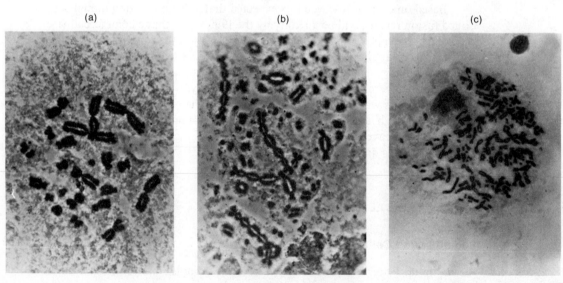

something like this:

$$H \overset{\times}{\underset{\underset{H}{\times}}{:}} \overset{..}{O} : + \text{ radiation} \longrightarrow : \overset{..}{\underset{.}{O}} \overset{\times}{} H + \overset{.}{H}$$

Hydroxide   Hydrogen
free radical   free radical

$$2 : \overset{..}{\underset{.}{O}} \overset{\times}{} H \longrightarrow \begin{array}{c} H \\ \overset{\times}{:\underset{..}{O}} \overset{..}{:\underset{\underset{H}{\times}}{O}} : \end{array}$$

Hydrogen
peroxide

The effects of radiation in cells can be divided into two stages. *Stage 1 effects* are the actual changes in the chemical makeup of the cells. *Stage 2 effects* result from the cells' inability to carry on their normal functions, which results in a body malfunction. For example, when iodine-131 is administered to an individual, it will localize in the thyroid gland. The radiation from the iodine-131 will destroy some cells in the thyroid gland (a stage 1 effect). The thyroid gland will not be able to produce as much thyroxine (a stage 2 effect). For an individual with a hyperactive thyroid, this represents an excellent therapeutic method.

The cells most sensitive to radiation are those undergoing rapid growth, for example, cells of the bone marrow and reproductive organs. Also, infants and children are very susceptible to radiation because a large number of cells in their bodies are undergoing rapid growth. Because cancer cells grow very rapidly compared to normal cells, radiation is very effective against these types of malignancies. Radiation directed at an area containing fast-growing cancer cells will have a greater devastating effect on those cells than on normal cells.

The Detection of Ionizing Radiation

Various instruments and devices are used for the detection of radiation. The best known device is the *Geiger–Müller counter*. This device consists of two parts, a detecting tube and a counter (Fig. 11-9). The heart of the system is the detecting tube, which consists of a pair of electrodes surrounded by an ionizable gas. As radiation enters the tube, it ionizes the gas. The ions produced travel toward the electrodes, between which there is a high voltage. The ions cause pulses of current at the electrodes, which are picked up and recorded on the counter. (Fig. 11-10).

The Geiger tube is most sensitive to beta radiation. Gamma radiation can pass right through the tube without being counted, and some alpha radiation can't make it through the window of the tube. The Geiger–Müller counter indicates the counts per minute of radiation entering the tube, but it doesn't tell you the energy of the radiation.

Figure 11-9 A Geiger-Müller counter. (Dosimeter Corporation of America)

A scintillation counter is a device that not only counts radioactivity, but also enables the operator to determine the energy of the radiation. The principle of operation involves the radiation reacting with a crystal containing sodium iodide and thallium iodide, which produces a series of flashes of varying intensity. The intensity of the flashes is proportional to the energy of the radiation (Fig. 11-11).

Figure 11-10 A Geiger tube.

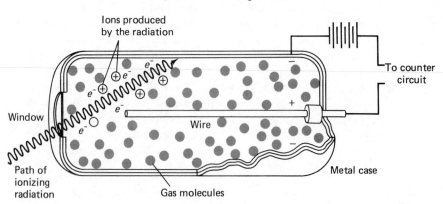

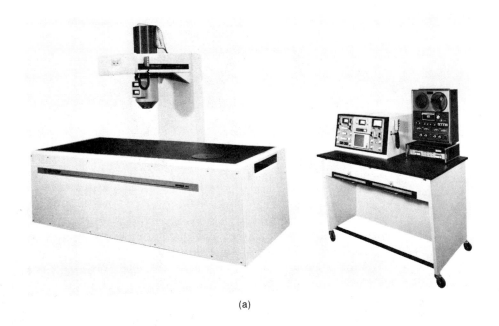

(a)

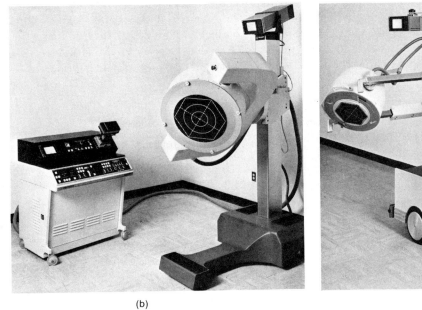

(b)

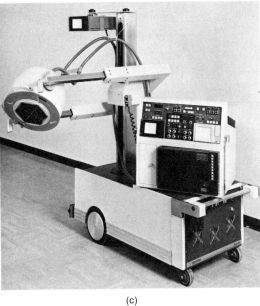

(c)

Figure 11-11 Some scintillation detectors: (a) scanner, (b) wide-field gamma-ray camera, (c) mobile gamma-ray camera. (Ohio Nuclear)

Figure 11-12   A film badge. (American Cancer Society)

Film badges are small portable devices that are worn by people such as x-ray technicians and nurses, who may be exposed to radiation. The badge contains a piece of photographic film that is removed monthly and developed. The darker the film badge, the greater the degree of exposure (Fig. 11-12).

Another device called a *dosimeter* is quickly replacing the film badge. One type of dosimeter works on the property of thermoluminescence, and is called a TLD for short. The TLD consists of a penlike device and a reading unit (Fig. 11-13). The penlike device, which is worn by the individual, contains a crystal such as lithium fluoride, which absorbs radiation. When the lithium fluoride crystal absorbs the radiation, its structure changes slightly. To determine the amount of radiation that the crystal has absorbed, the penlike device is placed in its reading unit, where it is heated quickly. This causes the lithium fluoride crystal to return to its original state. As it does, it gives off visible light. The visible light is proportional to the radiation absorbed by the lithium fluoride crystal.

Figure 11-13   A thermoluminescence dosimeter consists of a penlike device and a reading unit. (Dosimeter Corporation of America)

TABLE 11-2 *Unit used for expressing the level of ionizing radiation*

UNIT	WHAT IT MEASURES	EXPLANATION
curie (Ci)	The activity of the radiation source.	One curie equals 3.7×10^{10} disintegrations per second.
roentgen (r)	Measures the ionizing ability of x rays and gamma rays.	One roentgen of radiation produces 2×10^9 ion pairs in 1 cm^3 of air.
rad	Radiation absorbed *dose* (i.e., the amount of energy absorbed by living tissue).	One rad is equal to the 1 g of irradiated tissue absorbing 100 ergs of energy.
rem	Roentgen equivalent *man* (i.e., a *weighted* absorbed dose of radiation).	One rem equals 1 rad for beta and gamma radiation. One rem equals 0.1 rad for alpha radiation.

Now that we've looked at some devices used to detect radiation, let's learn about the *units* used to express the concentration or dose of radiation.

Units Used for Expressing Dosages of Radiation

There are several units used for expressing the level of ionizing radiation (Table 11-2). Some of these units simply express the activity of the radiation source; others relate the effect of a specific type of radiation to its effect on living tissue. Let's see how this works.

The *curie*, abbreviated Ci, is the unit used to describe the activity of the radiating source. One curie is equal to 3.7×10^{10} disintegrations per second, which happens to be the disintegration rate of 1 g of radium.

Example 11-2   A gold-198 source rated at 0.50 Ci is used in the treatment of widespread abdominal cancers. What is the activity of this gold in disintegrations per second (dps)?

Solution   We know that one curie (1 Ci) equals 3.7×10^{10} dps; therefore,

$$? \text{ dps} = (0.50 \text{ Ci})\left(\frac{3 \times 10^{10} \text{ dps}}{1 \text{ Ci}}\right) = 1.5 \times 10^{10} \text{ dps}$$

The *roentgen*, abbreviated r, measures the ionizing ability of x rays and gamma rays. One roentgen is the amount of radioactivity that produces 2×10^9 ion pairs in 1 cubic centimeter of air.

The *rad* (radiation *a*bsorbed *d*ose) measures the amount of energy absorbed by living tissue, regardless of the type of radiation. (By the way, the

TABLE 11-3 *Lethal dose values for various organisms*

ORGANISM	LD$_{50}^{30}$(r)
Guinea pig	200
Rabbit	300
Dog	325
Human being	400
Monkey	450
Rat	850
Bacteria	5,000–13,000
Viruses	100,000–200,000

energy absorbed is measured in units called ergs.* When 1 g of irradiated tissue absorbs 100 ergs of energy, that is 1 rad.) In general, the exposure to radiation in roentgen units is numerically equal to the absorbed dose in rads.

Example 11-3 Most authorities on radiation feel that a person exposed to 600 roentgens of radiation over his or her entire body will probably die. What is this value in rads?

Solution In general, 1 roentgen equals 1 rad; therefore, 600 roentgens equals 600 rads.

The toxicity of radiation is sometimes expressed in terms of a *lethal dose*, which is the amount of radiation that will kill 50% of those organisms exposed to it within 30 days. The symbol for this 30-day, 50% kill lethal dose is LD$_{50}^{30}$. The units of radiation are usually expressed in roentgens. Table 11-3 is a summary of lethal dose values for various organisms.

Different types of radiation have different effects on body tissue. In other words, 1 rad of alpha radiation will have a different effect on the body than will 1 rad of beta radiation. To account for this difference, scientists have developed a unit called the *rem* (roentgen equivalent man). The rem is calculated by multiplying the number of rads by a weighting factor. The weighting factor for alpha radiation is 10, whereas that for beta and gamma radiation is 1.

$$rem = rad \times weighting\ factor$$

When a dose of radiation is expressed in rems, it doesn't matter what type of radiation it is since it has already been weighted (Fig. 11-14). The rem is a

* An erg is a unit of work in the metric system.

Figure 11-14 Sometimes it doesn't matter what the units are when measuring radiation.

useful unit for expressing amounts of radiation that individuals receive each year (Table 11-4).

Example 11-4 A recommended dose limit for the general public is 0.5 rem/year.

(a) If you received 0.5 rem of radiation during the year, and it was all in the form of alpha rays, what is your dose in rads?

(b) If you received 0.5 rem of radiation during the year, and it was all in the form of beta rays, what is your dose in rads?

Solution We know that: rem = rad × weighting factor. Therefore,

$$\text{rad} = \frac{\text{rem}}{\text{weighting factor}}$$

TABLE 11-4   *Annual per capita radiation doses in the United States*

SOURCE OF RADIATION	DOSE PER YEAR (mrem)
Environmental	134
Medical	74
Occupational	1
Miscellaneous (e.g., television, etc.)	3

(a) The weighting factor for alpha radiation is 10. Therefore,

$$? \text{ rad} = \frac{0.5 \text{ rem}}{10} = 0.05 \text{ rad}$$

(b) The weighting factor for beta radiation is 1. Therefore,

$$? \text{ rad} = \frac{0.5 \text{ rem}}{1} = 0.5 \text{ rad}$$

Radioisotopes:
Their Uses in Medical Diagnosis and Therapy

Radiation is used to both detect and treat abnormalities in the body. We've already discussed the use of iodine-131 therapy for hyperthyroidism. Iodine-131 can also be used in a diagnostic procedure to monitor the function of the thyroid. The rate at which the thyroid takes up the iodine-131 can be monitored with a scanning device to see if it is functioning properly.

What makes radioisotopes so useful in diagnostic procedures is that the body treats the tagged isotope in the same way that it treats the nonradioactive element. Therefore, the tagged isotope goes right to the area of the body where you want it to go. For example, iodine, whether it's radioactive or not, goes right to the thyroid, where it is incorporated into the amino acid thyroxine (the only molecule in the body that contains iodine). Therefore, iodine-131 is perfect for monitoring the thyroid gland.

Chromium, in the form of sodium chromate, attaches strongly to the hemoglobin of red blood cells. This makes radioactive chromium-151 an excellent isotope for determining the flow of blood through the heart. This isotope is also useful for determining the lifetime of red blood cells, which can be of great importance in the diagnosis of anemias.

Radioactive cobalt (cobalt-59 or cobalt-60) is used to study defects in vitamin B_{12} absorption. Cobalt is the metallic atom at the center of the B_{12} molecule. By injecting a patient with vitamin B_{12}, labeled with radioactive cobalt, the physician can study the path of the vitamin through the body and discover any irregularities.

We've already discussed how radiation therapy can be used to destroy cancer cells. Radiation can be delivered to a malignant area in three ways.

One method is *teletherapy*, in which a high-energy beam of radiation is aimed at the cancerous tissues (Fig. 11-15). A second method is *brachytherapy*, in which a radioactive isotope is placed into the area to be treated. This is usually done by means of a seed, which could be a glass bead containing the isotope. In this way the isotope delivers a constant beam of radiation to the affected area.

The third method is called *radiopharmaceutical therapy*. This method involves oral or intravenous administration of the isotope. The isotope then

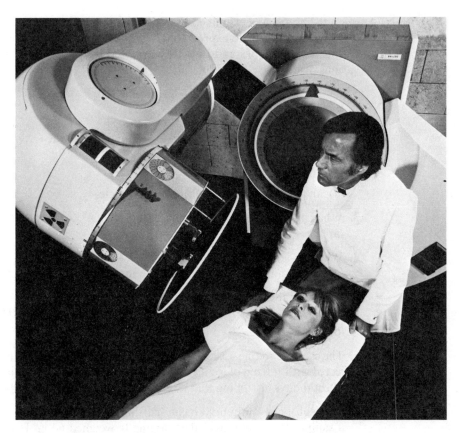

Figure 11-15 A rotational teletherapy unit for cobalt-60 therapy. (Philips Medical Systems)

uses the normal body pathways to seek its target. This is the method that is used to get iodine-131 to the thyroid gland.

SUMMARY

In this chapter we surveyed the topic of radiation. We began with the basics and read about the discovery of radiation. We discovered that there are three major types of ionizing radiation: alpha rays, beta rays, and gamma rays. We learned a little about nuclear transformation, and how one radioactive isotope can change into another isotope by emitting nuclear particles. We also studied half-life, the amount of time it takes for one-half of the radioactive material to decay. We also learned about the processes of nuclear fission and fusion and how these processes can be used for humankind's benefit or destruction.

In the second half of the chapter we looked at radiation and its medical uses. We discovered how radiation damages cells, and what effects this can have on the body. We examined the various methods used to detect radiation as well as the units used to express the dosages of radiation. Finally, we looked at various radioisotopes and how they are used in the diagnosis and therapy of various disorders.

EXERCISES

1. Name the three type of nuclear radiation and give the Greek symbols for each.

2. Show the charge, mass, and symbol of
   (a) an alpha particle   (b) a beta particle

3. When an isotope emits an alpha particle, what happens to its
   (a) mass number?   (b) atomic number?

4. When an isotope emits a beta particle, what happens to its
   (a) mass number?   (b) atomic number?

5. Complete the following nuclear equations.
   (a) $^{14}_{6}C \rightarrow ^{0}_{-1}e + ?$ (b) $? \rightarrow ^{0}_{-1}e + ^{24}_{12}Mg$ (c) $? \rightarrow ^{4}_{2}He + ^{234}_{90}Th$
   (d) $^{87}_{36}Kr \rightarrow ^{1}_{0}n + ?$ (e) $^{212}_{84}Po \rightarrow ^{4}_{2}He + ?$

6. The isotope iodine-131 has a half-life of 8 days. If an individual is injected with $1\overline{00}$ mg of iodine-131 today, how much will remain in her body after 24 days (assuming that none is lost through normal body channels)?

7. The isotope strontium-90 has a half-life of 28 years. Suppose you have a sample of strontium-90 that originally weighed 64 g. How old is this sample if today it weighs 8 g?

8. Explain the following terms.
   (a) chain reaction   (b) nuclear fission   (c) nuclear fusion

9. Explain the two ways in which radiation can damage or destroy cells.

10. Write the Lewis dot structure for
    (a) a hydroxide free radical   (b) a hydrogen free radical
    (c) hydrogen peroxide

11. Determine whether each of the following statements refers to a stage 1 or a stage 2 effect.
    (a) Yttrium-90 encased in glass beads is placed into the pituitary gland, where it destroys some of its cells.
    (b) Iodine-131 therapy decreases the amount of thyroxine produced by the thyroid gland.
    (c) A patient with inoperable brain tumors receives cobalt-60 therapy, which shrinks the size of the tumors.

12. Name two types of cells that are highly sensitive to radiation, and explain what makes them this way.

13. Name four types of radiation detectors, and explain the use of each.

14. Define the following terms.
    (a) curie   (b) roentgen   (c) rad   (d) rem   (e) LD_{50}^{30}

15. A $\overline{3000}$ curie cobalt-60 source is sometimes used to shrink inoperable tumors. What is the activity of this cobalt-60 source in disintegrations per second?

16. The LD_{50}^{30} for human beings is about 400 roentgens. What is this in rads?

17. The U.S. Environmental Protection Agency estimates that each of us receives about 4 millirems (0.004 rem) of radiation from fallout each year. If all of this were in the form of alpha rays, what would be the dose in rads?

18. Name the three ways that radiation can be delivered to malignant areas. Give an example showing how each method is used.

19. When iodine is administered to an individual, it goes straight to the thyroid. Why?

20. What type of shielding is needed to stop
    (a) alpha rays?   (b) beta rays?   (c) gamma rays?

21. Discuss the advantages and disadvantages of nuclear power (fission) plants as a source of electricity.

22. Tell how a breeder reactor works.

23. Write the principal reactions for atomic fusion.

24. Discuss the advantages and disadvantages of fusion power as a source of electricity.

Chapter 12

Organic Chemistry

Some Things You Should Know After Reading This Chapter

You should be able to:

1. State whether a compound is organic or inorganic when given its chemical formula.
2. Write the electron dot diagram for carbon.
3. Write the electron dot diagram for a methane molecule (CH_4).
4. Define and give an example of a tetrahedron.
5. Show how two carbon atoms can bond to form a carbon–carbon single bond, double bond, or triple bond.
6. Define and give an example of a molecular formula and a structural formula for an organic compound.
7. Define an R group.
8. State the properties of alkanes, and give some examples.
9. State the properties of alkenes, and give some examples.
10. State the properties of alkynes, and give some examples.
11. Define and give an example of a saturated hydrocarbon, an unsaturated hydrocarbon, a cyclic hydrocarbon, and an aromatic hydrocarbon.
12. State what is meant by a homologous series, and give an example of such a series.
13. Define and illustrate the term isomer.
14. Give an example of hydrocarbon combustion.
15. Write the products of alkene and alkyne hydrogenation given the initial reactants.
16. Explain the meaning of the term functional group.
17. Write the general formula for an alcohol, ether, aldehyde, ketone, carboxylic acid, ester, amine, and amide.
18. Give an example of each of the types of organic compounds listed in objective 17 and state some uses for each compound.
19. Write an equation for the oxidation of an aldehyde, and the reduction of an aldehyde.
20. Write an equation for the formation of an ester.
21. State whether an amine is primary, secondary, or tertiary when given its structural formula.

Part 1 The Basics

During the 1700's and early 1800's, most chemists believed that there were two distinct classes of chemical compounds: organic and inorganic. They said that organic compounds were derived from living or once-living organisms and that inorganic compounds were part of the nonliving world. Organic compounds were substances such as sugar, fats, oils, and wood. Inorganic compounds were substances such as salt and iron. And although many chemists knew that organic substances could be converted into inorganic substances by heating, the reverse was not true—inorganic substances could not be converted into organic ones. It was these observations that gave rise to the theory of *vitalism*. This theory stated that life and substances associated with life were not bound by the laws of science and that these life-connected substances had a *vital force* that human beings could not control. This, of course, meant that chemists would never be able to synthesize organic compounds from inorganic ones.

But in 1828, the German chemist Friedrich Wöhler astounded the chemical world by having synthesized the organic compound urea from what was then considered the inorganic compound ammonium cyanate. Urea is an organic compound found in urine, and until Wöhler synthesized it in his laboratory, urea was thought by most scientists to be associated with living things. The belief in the theory of vitalism was on its way out.

Today, a little more than 150 years after Wöhler's discovery, the synthesis of organic compounds is commonplace. In fact, many of the complex molecules associated with life itself, for example insulin, have been produced artificially in the laboratory.

ORGANIC COMPOUNDS DEFINED

Organic compounds are primarily those which contain carbon atoms. The old definition that organic substances are those from living or once-living sources no longer applies. For example, plastics and pharmaceuticals are organic compounds, because their molecules contain numerous carbon atoms. However, not all carbon-containing compounds are organic. For example, carbonate compounds such as $CaCO_3$ and Na_2CO_3, as well as the oxides of carbon (CO and CO_2), are considered to be inorganic compounds. Although a precise definition of organic compounds is difficult to give, we can say that for the most part,

> *organic compounds are those which contain carbon together with elements such as hydrogen, oxygen, nitrogen, sulfur, and the group VIIA elements (Fig. 12-1).*

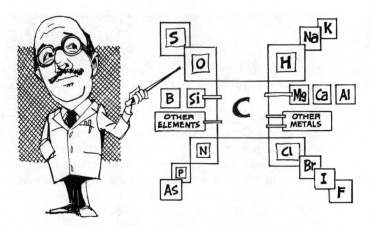

Figure 12-1 For the most part, organic compounds are those that contain carbon together with elements such as hydrogen, oxygen, nitrogen, sulfur, and the group VIIA elements.

After we've looked at some organic compounds, you'll see that we really don't need a precise definition anymore.

THE AMAZING CARBON ATOM

Perhaps the most unique thing about the carbon atom is its ability to combine with itself. The other elements in the periodic table don't have this ability to any significant extent. But it is this unusual ability that allows carbon atoms to form all kinds of chainlike and ring-shaped molecules, making the number of potential organic compounds almost infinite (Fig. 12-2). Just think, over 3 million organic compounds are known today, and the possibility exists for countless more to be discovered or synthesized.

Carbon is a group IVA element, which means that a carbon atom has four electrons in its outermost energy level. The carbon atom can use its four electrons to form covalent bonds with other carbon atoms of other elements. This is exactly what happens in organic compounds.

When a carbon atom shares each of its outer electrons with four other atoms, the result is a compound such as CH_4, the compound *methane*, which we know as natural gas (cooking gas). In this compound, the central carbon atom is bonded to each of the four hydrogen atoms. We can write the electron dot structure in the usual way.

$$\begin{array}{ccc} & H & & H \\ & | & & | \\ H : \overset{..}{\underset{..}{C}} : H & \text{or} & H - C - H \\ & | & & | \\ & H & & H \end{array}$$

As you can see, this diagram gives the impression that the carbon–hydrogen bonds are at 90° angles to each other.

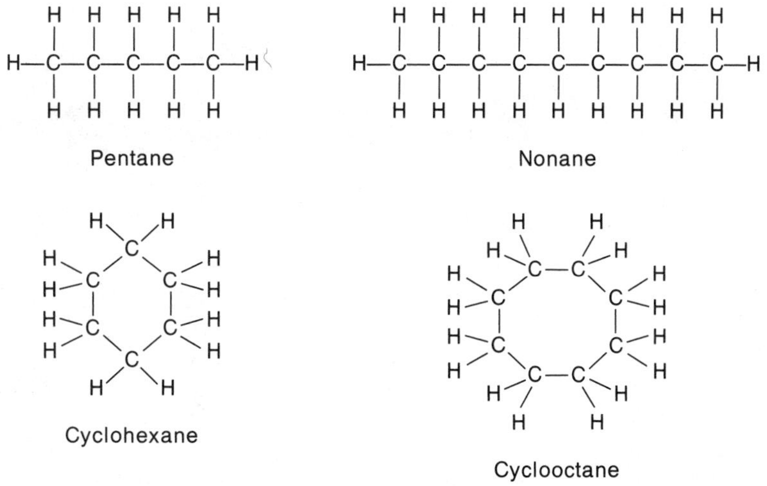

H H H H H
| | | | |
H—C—C—C—C—C—H
| | | | |
H H H H H

Pentane

H H H H H H H H H
| | | | | | | | |
H—C—C—C—C—C—C—C—C—C—H
| | | | | | | | |
H H H H H H H H H

Nonane

Cyclohexane

Cyclooctane

Figure 12-2 Carbon atoms can form chainlike and ring-shaped molecules.

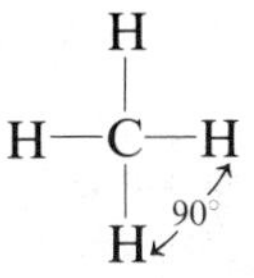

$$
\begin{array}{c}
H \\
| \\
H-C-H \\
| \quad 90° \\
H
\end{array}
$$

However, this is not the way the molecule actually exists. Molecules are *three-dimensional*, and the four outer electrons in carbon move as far apart from each other as possible, because electrons repel each other. (Remember that particles with similar charges always repel one another.) The electrons can put the greatest distance between themselves if the carbon–hydrogen bonds are in the shape of a tetrahedron. The tetrahedral shape allows the four carbon electrons to get as far apart as possible (Fig. 12-3). A tetrahedron ("tetra" is from the Greek word "four") is a pyramid-shaped figure with three upper sides plus a bottom side. You can make a tetrahedron this way. Take six pencils of equal height. Place three of them on a table in the form of a triangle. Hold the other three pencils upright, placing one at each corner of the triangle. Now tilt the three upright pencils inward until they touch. You have formed a tetrahedron (Fig. 12-4). The tetrahedral shape means that the carbon–hydrogen bonds in methane are 109.5° apart. Methane is only one example; this tetrahedral structure is common to many organic compounds.

The Bonding Between Carbon Atoms

When two carbon atoms bond with each other, they can do it in three ways:

1. They can form a single bond: ·C—C·

2. They can form a double bond: ·C=C·

3. They can form a triple bond: ·C≡C·

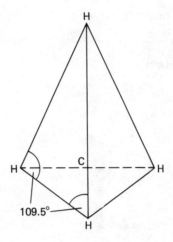

Figure 12-3 The tetrahedral structure of a CH_4 molecule.

Figure 12-4 How to make a tetrahedron using six pencils.

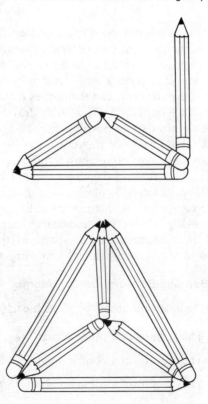

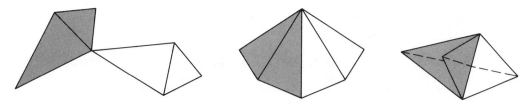

Figure 12-5 Two carbon atoms joined together by a single bond (left), a double bond (center), and a triple bond (right).

In the case of a single bond, we can visualize the two carbon atoms joining in such a way that the compound looks like two tetrahedrons with their corners touching. The lines in the figure represent the bonds, and the corners represent the atoms. For the double bond, we can visualize two tetrahedrons with a common *edge*. For a triple bond, we can visualize two tetrahedrons with a common *face* (Fig. 12-5).

STRUCTURAL FORMULAS OF ORGANIC COMPOUNDS

In earlier chapters we wrote the *molecular formulas* for various compounds. For example, barium chloride is $BaCl_2$. The molecular formula tells us the actual number of atoms of each element in a molecule of the compound.

For the organic chemist, however, the molecular formula is not enough. As we have seen, carbon atoms can bond together in different ways. So there can be many different compounds with the same molecular formula. For example, the molecular formula C_4H_{10} represents two distinct compounds; and the formula C_8H_{18} represents 18 distinct compounds! This is why, when organic chemists are writing the formula of an organic compound, they use a *structural formula*. The structural formula indicates *in what way* the carbon atoms are bonded to each other. We can write the two compounds represented by the formula C_4H_{10} in this way:

$$
\begin{array}{cccc}
& \text{H} & \text{H} & \text{H} & \text{H} \\
& | & | & | & | \\
\text{H}- & \text{C} & \text{C} & \text{C} & \text{C} -\text{H} \\
& | & | & | & | \\
& \text{H} & \text{H} & \text{H} & \text{H}
\end{array}
\quad\text{and}\quad
\begin{array}{ccc}
\text{H} & \text{H} & \text{H} \\
\diagdown & | & \diagup \\
\text{H} & \text{C} & \text{H} \\
| & | & | \\
\text{H}-\text{C} & \text{C} & \text{C}-\text{H} \\
| & | & | \\
\text{H} & \text{H} & \text{H}
\end{array}
$$

Structural formulas are sometimes written in a more convenient form. For example, the two formulas we just wrote can also be written like this:

$$
CH_3-CH_2-CH_2-CH_3 \quad\text{and}\quad
\begin{array}{c}
CH_3 \\
| \\
CH_3-CH-CH_3
\end{array}
$$

Take a minute to see what we've done. Compare the two ways of writing the formulas and you'll soon get the idea. The second way—the more convenient one—still shows how each carbon atom is bonded to its neighbors.

Example 12-1 Write the following structures in the more convenient form, sometimes called the *condensed form*.

(a) 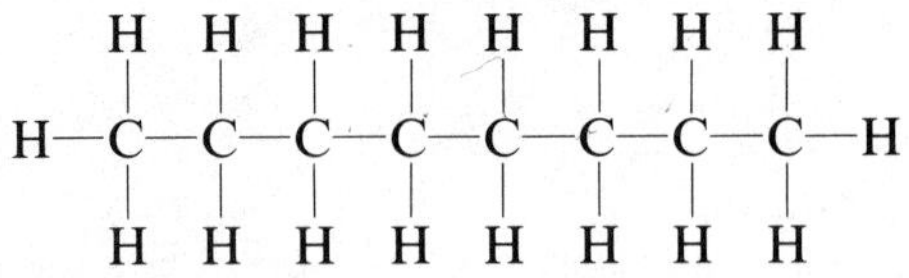
$$
\begin{array}{c}
\;\;\;\;\text{H}\;\;\text{H}\;\;\text{H}\;\;\text{H}\;\;\text{H}\;\;\text{H}\;\;\text{H}\;\;\text{H} \\
\;\;\;\;|\;\;\;\;|\;\;\;\;|\;\;\;\;|\;\;\;\;|\;\;\;\;|\;\;\;\;|\;\;\;\;| \\
\text{H}-\text{C}-\text{C}-\text{C}-\text{C}-\text{C}-\text{C}-\text{C}-\text{C}-\text{H} \\
\;\;\;\;|\;\;\;\;|\;\;\;\;|\;\;\;\;|\;\;\;\;|\;\;\;\;|\;\;\;\;|\;\;\;\;| \\
\;\;\;\;\text{H}\;\;\text{H}\;\;\text{H}\;\;\text{H}\;\;\text{H}\;\;\text{H}\;\;\text{H}\;\;\text{H}
\end{array}
$$

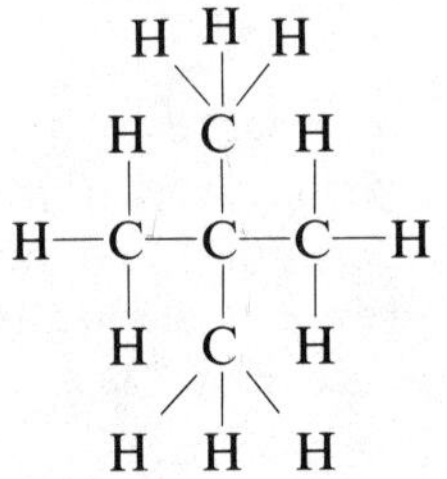

(b)
$$
\begin{array}{c}
\;\;\;\;\text{H}\;\;\text{H}\;\;\text{H} \\
\;\;\;\;\text{H}\;\;\text{C}\;\;\text{H} \\
\;\;\;\;|\;\;\;\;|\;\;\;\;| \\
\text{H}-\text{C}-\text{C}-\text{C}-\text{H} \\
\;\;\;\;|\;\;\;\;|\;\;\;\;| \\
\;\;\;\;\text{H}\;\;\text{C}\;\;\text{H} \\
\;\;\;\;\text{H}\;\;\text{H}\;\;\text{H}
\end{array}
$$

(c)
$$
\begin{array}{c}
\;\;\;\;\text{H}\;\;\text{H}\;\;\text{H}\;\;\text{H} \\
\;\;\;\;|\;\;\;\;|\;\;\;\;|\;\;\;\;| \\
\text{H}-\text{C}-\text{C}-\text{C}-\text{C}-\text{H} \\
\;\;\;\;|\;\;\;\;|\;\;\;\;|\;\;\;\;| \\
\;\;\;\;\text{H}\;\;\text{H}\;\;\text{C}\;\;\text{H} \\
\;\;\;\;\text{H}\;\;\text{H}\;\;\text{H}
\end{array}
$$

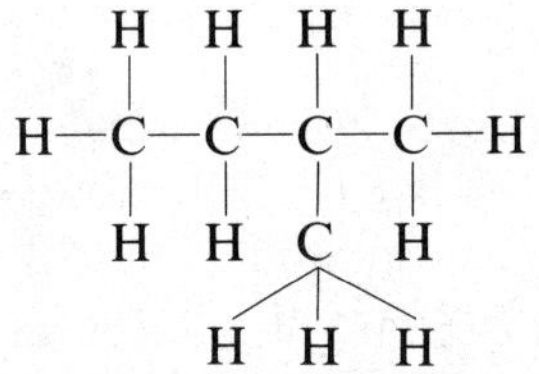

Solution Check each structure carefully. Be sure that you have the same number of carbon atoms and hydrogen atoms as in the original structure.

(a) $CH_3-CH_2-CH_2-CH_2-CH_2-CH_2-CH_2-CH_3$

(b) 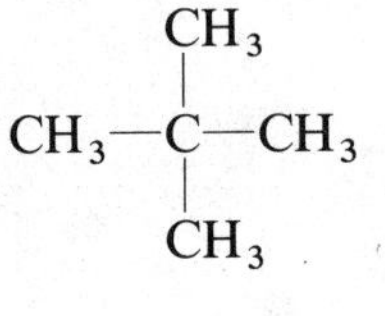
$$
\begin{array}{c}
\;\;\;\;\;\;CH_3 \\
\;\;\;\;\;\;\;| \\
CH_3-C-CH_3 \\
\;\;\;\;\;\;\;| \\
\;\;\;\;\;\;CH_3
\end{array}
$$

(c)
$$
\begin{array}{c}
\;\;\;\;\;\;\;\;H \\
\;\;\;\;\;\;\;\;| \\
CH_3-CH_2-C-CH_3 \\
\;\;\;\;\;\;\;\;| \\
\;\;\;\;\;\;\;CH_3
\end{array}
$$

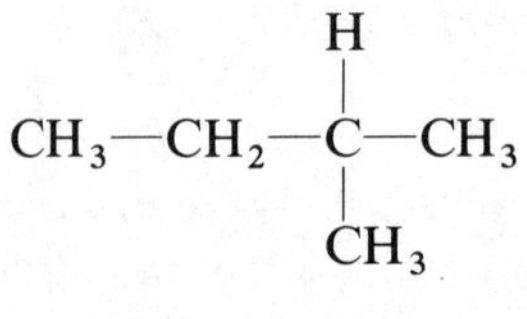

THE CLASSIFICATION OF ORGANIC COMPOUNDS

We have said that there are more than 3 million organic compounds known to modern chemists. Is each of these compounds completely unique, or do they have similarities? In other words, can we classify this great number of compounds into groups?

Fortunately, we can. We can classify most of them into a few simple groups. The categories are set up so that compounds that react similarly are placed together. Let's take a general look at organic compounds and see what we can learn about the various groups.

THE R GROUP

Suppose that an organic chemist begins by studying a compound with the formula

$$CH_3-C\equiv C-CH_3$$

The chemist then extends the study to cover other compounds with the same type of structure:

$$CH_3-CH_2-C\equiv C-CH_2-CH_3$$

$$CH_3-CH_2-CH_2-C\equiv C-CH_2-CH_2-CH_3$$

$$CH_3-CH_2-CH_2-CH_2-CH_2-C\equiv C-CH_2-CH_2-CH_2-CH_3$$

(By the way, this technique is used by many pharmaceutical companies to find new drugs that have certain similarities to existing ones.) The main point, however, is that the chemists's study shows that, in all these compounds, the *reactive* part of the molecule is the triple bond. Also, all these compounds behave in the same way. In other words, the chemist finds that all these compounds *belong in the same group*. We can write the general formula for this group as

$$R-C\equiv C-R'$$

The R and R' (pronounced "R prime") in this general formula are called *alkyl groups*, and they represent one, two, three, or any number of carbon atoms attached to the reactive part of the molecule. We'll use the R-group notation to represent classes of organic compounds later in this chapter.

THE ALKANES

Let's begin our study of particular organic compounds with the *alkanes*. This group of compounds is one of the simplest in composition and structure, for the following reasons.

1. All alkanes are *hydrocarbons*, which means that they contain only carbon and hydrogen.

2. The bonds between carbon atoms in alkanes are all single bonds. When all the carbon–carbon bonds in an organic compound are single bonds, we say that the compound is *saturated*.

The simplest member of this group of hydrocarbons is methane, CH_4, which we're already familiar with:

$$
\begin{array}{c}
\text{H} \\
| \\
\text{H}-\text{C}-\text{H} \\
| \\
\text{H}
\end{array}
$$

Methane

The next-to-the-simplest alkane has two carbon atoms and is called ethane, CH_3—CH_3, or

$$
\begin{array}{cc}
\text{H} & \text{H} \\
| & | \\
\text{H}-\text{C}-\text{C}-\text{H} \\
| & | \\
\text{H} & \text{H}
\end{array}
$$

Ethane

Then we come to the member of this group that has three carbon atoms—and many of us have had some experience with this one—*propane* (also known as bottled gas), CH_3—CH_2—CH_3, or

$$
\begin{array}{ccc}
\text{H} & \text{H} & \text{H} \\
| & | & | \\
\text{H}-\text{C}-\text{C}-\text{C}-\text{H} \\
| & | & | \\
\text{H} & \text{H} & \text{H}
\end{array}
$$

Propane

Note that each member of this group differs from the member before it by having one more CH_2 group. We can see this more clearly if we write the molecular formulas of these compounds.

$$CH_4 \qquad C_2H_6 \qquad C_3H_8$$
Methane   Ethane   Propane

So we say that these compounds are members of a *homologous series*. This means that all the compounds in this series are of the same chemical type and differ only by fixed increments (an increment is an increase in quantity). In the alkane series, the fixed increment is CH_2. The general formula for this series of compounds can be written C_nH_{2n+2}, where n represents the number of carbon atoms (Table 12-1). Think about this formula for a minute and look at the formulas in Table 12-1, and you'll understand.

TABLE 12-1 *The first 10 members of the alkane series*

NUMBER OF CARBON ATOMS	MOLECULAR FORMULA	STRUCTURAL FORMULA	NAME	COMMENTS
1	CH_4	CH_4	Methane	Main ingredient of natural gas
2	C_2H_6	CH_3-CH_3	Ethane	Minor ingredient of natural gas
3	C_3H_8	$CH_3-CH_2-CH_3$	Propane	Minor ingredient of natural gas; usually separated from other ingredients and sold as bottled gas
4	C_4H_{10}	$CH_3-(CH_2)_2-CH_3$	Butane	Also found in natural gas; usually separated from other ingredients and used as a fuel—for example, in butane lighters
5	C_5H_{12}	$CH_3-(CH_2)_3-CH_3$	Pentane	
6	C_6H_{14}	$CH_3-(CH_2)_4-CH_3$	Hexane	
7	C_7H_{16}	$CH_3-(CH_2)_5-CH_3$	Heptane	Ingredients of gasoline
8	C_8H_{18}	$CH_3-(CH_2)_6-CH_3$	Octane	
9	C_9H_{20}	$CH_3-(CH_2)_7-CH_3$	Nonane	
10	$C_{10}H_{22}$	$CH_3-(CH_2)_8-CH_3$	Decane	Ingredient of kerosene

ISOMERS

A little while ago we said that the molecular formula for an organic compound could actually represent many different compounds. As an example we showed that the formula of C_4H_{10} represented two compounds. This phenomenon is known as *structural isomerism*. Isomers are compounds with the same molecular formula, but different *structural* formulas.

The title of Table 12-1 states that it shows the first 10 members of the alkane series. However, it really shows the first 10 *straight-chain* members of the alkane series. This is because, starting with C_4H_{10}, there are isomers for the rest of the alkane series (Table 12-2). Let's write the structures for the isomers of C_4H_{10} once again.

$$CH_3-CH_2-CH_2-CH_3$$

Normal butane

$$CH_3-\overset{\overset{\displaystyle CH_3}{|}}{\underset{\underset{\displaystyle H}{|}}{C}}-CH_3$$

Isobutane

267 *Organic Chemistry*

TABLE 12-2 *Number of isomers of some compounds in the alkane series*

MOLECULAR FORMULA	NUMBER OF ISOMERS
C_4H_{10}	2
C_5H_{12}	3
C_6H_{14}	5
C_7H_{16}	9
C_8H_{18}	18
C_9H_{20}	35
$C_{10}H_{22}$	75
$C_{15}H_{32}$	5000 (approx.)
$C_{40}H_{82}$	60 trillion (approx.)

The first structure is the *straight-chain* isomer, which we call normal butane (or *n*-butane). The second structure is the *branched-chain* isomer, which is commonly called isobutane. Look at these two structures carefully and be sure that you understand the differences between them.

The three isomers of pentane (C_5H_{12}) would look this way.

$$CH_3{-}CH_2{-}CH_2{-}CH_2{-}CH_3$$

n-Pentane

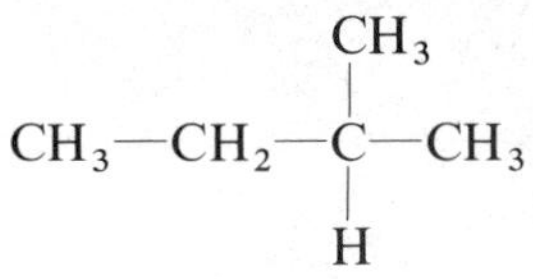

$$CH_3{-}CH_2{-}\underset{\underset{H}{|}}{\overset{\overset{CH_3}{|}}{C}}{-}CH_3$$

Isopentane

$$CH_3{-}\underset{\underset{CH_3}{|}}{\overset{\overset{CH_3}{|}}{C}}{-}CH_3$$

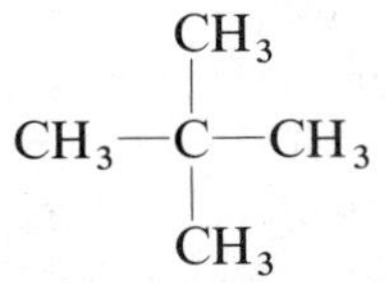

Neopentane

Again be sure that you see the differences among the three structures.

At this point you might be asking whether there are still more structures that can be written for the formula C_5H_{12}. You might try writing the following structure:

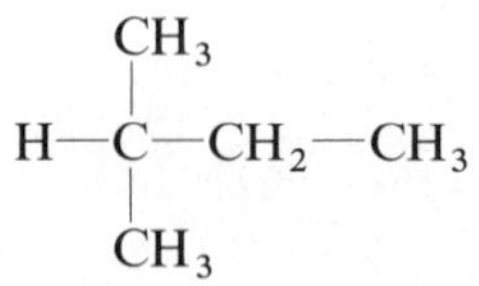

$$H{-}\underset{\underset{CH_3}{|}}{\overset{\overset{CH_3}{|}}{C}}{-}CH_2{-}CH_3$$

Is this structure different from that of isopentane? The ~~simply the molecule of isopentane flipped around. The b~~ It is
still the same. ~~e is~~

Example 12-2   Write the structural formulas for the five isomers of C_6H_{14}

Solution   Try to obtain five different bonding sequences. Remember carbon atom can be bonded to only four other atoms.

(a) $CH_3-CH_2-CH_2-CH_2-CH_2-CH_3$

(b) $CH_3-CH_2-CH_2-\underset{\underset{\displaystyle H}{|}}{\overset{\overset{\displaystyle CH_3}{|}}{C}}-CH_3$

(c) $CH_3-CH_2-\underset{\underset{\displaystyle H}{|}}{\overset{\overset{\displaystyle CH_3}{|}}{C}}-CH_2-CH_3$

(d) $CH_3-CH_2-\underset{\underset{\displaystyle CH_3}{|}}{\overset{\overset{\displaystyle CH_3}{|}}{C}}-CH_3$

(e) $CH_3-\underset{\underset{\displaystyle H}{|}}{\overset{\overset{\displaystyle CH_3}{|}}{C}}-\underset{\underset{\displaystyle H}{|}}{\overset{\overset{\displaystyle CH_3}{|}}{C}}-CH_3$

THE ALKENES AND ALKYNES

The alkenes and alkynes, like the alkanes, are hydrocarbons. But the alkenes have a carbon–carbon double bond and the alkynes have a carbon–carbon triple bond (Fig. 12-6).

The simplest alkene has the formula C_2H_4 and the structure

$$\underset{H}{\overset{H}{\diagdown}}C=C\underset{H}{\overset{H}{\diagup}}$$

Ethylene

Notice the double bond linking the carbon atoms. The common name for this compound is ethylene. It is the substance from which we make polyethylene,

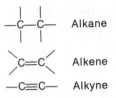

	Alkane
	Alkene
	Alkyne

Figure 12-6 Alkanes have carbon-carbon single bonds; alkenes have a carbon-carbon double bond; and alkynes have a carbon-carbon triple bond.

TABLE 12-3 *The alkene and alkyne series*

	ALKENES	ALKYNES
Structural formula of the first member of the series	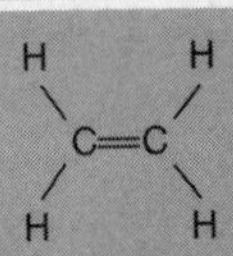 $\begin{array}{c}\text{H}\qquad\text{H}\\ \diagdown\ /\\ \text{C}=\text{C}\\ /\ \diagdown\\ \text{H}\qquad\text{H}\end{array}$	H—C≡C—H
Molecular formula	C_2H_4	C_2H_2
Common name	Ethylene	Acetylene
IUPAC name	Eth*ene*	Eth*yne*
General formula for the series	C_nH_{2n}	C_nH_{2n-2}

a plastic used for containers, electrical insulation, and packaging. (By the way, the official IUPAC name for this compound is eth*ene*.)

The simplest *alkyne* has the formula C_2H_2 and the structure

$$\text{H}—\text{C}\equiv\text{C}—\text{H}$$
Acetylene

Notice the triple bond linking the carbon atoms. The common name for this compound is acetylene. Today it is used as a fuel, for example in the acetylene torch. But in the past it has been used as a surgical anesthetic. (By the way, the official IUPAC name for this compound is eth*yne*.)

Each of these compounds is the first compound in its homologous series (Table 12-3). Both of these series of hydrocarbons are said to be *unsaturated*. This means that there can be double or triple bonds in the compounds.

THE CYCLIC HYDROCARBONS

There is a whole series of hydrocarbons that form *cyclic structures*—that is, structures that join to form a closed chain, like a snake biting its tail. An example of this is:

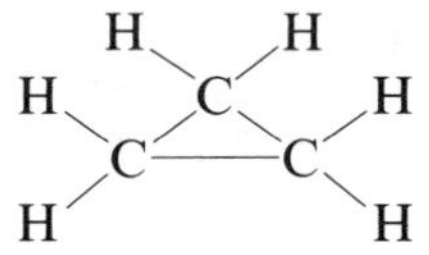

Cyclopropane

or C_3H_6

The name of this compound—cyclopropane—is derived from the name of the three-carbon alkane (propane) and the prefix "cyclo." Many cyclic hydrocarbons have useful properties. For example, cyclopropane is an anesthetic that is sometimes used in surgery. However, cyclopropane is very flammable and great precautions must be taken in the operating room when it is used.

Another example of a cyclic hydrocarbon is

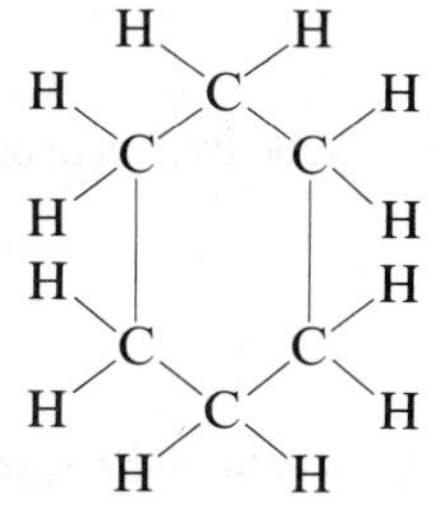

Cyclohexane

or C_6H_{12}

The name of this compound—cyclohexane—is derived from the name of the six-carbon alkane (hexane) and the prefix "cyclo." Cyclohexane is used to remove paint and varnish. To make the writing of cyclic hydrocarbons simpler, we usually do this:

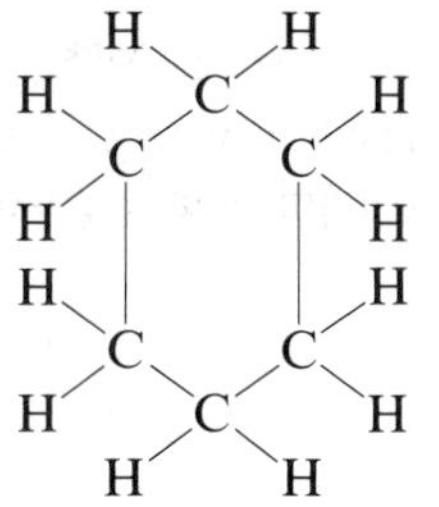

we write as

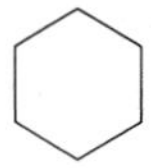

The shape of the figure tells you the name of the compound. In this example the figure is a hexagon (six sides). Each corner of the hexagon represents a CH_2 group.*

---

* In general, when writing cyclic hydrocarbons, each corner of the figure represents a C atom with the appropriate number of hydrogen atoms. Remember, each carbon atom can have only four bonds.

Example 12-3   Name the following cyclic hydrocarbons.

(a) ▢   (b)  ⬠

(c) 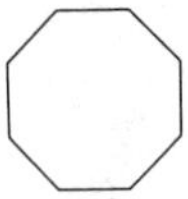 ⬡   (d) 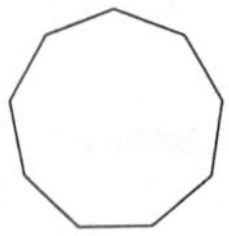 ⬣

Solution   (a) ▢ This is a cyclic hydrocarbon with four carbon atoms and is called cyclobutane.

(b) 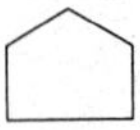 ⬠ This is a cyclic hydrocarbon with five carbon atoms and is called cyclopentane.

(c) 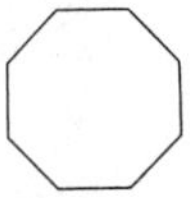 ⬡ This is a cyclic hydrocarbon with eight carbon atoms and is called cyclooctane.

(d) 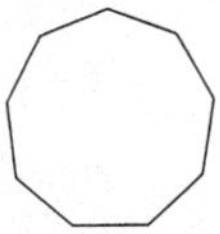 ⬣ This is a cyclic hydrocarbon with nine carbon atoms and is called cyclononane.

Some cyclic hydrocarbons have a double or triple bond: for example,

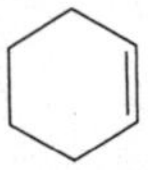

Cyclohexene

This six-carbon cyclic hydrocarbon with a double bond is called cyclohex*ene* and is used as a stabilizer for high-octane gasolines. And the very interesting compound

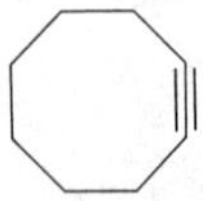

Cyclooctyne

which is called cyclooct*yne*, has no practical uses as yet.

Aromatic Hydrocarbons: They Don't All Smell Sweet

One series of cyclic hydrocarbons has alternating single and double bonds. The members of this series are called *aromatic hydrocarbons*, because early organic chemists noticed that these compounds had pleasant smells. (However, later organic chemists found that some of these compounds were foul smelling!) The parent compound of the aromatic hydrocarbons is benzene, C_6H_6, whose structure is

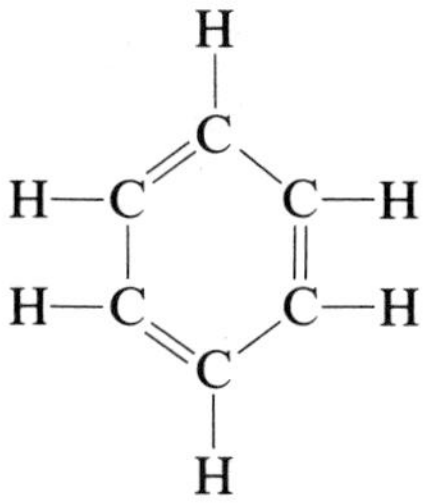

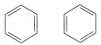

Benzene

Although only three of the carbon–carbon bonds in benzene are written as double bonds, in reality all the bonds seem to be equivalent. At one time chemists thought that the double bonds switched back and forth between the carbon atoms. If that were true, either of these two diagrams would give a valid picture of the benzene molecule:

Today, chemists represent the structure of benzene using any one of the following representations:

The third structure is meant to show that all the carbon–carbon bonds in benzene are equivalent, which is known to be true.

Benzene, which can be obtained from coal tar, is one of the most important organic compounds. It is the starting material for countless products that make life more comfortable and pleasant, including many pesticides, pharmaceuticals, and synthetic fibers such as nylon, to say nothing of perfumes and synthetic dyes.

Other aromatic hydrocarbons seem to be made of benzene rings joined together (Fig. 12-7). As you might have guessed, there are many compounds that consist of these aromatic compounds with various side chains. An example is

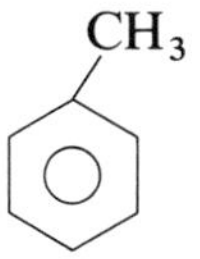

Toluene

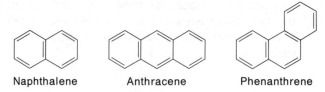

Naphthalene   Anthracene   Phenanthrene

Figure 12-7 Some aromatic hydrocarbons.

One name for this compound is methylbenzene, but organic chemists usually call it *toluene*. It is used in making explosives and dyes.

You'll be seeing a lot more of these aromatic compounds later in the chapter when we look at organic compounds other than hydrocarbons.

REACTIONS OF HYDROCARBONS

Hydrocarbon compounds undergo many different kinds of chemical reactions to produce some very interesting and useful products. Many of these products are very useful to us in our daily lives. Let's briefly examine some of the reactions of hydrocarbons.

Alkanes

Alkanes are a fairly inert group of organic compounds. However, they do react with oxygen in a process known as *combustion*. The resulting products of hydrocarbon combustion are carbon dioxide and water. For example, in the burning of natural gas, methane, the reaction looks something like this:

$$CH_4 + 2\,O_2 \longrightarrow CO_2 + 2\,H_2O$$

Now you can see why people who spend all day cooking with their gas oven turned on should also keep a window open. If they don't, the oxygen content in the kitchen may drop and the carbon dioxide level increase, causing the individuals to become dizzy or drowsy. (By the way, incomplete combustion can also result in poisonous carbon monoxide, CO, also forming.)

Alkenes

Alkenes are more reactive than alkanes because these hydrocarbons have double bonds (they're unsaturated). The carbon atoms joined by double bonds can react with other substances in what is called an *addition reaction*. For example, an alkene can react with hydrogen gas to form an alkane. This process is known as *hydrogenation*.

$$H_2C{=}CH_2 + H_2 \longrightarrow H_3C{-}CH_3 \quad \text{(hydrogenation)}$$

Ethene   Ethane

Hydrogenation is used to convert vegetable oils that contain double bonds in their molecules into solid shortenings such as Spry and Crisco.

Another type of addition reaction involves the reaction of an alkene with water to produce another type of organic compound, known as an *alcohol*. This type of reaction is known as *hydration*.

$$H_2C{=}CH_2 + H{-}OH \longrightarrow CH_3CH_2{-}OH \quad \text{(hydration)}$$

Ethene   Ethanol

We'll learn more about alcohols later in the chapter.

Alkynes

Alkynes are very reactive hydrocarbons. Like alkenes, they undergo addition reactions. In the case of hydrogenation, four atoms of hydrogen are added to the molecule.

$$HC{\equiv}CH + 2H_2 \longrightarrow CH_3{-}CH_3 \quad \text{(hydrogenation)}$$

Ethyne   Ethane

Cycloalkanes

Cyclic hydrocarbons are also reactive. This is especially true in the case of the smaller molecules, where bond angles are strained. For example, the hydrogenation of cyclopropane yields *n*-propane (the straight-chain propane).

$$\begin{array}{c} CH_2 \\ \diagup \quad \diagdown \\ CH_2{-}CH_2 \end{array} + H_2 \longrightarrow CH_3CH_2CH_3$$

Cyclopropane   *n*-Propane

Cyclopropane also reacts very readily with oxygen in an explosive reaction. All that's necessary is a spark.

$$2 \begin{array}{c} CH_2 \\ \diagup \quad \diagdown \\ CH_2{-}CH_2 \end{array} + 9O_2 \xrightarrow{\text{spark}} 6CO_2 + 6H_2O$$

We mention this because cyclopropane, a sweet-smelling gas, is used in hospitals as an anesthetic. Therefore, it's very important that precautions be taken to avoid sparks or static electricity in an operating room where this material is being used.

The list of organic reactions goes on and on, but we will not. Our purpose in the preceding pages was to give you a brief introduction into the reactions of hydrocarbons so that you could learn about some of the interesting reactions that they undergo.

In Part 1 of the chapter we studied only hydrocarbon compounds. However, there are many classes of organic compounds. We can write the general formula of these compounds by using the R-group notation that we discussed earlier. Each class of organic compounds is denoted by a *functional group*. The functional group is what the R group is attached to: it is the reactive part of the molecule (Fig. 12-8). Table 12-4 lists some of the common classes of organic compounds, together with their general formulas.

Each class of compounds has very different properties and uses. In this part of the chapter we survey the various classes of organic compounds and learn about some of their properties and uses.

ALCOHOLS: THE —OH FUNCTIONAL GROUP

The general formula for an alcohol is

$$R—OH$$

The —OH is called the functional group. The R is usually a chain of carbon atoms. If the R group consists of only a few carbon atoms, the alcohol tends

Figure 12-8 The functional group is what the R group is attached to; it is the reactive part of the molecule.

TABLE 12-4 *Some common classes of organic compounds*

CLASS	GENERAL FORMULA[a]	EXAMPLE	COMMON NAME	IUPAC NAME
Alkane	C_nH_{2n+2}	CH_4	Methane (swamp gas)	Methane
Alkene	C_nH_{2n} (n = 2 or more)	$CH_2{=}CH_2$	Ethylene	Ethene
		$CH_3{-}CH{=}CH_2$	Propylene	Propene
Alkyne	C_nH_{2n-2} (n = 2 or more)	$H{-}C{\equiv}C{-}H$	Acetylene	Ethyne
		$CH_3{-}C{\equiv}C{-}H$	Methylacetylene	Propyne
Alcohol	$R{-}OH$	$CH_3{-}OH$	Methyl alcohol (wood alcohol)	Methanol
		$CH_3{-}CH_2{-}OH$	Ethyl alcohol (grain alcohol)	Ethanol
Ether	$R{-}O{-}R$	$CH_3{-}O{-}CH_3$	Dimethyl ether	Methoxymethane
		$CH_3{-}CH_2{-}O{-}CH_2{-}CH_3$	Diethyl ether	Ethoxyethane
Aldehyde	$R{-}\overset{\displaystyle }{C}{=}O$ with H below	$H{-}C{=}O$ with H below	Formaldehyde	Methanal
		$CH_3{-}CH_2{-}C{=}O$ with H below	Propionaldehyde	Propanal
Ketone	$R{-}\underset{O}{\overset{\|}{C}}{-}R$	$CH_3{-}\underset{O}{\overset{\|}{C}}{-}CH_3$	Dimethyl ketone (acetone)	Propanone
		$CH_3{-}CH_2{-}\underset{O}{\overset{\|}{C}}{-}CH_2{-}CH_3$	Diethyl ketone	3-Pentanone
Carboxylic acid	$R{-}\underset{O}{\overset{\|}{C}}{-}OH$	$H{-}\underset{O}{\overset{\|}{C}}{-}OH$	Formic acid	Methanoic acid
		$CH_3{-}\underset{O}{\overset{\|}{C}}{-}OH$	Acetic acid	Ethanoic acid
Ester	$R{-}\underset{O}{\overset{\|}{C}}{-}O{-}R$	$H{-}\underset{O}{\overset{\|}{C}}{-}O{-}CH_3$	Methyl formate	Methylmethanoate
		$CH_3{-}\underset{O}{\overset{\|}{C}}{-}O{-}CH_2{-}CH_3$	Ethyl acetate	Ethylethanoate
Amine	$R{-}NH_2$	$CH_3{-}NH_2$	Methylamine	Aminomethane
		$CH_3{-}CH_2{-}NH_2$	Ethylamine	Aminoethane
Amides	$R{-}\underset{O}{\overset{\|}{C}}{-}NH_2$	$CH_3{-}\underset{O}{\overset{\|}{C}}{-}NH_2$	Acetamide	Ethanamide
		$CH_3CH_2{-}\underset{O}{\overset{\|}{C}}{-}NH_2$	Propionamide	Propanamide

[a] Remember that n = number of carbon atoms.

Source: Alan Sherman, Sharon Sherman, and Leonard Russikoff, *Basic Concepts of Chemistry*, 2nd ed., Houghton Mifflin Company, Boston, 1980. By permission of the publisher.

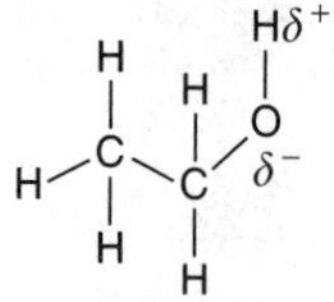

Figure 12-9 If the R group of an alcohol consists of only a few carbon atoms, the alcohol tends to be polar, as a result of the polarity of the —OH group.

to be polar, because of the polarity of the —OH group (Fig. 12-9). This allows the alcohol to dissolve in polar solvents such as water. For example, the alcohol CH_3CH_2OH, ethanol, dissolves in water. However, if the R group consists of a long chain of carbon atoms, the polarity of the —OH group is less pronounced and the alcohol appears to be nonpolar (Fig. 12-10). The alcohol will dissolve in nonpolar solvents such as benzene and carbon tetrachloride. For example, the alcohol $CH_3CH_2CH_2CH_2CH_2CH_2OH$, hexanol, dissolves well in benzene.

Some Important Alcohols and Their Uses

Methanol, CH_3OH, whose common name is *methyl alcohol*, is the simplest alcohol. This very poisonous compound is also known as wood alcohol because it can be obtained from the burning of wood in the absence of air (a process called destructive distillation). Death can result from drinking just 30 ml of it: smaller amounts can produce nausea, convulsions, blindness, and respiratory failure. During prohibition in this country, many people drank beverages containing methanol, not realizing the dangers involved.

Ethanol, CH_3CH_2OH, whose common name is *ethyl alcohol*, is the alcohol we all consume in our favorite wine, liqueur, or beer. The ethanol found in these beverages is produced through the process of fermentation, which involves the breakdown of the sugar glucose.

$$C_6H_{12}O_6 \xrightarrow{\text{yeast}} 2CH_3CH_2OH + 2CO_2 + \text{energy}$$
Glucose

A solution that is 70% ethanol by volume acts as an antiseptic by coagulating the proteins of bacteria. Ethanol is also used to make *tinctures*, which are medicinal substances dissolved in this alcohol. For example, tincture

Figure 12-10 If the R group of an alcohol consists of a long chain of carbon atoms, the alcohol tends to be nonpolar, because the compound is mostly an alkane chain and behaves like a nonpolar alkane.

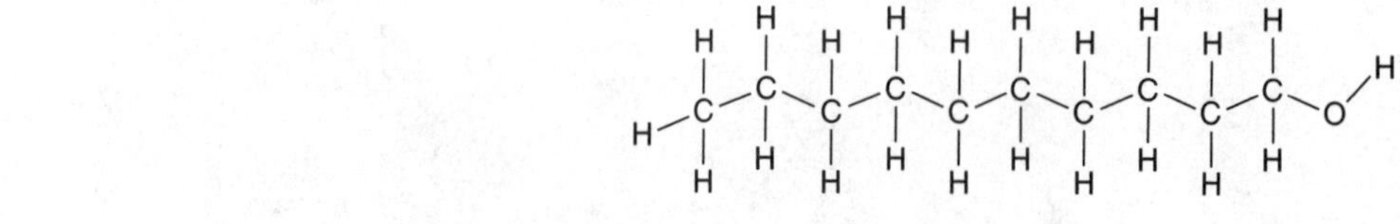

of iodine consists of 2% I_2, 2.4% NaI in a 50% weight/volume alcohol water solution.

2-Propanol, CH_3CHCH_3, whose common name is *isopropyl alcohol*, is
$$\underset{\displaystyle OH}{|}$$
used to give sponge baths to reduce high fevers. It is for this reason that this substance is also called rubbing alcohol.

1,2,3-Propanetriol, $CH_2\!-\!CH\!-\!CH_2$, which is commonly known as
$$\underset{\displaystyle OH\quad OH\quad OH}{|\quad\ |\quad\ |}$$
glycerol or *glycerine*, is a trihydroxy alcohol (in other words, it has three OH groups). Glycerol is a very thick, sweet-tasting compound that is found as part of natural fats and oils. It is nontoxic, which makes it an excellent carrier for certain medicines. The cosmetics industry uses glycerol in the formulation of hand and skin creams because it acts as a good lubricant.

5-Methyl-2-isopropyl-1-cyclohexanol,

commonly called *menthol*, is a substituted cyclohexanol. This substance has many commercial uses, for example in cough drops, nasal inhalers, shaving creams, liqueurs, and cigarettes. Menthol is found naturally in peppermint oil. It has a cooling sensation when rubbed on the skin. When placed in cough drops, menthol causes the mucous membranes of the throat to increase their secretions. This has a soothing effect on inflamed tissues.

2,6-di-tert-Butyl-4-methyl-1-phenol,

commonly known as *butylated hydroxytoluene*, or BHT for short, is used in many food substances as an antioxidant. This means that it doesn't allow the food to spoil. BHT is also put into certain types of rubbers and plastics to prevent them from undergoing oxidation.

As you can see by these brief examples, alcohols are an important class of compounds. Some additional alcohols and their uses are listed in Table 12-5.

ETHERS: THE R—O—R COMPOUNDS

The general formula for an ether is

$$R—O—R'$$

TABLE 12-5 *Some alcohols and their uses*

NAME	STRUCTURE	USE
ortho-Cresol		Antiseptic and disinfectant (found in products such as Lysol)
meta-Cresol		Antiseptic and disinfectant
para-Cresol		Antiseptic and disinfectant
Diethylstilbestrol		Estrogenic hormone therapy (this compound has been linked to uterine cancer in females whose mothers received this compound during pregnancy)
Hexylresorcinal		Antiseptic and germicide
Terpin hydrate		An expectorant

If the R and R′ groups are the same, we say that the ether is symmetrical. If the R and R′ groups are different, we say that the ether is unsymmetrical.

Symmetrical ethers can be synthesized from alcohols in the presence of sulfuric acid. For example, the simplest ether, methoxymethane (dimethyl ether), can be synthesized as follows:

$$CH_3OH + HOCH_3 \xrightarrow{H_2SO_4} CH_3OCH_3 + H_2O$$

Methyl alcohol   Dimethyl ether

The sulfuric acid acts as a dehydrating agent. Two molecules of methyl alcohol condense to form one molecule of dimethyl ether.

Some Properties of Ethers

Because ethers have two alkyl groups bonded covalently to an oxygen atom, ethers tend to have little polarity. This means that they don't dissolve well in water. Also, unlike their corresponding alcohols, ethers cannot hydrogen-bond with other ether molecules. This means that they have much lower boiling points than the alcohols from which they are derived. Ethers tend to be very flammable, and many of them have some type of anesthetic properties.

Some Important Ethers and Their Uses

Ethoxyethane, $CH_3CH_2OCH_2CH_3$, also called diethyl ether or sometimes just "ether," is probably the best known to medical personnel. It is used in hospitals as an anesthetic for long surgical procedures (Fig. 12-11). It is an extremely safe anesthetic because there is a large difference in dosage between that needed to anesthetize and that needed to kill. Also, it has little effect on blood pressure or respiration. However, great care must be taken in the operating room because of the high flammability of diethyl ether.

Figure 12-11   The Hinckley painting depicting the first public demonstration of the use of ether as a surgical anesthetic. (Massachusetts General Hospital)

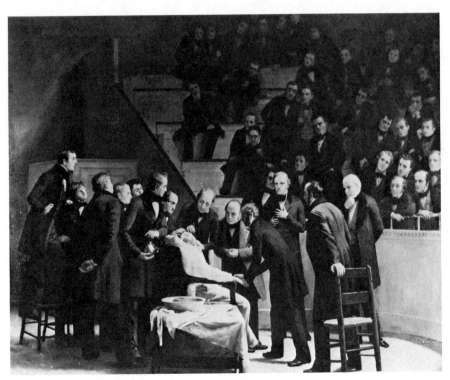

TABLE 12-6 *Some ethers and their uses*

NAME	STRUCTURE	USE
Furazolidone		Antimicrobial
Divinyl ether	$CH_2=CH-O-CH=CH_2$	Quick-acting anesthetic
Mephenesin		Skeletal muscle relaxant

Ethylene oxide,

$$\begin{array}{c} O \\ / \backslash \\ CH_2-CH_2 \end{array}$$

which is a gas at room temperature, is a cyclic ether, used as a fumigant (disinfectant) for foodstuffs and as a fungicide in agriculture. It is also used to sterilize surgical instruments.

Some additional ethers and their uses are listed in Table 12-6.

ALDEHYDES: THE $-\overset{\overset{\displaystyle O}{\|}}{C}-H$ FUNCTIONAL GROUP

The general formula for an aldehyde is

$$R-\overset{\overset{\displaystyle}{\|}}{\underset{O}{C}}-H$$

The $-\overset{\|}{\underset{O}{C}}-H$ is called the *aldehyde group*. Many aldehydes have attractive odors and are used in perfumes. And some aldehydes have pleasant tastes and are used in flavoring agents. The $-\overset{\|}{\underset{O}{C}}-H$ group makes aldehydes chemically reactive. For example, many aldehydes can undergo the process of reduction and be converted to alcohols. For example,

$$CH_3CH_2\overset{\|}{\underset{O}{C}}H + H_2 \xrightarrow{Pt} CH_3CH_2CH_2OH$$

Propan*al*   Propan*ol*

Aldehydes can also undergo the process of oxidation and be converted to carboxylic acids. For example,

$$CH_3CH_2\underset{\underset{O}{\|}}{C}H \xrightarrow{KMnO_4} CH_3CH_2\underset{\underset{O}{\|}}{C}-OH$$

Propanal   Propanoic acid

Some Important Aldehydes and Their Uses

Methanal, $H-\underset{\underset{O}{\|}}{C}-H$, commonly called *formaldehyde*, is the simplest and best known aldehyde. Although it is a gas at room temperature, it is usually dissolved in water and commonly sold in a 40% weight/volume solution. In this form it is used as a germicide, disinfectant, and preservative, and is commonly called *formalin*. Biologists use formalin solution to preserve tissues of various organisms. Many of you are probably familiar with the pungent odor of this compound. Formaldehyde is also very irritating to your eyes, nose, and throat.

3-Methoxy-4-hydroxybenzaldehyde,

better known as vanillin, is used both commercially and medically as a flavoring agent. It has a very pleasant odor and taste that can cover up the harsh taste of many medicines. The vanilla taste in vanilla ice cream is due to the compound vanillin.

Some additional aldehydes and their uses are listed in Table 12-7.

TABLE 12-7 *Some aldehydes and their uses*

NAME	STRUCTURE	USE
Benzaldehyde		Almond flavoring
Cinnamalaldehyde		Cinnamon flavoring
Propionaldehyde	$CH_3CH_2-\underset{\underset{O}{\|}}{C}H$	Disinfectant

KETONES: THE R—C—R′ COMPOUNDS

(with O double-bonded to C above)

The general formula for a ketone is

$$R\text{—}\underset{\underset{O}{\|}}{C}\text{—}R'$$

The —C— group (with ‖O below) is called the *carbonyl* or *keto group*. Ketones are very similar in structure to aldehydes. Ketones, however, have two R groups attached to the carbonyl group, whereas aldehydes have only one.

$$R\text{—}\underset{\underset{O}{\|}}{C}\text{—}R' \qquad R\text{—}\underset{\underset{O}{\|}}{C}\text{—}H$$

               Ketone   Aldehyde

Some Important Ketones and Their Uses

Propanone, $CH_3\text{—}\underset{\underset{O}{\|}}{C}\text{—}CH_3$, also called *dimethyl ketone* or *acetone*, is

the simplest and best known ketone. Acetone is an excellent solvent for many organic substances. It is found in paint removers and in nail polish removers. However, small amounts of acetone are also found in the blood, where it is called a *ketone body*. Large amounts of acetone in the blood indicate a body malfunction in the breakdown of fats. This is usually a sign of diabetes mellitus. Acetone has a very unique odor and sweetish taste. An individual suffering from uncontrolled diabetes mellitus will give off an acetone smell when he or she exhales. This is called *acetone breath*, which a physician can easily detect in an examination of such an individual.

Progesterone,

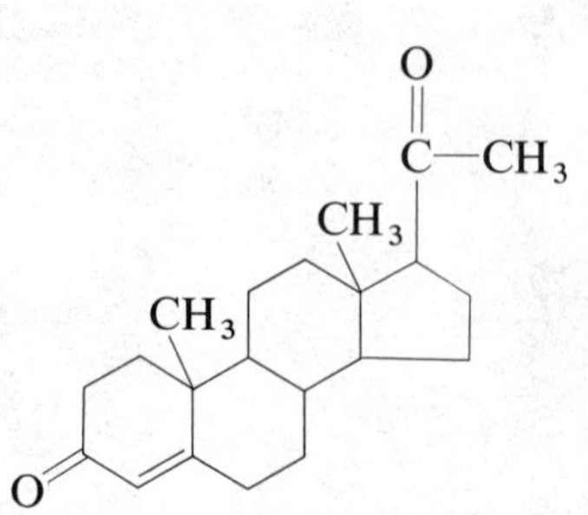

also called the *pregnancy hormone*, is a diketone. This substance is one of the major female hormones. It is secreted during the second half of the menstrual

TABLE 12-8 *Some ketones and their uses*

NAME	STRUCTURE	USE
Camphor		Anesthetic and mild antiseptic
Carvone		Oil of spearmint
Testosterone		Male sex hormone

cycle and prepares the uterine lining to receive the embryo. Progesterone will also prevent ovulation if it is given during the fifth to twenty-fifth days of the menstrual cycle. This is, of course, the basis of oral contraceptives.

Some additional ketones and their uses are listed in Table 12-8.

CARBOXYLIC ACIDS: THE ORGANIC ACIDS WITH THAT $-\overset{\displaystyle O}{\overset{\displaystyle \|}{C}}-OH$ GROUP

The general formula for a carboxylic acid is

$$R-\overset{}{\underset{\underset{O}{\|}}{C}}-OH$$

The $-\overset{}{\underset{\underset{O}{\|}}{C}}-OH$ is called the *carboxyl group*. This group acts like an acid because it releases hydrogen ions in aqueous solutions.

$$R-\overset{}{\underset{\underset{O}{\|}}{C}}-OH + H_2O \; \rightleftharpoons \; R-\overset{}{\underset{\underset{O}{\|}}{C}}-O^- + H_3O^+$$

The ionization occurs only to a slight degree, so carboxylic acids are weak acids. Some carboxylic acids are found in fats and are called *fatty acids*. We discuss fatty acids in more detail when we discuss lipids in Chapter 15. Still

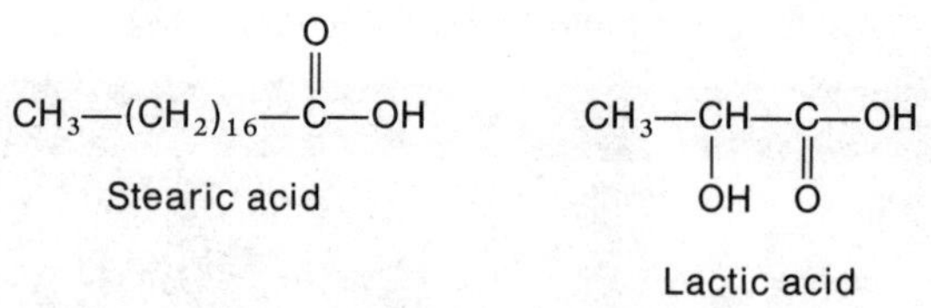

$$CH_3-(CH_2)_{16}-\overset{\displaystyle O}{\overset{\|}{C}}-OH \qquad\qquad CH_3-\underset{\underset{\displaystyle OH}{|}}{CH}-\overset{\displaystyle O}{\overset{\|}{C}}-OH$$

Stearic acid

Lactic acid

Figure 12-12   Stearic acid is a fatty acid used to make soaps, and lactic acid is a hydroxy acid found in sour milk.

other carboxylic acids contain an extra hydroxyl group and are called *hydroxy acids*. These hydroxy acids play an important part in certain metabolic reactions. Figure 12-12 shows an example of a fatty acid and a hydroxy acid.

Some Important Carboxylic Acids and Their Uses

Methanoic acid, $H-\overset{\|}{\underset{O}{C}}-OH$ whose common name is *formic acid*, is the simplest of the carboxylic acids. It is the acid that is produced by ants and bees and is injected into their victims when they bite or sting. If you were ever stung by a bee, you may be interested in knowing that it was formic acid that caused your skin to become inflamed.

By the way, it is formic acid that is formed by the liver when an individual consumes methanol.

$$CH_3-OH \xrightarrow[\text{by liver}]{\text{oxidation}} H-\overset{\|}{\underset{O}{C}}-OH$$

The formic acid, once formed, can cause severe disruptions in the body's blood chemistry, which can lead to death.

Ethanoic acid, $CH_3-\overset{\|}{\underset{O}{C}}-OH$, also known as *acetic acid*, is a substance found in vinegar. Acetic acid plays an important role in living cells, where it helps to produce long-chain fatty acids (see Chap. 15).

Butanoic acid, $CH_3CH_2CH_2-\overset{\|}{\underset{O}{C}}-OH$, also known as *butyric acid*, is produced when butter turns rancid. In its pure form it has a very pungent odor. It is also one of the substances that causes body odor.

2-Hydroxypropanoic acid, $CH_3CH-\overset{\|}{\underset{O}{C}}-OH$ with OH, better known as *lactic acid*, is a hydroxy acid. It is produced in your muscles when you exercise. Lactic acid is formed in milk when it turns sour, as a result of the action of lactic acid bacteria.

TABLE 12-9 *Some carboxylic acids and their uses*

NAME	STRUCTURE	USE
Citric acid	$O=C-OH$ $HO-C-CH_2-C-CH_2-C-OH$ $\quad\ \ \, O \qquad\ \ \, OH \qquad\ \ O$	The taste of citrus fruits, also used for rickets and as a mild astringent
Lactic acid	$CH_3-CH-C-OH$ $\qquad\ \ OH\ \ O$	Found in sour milk
Nicotinic acid	(pyridine ring with) $C-OH$ over O	Part of the essential B vitamins; also used as a vasodilator
Salicylic acid	(benzene ring with) $C{\diagup}^{OH}$ over O and OH	Used extensively as an antimicrobial agent; also used to make aspirin
Stearic acid	$CH_3-(CH_2)_{16}-C-OH$ $\qquad\qquad\qquad\qquad\ O$	Used for suppositories and for coating various types of pills, especially those which have a bitter taste; also used to make soap

Benzoic acid,

$$\underset{\text{(benzene ring)}}{\overset{O}{\underset{\|}{C}}-OH}$$

which can be thought of as benzenecarboxylic acid, is the simplest aromatic acid. It is used an an antiseptic and antifungal agent, as well as a preservative. Other acids related to benzoic acid have numerous practical uses.

Some additional carboxylic acids and their uses are listed in Table 12-9.

ESTERS: THE $R-\overset{O}{\overset{\|}{C}}-OR'$ COMPOUNDS

The general formula for an ester is

$$R-\overset{}{\underset{\|}{\overset{}{C}}}-OR'$$
$$\quad\ \ O$$

Esters can be formed by the reaction of an alcohol with a carboxylic acid. For example,

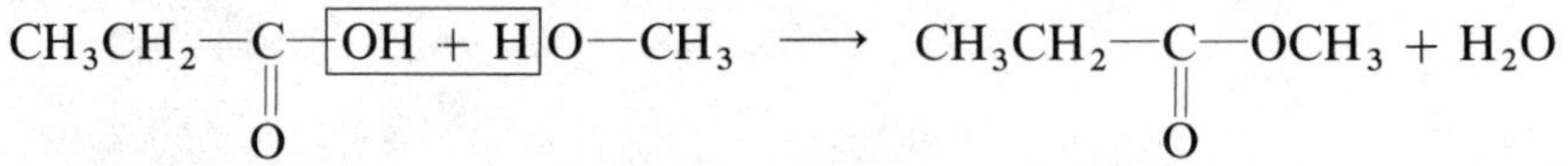

$$CH_3CH_2-\underset{\underset{O}{\|}}{C}\boxed{OH + H}O-CH_3 \longrightarrow CH_3CH_2-\underset{\underset{O}{\|}}{C}-OCH_3 + H_2O$$

Propanoic acid   Methanol   An ester

Notice how the acid and alcohol react to produce the ester.

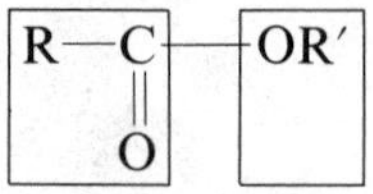

$$\boxed{R-\underset{\underset{O}{\|}}{C}}\boxed{OR'}$$

Acid   Alcohol
part   part

Water is also produced in this reaction, which is called *esterification*. Some of the most common flavors and fragrances are due to esters (Table 12-10).

Some Important Esters and Their Uses

    Methyl salicylate,

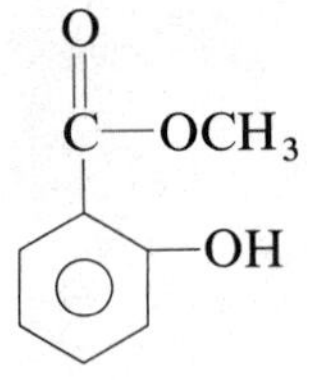

commonly known as *oil of wintergreen*, is the methyl ester of salicylic acid,

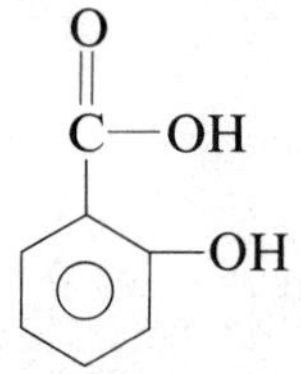

It has a pleasant spearmintlike odor, and for this reason is used (in very low concentrations) in candies and perfumes. Because it also creates a mild burning sensation on your skin, which makes sore muscles feel good, methyl salicylate is found in many liniments.

    Acetylsalicylic acid,

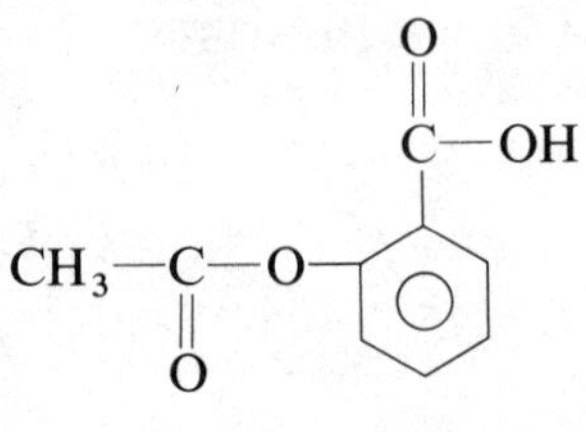

TABLE 12-10 *Some esters and their flavoring and fragrances*

NAME	FORMULA	FLAVOR OR FRAGRANCE
Amyl acetate	$CH_3-(CH_2)_4-O-\overset{\displaystyle O}{\underset{\|}{C}}-CH_3$	Banana
Amyl butyrate	$CH_3-(CH_2)_4-O-\overset{\displaystyle O}{\underset{\|}{C}}-CH_2CH_2CH_3$	Apricot
Ethyl butyrate	$CH_3CH_2-O-\overset{\displaystyle O}{\underset{\|}{C}}-CH_2CH_2CH_3$	Pineapple
Ethyl formate	$CH_3CH_2-O-\overset{\displaystyle O}{\underset{\|}{C}}-H$	Rum
Isoamyl acetate	$CH_3-\underset{\underset{\textstyle CH_3}{\|}}{CH}-CH_2CH_2-O-\overset{\displaystyle O}{\underset{\|}{C}}-CH_3$	Pear
Isobutyl formate	$CH_3-\underset{\underset{\textstyle CH_3}{\|}}{CH}-CH_2-O-\overset{\displaystyle O}{\underset{\|}{C}}-H$	Raspberry
Octyl acetate	$CH_3-(CH_2)_7-O-\overset{\displaystyle O}{\underset{\|}{C}}-CH_3$	Orange

TABLE 12-11 *Some esters and their uses*

NAME	STRUCTURE	USE
m-Cresyl acetate	(benzene ring with $\overset{\underset{\|}{O}}{C}-OCH_3$ and CH_3 substituent)	Topical antiseptic and fungicide
Methylparaben	(benzene ring with $\overset{\underset{\|}{O}}{C}-OCH_3$ and OH substituent)	Used as a preservative for certain drugs
Nitroglycerin	H_2C-ONO_2 $HC-ONO_2$ H_2C-ONO_2	Nitro ester, used as a vasodilator for certain heart conditions

more commonly known as aspirin, is probably the best known and safest pain reliever and fever reducer. If you look closely at the structure of this compound, you'll see that it is the ester formed from acetic acid and salicylic acid.

Some additional esters and their uses are listed in Table 12-11.

AMINES: THE —$\overset{|}{N}$— COMPOUNDS

There are three general formulas for amines:

1. R—NH_2 (primary amine)
2. R—$\underset{R'}{\overset{|}{N}H}$ (secondary amine)
3. R—$\underset{R'}{\overset{|}{N}}$—R'' (tertiary amine)

Amines are related to the compound ammonia, NH_3, except that the hydrogen atoms in the ammonia molecule have been replaced by R groups. If one R group has replaced a hydrogen atom, the compound is a primary amine. If two R groups have replaced two hydrogen atoms, the compound is a secondary amine. And if three R groups have replaced three hydrogen atoms, the compound is a tertiary amine.

Amines in general have very strong ammonia-like odors. Many of them act as organic bases.

$$RNH_2 + H_2O \rightleftharpoons RNH_3^+ OH^-$$

Amines are produced when living organisms decay. And amines are found in our bodies as amino acids. (Amino acids are compounds that have both an amine group and a carboxylic acid group. For example, H_2N—$\underset{CH_3}{\overset{|}{C}H}$—$\overset{\overset{O}{||}}{C}$—OH

is the amino acid alanine. We have more to say about amino acids when we discuss proteins in Chap. 16.)

Some Important Amines and Their Uses

1,5-Diaminopentane, H_2N—$CH_2CH_2CH_2CH_2CH_2$—NH_2, also called 1,5-pentanediamine but better known as *cadaverine*, is a compound produced by decaying organisms. Such amines are called *ptomaines* and, in fact, can also be formed by the action of bacteria on meat and fish (thus the name ptomaine poisoning). Cadaverine has a very pungent odor; it also acts as a very strong base.

4-Aminobenzoic acid,

$$
\begin{array}{c}
O \\
\parallel \\
C-OH \\
| \\
\text{(benzene ring)} \\
| \\
NH_2
\end{array}
$$

also known as *para-aminobenzoic acid,* or *PABA* for short, is used today in suntan lotions as a sunscreen (in other words, it prevents ultraviolet light from reaching the skin). PABA has also been used as an antirickettsial agent (rickettsia are bacterialike organisms that cause diseases such as Rocky Mountain spotted fever). PABA is also found in foods as part of the B complex vitamins. For example, baker's yeast contains 5 to 6 parts per million PABA, and brewer's yeast contains from 10 to 100 parts per million.

1-Phenyl-2-aminopropane,

$$
\text{(benzene ring)} - CH_2-CH-CH_3 \\
 | \\
 NH_2
$$

also called (*phenylisopropyl*)*amine,* but better known as *amphetamine* (or *benzedrine*), acts as a central nervous system stimulant. This compound and others related to it are called "uppers," because of their ability to keep you active and awake. Unfortunately, they can cause drug dependence, which means that after you've been taking them for a while, you need to continue to take them to stay alert. Amphetamine has also been used in nasal inhalers to relieve congestion in the nose.

Some additional amines and their uses can be found in Table 12-12.

AMIDES: THE $R-\overset{\overset{\displaystyle O}{\parallel}}{C}-\overset{|}{N}-$ COMPOUNDS

There are three general formulas for amides:

1. $R-\underset{\underset{\displaystyle O}{\parallel}}{C}-NH_2$

2. $R-\underset{\underset{\displaystyle O}{\parallel}}{C}-\underset{\underset{\displaystyle R'}{|}}{N}H$

TABLE 12-12 *Some amines and their uses*

NAME	STRUCTURE	USE
Ephedrine	$\bigcirc$—CH—CH—NH—CH$_3$ with OH and CH$_3$ below	Used as a vasoconstrictor, it also opens stuffed nasal passages
Methadone	CH$_3$CH$_2$—C—C—CH$_2$—CH—N—CH$_3$ (with phenyl groups, O, CH$_3$, CH$_3$)	Substitute for heroin in the treatment of heroin addiction
Piperazine	piperazine ring with N—H groups	Kills pinworms and roundworms in many types of domestic animals

3. $\text{R}-\underset{\underset{\text{O}}{\|}}{\text{C}}-\underset{\underset{\text{R}'}{|}}{\text{N}}-\text{R}''$

The first example represents the formula for simple amides. The only R group is off the carbon atom. This type of amide can be thought of as being analogous to a carboxylic acid, with the OH group replaced by an NH$_2$ group.

$$\text{R}-\underset{\underset{\text{O}}{\|}}{\text{C}}-\text{OH} \qquad \text{R}-\underset{\underset{\text{O}}{\|}}{\text{C}}-\text{NH}_2$$

Acid   Amide

The second and third types of amides are more complex, in that they have one or two R groups off the nitrogen atom.

Amides can be formed by the reaction of a carboxylic acid with an amine (or ammonia). The general formula for such a reaction is

$$\text{R}-\underset{\underset{\text{O}}{\|}}{\text{C}}-\text{OH} + \text{H}-\underset{\underset{\text{R}'}{|}}{\text{N}}-\text{R}'' \longrightarrow \text{R}-\underset{\underset{\text{O}}{\|}}{\text{C}}-\underset{\underset{\text{R}'}{|}}{\text{N}}-\text{R}'' + \text{H}_2\text{O}$$

An understanding of the amide group is very important in the study of proteins. Proteins are composed of amino acids, connected by an amide linkage (called

a *peptide bond*). For example,

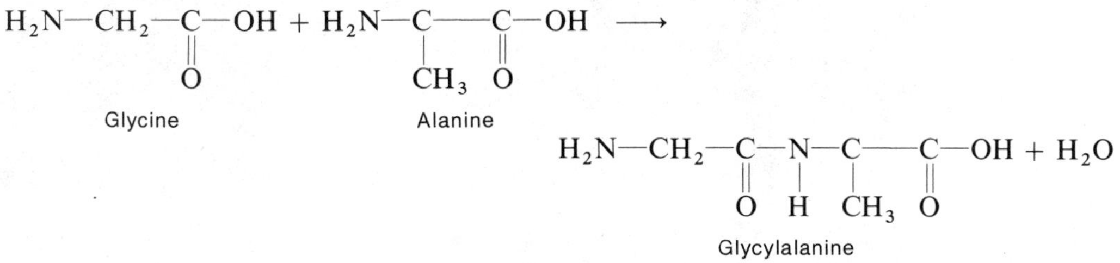

$$H_2N-CH_2-\underset{\underset{O}{\|}}{C}-OH + H_2N-\underset{\underset{CH_3}{|}}{C}-\underset{\underset{O}{\|}}{C}-OH \longrightarrow$$

Glycine   Alanine

$$H_2N-CH_2-\underset{\underset{O}{\|}}{C}-\underset{\underset{H}{|}}{N}-\underset{\underset{CH_3}{|}}{C}-\underset{\underset{O}{\|}}{C}-OH + H_2O$$

Glycylalanine

Protein bonds hold the amino acids together to form the long protein chains.

Some Important Amides and Their Uses

Carbamide, $H_2N-\overset{\overset{O}{\|}}{C}-NH_2$, better known as urea, is not a simple amide, but a diamide. Urea is the major end product of nitrogen metabolism in the body.

Para-ethoxy-N-phenylacetamide,

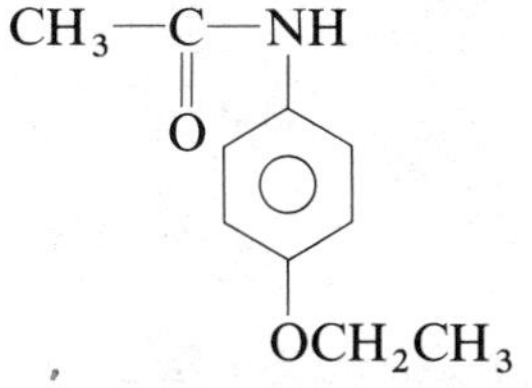

$$CH_3-\underset{\underset{O}{\|}}{C}-NH$$

OCH$_2$CH$_3$

also called *para-ethoxyacetanilide* but better known as *phenacetin*, is a pain reliever and fever reducer. This compound has few side effects and is used as an aspirin substitute.

N,N-Diethyllysergamide,

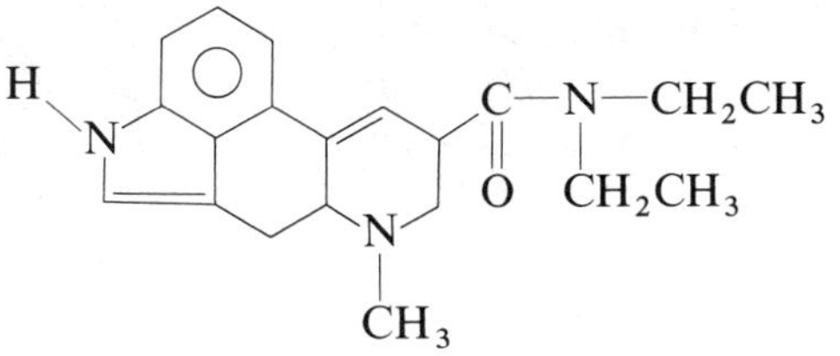

$$C-N-CH_2CH_3$$
$$O \quad CH_2CH_3$$

CH$_3$

better known as *lysergic acid diethylamide* (LSD, for short), has been used in the study and treatment of mental disorders. LSD gained notoriety in the 1960's and 1970's as a powerful hallucinogen. It appears that this drug can block the action of a chemical in the brain called serotonin, which controls nerve impulses in the brain. When the function of serotonin is blocked, nerve

293 *Organic Chemistry*

TABLE 12-13 *Some amides and their uses*

NAME	STRUCTURE	USE
Acetominophen	A benzene ring with OH at bottom and $NH-\overset{\overset{\displaystyle O}{\|}}{C}-CH_3$ at top	Aspirin substitute
Lidocaine	A benzene ring with CH_3 groups, $NH-\overset{\overset{\displaystyle}{\|}}{\underset{\underset{\displaystyle}{\|}}{C}}-CH_2-N-CH_2CH_3$ with O and CH_2-CH_3	Local anesthetic
Oxanamide	$CH_3-CH_2-CH_2-CH-\overset{\overset{\displaystyle CH_2CH_3}{\|}}{C}-\overset{\overset{\displaystyle}{\|}}{\underset{\underset{\displaystyle O}{\|}}{C}}-NH_2$ with epoxide O	Tranquilizer
Xylocaine	$\overset{\displaystyle CH_3CH_2}{\underset{\displaystyle CH_3CH_2}{}}N-CH_2-\overset{\overset{\displaystyle}{\|}}{\underset{\underset{\displaystyle O}{\|}}{C}}-NH-$ benzene ring with CH_3 groups	Local anesthetic

impulses in the brain become uncontrollable. This results in hallucinations and certain types of abnormal behavior.

Some additional amides and their uses can be found in Table 12-13.

SUMMARY

In this chapter we discussed the differences between organic and inorganic compounds. We looked at the properties of carbon and discovered why there are so many organic compounds. We learned how to write the structural formulas for organic compounds and how to tell the differences between alkane, alkene, alkyne, and cyclic hydrocarbons. We also studied isomers and practiced writing the structures of isomers given the molecular formula of an organic compound. We learned the meaning of the words saturated, unsaturated, and homologous series. We also looked at some of the reactions that hydrocarbons undergo.

In the second part of the chapter we looked at eight major functional groups of organic compounds. We studied a little about the chemistry of each group of compounds. We also surveyed some important members of each class of compounds and learned about their uses.

EXERCISES

1. In your own words, define an organic compound.

2. State the three-dimensional shape of a methane molecule, and then write the Lewis dot structure for methane.

3. Describe in terms of three-dimensional geometry what happens when
   (a) two carbon atoms form a single bond
   (b) two carbon atoms form a double bond
   (c) two carbon atoms form a triple bond

4. Give an example of
   (a) a saturated hydrocarbon   (b) an unsaturated hydrocarbon

5. Write the names and formulas for the first 10 straight-chain members of the alkane series. What is such a series called?

6. Write the following structural formulas in the more convenient form, sometimes called the condensed form.

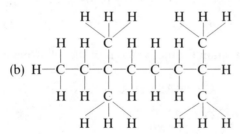

7. Write the structures for the nine isomers of heptane, C_7H_{16}.

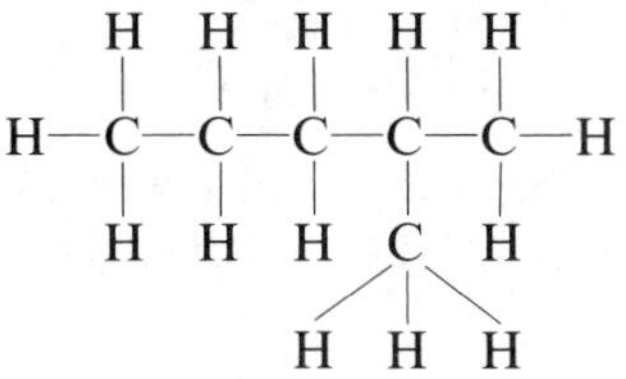

8. Rewrite each of the following structures by adding hydrogen atoms to the carbon skeleton.

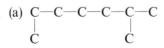   (b) C—C—C=C—C—C

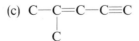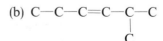

(c) C—C=C—C≡C

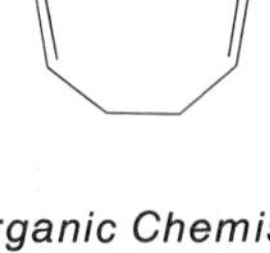

9. Name the following cyclic hydrocarbons.

(a)

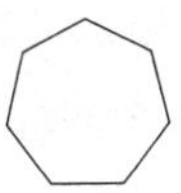

(b)

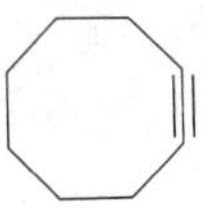

(c)

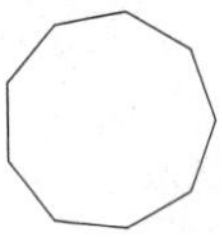

(d)

10. Write the products of the following hydrocarbon reactions.
    (a) $C_8H_{18} + O_2 \rightarrow$
    (b) $CH_3CH_2-CH=CHCH_3 + H_2 \rightarrow$
    (c) $CH\equiv C-CH_2CH_2CH_3 + H_2 \rightarrow$

11. Write the general formula for
    (a) an alcohol   (b) an ether
    (c) an aldehyde   (d) a ketone
    (e) a carboxylic acid   (f) an ester
    (g) an amine   (h) an amide

12. Write an equation for the synthesis of the following symmetrical ethers.
    (a) $CH_3CH_2OCH_2CH_3$ (b) $CH_3CH_2CH_2CH_2OCH_2CH_2CH_2CH_3$

13. Write an equation for the
    (a) reduction of CH_3CH (b) oxidation of CH_3CH
                  $\overset{\|}{O}$ $\overset{\|}{O}$

14. Write an equation for the synthesis of the following esters.

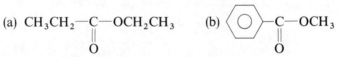

(a) $CH_3CH_2-\overset{\overset{\textstyle O}{\|}}{C}-OCH_2CH_3$ (b) $\langle O \rangle -\overset{\overset{\textstyle O}{\|}}{C}-OCH_3$

15. Write an equation for the synthesis of the following amides.
    (a) $CH_3-\overset{\overset{\textstyle O}{\|}}{C}-NH_2$ (b) $CH_3CH_2-\overset{\overset{\textstyle O}{\|}}{C}-\overset{\overset{\textstyle }{}}{\underset{\underset{\textstyle CH_3}{|}}{N}}H$

16. State whether each of the following amines is a primary, secondary, or tertiary amine.
    (a) $CH_3CH_2CH_2\underset{\underset{\textstyle NH_2}{|}}{C}HCH_2CH_3$ (b) $CH_3CH_2-\underset{\underset{\textstyle CH_2CH_3}{|}}{N}-CH_2CH_3$

    (c) 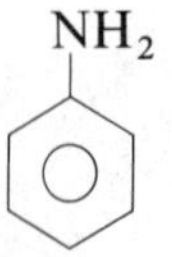 NH_2 (d) $CH_3CH_2CH_2CH_2-\underset{\underset{\textstyle CH_2CH_2CH_3}{|}}{N}H$

17. Give an example of each of the following types of organic compounds, and state some uses for each compound.
    (a) an alcohol   (b) an ether
    (c) an aldehyde   (d) a ketone
    (e) a carboxylic acid   (f) an ester
    (g) an amine   (h) an amide

18. Discuss the theory of *vitalism*. Why do you think it was such a strongly held theory during the 1700's and early 1800's?

Chapter 13

Plastics and Polymers

Some Things You Should Know After Reading This Chapter

You should be able to:

1. Explain how polymers are built up from smaller molecules.
2. Explain how crystalloids are different from colloids.
3. Explain the process of osmosis.
4. Discuss the importance of the discovery of nitrocellulose, pyroxylin, and celluloid.
5. Explain how celluloid film and cellulose acetate film played an important part in the development of photography.
6. Explain how condensation polymers differ from addition polymers.
7. State some of the properties and uses of Bakelite.
8. Write the structural formula of isoprene and show how it is related to other monomers used in the synthesis of synthetic rubberlike materials.
9. Explain the process of vulcanization, and explain why it's so important.
10. Define the following terms: copolymer, thermosetting polymer, and thermoplastic polymer.
11. Discuss some of the properties of nylon and explain the reason for the development of this polymer.
12. State some of the properties and uses of high- and low-density polyethylene.
13. Explain how PVC and polyacrylonitrile are similar to polyethylene in terms of their monomers.
14. Explain why polyester fabrics are used so extensively in clothing.
15. State five environmental problems associated with polymers.

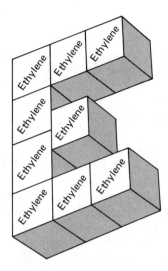

Figure 13-1 Polymers are never composed of single units but are made up of small building blocks. For example, polyethylene is made up of ethylene molecules bonded one to another.

INTRODUCTION

The organic substances that we studied in Chapter 12 were for the most part composed of single units containing not more than 50 or so atoms in their molecules. However, there exist organic substances whose molecules contain thousands or even millions of atoms. These molecules are never composed of single units but are made up of small "building blocks" (Fig. 13-1). Some of these giant* molecules, such as starch and proteins (see Chap. 16), are necessary for life processes; other giant molecules, such as natural rubber, have important commercial uses.

Some of these giant molecules, or *macromolecules* as they are called, have been used by our ancestors since their appearance on the planet. Some common examples of these naturally occurring macromolecules are wood, cotton, wool, rubber, and silk. In addition, foods such as proteins and starches are composed of these giant molecules. For centuries human beings used these molecules just as nature provided them. But beginning in the mid-1800's organic chemists turned their attention to the study of these molecules. Initial studies attempted to discover the properties of these molecules. Further research attempted to discover how these molecules were put together. As research continued into the 1900's, it soon became apparent that chemists would be able to synthesize these molecules in the laboratory and even improve on them. It wouldn't be

---

* The word "giant" is used here in a relative sense. These *giant* molecules are much larger than molecules such as H_2O, CH_4, and even ⬡. However, they are still too small to see with optical microscopes.

long before chemists would be able to tailor-make a macromolecule with a particular set of properties to meet a specific need. The development of a whole new class of compounds, which came to be called *plastics*, resulted from this research. Here's how it all began.

A SHORT HISTORY OF BIG MOLECULES

In the mid-1800's the Scottish chemist Thomas Graham devised an experiment to compare the rates of diffusion of different *dissolved* substances through a piece of parchment. (The parchment was thought to contain submicroscopic holes.) Graham found that certain *dissolved* substances, such as salt and sugar, would find their way through the parchment, but that other dissolved substances, such as glue and gelatin, would not. Graham called those materials that could pass through the parchment *crystalloids* (probably because he was able to obtain them as crystals after evaporation of the solvent). Those substances that did not pass through the parchment Graham called *colloids*. (*Glue*, which Graham determined was a colloid, is *kolla* in Greek; thus the name *colloid*.) The physical study of colloids, which Graham initiated, played an important part in the understanding of giant molecules.

In 1877, a German botanist named Wilhelm Pfeffer devised a technique to determine the molecular weights of colloidal substances in solution. Here's what he did. Pfeffer placed a sheet of parchment in a beaker in such a way that the beaker was divided into two compartments. In one compartment Pfeffer placed pure water. In the other compartment, Pfeffer placed a colloidal solution. The solutions in the beaker are allowed to remain undisturbed (Fig. 13-2). A process known as *osmosis* begins. In this process, water moves from an area of low *solute concentration* to an area of *high solute concentration*. (In Pfeffer's apparatus water moves to the side of the beaker containing the

Figure 13-2 In Pfeffer's experiment, pure water was placed in one compartment and a colloidal solution in the other.

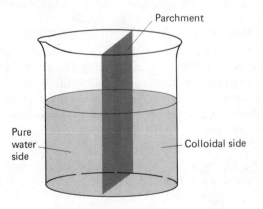

Parchment

Pure water side

Colloidal side

colloid.) This process occurs naturally and is nature's way of trying to equalize the *solute concentration* (colloid) on both sides of the compartment. However, in this situation, this result cannot be achieved, but the process of osmosis does eventually stop. This occurs when the number of water molecules passing from compartment A (the pure water side) to compartment B (the colloid side) becomes equal to the number of water molecules passing from side B to side A. In other words, a dynamic equilibrium results. But why does this occur? The answer lies in looking at the heights of the water in both compartments. The height of the water in compartment B is greater than that in A (Fig. 13-3). At first, water molecules move from compartment A to compartment B in an attempt to equalize the concentration of the two solutions. But eventually the height of water in compartment B causes a pressure to be exerted on the water molecules in that compartment, pushing them back into compartment A. When these two opposing forces become equal, the process of osmosis stops. The pressure exerted by the water in compartment B at this point is called the *osmotic pressure*. Knowing the osmotic pressure and the original concentration of the colloid particles, Pfeffer was able to determine the molecular weight of the colloid in compartment B. By performing the experiment with various colloids, Pfeffer was able to show that the molecular weights of these molecules were in the thousands of atomic mass units. (Later experiments showed that some of these giant molecules could have molecular weights in the millions.)

In 1907, the Swedish chemist Arne Tiselius devised a method for separating giant molecules in solution using an electric current. This technique,

Figure 13-3 The process of osmosis stops when the number of water molecules moving from compartment A to compartment B equals the number of water molecules moving from compartment B to compartment A. Osmotic pressure is due to the height of the water column in compartment B.

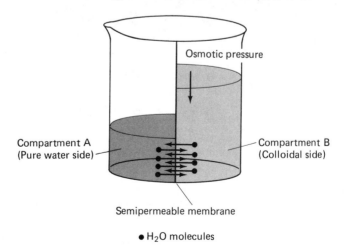

known as *electrophoresis*, became an important method for separating and purifying proteins (one of the types of giant molecules). Using these and other techniques it soon became apparent that these giant molecules were of two types. Substances such as starch, cellulose, and wood were built up of single building blocks which repeated endlessly in the molecule. Other substances, such as proteins, were composed of building blocks that were different (Fig. 13-4). In the case of proteins the building blocks are amino acids. (There are 21 different amino acids that form proteins. These amino acids join together in different numbers and combinations to form the various protein molecules.) These giant molecules eventually came to be called polymers (*poly* is Greek for *many* and *meros* is Greek for *parts*). Today, polymers not only include all of nature's giant molecules but all of the synthetic plastics and fibers that chemists have created in the laboratory.

Figure 13-4 Proteins are composed of building blocks that are different, whereas other substances, like cellulose, are composed of building blocks that repeat in the molecule again and again.

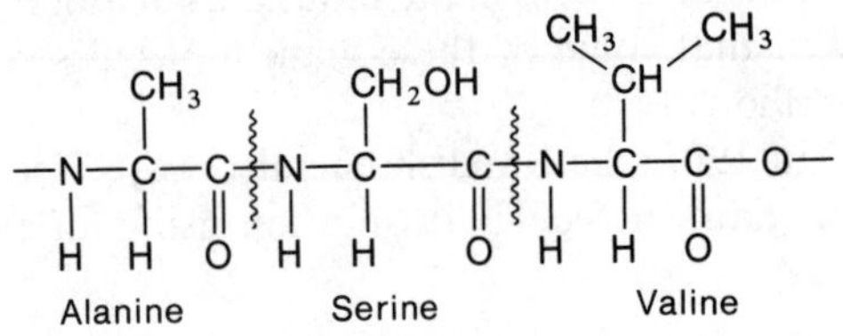

A section of a protein showing three different amino acids, alanine, serine, and valine.

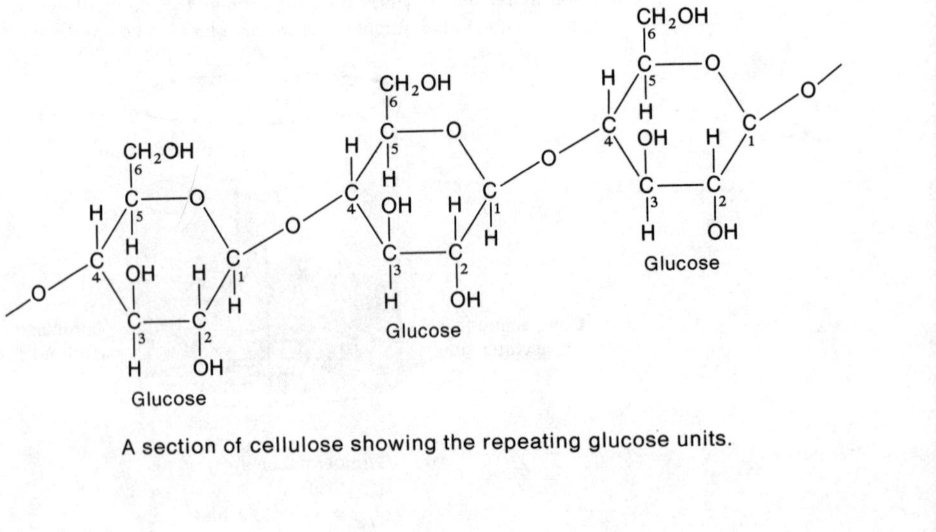

A section of cellulose showing the repeating glucose units.

THE NITROCELLULOSE STORY:
STARTING THINGS OFF WITH A BANG

Even as the properties of macromolecules were being investigated, a German-Swiss chemist, Christian Schönbein, made an important accidental discovery. In 1845, Schönbein was working in his home with a mixture of nitric and sulfuric acids when he accidentally spilled some of the mixture. He used his wife's cotton apron to mop up the mixture. (Keep in mind that cotton is composed of cellulose, a natural polymer.) Schönbein hung the apron up to dry over a stove. But as the moisture evaporated from the apron, the entire apron went "poof"! You might say it went off with a bang. What Schönbein had done was to turn the cellulose into nitrocellulose (the nitro groups adding to the cellulose from the nitric acid). This nitrocellulose was a powerful explosive.

Schönbein saw the enormous potential of his discovery. Nitrocellulose was not only powerful but smokeless. It exploded a lot more cleanly than did black gunpowder. This meant that it wouldn't foul up artillery. It wasn't long before nitrocellulose, which came to be known as *guncotton*, was manufactured for military use. However, a few problems had to be worked out because of the unstable nature of guncotton. But in 1891, two chemists, James Dewar and Frederick Abel, were able to formulate a stable mixture of guncotton that could be molded into long cords. This material, called *cordite*, was safe for military use.

CELLULOID:
THE DISCOVERY OF A SYNTHETIC PLASTIC

Soon after the discovery of nitrocellulose, the fully nitrated derivative of cellulose, it was discovered that a partially nitrated derivative was safer to handle. This partially nitrated derivative was called *pyroxylin*, and it was to play an important part in the synthesis of new polymers.

During the mid-1800's a prize was being offered to anyone who could produce an ivory substitute for billiard balls. Ivory, after all, was a very expensive material, and billiards was a rich man's game. The American inventor John Wesley Hyatt set about to produce such a material. He began with pyroxylin, which he dissolved in a mixture of ethyl alcohol and ethyl ether. Next, he added camphor, which made the resultant material soft and malleable. Hyatt called the material *celluloid* and it won him the prize (Fig. 13-5). Celluloid is considered to be the first synthetic plastic. (A plastic is a material that can be molded into shape.) Hyatt helped to make billiards a very popular game and, more important, opened the door to further research for new polymers.

However, there was more to pyroxylin than billiard balls. The French chemist Louis Bernigaud, Count of Chardonnet, produced fibers from pyroxylin by forcing solutions of the material through tiny holes. As the solution was forced through the small holes the solvent evaporated, leaving a fine thread

Figure 13-5 A daylight-loading Kodak camera and celluloid film. (Eastman-Kodak Company, Rochester, New York)

Figure 13-6 Some photographs of early Kodak cameras: (a) the original Kodak camera disassembled, (b) a folding Kodak camera, (c) a folding pocket Kodak camera, vintage 1898. (Eastman-Kodak Company, Rochester, New York)

(a)

behind. These threads could be woven into a silklike material. In 1884, Chardonnet patented this material, which he called rayon.

CELLULOSE ACETATE:
THE GENIUS OF GEORGE EASTMAN

George Eastman of Eastman-Kodak fame also worked with polymers. One of George Eastman's great contributions to photography was the preparation of a stable photographic emulsion that did not have to be prepared on the spot. Eastman prepared his emulsion by mixing various silver compounds with gelatin. However the emulsion had to be spread over a glass plate, which made picture taking difficult. In 1884, Eastman replaced the glass plate with celluloid film. With a stable emulsion on celluloid film, picture taking entered the realm of the novice. Now everyone would be able to take pictures (Fig. 13-6).

But Eastman was not satisfied. He was troubled because celluloid, although it was not explosive, was very flammable. Eastman tried to find a less flammable material. He discovered that if acetate groups rather than nitrate groups were added to cellulose, the resulting product would still be a plastic. This new material, cellulose acetate, was much less flammable than nitro-cellulose. (Remember, it is nitrocellulose that makes up celluloid.) Eastman accomplished his task, and just in time, because the motion picture industry was just getting started. Now motion picture producers could use a film with a much lower fire hazard.

THE DISCOVERY OF BAKELITE:
THE FIRST COMPLETELY SYNTHETIC PLASTIC

The Belgian-American chemist Leo Baekeland was searching for a shellac substitute. He wanted to synthesize a polymer by combining small molecules

(b)   (c)

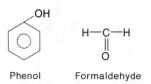

Phenol   Formaldehyde

Figure 13-7   The compounds phenol and formaldehyde, the monomers used to produce Bakelite.

(monomers) into large ones (polymers). Baekeland attempted to combine phenol and formaldehyde (Fig. 13-7) as the monomer units. What he obtained was not a shellac, but a polymer for which he could not find a solvent. However, this material had some very interesting properties. It was extremely strong and solvent resistant, but it could be molded as it formed. It 1909, Baekeland announced the discovery of this completely synthetic plastic material, which he called Bakelite (Fig. 13-12). Because of its inertness, Bakelite found many uses as a plastic for radio cabinets and telephone receivers, electric insulators, and molded plastic ware. It could also be machined if necessary.

Bakelite was not only the first completely synthetic plastic but also the first synthetically produced *condensation polymer*. In a condensation polymer the monomer units (which in Bakelite are phenol and formaldehyde) join together by splitting out water to form an intricate three-dimensional network

Figure 13-8   Part of the polymer structure of Bakelite. Notice how the CH_2 units of formaldehyde connect the phenol groups, forming a repeating network. Although shown here in two dimensions, this network actually exists in three dimensions.

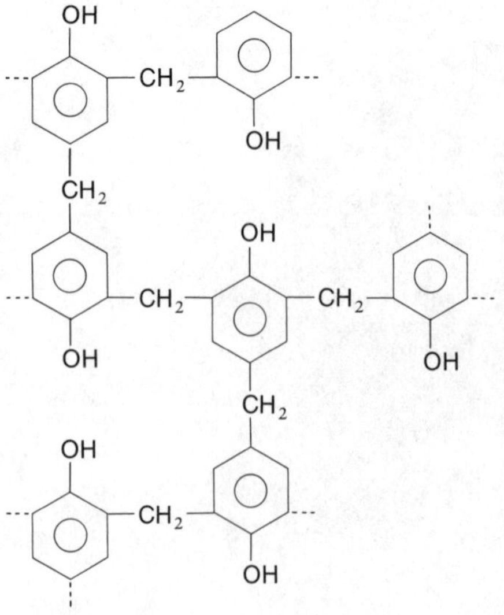

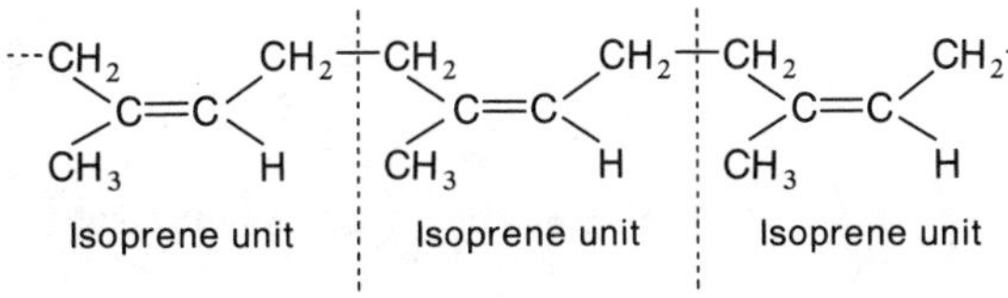

$$\cdots CH_2 \quad CH_2 \overset{|}{\underset{|}{}} CH_2 \quad CH_2 \overset{|}{\underset{|}{}} CH_2 \quad CH_2 \cdots$$

Isoprene unit | Isoprene unit | Isoprene unit

Figure 13-9 In natural rubber the isoprene unit is repeated over and over.

as shown in Figure 13-8. We look at some other condensation polymers later in the chapter.

THE RUBBER REVOLUTION: THE ELASTOMERS

Natural rubber is a polymer obtained from various tropical plants. The monomer from which natural rubber is composed is isoprene:

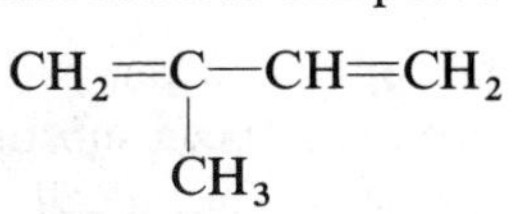

$$CH_2{=}C{-}CH{=}CH_2$$
$$\overset{|}{CH_3}$$

In rubber this isoprene unit is repeated over and over (Fig. 13-9).

Rubber, as produced by nature, is not a very useful substance. That's because it's very sticky in warm weather and hard (or brittle) in cold weather. But the American inventor Charles Goodyear accidentally discovered that rubber took on more desirable elastic properties over a wide range of temperatures when it was heated with sulfur. In 1844, Goodyear patented his *vulcanized rubber*. What the vulcanization process had done was allow the long polymer chains of rubber to cross-link (Fig. 13-10). This cross linking gave the rubber the properties that were necessary to make it a useful substance.

However, as the decade of the 1930's approached, chemists were looking for ways to create synthetic rubberlike materials that would have better durability than natural rubber, and also be resistant to damage by gasoline and oil.

Figure 13-10 In vulcanized rubber, sulfur atoms cross-link the polymer chain. The S_x notation is used to indicate an indefinite number of sulfur atoms cross linking each chain.

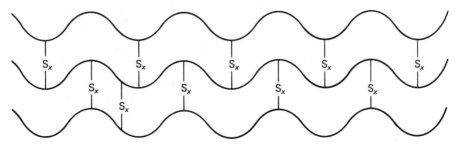

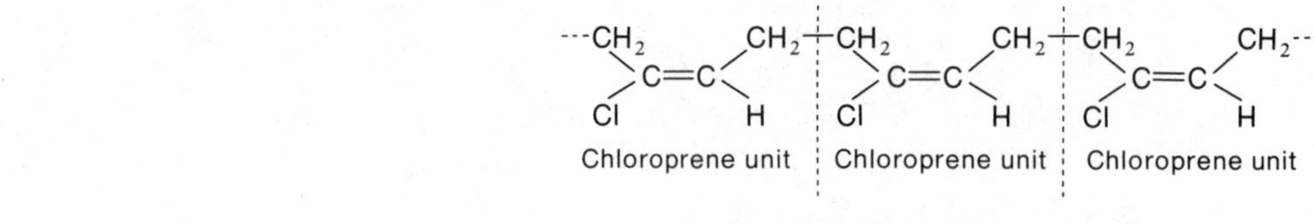

Figure 13-11 In neoprene, technically known as polychloroprene, the chloroprene unit is repeated over and over.

One such material was synthesized in 1932 by two chemists, Wallace Carothers and Julius Nieuland. The material they produced was named neoprene (Fig. 13-11). It was produced from the monomer chloroprene:

$$CH_2{=}C{-}CH{=}CH_2$$
$$\quad\;\; | $$
$$\quad\;\; Cl$$

Notice that chloroprene is very similar to isoprene, the only difference being that a chloro group has been substituted for a methyl group.

The neoprene (or technically speaking, *polychloroprene*) had excellent elastic properties. Because of these properties, neoprene and synthetic rubbers like it were called *elastomers*. Upon testing, neoprene showed excellent resistance to deterioration by gasoline and for that reason was used to make gasoline hoses for service stations (Fig. 13-12).

It wasn't long before other rubber substitutes were synthesized. One example was the material polybutadiene (Fig. 13-13), made from polymerizing the monomer butadiene:

$$CH_2{=}CH{-}CH{=}CH_2$$

Butadiene is also similar to isoprene, except that a hydrogen atom has replaced a methyl group. Although polybutadiene was inexpensive to produce, it was not as versatile as natural rubber or neoprene. However, it was discovered that a mixture of 75% butadiene and 25% styrene could be polymerized to produce a very useful polymer. This resultant material was called SBR, and proved to be a very good elastomer. It had excellent anticracking properties, and found extensive use by tire manufacturers.

The SBR is known as a *copolymer* since the molecular chains contain a mixture of two monomers: styrene and butadiene (Fig. 13-14).

All of the substances that we've discussed in this section, including natural rubber, are called *addition polymers*. An addition polymer is one in which a small molecule *adds* to itself, over and over, to form the polymer. Look once more at the structure of isoprene and at the structure of natural rubber. See how the isoprene unit repeats over and over in the rubber. We look at other addition polymers later in the chapter.

(a)

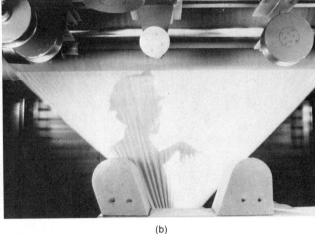

(b)

Figure 13-12 (a) Dr. Wallace H. Carothers, shown here with neoprene, the first commercially successful general purpose synthetic rubber, came to the Du Pont Company the year after it launched its program of fundamental research in 1927. Dr. Carothers is remembered primarily for his direction of the program of fundamental research from which came neoprene and nylon, the world's first synthetic fibers comparable to nature's fibers. He died in 1937. (b) A web of neoprene synthetic rubber is gathered into a hawserlike strand as part of the final manufacturing stage at Du Pont's Pontchartrain, Louisiana, plant. Although neoprene was the first commercially successful synthetic elastomer, its all-round superior characteristics have earned it steadily growing markets over the years. (c) Hot water spray washes a moving web of neoprene synthetic rubber at a Du Point Company plant near Laplace, Louisiana. (Photographs courtesy of E. I. du Pont de Nemours & Company, Wilmington, Delaware)

(c)

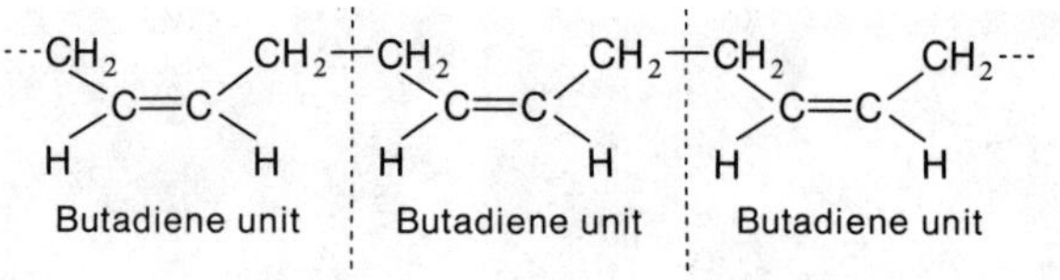

Figure 13-13 In the polymer polybutadiene the monomer butadiene is repeated over and over.

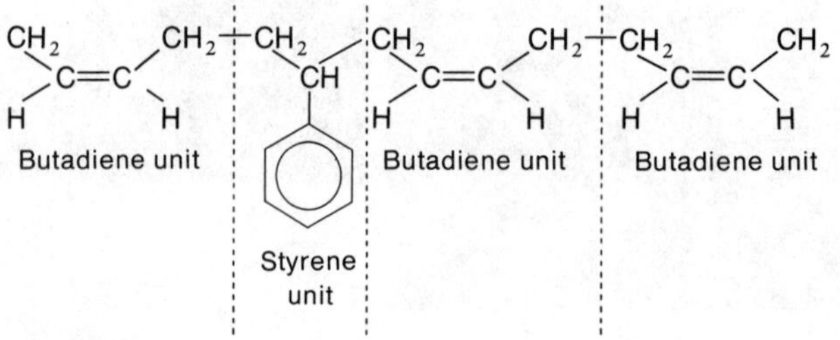

Figure 13-14 The copolymer known as SBR is composed of butadiene and styrene units.

THE SYNTHESIS OF NYLON: ANOTHER CONDENSATION POLYMER

Carothers not only developed synthetic rubber but he also helped to develop synthetic fibers. With the onset of World War II in Europe, it appeared that the American supply of silk was in jeopardy. Carothers had been experimenting with two types of organic compounds, amines and carboxylic acids. The amine

Figure 13-15 Carothers reacted a diamine and a dicarboxylic acid with each other and produced a polymer in which the monomer units are held together by peptide bonds. These are the bonds that hold amino acids together in proteins. Water is also produced in this reaction.

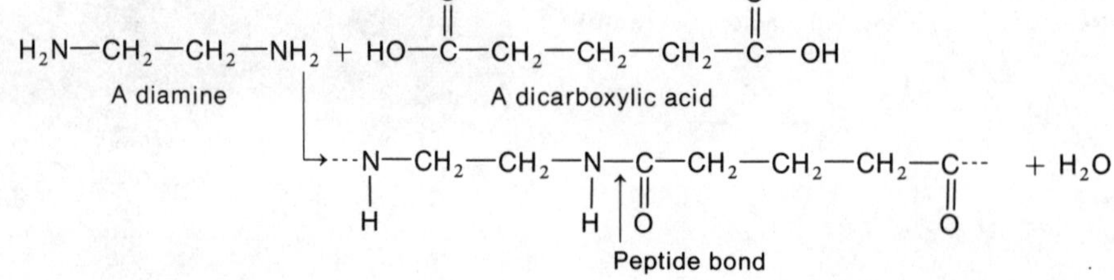

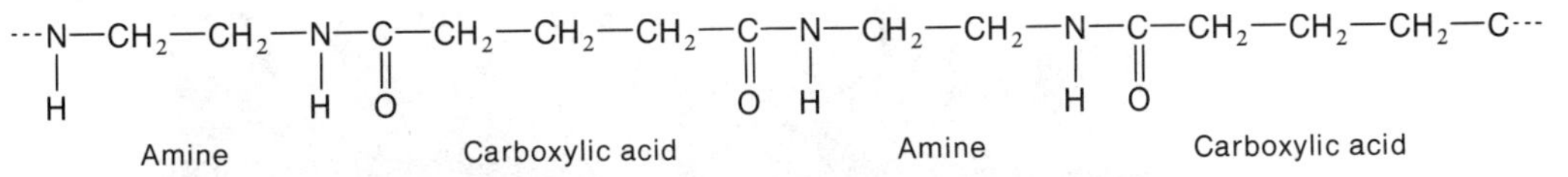

···N—CH$_2$—CH$_2$—N—C—CH$_2$—CH$_2$—CH$_2$—C—N—CH$_2$—CH$_2$—N—C—CH$_2$—CH$_2$—CH$_2$—C···
　　　　　　　|　 |　‖　　　　　　　　 ‖　 |　　　　　　 |　 ‖
　　　　　　　H　 H　O　　　　　　　　 O　 H　　　　　　 H　 O

　　　　Amine　　　　　Carboxylic acid　　　　Amine　　　　Carboxylic acid

Figure 13-16 The reaction of a dicarboxylic acid with a diamine always leaves a free amino group on one end of the molecule and a free carboxyl group on the other end of the molecule. In this way the polymer chain can be extended.

compounds he was using were actually diamines. In other words, there were two amine groups per molecule. Also, the carboxylic acid compounds were actually dicarboxylic acids, which had two carboxyl groups per molecule (Fig. 13-15).

Carothers hoped that he would be able to get the amine group of his diamine to react with a carboxyl group of his dicarboxylic acid. The bond that formed would be similar to the *peptide link* found in silk protein (Fig. 13-16). (The *peptide link* or *peptide bond* is what holds amino acids together in proteins. We have more to say about this in Chap. 16.) Of course, since this new molecule would still have a free amine group on one end and a free carboxyl group on the other end, it could combine with another dicarboxylic acid molecule and another diamine molecule (Fig. 13-17). In this way a long polymer chain could be produced. Carothers was right! The results of his work led to the development of a synthetic silklike material called *nylon*.

The best known form of nylon is nylon 66. It is synthesized from hexamethylenediamine and adipic acid (Fig. 13-18). When these two materials are allowed to react under the proper conditions, they join together and split out water, forming the nylon.

The first use of nylon was for hosiery, replacing silk stockings. But today nylon has many other uses, including many types of clothing items, rugs and carpets, small items such as combs and hair brushes, and various types of rope and hardware.

Figure 13-17 Nylon 66 is synthesized from hexamethylenediamine and adipic acid.

H$_2$N—CH$_2$CH$_2$CH$_2$CH$_2$CH$_2$CH$_2$—NH$_2$ + HO—C—CH$_2$CH$_2$CH$_2$CH$_2$—C—OH
　　　　　　　　　　　　　　　　　　　　　　　　‖　　　　　　　　 ‖
　　　　　　　　　　　　　　　　　　　　　　　　O　　　　　　　　 O

　　Hexamethylenediamine　　　　　　　　　Adipic acid

···N—CH$_2$CH$_2$CH$_2$CH$_2$CH$_2$CH$_2$—N—C—CH$_2$CH$_2$CH$_2$CH$_2$—C···
　 |　　　　　　　　　　　　　 |　 ‖　　　　　　　　　　　 ‖
　 H　　　　　　　　　　　　　 H　 O　　　　　　　　　　　 O

　　　　　　　　　　　Nylon 66

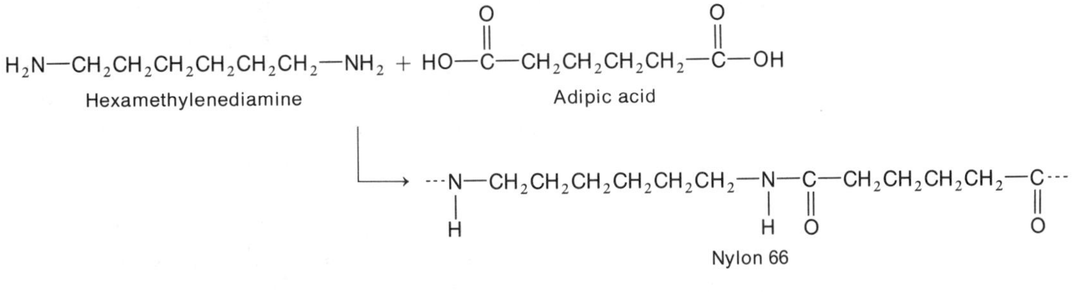

Figure 13-18 Probably the most dramatic moment in Du Pont history is reenacted above—the birth of the first completely synthetic fiber, impractical for commercial use but a true forerunner of nylon itself. Here chemist Julian Hill shows how he pulled the molten sample of material from a laboratory test tube at the company's Experimental Station near Wilmington, Delaware. The molasseslike mass stuck to the glass stirring rod and was drawn out into a thin fiber. (E. I. du Pont de Nemours & Company, Wilmington, Delaware)

POLYETHYLENE:
ONE OF OUR MOST IMPORTANT POLYMERS

One of the major compounds obtained from the cracking of petroleum is ethylene, $CH_2=CH_2$, a gas at room temperature. Ethylene, you might remember from Chapter 12, is the first member of the alkene series of hydrocarbons. When subjected to high temperatures and pressures in the presence of a catalyst, ethylene molecules can be made to combine with each other to form long chains in which $-CH_2-CH_2-$ units are repeated over and over (Fig. 13-19).

The polyethylene that forms is a *thermoplastic material*. That means that it can be softened by heat and remolded into various forms. (Other plastics, such as Bakelite, are called *thermosetting polymers* because they can't be

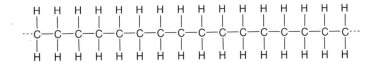

Figure 13-19 A two-dimensional representation of polyethylene. The ethylene units are repeated over and over.

re-formed once they are made.) The uses of polyethylene range from the production of plastic bags of all kinds, to the manufacture of toys, automobile parts, and electrical wiring insulation (Fig. 13-20).

Polyethylene can be produced in two forms; high-density polyethylene and low-density polyethylene. In high-density polyethylene the molecules have an almost crystalline arrangement along the polymer chain. In low-density polyethylene the molecules are more randomly distributed along the polymer

Figure 13-20 As a continuing line of bottles passes by, a filling line operator at Union Carbide Home and Automotive Division's antifreeze plant in Freehold, New Jersey, checks Prestone II containers, which are molded at the plant from Union Carbide polyethylene. (Union Carbide)

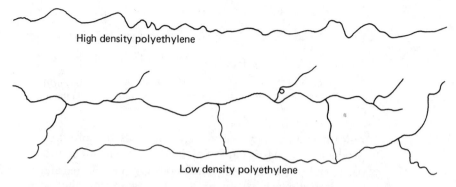

High density polyethylene

Low density polyethylene

Figure 13-21 In high-density polyethylene the molecules have an *almost* crystalline arrangement along the polymer chain. In low-density polyethylene the molecules are more randomly distributed and may also be branched or cross linked.

chain (Fig. 13-21). Because the carbon chain in high-density polyethylene is fairly regular, the material is rigid. High-density polyethylene is used in the production of toys, bottle caps, and plastic plumbing pipes, or wherever a rigid plastic is needed. Low-density polyethylene is a semirigid, translucent plastic which has a high degree of flexibility. Low-density polyethylene is also resistant to many types of chemicals, which makes it an excellent packaging material for these substances. Some other uses of low-density polyethylene are electrical wiring insulation, plastic bags and bottles, and any other instance where a flexible, fairly inert material is needed.

POLYVINYL CHLORIDE (PVC): A REMARKABLE POLYMER

Polyethylene is a very useful material. However, an even more versatile material is polyvinyl chloride or PVC, for short. PVC is synthesized from the monomer vinyl chloride.

$$\begin{array}{c} H \\ \diagdown \\ C = C \\ \diagup \qquad \diagdown \\ H \qquad\qquad Cl \end{array}$$

Vinyl chloride

Notice the similarity between this molecule and ethylene.

$$\begin{array}{c} H \qquad\qquad H \\ \diagdown \qquad \diagup \\ C = C \\ \diagup \qquad \diagdown \\ H \qquad\qquad H \end{array}$$

Ethylene

The only difference is that a chlorine atom has been substituted for a hydrogen atom. Vinyl chloride is a water-insoluble gas at room temperature, which is

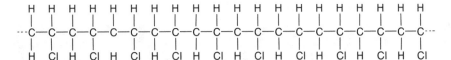

Figure 13-22 A section of the molecular chain of polyvinyl chloride. The vinyl chloride monomer is repeated over and over in the molecule.

readily polymerized under pressure and heat (Fig. 13-22). The polymer produced is a tough synthetic material which can be used for floor coverings, unbreakable bottles, clear plastic wraps, and synthetic leather. PVC can be easily molded into various shapes, and it can be colored and textured to simulate leather (Fig. 13-23).

(a)

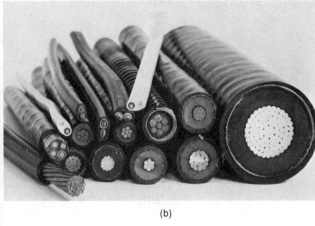

(b)

Figure 13-23 (a) Rigid polyvinyl chloride has gained a substantial share of several pipe and fitting markets, including agriculture, residential, industrial manufacturing, and municipal water supply systems. PVC pipe is lightweight, resistant to chemicals, and easily installed. (b) Flexible polyvinyl chloride compounds are used to extrude wire and cable sheath and insulation in applications ranging from speaker wire to 35,000 volt cable. Vinyl provides good abrasion resistance and low temperature flexibility. (c) Rigid polyvinyl chloride can be used for such construction applications as house siding, gutter systems, and window frames, to provide a low maintenance product with excellent weatherability. (Photographs courtesy of BF Goodrich Chemical Group.)

(c)

TABLE 13-1 *Polymers that are derivatives of, or are similar to, polyethylene*

NAME OF POLYMER	STRUCTURE OF MONOMER UNIT	USES OF POLYMER
Polypropylene	CH_3 and H on one carbon, H and H on the other, C=C (propylene)	As an indoor–outdoor carpet fabric; also for clear plastic bottles, plastic lab ware, and kitchenware
Polystyrene	H and H, C=C with H and phenyl ring (styrene)	As a packaging material; in its foam form (known as Styrofoam) it is used to make inexpensive cups and plates
Polytetrafluoroethylene (trade name, Teflon)	F, F, C=C, F, F (tetrafluoroethylene)	As a nonstick coating material for pots and pans
Polymethyl methacrylate (trade name, Lucite or Plexiglas)	H, H, C=C, CH_3, C—OCH_3, O double bond (methyl methacrylate)	As a transparent, unbreakable glass
Polyvinylidene (trade name, Saran)	H, H, C=C, Cl, Cl (vinylidene chloride)	As a self-adhering wrap, known commercially as Saran Wrap

There are a number of other polymers which are derivatives of, or are similar to, polyethylene. These polymers have thousands of uses in our technological society, from indoor–outdoor carpeting to transparent plastic glass such as Lucite and Plexiglas. A number of these substances and their uses are listed in Table 13-1.

THE WORLD OF SYNTHETIC FIBERS

Earlier in this chapter when we discussed the history of polymers we learned how rayon, a semisynthetic fiber, was produced from a derivative of cellulose. Another cellulose fiber was produced by treating cellulose with acetic anhydride. A reaction known as esterification takes place (see Chap. 12 for a discussion of esterification) producing a polymer known as rayon acetate. When the rayon acetate fibers are spun into a cloth, they produce a soft expensive-looking fabric.

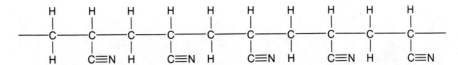

Figure 13-24 A section of the molecular chain of polyacrylonitrile. The acrylonitrile molecule is repeated over and over in the molecule.

Other synthetic fibers have been produced over the years. A very useful synthetic fiber is polyacrylonitrile (Fig. 13-24). This material is usually woven into fabrics for sweaters, blankets, and carpeting, and goes under the trade names of Acrilan and Orlon. Polyacrylonitrile is produced from the monomer acrylonitrile. It has a structure very similar to that of vinyl chloride, except that a —C≡N group has replaced the Cl atom in vinyl chloride.

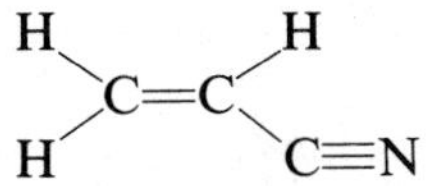

$$
\begin{array}{cc}
\text{H} \quad\quad \text{H} & \text{H} \quad\quad \text{H} \\
\diagdown \text{C=C} \diagup & \diagdown \text{C=C} \diagup \\
\text{H} \diagup \quad \diagdown \text{C≡N} & \text{H} \diagup \quad \diagdown \text{Cl} \\
\text{Acrylonitrile} & \text{Vinyl chloride}
\end{array}
$$

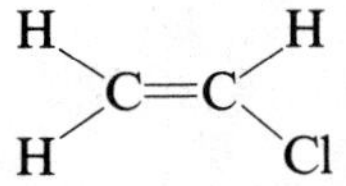

Synthetic fibers known as polyesters have found wide use in the production of permanent-press fabrics. These polyesters are quite strong, extremely resilient, and resist stretching and shrinking. One type of polyester, with the trade name Dacron, is synthesized from the reaction of ethylene glycol and terephthalic acid (Fig. 13-25). It is a condensation polymer. Clothes made of dacron have excellent wear characteristics (Fig. 13-26).

Figure 13-25 Dacron is made from the reaction of ethylene glycol and terephthalic acid. Notice how the ethylene glycol units and terephthalic acid units alternate in the chain. These units are repeated over and over, in this sequence, throughout the polymer.

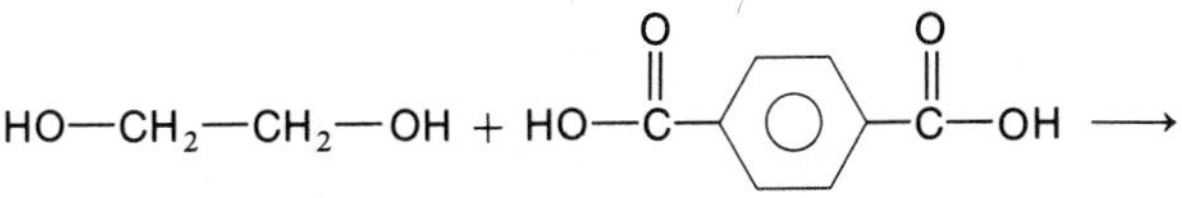

$$
\text{HO—CH}_2\text{—CH}_2\text{—OH} + \text{HO—}\overset{\overset{\text{O}}{\|}}{\text{C}}\text{—}\langle\bigcirc\rangle\text{—}\overset{\overset{\text{O}}{\|}}{\text{C}}\text{—OH} \longrightarrow
$$

Ethylene glycol   Terephthalic acid

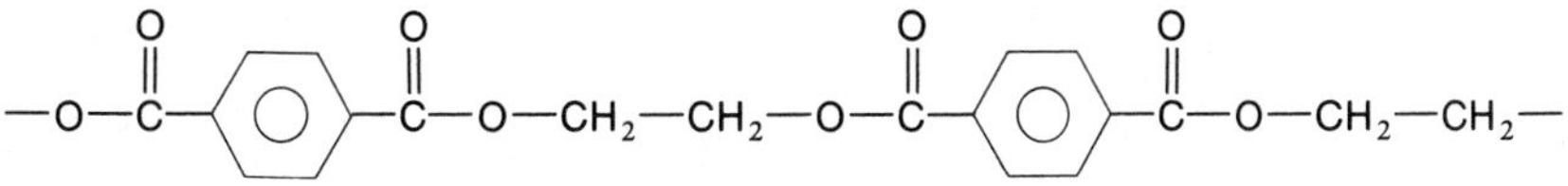

$$
\text{—O—}\overset{\overset{\text{O}}{\|}}{\text{C}}\text{—}\langle\bigcirc\rangle\text{—}\overset{\overset{\text{O}}{\|}}{\text{C}}\text{—O—CH}_2\text{—CH}_2\text{—O—}\overset{\overset{\text{O}}{\|}}{\text{C}}\text{—}\langle\bigcirc\rangle\text{—}\overset{\overset{\text{O}}{\|}}{\text{C}}\text{—O—CH}_2\text{—CH}_2\text{—}
$$

Dacron

(b)

(a)

Figure 13-26 (a) Dacron fibers being spun. (b) Dacron fibers being inspected at a Du Pont Plant. (E. I. du Pont de Nemours & Company, Wilmington, Delaware)

POLYMERS AND POLLUTION

Each year the demand for polymers increases. The present demand is about 100 million tons per year. How do these substances affect our environment? Table 13-2 lists some of the problems caused by the abundant use of these materials. Let's take a brief look at some of these problems.

Shortage of Raw Materials

A major source of raw materials for polymers is petroleum. As our supply of this raw material decreases, the availability of many polymers may also decrease. The decade of the 1970's saw the beginning of the petroleum shortage,

TABLE 13-2 *Environmental problems caused by excessive use of synthetic polymers*

1. Overconsumption of nonrenewable resources, especially petroleum.
2. High energy consumption needed to produce synthetic fibers and plastics.
3. Solid waste disposal problems.
4. Fire hazards from various synthetic fibers and plastics that burn.
5. Harmful effects on animals and human beings from the use of plasticizers in the production of plastics.

and future shortages are probable. Our world depends on polymers to a great extent. What will happen when the oil wells dry up?

High Energy Consumption

Numerous production steps are necessary for the manufacture of most polymers. Each production step requires huge amounts of energy. How long can we continue to supply this energy? Today, most of our energy comes from fossil fuels—especially oil. Can we continue to produce synthetic fibers using nonrenewable energy sources? Remember, natural fibers such as cotton are not dependent on an energy-intensive technology; instead they are produced by solar energy through the process of photosynthesis.

Disposal Problems

The biggest problem with many polymers, especially plastics, is: What do we do with them after we're finished using them? Most plastics don't decompose. They don't get recycled back into the environment. A toy made out of polyethylene or a bottle made out of PVC may last for thousands upon thousands of years. Sure, we could bury these materials in a landfill and get them out of sight. But once buried, these materials remain essentially unchanged.

How about disposing of these plastic materials by burning them? Well, most plastics will burn, but many of them give off toxic gases. For example, PVC produces hydrogen chloride gas upon incineration. Our only solution to this problem seems to be to develop a plastic that will degrade after a reasonable amount of time. Such materials are now being investigated.

Fire Hazards

Because plastics and other polymers do burn, they present a fire hazard. About 200,000 people a year (according to government statistics) are injured or burned to death by accidents associated with fabrics that burn. Then, too, there is the problem of fires in buildings that have plastic panels and other such building materials. In August 1973 there was a fire in the Summerland Amusement Park on the Isle of Man, off the coast of Britain. Fifty people died, many of them *not* because they burned to death, but because of the toxic fumes produced by the burning plastic panels.

Plasticizers

Finally, certain plastics, such as PVC, have additives in them to make them more pliable. These additives are called *plasticizers*. Compounds such as polychlorinated biphenyls (PCBs for short) and phthalates are used as plasticizers (Fig. 13-27). PCBs were first thought to be safe by most scientists.

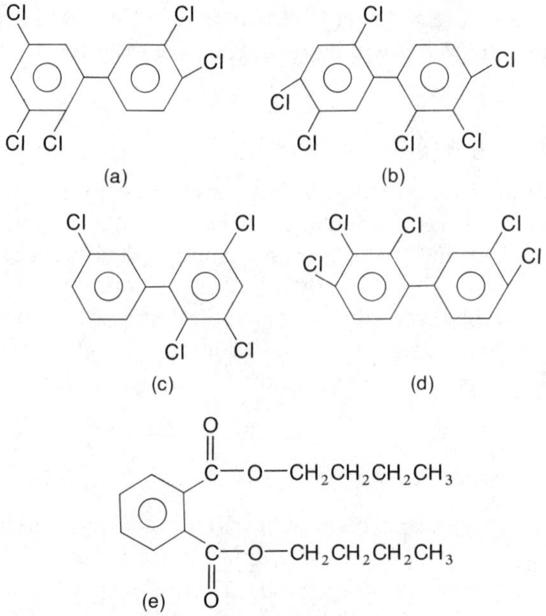

Figure 13-27 (a), (b), (c) and (d) Some of the PCBs; many others also exist. (e) One of the phthalates, known as dibutyl phthalate (DBP, for short). All these materials have been used as plasticizers.

However, they have been shown to be a health hazard to human beings. PCBs have molecular structures similar to DDT (Fig. 13-28), and they can get leached out of a plastic and end up in the environment. Once in the environment, PCBs usually get incorporated into the food chain and end up in the fatty tissues of animals and people. Because of these dangers, PCBs are no longer used in the manufacture of plastics. But just think of all the plastic that is out there already, loaded with PCBs.

Phthalates, like PCBs, have also been found to get incorporated into the food chain. Phthalates have also been found to have adverse health effects on

Figure 13-28 PCBs have molecular structures similar to that of DDT. Compare the structure of DDT shown here to those of the PCBs shown in Figure 13-27.

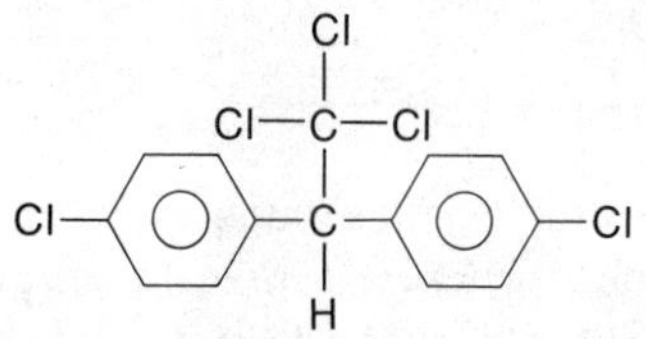

DDT

animals. But their effects on human beings is still open to question, so phthalates are still being used as plasticizers.

Even with all these problems, plastics will still play an important part in our future. Perhaps science and technology will solve most of the problems with plastics and other polymers, and "Better Living Through Chemistry" will become a reality (Figs. 13-29 and 13-30).

Figure 13-29 (a) The Reigning Resin: A rain of high purity Cleartuf polyester resin is packed for shipment at Goodyear's Point Pleasant, West Virginia, plant. Goodyear is the world's leading producer of high purity polyester resin used to make food packaging, soft drink bottles, coatings for paper cooking trays, and film wraps for lunch meats. (b) The Suitable Bottle: Just 26 soft drink bottles made of polyester are tailor-made to be recycled into this coat and pants, says Goodyear, maker of the polyester used in the bottles. The bottles—now being used by Pepsi-Cola and Coca-Cola—can be converted into hundreds of useful products from adhesive tape to zippers, Goodyear says. (P.S. The tie was made from just two more bottles.) (Photographs courtesy of Goodyear Tire and Rubber Company, Akron, Ohio.)

(a)   (b)

(a)

(b)

Figure 13-30 (a) Grand Canyon: A tractor-trailer truck is dwarfed in a canyon of earthmover tires in Goodyear's unique tire storage "bubble" at its Topeka, Kansas, production facility. The air-inflated structure provides a 125 × 600-foot storage area where Goodyear inventories more than 4000 earthmover tires.
(b) Once Over Carefully: An inspector at Goodyear's tire plant in Topeka, Kansas, makes a visual inspection of a Wrangler Radial off-road tire still hot from the curing press. Developed in four-wheel-drive off-road racing, the tire was on the winning vehicles in six classifications of the 1979 World Championship Off-Road Races at Riverside, California. (Photographs courtesy of Goodyear Tire and Rubber Company, Akron, Ohio.)

SUMMARY

In this chapter we studied the macromolecules known as polymers. We began with a short history of these big molecules and learned about their structures. We read about the accidental discovery of nitrocellulose, which we could say "began the polymer revolution."

We studied the many different types of polymers—plastics and fibers. We learned the difference between an addition polymer and a condensation polymer. We reviewed the properties of elastomers. We also learned about thermoplastic materials and thermosetting materials. And of course we reviewed the many uses of these materials in our daily lives.

Finally, we studied some of the environmental problems associated with the tremendous world demand for polymers, and we also looked at some possible solutions.

EXERCISES

1. Name four common examples of naturally occurring macromolecules.

2. Explain the differences between crystalloids and colloids.

3. Explain how the process of osmosis was used in the study of polymers.

4. "As research on polymers progressed it became apparent that these giant molecules were of two types, in terms of their building blocks." Explain this statement.

5. Explain the derivation of the word polymer.

6. Discuss the importance of the discovery of nitrocellulose, pyroxylin, and celluloid.

7. Explain how celluloid film and cellulose acetate film played an important part in the development of modern photography.

8. Show how the two monomers phenol and formaldehyde react to form the condensation polymer Bakelite.

9. State some of the properties and uses of Bakelite.

10. Using examples, show how a condensation polymer differs from an addition polymer.

11. Write the structural formula of isoprene. Show how it is similar to chloroprene and butadiene.

12. Explain the process of vulcanization, and tell why it's so important.

13. Define the following terms: copolymer, thermosetting polymer, and thermoplastic polymer. Give examples of each.

14. Discuss some of the properties of nylon and explain one of the reasons for its development.

15. Explain Carother's rationale in using dicarboxylic acids and diamines to produce a silklike material.

16. Show by writing a chemical equation how polyethylene is formed from ethylene.

17. State some of the properties and uses of high-density polyethylene and low-density polyethylene.

18. What is the difference in molecular structure between high- and low-density polyethylene?

19. Explain how PVC and polyacrylonitrile are similar to polyethylene, in terms of their molecular structure.

20. Why are polyester materials such great fabrics for clothing?

21. Discuss five environmental problems associated with polymers. Can you find solutions for these problems?

22. What do you see as the future for plastics and other polymers?

Chapter 14

Agricultural Chemistry

Some Things You Should Know After Reading This Chapter

You should be able to:

1. Explain the process of photosynthesis.
2. Describe the ways in which nitrogen can be supplied to plants.
3. Write an equation for the Haber process for the formation of ammonia.
4. List the various forms of fixed nitrogen.
5. Name the sources of potassium and phosphorus fertilizers.
6. Explain the meaning of the numbers on a bag of fertilizer.
7. Discuss the Green Revolution.
8. Describe the initial effects of using DDT worldwide.
9. Explain how DDT is made.
10. Discuss the environmental impact of DDT.
11. Discuss the various ways in which insect populations can be controlled.
12. Explain the meaning of the term herbicide.
13. Discuss the positive and negative aspects of the use of herbicides.
14. Explain the meaning of the term defoliant.
15. Discuss the advantages of organically grown food.

INTRODUCTION

There are more than 4 billion people in the world today. The United Nations predicts that world population will reach 12 billion by 2075, depending on the fertility rate. Today many of these people are undernourished. Many do not have enough to eat. Those who live on rice, wheat, or potatoes don't get enough protein. Modern agriculture is responsible for providing food for the world's people. Farming today depends on certain fundamental techniques. These include the use of chemicals to fertilize the soil, machines to do the heavy work, water for plant growth, and chemicals to control disease, insects, and weeds. In this chapter we take a look at the status of agriculture today.

HOW DO PLANTS GROW?

All plants undergo a process called *photosynthesis*. This involves the chemical reaction of carbon dioxide (from air) and water (from soil) in the presence of sunlight, to produce a simple sugar (a carbohydrate) and oxygen. The plant stores the energy taken from the sun in the form of an energy-rich carbohydrate molecule (Fig. 14-1). The balanced reaction for the process of photosynthesis is

$$6\,CO_2 + 6\,H_2O \xrightarrow[\text{chlorophyll}]{\text{sunlight}} C_6H_{12}O_6 + 6\,O_2 \quad \text{(photosynthesis)}$$

Figure 14-1 Plants store energy taken from the sun in the process of photosynthesis. We consume the plants to obtain energy for ourselves.

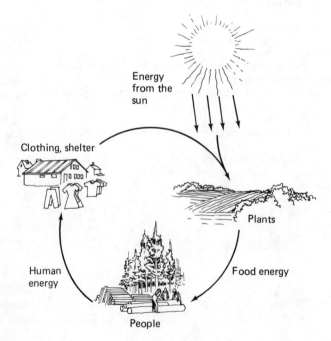

Energy from the sun

Clothing, shelter

Plants

Human energy

Food energy

People

Carbohydrates are produced by plants and by certain microorganisms, both of which contain the compound *chlorophyll*. This compound is essential for the process of photosynthesis to take place.

It is interesting to see how plants and animals rely on each other to provide an oxygen-carbon dioxide balance in nature. Plants can use the carbon dioxide from the air and water from the soil to synthesize carbohydrates. Animals can't form carbohydrates in this way, so they must rely on plants for their carbohydrate supply. But animals have the ability to utilize carbohydrates in their bodies in a process called metabolism. *Metabolism* is the process by which the food we eat is transformed in the body by a series of complex chemical reactions, catalyzed by enzymes, to provide material for cellular function and energy to power the body. Many of the chemical processes that take place in the body are classified under the name "metabolism." In this overall reaction, a carbohydrate molecule reacts with oxygen to produce carbon dioxide, water, and energy. The carbon dioxide and water are returned to the environment and the energy is used to power the body. The balanced equation for this reaction is

$$C_6H_{12}O_6 + 6O_2 \longrightarrow 6CO_2 + 6H_2O + energy \quad \text{(animal metabolism)}$$

You will notice that this is the reverse of the equation for photosynthesis, where carbohydrates are produced. So you see that a carbon dioxide–oxygen cycle exists in nature. Plants take in carbon dioxide from the air during photosynthesis and give off oxygen. Animals take in oxygen from the air and give off carbon dioxide during metabolism. This is a mutually beneficial cycle.

SOIL FERTILIZATION

In days past, farmers used manure, straw, or dead fish to fertilize their crops. Fertile soil contains many important elements. Nitrogen, potassium, and phosphorus are essential in large amounts. Minerals such as calcium, zinc, magnesium, iron, and sulfur are also needed, but in smaller amounts (Table 14-1). Today farmers use fertilizers that are mined and manufactured. These are more expensive fertilizers, because energy is required for their production. As energy sources become more costly and less abundant, agricultural productivity is threatened.

Nitrogen Fertilizers

All plants and animals must have nitrogen in order to survive. The atmosphere consists of nearly 80% nitrogen, but plants cannot use atmospheric nitrogen. Atmospheric nitrogen must be *fixed* in order to be used by plants. To *fix* means to stabilize the nitrogen or turn it from a gaseous form to a solid or liquid form in which it is bound with another element or with more than one element. Some forms of fixed nitrogen are ammonia (NH_3), nitrite ions (NO_2^{-1}), and nitrate ions (NO_3^{-1}). Certain bacteria and algae are capable

TABLE 14-1 *Nutrients needed by plants*

PRIMARY NUTRIENTS (LARGEST AMOUNTS)	SECONDARY NUTRIENTS (SMALLER AMOUNTS)	MICRONUTRIENTS (VERY SMALL AMOUNTS)
Carbon	Calcium	Boron
Hydrogen	Magnesium	Copper
Oxygen	Sulfur	Iron
Nitrogen		Manganese
Phosphorus		Zinc
Potassium		Molybdenum
		Chlorine

of fixing atmospheric nitrogen. These organisms are found free in nature and some live in colonies on the roots of plants called *legumes* (Fig. 14-2).

There are three ways in which nitrogen can be supplied to plants. Most simply, dead plant matter and animal wastes can be used to fertilize the soil. Crop rotation is a second method supplying nitrogen by planting legumes and grain in the same spot in alternate years. Finally, chemical fertilizers such as ammonia can be used. The equation for the commercial production of ammonia is

$$3\,H_2 + N_2 \xrightarrow[\text{high pressure}]{\text{high temperature}} 2\,NH_3$$

The process of combining nitrogen and hydrogen to produce ammonia was developed by Fritz Haber in the early 1900's and is referred to as the Haber

Figure 14-2 Certain bacteria are able to turn atmospheric nitrogen into a form that can be used by plants. Some of these bacteria live in colonies on the roots of plants called *legumes*.

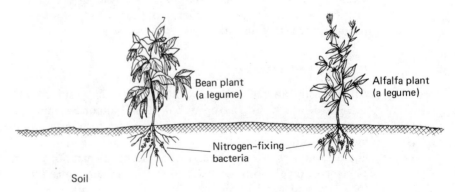

Bean plant (a legume)

Alfalfa plant (a legume)

Nitrogen-fixing bacteria

Soil

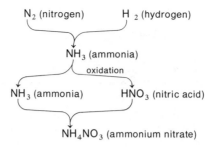

Figure 14-3 How the Germans produced ammonium nitrate in World War I.

process. Fritz Haber received the Nobel Prize in Chemistry in 1918 for his discovery. Haber was a German chemist and his discovery helped his country in World War I. Because of the Haber process used for producing ammonia, the Germans were able to make ammonium nitrate (NH_4NO_3), which is a very good fertilizer, but more important, is an excellent explosive. The Germans were able to carry on the war by making explosives even when their supplies of normal raw materials were cut off by the Allied forces (Fig. 14-3). (It should be pointed out that the reaction for the formation of ammonia from its elements is more properly represented as

$$N_2 + 3 H_2 \xrightleftharpoons[\text{high pressure}]{\text{high temperature}} 2 NH_3$$

This indicates that the reaction is reversible and that an equilibrium exists between the reactants and products. We discussed this type of reaction in Chapter 8. To shift this equilibrium to the product side of the reaction, high temperatures and pressures are used.)

Potassium and Phosphate Fertilizers

Potassium and phosphorus, together with nitrogen, are needed in large amounts for growing plants. Potassium is found in chemical fertilizers primarily as potassium chloride (KCl). There are large reserves of potassium chloride in mines in the United States and in Canada. Since this nutrient is mined and it eventually ends up in the ocean (after being scattered throughout the environment), it is a *nonrenewable* resource. Nutrients that wind up in the ocean require tremendous amounts of energy for retrieval, so they're essentially lost for reuse at a later date.

Phosphorus is also mined. Bones and teeth of fish and other animals that died in centuries past make up the deposits of phosphorus-containing ore. Phosphorus, which is commercially mined as phosphate ores, is highly insoluble. To be used as plant fertilizers, these phosphate ores must be treated with sulfuric acid so that they can be converted to more soluble forms of phosphates called superphosphates (Fig. 14-4). Since phosphate ores must be mined, they too are a *nonrenewable* resource.

$$Ca_3(PO_4)_2 + 2\,H_2SO_4 \longrightarrow Ca(H_2PO_4)_2 + 2\,CaSO_4$$

Calcium   A superphosphate
phosphate

Figure 14-4 Phosphate ores must be treated with sulfuric acid so that they can be converted to more soluble forms of phosphates called superphosphates, which are then used as fertilizers.

Figure 14-5 A fertilizer that is labeled 10-6-4 contains 10% nitrogen, 6% phosphorus, and 4% potassium.

10–6–4: What Does It All Mean?

When farmers or home gardeners purchase bags of chemical fertilizer, there are three numbers on the bag. This indicates the percentage of the nutrients nitrogen, phosphorus, and potassium, in that order (N–P–K). So a 10–6–4 mixture is 10% nitrogen, 6% phorphorus (as P_2O_5), and 4% potassium (as K_2O) (Fig. 14-5).

Fertilizers are spread on plants in gardens or on lawns. The nutrients are water soluble and when it rains they are washed into lakes and streams. This causes the algae found in the bodies of water to grow too much. These chemicals also penetrate the groundwater and sometimes end up in wells, where they can reach toxic concentrations. However, on the brighter side, they enrich the soil and foster plant growth. This has allowed farmers to grow substantially more food than ever before.

THE "GREEN REVOLUTION": NEW WONDER SEEDS

Have you ever looked at wild grass growing in a field or meadow? If you did, you probably noticed that the stalk is thin with only a few leaves, and there are just a few seeds at the top. The seeds contain proteins, carbohydrates, and vitamins. But if there are only a few seeds, there isn't much to harvest and little food would be provided.

Scientists have recently been able to develop new "wonder seeds" which produce much more grain than can be produced by the wild plants. So much food can be produced by these seeds that their presence has been termed the start of the "Green Revolution." They provide a greater amount of food per hectare of land. (A hectare is a metric unit for measuring surface area. One hectare equals 10,000 square meters or 2.47 acres.)

The "wonder seeds" require more careful planting than do older grain varieties. Farmers must invest in chemical fertilizers, irrigation systems, and pesticides in order to get the higher yield. However, all of this requires the use of fuel, and with escalating fuel costs, farmers can't always afford what's needed. So the results of the Green Revolution are twofold. Wealthy farmers have been able to produce more food. Starvation has decreased in the developing nations. But poorer farmers in some developing nations have been driven out of business and many have been forced to live in poverty in urban areas.

HUMANKIND VERSUS PESTS: PROTECTING OUR FOOD SUPPLY

Farmers have always had to protect their crops from pests. What are pests? They include insects, field mice, worms, and fungi (Fig. 14-6). Not all insects, rodents, and microorganisms found in soil are pests however. Most of them are important and necessary to help maintain the balance of nature. Many insects eat other insects which are pests that destroy plant life. Bees are important because they spread out plant pollen as they search for food and are responsible for fertilizing many plants.

Nevertheless, many harmful pests do exist. Driver ants, for example, are responsible for devastating some villages in Africa. When they attack, they chase away the people and eat everything in sight. But besides destroying the food supply and property, they kill rats, cockroaches, and other pests along the way. This is an example of a natural cycle. When the people return, their village is rid of pests which threatened their own health.

DDT: The Start of the Modern Pesticide Era

In 1940, a chemical called DDT was discovered. DDT stands for dichloro-diphenyltrichloroethane (Fig. 14-7). This chemical was quite inexpensive and a potent *insecticide*. An insecticide is a chemical that kills insects. DDT killed more insects than any chemical known previously.

During World War II, the Americans and British needed a powerful in-secticide to kill lice and ticks (which cause typhus fever) which were threatening to cause an epidemic among the soldiers. Mosquitoes and other pests threatened those fighting in the South Pacific. DDT was mixed with talcum

Figure 14-6 A boll weevil attacking a cotton plant. (U.S. Department of Agriculture)

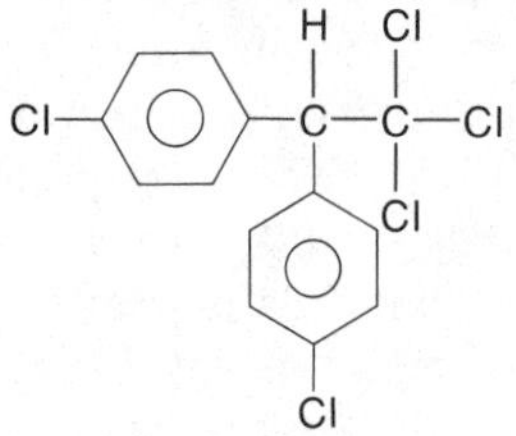

DDT

Figure 14-7 The insecticide DDT, whose full name is dichlorodiphenyl-trichloroethane.

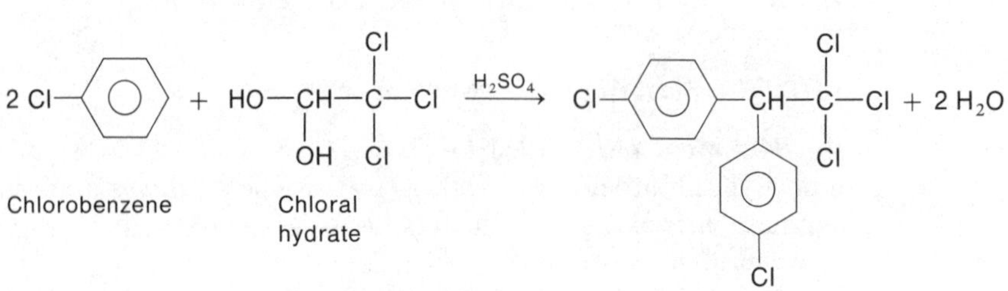

Chlorobenzene   Chloral hydrate   DDT

Figure 14-8 DDT is made by reacting chlorobenzene and chloral hydrate in the presence of sulfuric acid.

to form a powder. It was great for killing lice. People and animals didn't seem to be affected by the DDT. It was a miracle insecticide—cheap, effective, and seemingly safe.

How is DDT made? Two chemicals, chlorobenzene and chloral hydrate, are mixed and warmed in the presence of sulfuric acid (Fig. 14-8). After the reaction is over, the mixture is poured into water. DDT is a *chlorinated hydrocarbon* and it is insoluble in water. The DDT thus separates out.

The Effects of DDT

The World Health Organization (WHO) estimates that the use of DDT and other chlorinated hydrocarbons has saved 25 million lives. In tropical countries malaria cases have been dramatically reduced and life spans have been increased. In India, for example, the incidence of malaria dropped from 75 million cases a year a few decades ago, to 5 million cases a year at present.

Paul Müller is the chemist who discovered that DDT had insecticidal properties. He won the Nobel Prize in Medicine and Physiology in 1948 for his work.

After things seemed to be going so well, the bottom started to fall out of the DDT miracle. Resistant strains of insects evolved. Bird populations began to decline. Fish were dying. Government officials were afraid of future catastrophes and in 1972 the use of DDT was banned by the Environmental Protection Agency (EPA).

What's so bad about DDT and other chlorinated hydrocarbons? These chemicals are relatively stable in the environment and undergo few chemical reactions. The estimated half-life of DDT is about 15 years. So DDT stays around for a long, long time.

DDT, as we said, is insoluble in water. But it is soluble in fatty tissues. It's hard for the body to get rid of DDT once it's inside.

DDT has also gotten into the food chain. DDT sprayed into the air gets into the soil. Once in the soil, the DDT gets into earthworms. Robins eat the earthworms and ingest the DDT. In many cases, DDT ingested by robins has been so high that it constituted a lethal dose.

After so much DDT was sprayed around (over 500 million kilograms, it is estimated), it was no longer as effective as before. Resistant strains of insects appeared. They were capable of detoxifying DDT. Body lice in Korea which are carriers of typhus became DDT-resistant. The same held true for many mosquitoes that carry malaria and yellow fever, and fleas, which carry bubonic plague. Some of the natural predators of these pests were killed off by DDT, so their populations could not even be controlled by natural means.

Other Ways of Killing Insects

Insects can be eliminated by using their natural enemies. In Japan, the population of Japanese beetles is controlled by a type of wasp which is its natural enemy. The female wasp attaches her eggs to a Japanese beetle grub. When the eggs hatch, the young wasps eat the grub. The beetles' population is thus controlled.

Bacteria and viruses are also effective against certain pests. In the United States a virus that kills cotton bollworm and corn earworm has been manufactured. These classes of bacteria and viruses reproduce naturally and need not be "reapplied" for years.

Sterilization of adult pests is another method of population control. For example, the screwworm population was controlled for a time by sterilizing male screwworm flies and having them mate with normal females. Eggs were laid, but they did not hatch. However, females eventually stopped mating with sterile males and the screwworm population rose once again.

Hormones can be used to kill pests. Since the life cycles of insects are controlled by hormones, artificial hormones can be sprayed on insects to upset their life cycles (Fig. 14-9). The environmental impact of these hormones is minimal and predators are not killed (Fig. 14-10).

Sex attractants are another group of chemicals used to kill insects. When a virgin female is ready to mate, she emits a sex attractant. These chemicals are called *pheromones*. Traps have been baited both with live females and with artificial pheromones. The males are attracted to the traps by the chemicals. Little environmental disturbance is brought about by this method and the

Figure 14-9 The life cycle of an insect. Hormones regulate this life cycle. Application of juvenile hormone during the insect's larva or pupa stage can trap the insect in this stage and prevent it from maturing into an adult.

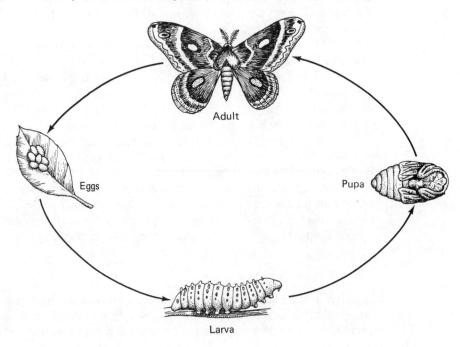

Adult

Eggs

Pupa

Larva

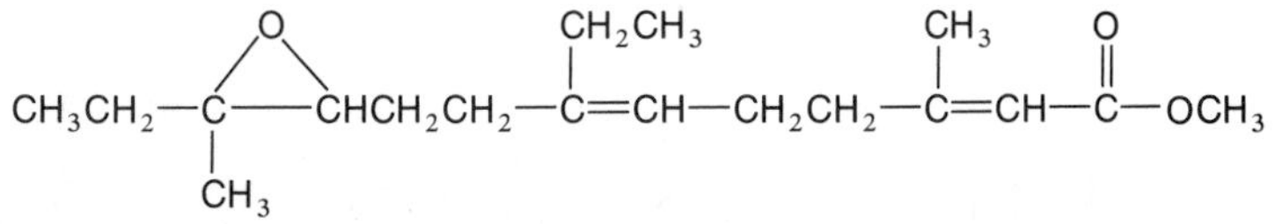

Figure 14-10 The structure of mosquito juvenile hormone.

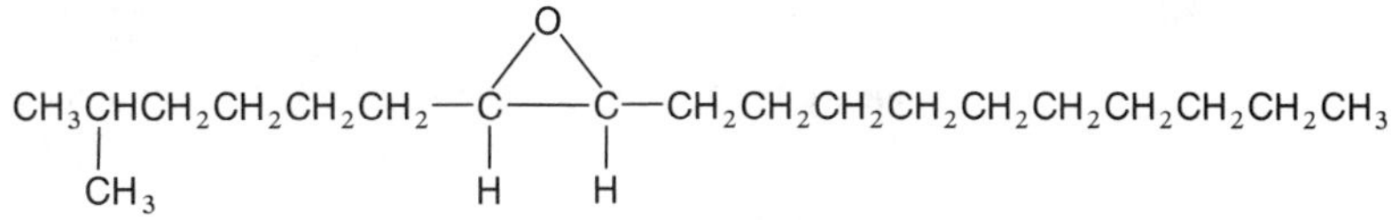

Figure 14-11 The chemical structure of the gypsy moth pheromone.

only drawback is that large numbers of traps must be set to control populations (Fig. 14-11).

Resistant strains of crops are yet another way to solve the insect problem. Certain plant varieties produce their own insecticides. Researchers have been able to develop alfalfa that is resistant to alfalfa weevil. Some cereal crops that are resistant to rust infections have also been developed. It is a very big job to produce a resistant strain of plant which maintains a high yield as well. And, of course, there is nothing to stop insects from undergoing an evolutionary change which allows them to overcome the plant's defense system.

HERBICIDES: THE WEED KILLERS

Weeds are another problem with which farmers and gardeners must deal. Weeds can grow so thick that they steal nutrients from the plants and choke them out. Chemicals called *herbicides* have been developed which kill weeds but don't harm plants. In the United States alone, about 400 million kilograms of herbicides were produced in 1978.

Herbicides, like all chemicals sprayed in the environment, have some adverse affects. Earthworms and other soil microorganisms have been killed by some herbicides. Others have killed the much needed insect predators.

The U.S. Forest Service has used the herbicide 2,4,5-T to kill weeds which hinder the growth of timber (Fig. 14-12). But 2,4,5-T is toxic to human beings and it causes birth defects in combination with a *dioxin*, which is a contaminant in the commercial preparation of the herbicide (Fig. 14-13). A very large number of women living in an area of Minnesota that had been sprayed with 2,4,5-T suffered miscarriages within months of the application. Farmers in the area became dizzy and had headaches, nausea, and diarrhea. The herbicide 2,4,5-T is currently on the list of dangerous chemicals.

The use of herbicides poses a problem for farmers. If they choose not to use herbicides, people must be hired to pick the weeds. This costs money and drives up the cost of food. The alternative is to continue using herbicides,

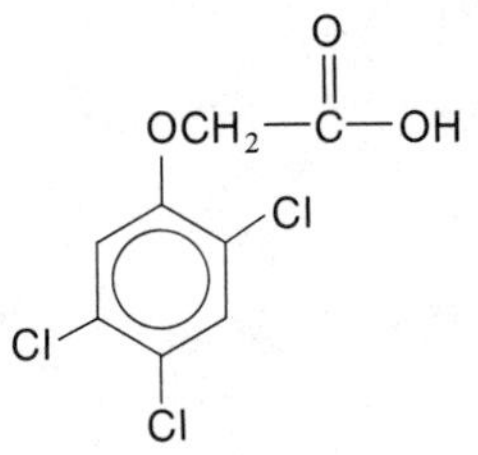

Figure 14-12 The structure of 2,4,5-T, a powerful herbicide.

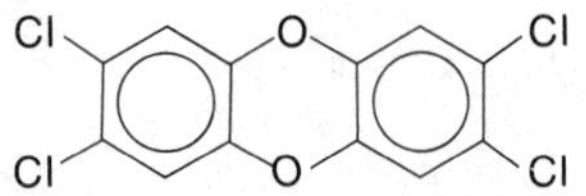

2,3,7,8-Tetrachlorodibenz-p-dioxin

Figure 14-13 The compound 2,3,7,8-tetrachlorodibenz-*p*-diozin, which is a contaminant in the preparation of 2,4,5-T, may be responsible for causing birth defects in the offspring of women exposed to this substance.

which is what is being done. The effect on the health of people and the environment remains to be seen.

Here are some other herbicides which are in use. *Atrazine* is a chemical that is capable of blocking an important step in the process of photosynthesis (Fig. 14-14). The root cells in corn and sorghum plants can deactivate this chemical, so atrazine can kill weeds but not corn or sorghum.

Paraquat is a preemergent herbicide (Fig. 14-15). It kills weeds before the seedlings emerge. It is broken down in the soil and doesn't stay around too long.

Another class of herbicide is the *defoliant*. These chemicals cause plants to lose their leaves. *Agent Orange* is a chemical defoliant which was used during the Vietnam war. Veterans of this war are now experiencing long-range effects of the use of this defoliant. This substance may have damaging effects on the kidney and liver and may even be a cancer-causing agent. There is also some indication that this substance can cause birth defects in children whose parents were exposed to it.

Figure 14-14 The compound atrazine is an important herbicide that can kill weeds but not corn or sorghum.

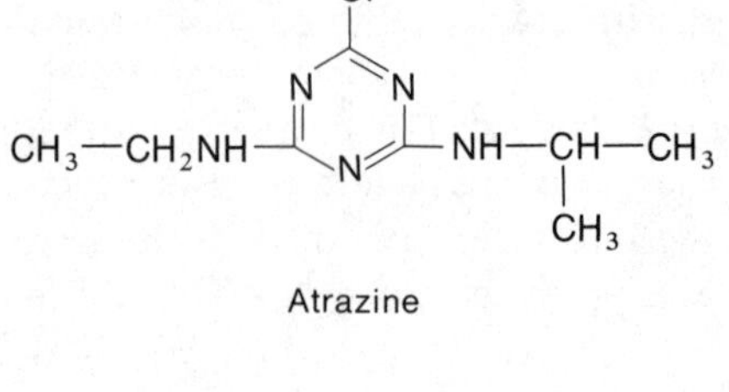

Atrazine

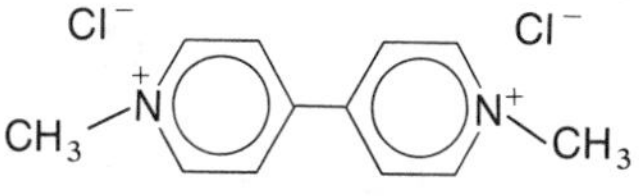

Cl⁻   Cl⁻

Paraquat

Figure 14-15 The compound paraquat is a preemergent herbicide. It kills weeds before the seedlings emerge.

FOOD THAT'S ORGANICALLY GROWN

Organic farming is accomplished without the use of chemical fertilizers, pesticides, or herbicides. The fertilizers used are natural fertilizers. This means that they are not the inorganic type of fertilizer that is manufactured in a factory. Manure, slaughterhouse waste, and other organic materials are used. Such fertilizer is bulkier to use than manufactured fertilizers. It's also cheaper and is a renewable resource. Fuel is conserved and mineral reserves aren't touched. More human labor is needed by organic farmers, but much more energy is saved (Fig. 14-16).

Are organically grown foods more nutritious to eat? Not really. Chemical analysis shows that the nutrient contents of foods are the same whether grown organically or inorganically. However, food that is grown organically has a less negative impact on the environment.

SUMMARY

In this chapter we discussed agricultural chemistry. We began by looking at the process of plant growth. We saw how photosynthesis and animal metabolism work hand in hand to produce a carbon dioxide–oxygen cycle in

Figure 14-16 Organic versus inorganic farming.

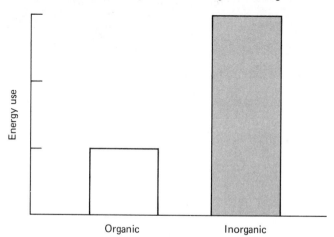

nature. Next, we reviewed the various ways that soil is fertilized. We began by looking at nitrogen fertilizers, and we examined three ways that nitrogen can be supplied to plants. We also examined potassium and phosphate fertilizers and discovered how they are produced. We also learned what the numbers on a bag of fertilizer mean.

In the next part of the chapter we looked at hybrid plants and discussed some of the implications of using these plants. Next, we looked at the use of pesticides and herbicides that have been developed to protect our food supply. We also examined some natural ways of eliminating pests.

Finally, we took a brief look at organically grown food.

EXERCISES

1. Explain the process of photosynthesis.

2. Describe the ways in which nitrogen can be supplied to plants.

3. Write the equation for the Haber process for the formation of ammonia.

4. List the various forms of fixed nitrogen.

5. Write an equation for animal metabolism.

6. What are legumes?

7. Name some sources of potassium and phosphorus fertilizers.

8. What is the meaning of the numbers 20–10–5 on a bag of fertilizer?

9. What is the Green Revolution?

10. What is a pest?

11. List some reasons for the use of DDT.

12. What chemicals are needed to synthesize DDT?

13. Discuss the long-range environmental impact of DDT.

14. Discuss various ways in which insect populations can be controlled.

15. What is a pheromone?

16. What is a herbicide? Discuss some of the positive and negative aspects of using herbicides.

17. What is a defoliant?

18. The defoliant Agent Orange was used during the Vietnam war. What were some of the reasons for its use?

19. List some of the advantages of growing food *organically*. Does organically grown food have any greater nutrient value than food not grown organically?

Chapter 15

Food: A Chemical Collection

Some Things You Should Know After Reading This Chapter

You should be able to:

1. Name the various nutrients found in foods.
2. Name the three classes of carbohydrates and give an example of each.
3. Explain the importance of starch in the diet.
4. Calculate the energy produced from the metabolism of a carbohydrate or fat given the necessary information.
5. Explain the difference in the terms calorie, Calorie, and kilocalorie.
6. List the properties of lipids.
7. Distinguish among monoglycerides, diglycerides, and triglycerides and give an example of each.
8. Distinguish between a saturated fat and an unsaturated fat.
9. Discuss the importance of protein in the diet, and explain the difference between a complete and an incomplete protein.
10. Discuss the basic building blocks of proteins—the amino acids.
11. Explain the difference between fat-soluble and water-soluble vitamins.
12. Discuss the functions in the body of the minerals calcium, iodine, and chlorine.
13. Discuss some of the pros and cons of eating processed foods.
14. Discuss some of the positive and negative aspects of consuming soft drinks.
15. Explain the importance of milk in the diet.
16. List constituents of breast milk and infant feeding formulas.
17. Name four categories of food additives.
18. Discuss the pros and cons of using food additives in the food we eat.

STARTING OFF RIGHT

It is the year 2005, a step into the twenty-first century. A baby has just been born and is given to her mother to be fed. The baby is being breast-fed and a natural process is taking place. The mother is relaxed and comfortable and the baby is being well nourished.

Toward the end of the twentieth century, a trend toward artificial feeding of infants was seen. In the industrialized countries of the world substitutes for breast milk were easily available. Millions of mothers gave their infants formulas made from pasteurized cow's milk. These babies showed a fine pattern of growth and weight gain. But experts learned that breast milk was better. It provides the infant with some forms of immunity to disease. It is more easily digested, and it rarely causes an allergic reaction. Breast milk also contains a natural laxative.

But problems related to artificial feeding soon arose in many of Middle Eastern and African countries and in South and Central America as well. Largely as a result of commercial advertising, urban women of these countries abandoned breast feeding and switched to milk formulas. In the absence of

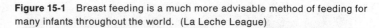

Figure 15-1   Breast feeding is a much more advisable method of feeding for many infants throughout the world. (La Leche League)

absolutely safe drinking water, good sanitary conditions, and refrigeration, baby formulas sometimes became the carriers of common infections. This caused many diseases, which were once seen mainly in older children, to be seen in infants. Another problem with artificial milk formulas is that some mothers tend to dilute them in order to save money; some of them carry it to an extreme, with the result that the baby does not get sufficient nourishment. Reeducation of the population as to the values of breast feeding reversed this trend. Even an undernourished mother is able to produce 400 to 600 ml of milk daily for her child. This is a much more advisable method of feeding for many infants throughout the world (Fig. 15-1).

INTRODUCTION: YOU ARE WHAT YOU EAT

From the moment of conception it is important that a human being be properly nourished. The amounts as well as the types of food that you eat have an effect on the ultimate size and strength that you achieve.

Foods are composed of water and a few organic compounds, which are proteins, carbohydrates, fats, vitamins, and minerals. When you eat, the food is digested and broken down by the various digestive organs of the body to form smaller molecules. These small molecules then get together to build new cells, which help us to grow and to replace old, worn-out cells when necessary. Energy is also produced to help power the body (Fig. 15-2). The human body is an amazing machine. It is terribly complex in its design and works with great

Figure 15-2 How food is used.

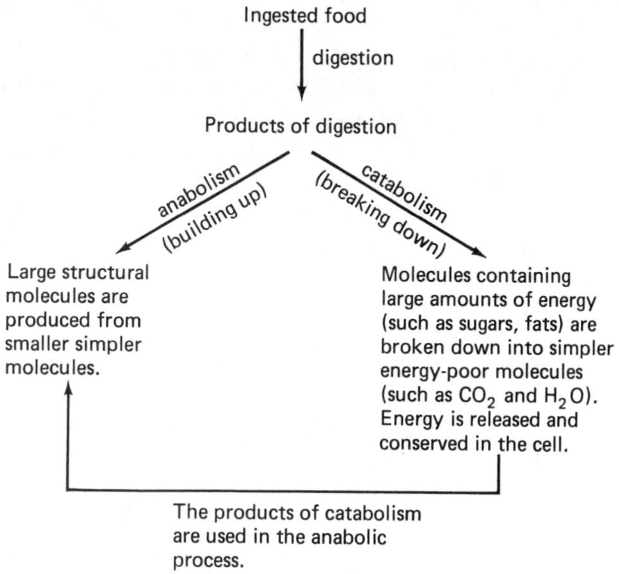

Ingested food

| digestion

Products of digestion

anabolism (building up)   catabolism (breaking down)

Large structural molecules are produced from smaller simpler molecules.

Molecules containing large amounts of energy (such as sugars, fats) are broken down into simpler energy-poor molecules (such as CO_2 and H_2O). Energy is released and conserved in the cell.

The products of catabolism are used in the anabolic process.

accuracy. For example, it is able to maintain itself at the proper working temperature of 37°C (98.6°F). The chemicals that make up the body work best at this temperature, so a regulatory mechanism works to maintain this temperature. Proper levels of various special chemicals called hormones and enzymes must be present for chemical reactions to take place in the body. Just the right amounts of these hormones and enzymes are produced and appear at the right place at the right time so that the numerous chemical reactions can occur. The body is a fantastic machine and all that we have to do to keep it working is to feed it right! Let's see what constitutes our food.

CARBOHYDRATES: HOW SWEET THEY ARE

Carbohydrates are an important part of our diets. The foods we eat furnish us with two basic types of carbohydrates: sugars and starches. The sugars are simple carbohydrates and the starches are complex carbohydrates. At the beginning of the twentieth century, Americans consumed *more* carbohydrates than we do today. Even though we have reduced our overall carbohydrate intake from about 492 g per person per day to about 390 g per person per day, the types of carbohydrates that we consume have changed drastically. In the early 1900's, 68.3% of our carbohydrate consumption was starches and only 31.7% was sugars. Now we consume 47.1% starches and a whopping 52.9% sugars (Table 15-1). What's so bad about this? Well, sources of starch, which include foods such as breads, cereals, whole grains, and potatoes, are also good sources of vitamins, proteins, and minerals. Sugars contain only calories—no nutrients to speak of, so they add empty calories to the diet. Starches are also good sources of dietary fiber. Such starches as whole grains, whole-grain breads, and cereal contain dietary fiber, which means that they are made of complex carbohydrates that can't be digested by the body. This means that they add bulk to the feces and are important for the proper functioning of the digestive system. Some people think that these foods also help reduce the risk of cancer of the colon, since they're constantly cleaning out the digestive system of solid wastes.

Now that we know a little about carbohydrates, let's look at their chemical composition (Fig. 15-3).

First, carbohydrate molecules are organic compounds which are generally composed of the elements carbon, hydrogen, and oxygen. Some carbohydrates

TABLE 15-1 *Sources of carbohydrates*

YEAR	PERCENT STARCH IN THE DIET	PERCENT SUGAR IN THE DIET
1909–1913	68.3	31.7
1957–1959	49.6	50.4
1976	47.1	52.9

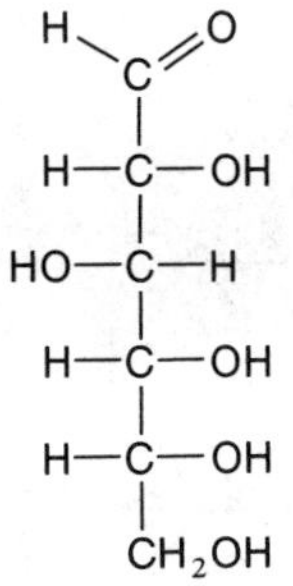

Glucose (open chain structure)

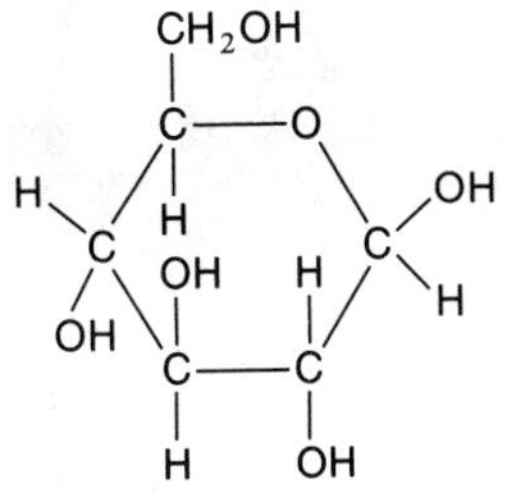

Glucose (ring structure)

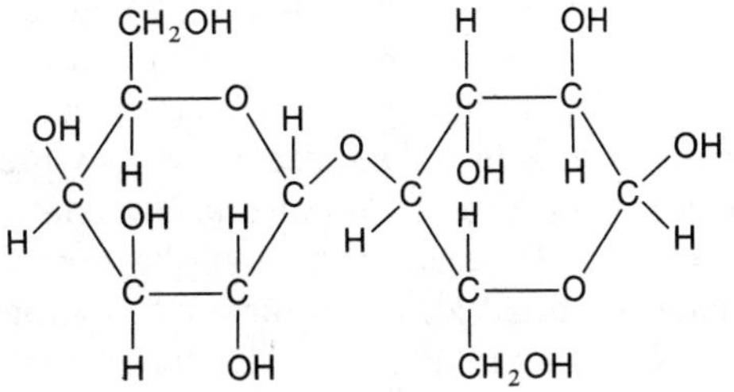

Lactose

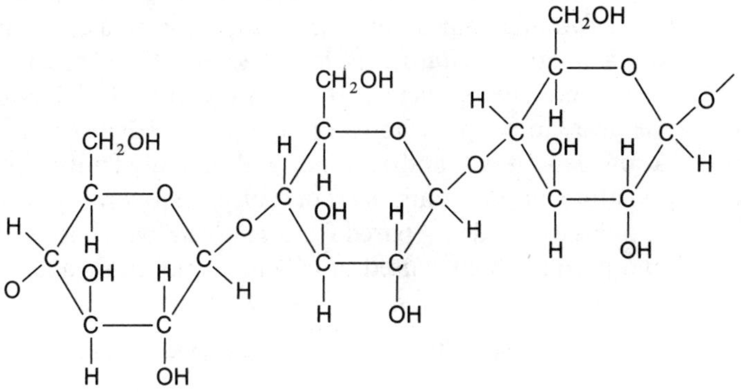

Cellulose

Figure 15-3 Some carbohydrate structures.

also contain nitrogen and sulfur. Chemically speaking, carbohydrates are defined as polyhydroxyaldehydes or polyhydroxyketones. You remember from our discussion of organic chemistry (Chap. 12) that polyhydroxy means that several —OH groups are present. Aldehydes and ketones both contain the carbonyl group $(-\underset{\underset{\displaystyle O}{\|}}{C}-)$. So carbohydrates are alcohols and are also

TABLE 15-2 *Hydrolysis products*

Disaccharides $\xrightarrow{\text{hydrolysis (H}_2\text{O) (acid)}}$ 2 monosaccharides

Polysaccharides $\xrightarrow{\text{hydrolysis (H}_2\text{O) (acid)}}$ many monosaccharides

aldehydes and ketones. Substances that form polyhydroxyaldehydes or polyhydroxyketones upon *reaction with water* (this is called hydrolysis) are also classified as carbohydrates.

Carbohydrates are classified according to the size of the molecule. There are three principal categories, based on the number of sugar or *saccharide* groups in the molecule. These are monosaccharides, disaccharides, and polysaccharides. *Monosaccharides*, which are also called *simple sugars*, contain only one saccharide group in the molecule. *Disaccharides* contain two saccharide groups in the molecule and they break down into simple sugars upon reaction with water. *Polysaccharides* contain many saccharide groups in the molecule, and these are the complex carbohydrates. Upon hydrolysis, polysaccharides break down to form many simple sugars (Table 15-2).

Now let's take a look at some important carbohydrates. Of the simple sugars, glucose, galactose, and fructose are the three most important. They all have the molecular formula $C_6H_{12}O_6$, but their structures are slightly different (Fig. 15-4). Because these three simple sugars have the same molecular formulas but different structural formulas, we say that they are *isomers* of each other. *Glucose* is found in the bloodstream and is also known as dextrose or blood sugar. It doesn't have to be digested, so it is given intravenously to people who can't take food orally. *Fructose* is the sweetest sugar known and it is found in honey and in many fruits. *Galactose* isn't found free in nature but is a component of such carbohydrates as lactose (*milk sugar*), *agar* (a polysaccharide prepared from seaweed, which is used to treat constipation), and pectin (which is used in jelly as a thickening agent).

Figure 15-4 The structures of glucose, galactose, and fructose.

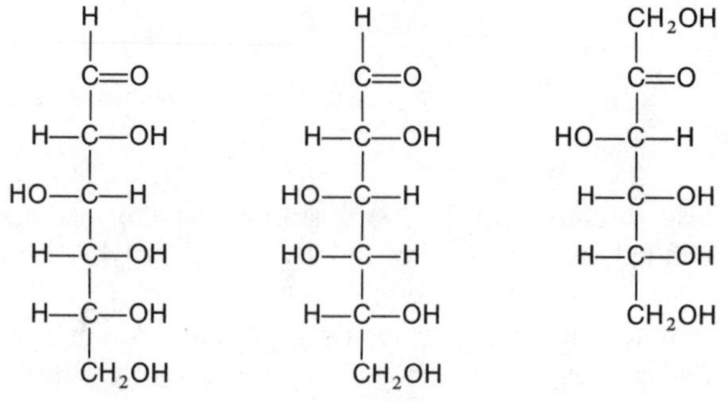

Glucose   Galactose   Fructose

Of the disaccharides, the three most important are sucrose, lactose, and maltose. The disaccharides are polymers of monosaccharides. In other words, certain monosaccharides link together to form the disaccharides. *Sucrose* is found in sugarcane and sugar beets. It is called *table sugar*. *Lactose* is known as milk sugar and is the sugar found in human milk as well as in the milk of many animals. *Maltose* is known as malt sugar and is used in the production of beer (Fig. 15-5).

Figure 15-5 The structure of maltose, sucrose, and lactose.

Glucose + glucose $\longrightarrow$ maltose + H_2O

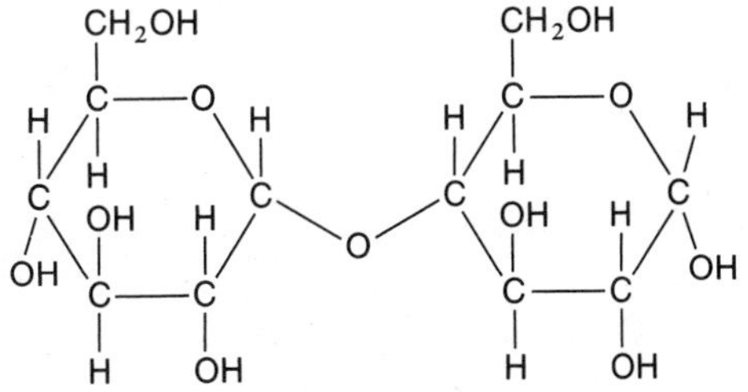

maltose

Glucose + fructose $\longrightarrow$ sucrose + H_2O

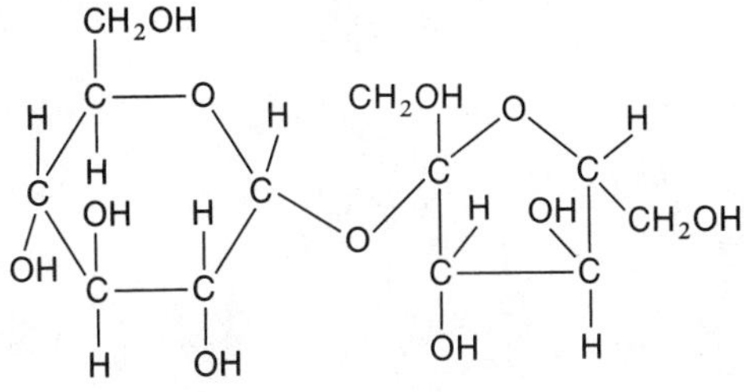

sucrose

Glucose + galactose $\longrightarrow$ lactose + H_2O

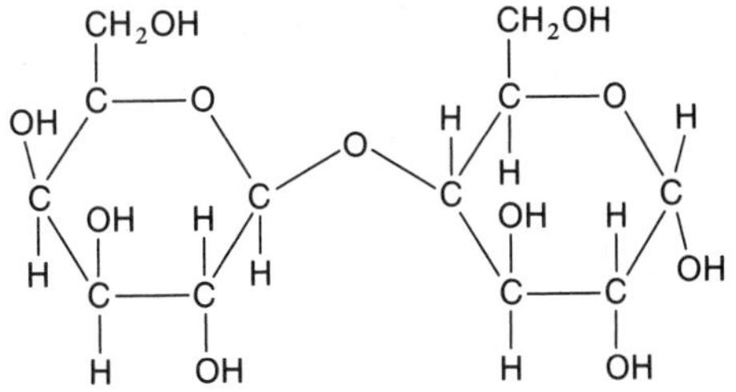

lactose

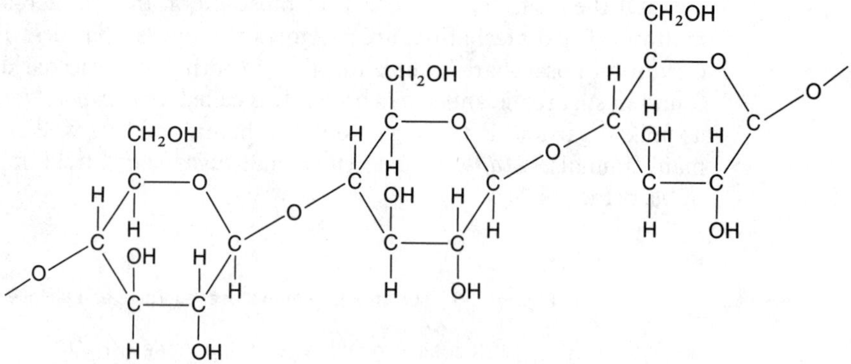

Cellulose consists of from 900 to 6000 glucose units linked together in a straight chain. This polymer consists of beta-glucose units.

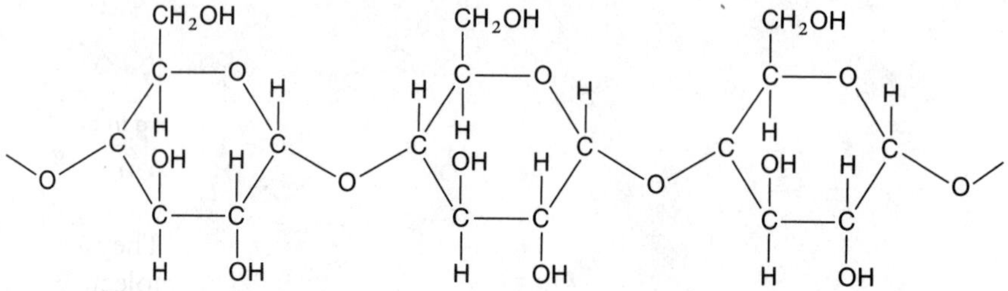

Amylose: a straight chain glucose polymer consisting of from 250 to 500 alpha-glucose molecules linked together is found in starch.

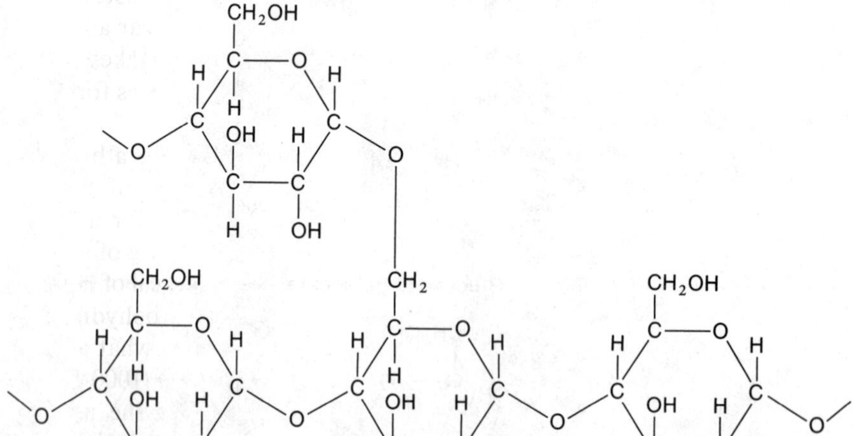

Amylopectin: a highly branched glucose polymer consisting of about 1000 alpha-glucose molecules is found in starch.

Figure 15-6 Some examples of polysaccharides.

Figure 15-7 Every gram of carbohydrate that we take in gives us 4 kcal of energy for our bodies.

There are several important polysaccharides. They are polymers made up of hundreds and even thousands of simple sugar molecules linked together in various arrangements. *Starch* is the most important carbohydrate that we eat. Such foods as vegetables, potatoes, wheat, corn, and rye contain large amounts of starch. *Cellulose* is important in the diet since it can't be digested and therefore serves as bulk. It aids in the excretion of solid wastes, as it helps move food through the intestinal tract. *Glycogen* is stored sugar and is found in the liver and muscles. *Dextrin* is the polysaccharide that makes up the glue found on postage stamps. During the baking of bread dextrins form and give the bread a golden brown color (Fig. 15-6).

Before we go on to the next section, let's take a brief look at the energy produced from carbohydrates. For every gram of carbohydrate that we utilize, 4 kilocalories (kcal) of energy is produced for use by our bodies (Fig. 15-7). What this really means is that if we burned 1 g of a carbohydrate in pure oxygen in a special device called a calorimeter, 4 kcal of heat would be released. So if your body burns (metabolizes) 1 g of carbohydrate, 4 kcal of energy would be provided to your body as a fuel. Now what about this term "kilocalorie"? The chemist uses the word *kilocalorie* (1000 calories) to mean the same thing as the nutritionist does when he or she uses the word *Calorie* (with a capital C). So 1 *kilocalorie* is the same as the nutritionist's 1 Calorie, sometimes called a *big Calorie*.

Example 15-1 How many kilocalories of energy are produced when 125 g of a carbohydrate is burned? How many *big Calories* are produced?

Solution
$$? \text{ kcal} = (125 \text{ g})\left(\frac{4 \text{ kcal}}{1 \text{ g}}\right) = 500 \text{ kcal}$$

Because 1 kcal = 1 big Calorie, 500 kcal equals 500 big Calories.

LIPIDS: THE CHEMICALS THAT CAUSE THE GREASY SPOON

It is very alarming to see that our diets are becoming higher in fat intake than ever before. Our fat intake has grown by a full 10% since the early 1960's. Where has this increased fat consumption come from? We are eating more margarine, vegetable oil, more shortening, and more salad and cooking oils than ever before. We have cut down on animal fat consumption and increased our consumption of vegetable fats. We have been bombarded with information which tells us that there seems to be a link between high fat intake and cardiovascular diseases and maybe even certain cancers. But it is still not known why some people who eat diets high in fats, especially saturated fats and cholesterol, develop atherosclerosis (clogged arteries) and why others, on the same diet, don't suffer from these problems. Over 1 million people in America die each year from cardiovascular disease. And we are told to eat polyunsaturated fats, yet we are just beginning to explore the possible hazards of consuming polyunsaturates, since some of them are suspected of promoting cancer.

In 1979, the Senate Select Committee on Nutrition and Human Needs, under Senator George McGovern, Democrat of South Dakota, suggested that we cut our cholesterol intake in half and reduce our fat consumption to 30% of our total food intake. We presently consume 42% fats in our diet. Let's take a look at the chemical composition of fats so that we can have a greater understanding of the terminology involved.

Fats and oils fall under the general category of *lipids* (Table 15-3). Lipids are an important group of biochemical substances found in both plant and animal tissues. The seeds and the fruit of plants contain most of the fat found in plants. *Vegetable fats*, called *oils*, are found mainly in the seeds. Some seeds have a very high oil content. For example, the coconut seed is about 65% oil. *Waxes*, which are also lipids, serve as protective agents on plant leaves. Waxes are there to prevent or reduce the loss of moisture from the plant. Plants found in dry desert regions develop a very thick wax coat on the underside of the leaf to reduce water loss. Carnauba wax, which is found on the leaves of a special type of Brazilian palm tree, is widely used in floor polishes, shoe creams, car wax, and in the manufacture of carbon paper.

TABLE 15-3 *Some examples of lipids*

Fats	Cholesterol	Oils
Waxes	Vitamin A	Cortisone

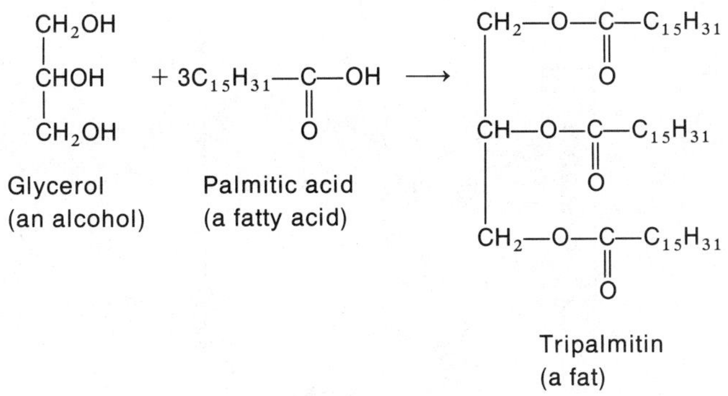

$$
\begin{array}{c}
\text{CH}_2\text{OH} \\
| \\
\text{CHOH} \\
| \\
\text{CH}_2\text{OH}
\end{array}
\quad + 3\text{C}_{15}\text{H}_{31}-\underset{\underset{\text{O}}{\|}}{\text{C}}-\text{OH} \longrightarrow
\quad
\begin{array}{c}
\text{CH}_2-\text{O}-\underset{\underset{\text{O}}{\|}}{\text{C}}-\text{C}_{15}\text{H}_{31} \\
| \\
\text{CH}-\text{O}-\underset{\underset{\text{O}}{\|}}{\text{C}}-\text{C}_{15}\text{H}_{31} \\
| \\
\text{CH}_2-\text{O}-\underset{\underset{\text{O}}{\|}}{\text{C}}-\text{C}_{15}\text{H}_{31}
\end{array}
$$

Glycerol   Palmitic acid

(an alcohol)   (a fatty acid)

Tripalmitin

(a fat)

Figure 15-8 Fats are esters of fatty acids and glycerol.

Let's look at some of the properties of lipids:

1. They are insoluble in water and soluble in one or more organic solvents, such as ethanol, acetone, benzene, chloroform, and carbon tetrachloride. These organic solvents are called "fat solvents," and they are nonpolar organic molecules.

2. They contain carbon, hydrogen, and oxygen atoms and sometimes phosphorus or nitrogen atoms.

3. They are almost all esters that form *fatty acids* on hydrolysis. (Fatty acids are usually straight-chain carboxylic acids that generally contain an even number of carbon atoms.)

4. They are important components of all living matter.

To understand what saturated and polyunsaturated fats are, we must first learn about fatty acids. Most *fatty acids* are found as essential components of other lipid molecules. When fats are reacted with water (hydrolysis), fatty acids can be obtained. To produce fats, fatty acids must bond to an alcohol called glycerol, in a special reaction called an *esterification reaction* (Fig. 15-8). Glycerol is an alcohol containing three —OH groups (Fig. 15-9). It is commonly called *glycerin*. It has a sweet taste and is a colorless, oily liquid. When one,

Figure 15-9 The structure of the polyhydroxy alcohol, glycerol.

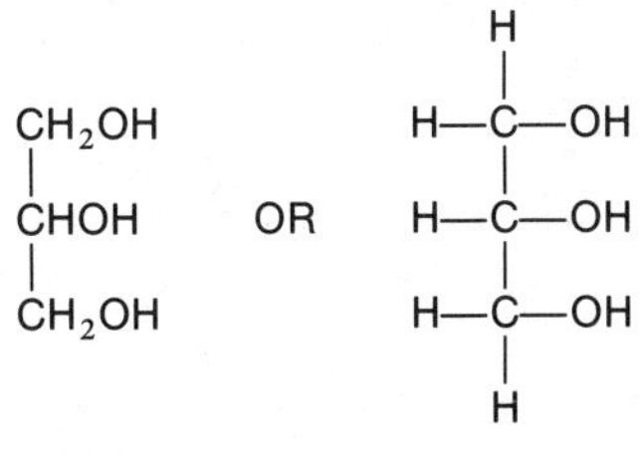

$$
\begin{array}{c}
\text{CH}_2\text{OH} \\
| \\
\text{CHOH} \\
| \\
\text{CH}_2\text{OH}
\end{array}
\qquad \text{OR} \qquad
\begin{array}{c}
\text{H} \\
| \\
\text{H}-\overset{}{\underset{}{\text{C}}}-\text{OH} \\
| \\
\text{H}-\overset{}{\underset{}{\text{C}}}-\text{OH} \\
| \\
\text{H}-\overset{}{\underset{}{\text{C}}}-\text{OH} \\
| \\
\text{H}
\end{array}
$$

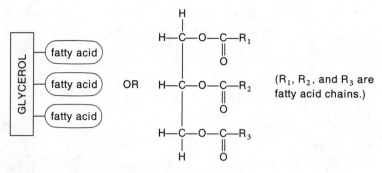

Figure 15-10 The general formula for a fat.

two, or three fatty acids bond to a molecule of glycerol, a fat is produced (Fig. 15-10).

Monoglycerides are fats that are produced when just *one* glycerol —OH group combines with one fatty acid in an ester linkage. *Diglycerides* are fats in which *two* glycerol —OH groups are combined with *two* fatty acids in ester linkages. The —OH groups that have not undergone esterification remain unchanged. *Triglycerides* are fats in which all *three* of the glycerol —OH groups have undergone esterification and are bonded to the fatty acid chains (Fig. 15-11).

To get the whole picture, let's take a look at the different types of fatty acids. Most fatty acids are composed of a straight chain of carbon atoms connected to a carboxyl group, $-\overset{\displaystyle \|}{\underset{\displaystyle O}{C}}-OH$, which is found at one end of the molecule. Most of the natural fatty acids have an even number of carbon atoms in their molecules. The fatty acids that compose natural fats and oils can be classified into two groups; the *saturated fatty acid series* and the *unsaturated fatty acid series*. The saturated fatty acids contain fatty acids with *no double bonds*. The unsaturated fatty acids have *one or more double bonds*

Figure 15-11 The general formula for a monoglyceride and a diglyceride.

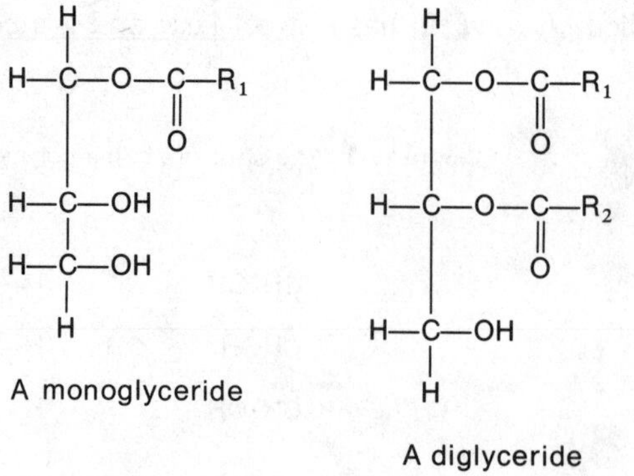

A monoglyceride

A diglyceride

TABLE 15-4 *Some important fatty acids*

ACID SERIES	NAME	TOTAL NUMBER OF CARBON ATOMS	FORMULA	OCCURRENCE
Saturated fatty acids	Butyric	4	C_3H_7COOH	Milk fat
	Caproic	6	$C_5H_{11}COOH$	Butter, coconut oil, palm nut oil
	Caprylic	8	$C_7H_{15}COOH$	Butter, coconut oil, palm nut oil
	Capric	10	$C_9H_{19}COOH$	Butter, coconut oil, palm nut oil
	Lauric	12	$C_{11}H_{23}COOH$	Laurel oil, coconut oil
	Myristic	14	$C_{13}H_{27}COOH$	Nutmeg oil
	Palmitic	16	$C_{15}H_{31}COOH$	Animal and vegetable fats
	Stearic	18	$C_{17}H_{35}COOH$	Animal and vegetable fats
	Arachidic	20	$C_{19}H_{39}COOH$	Peanut oil
Unsaturated fatty acids	Crotonic	4	C_3H_5COOH	Croton oil
	Oleic	18	$C_{17}H_{33}COOH$	Animal and vegetable fats
	Linoleic	18	$C_{17}H_{31}COOH$	Vegetable oils such as linseed and cottonseed oil
	Linolenic	18	$C_{17}H_{29}COOH$	Linseed oil
	Arachidonic	20	$C_{19}H_{31}COOH$	Nervous tissue, lecithin, cephalin

and are liquids at room temperature. Polyunsaturated fatty acids contain many double bonds (Table 15-4).

Vegetable fats, called oils, generally contain more unsaturated fatty acids than saturated fatty acids, so they are liquids at room temperature. Animal fats are usually solids at room temperature because they contain an abundance of saturated fatty acids (Fig. 15-12). Liquid vegetable oils, as we said, are generally high in polyunsaturated fat, except for palm and coconut oils, which are high in saturated fat, and peanut and olive oils, which are high in mono-unsaturated fat (just one double bond). But more and more vegetable oils are "hydrogenated," which means they are hardened by a process that turns polyunsaturated fat into saturated or mono-unsaturated fat. This is used for margarine, shortening, and in many processed foods.

Let's take a look at the energy produced by the burning of fat. For every gram of fat that is metabolized, 9 kcal of energy is produced. This is more energy production than any other nutrient provides. When we eat more food than we need and consume too many calories, our bodies produce fat. The fat is stored in various cells throughout the body and serves as an energy reservoir as well as an insulating material. Too much fat can pose a health problem and this is why so many people are trying to lose weight and decrease their fat reservoirs.

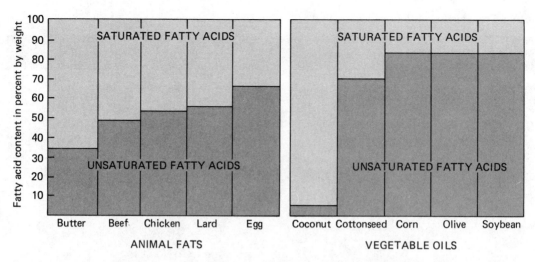

Figure 15-12 A comparison of saturated and unsaturated fatty acids in some foods.

PROTEINS: THE ESSENTIAL SUBSTANCES OF LIFE

Throughout the body we find a very important type of organic compound—the protein. Proteins are essential in sustaining normal life processes. Some of the functions of proteins include making the body's chemical reactions work right, defending the body against disease, helping with the digestion of foods, moving oxygen around in the blood, aiding in the clotting of blood, regulating the activities in our cells, and helping our genes function properly.

Proteins are made by living matter. Plants produce the greatest amount of protein. They synthesize carbohydrates and then combine these carbohydrates with nitrogen-containing compounds from the soil to produce proteins.

Proteins are composed of the elements carbon, hydrogen, oxygen, and nitrogen. Some proteins contain sulfur, iodine, phosphorus, or iron. Proteins

Figure 15-13 The general structure of an amino acid.

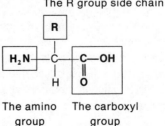

are large molecules with weights ranging from 5000 to millions of grams per mole. They are such large molecules because they are polymers composed of simpler units called *amino acids*. Let's take a brief look at amino acids.

Amino acids are the molecules that make up proteins. They have two functional groups, a carboxyl group ($-\overset{\underset{\|}{O}}{C}-OH$) on one end of the molecule and an amino group ($-NH_2$) on the other end of the molecule. Each amino acid has a different R group as a side chain, which is what makes each amino acid unique (Fig. 15-13 and Table 15-5). There are 20 major amino acids found in proteins. Because amino acids are the building blocks of proteins, they are very important to us biologically. Some amino acids can be synthesized by the body and others can be interconverted from closely related amino acids. There are eight amino acids that *can't* be synthesized by the body in any way and must be included in the diet. These are the *essential amino acids*: lysine,

TABLE 15-5   *The 20 amino acids commonly found in proteins*

THOSE CONTAINING NONPOLAR R GROUPS

Alanine (Ala)

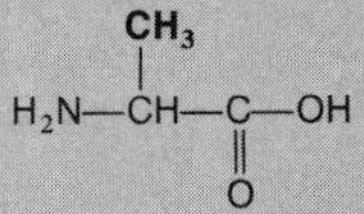

Valine (Val)

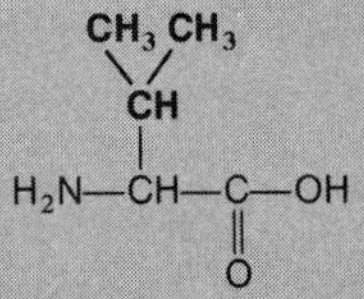

Leucine (Leu)

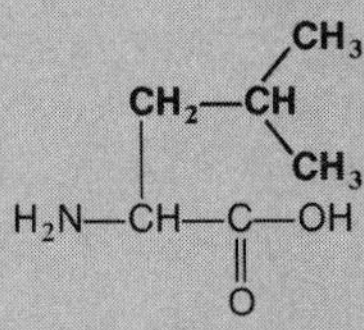

Isoleucine (Ile)

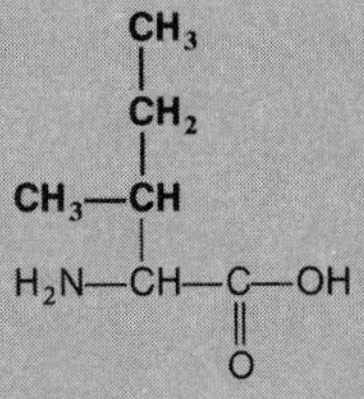

Methionine (Met)

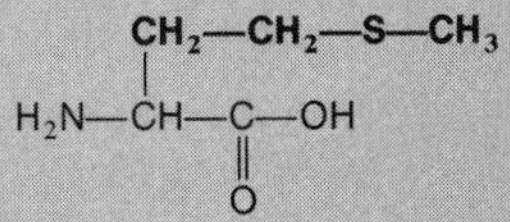

Phenylalanine (Phe)

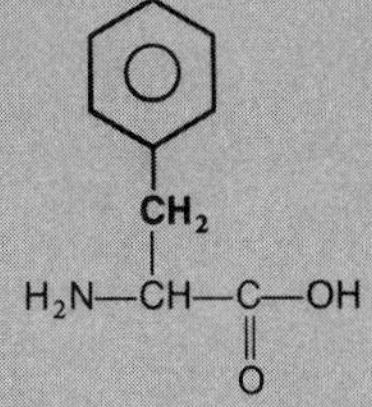

Proline (Pro)

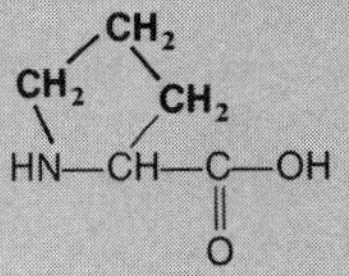

Tryptophan (Trp)

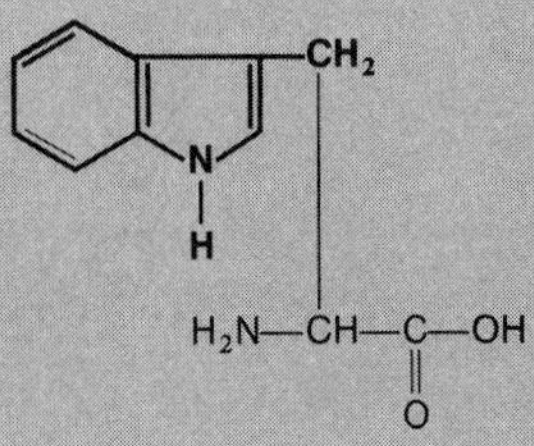

TABLE 15-5 *The 20 amino acids commonly found in protein (cont.)*

THOSE CONTAINING POLAR R GROUPS

Glycine (Gly)

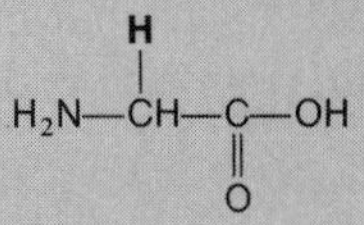

$$H_2N-CH-C-OH$$

with H above CH and O below C

Serine (Ser)

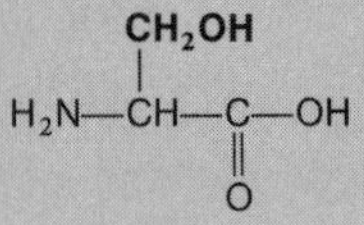

$$H_2N-CH-C-OH$$

with CH_2OH above CH and O below C

Threonine (Thr)

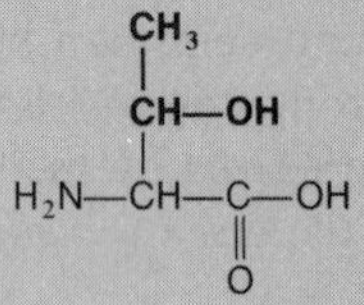

$$H_2N-CH-C-OH$$

with CH_3 / $CH-OH$ above CH and O below C

Tyrosine (Tyr)

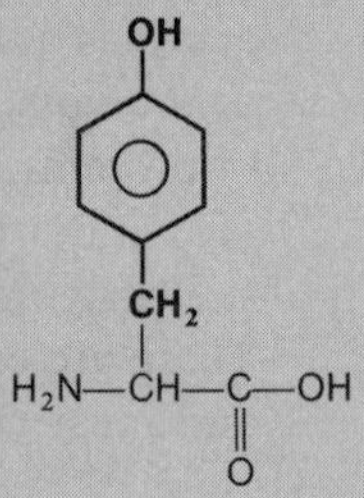

$$H_2N-CH-C-OH$$

with OH—benzene ring—CH_2 above CH and O below C

Cysteine (Cys)

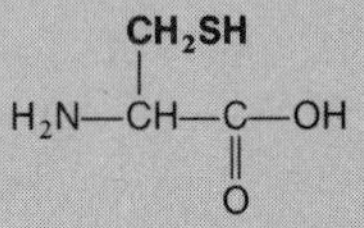

$$H_2N-CH-C-OH$$

with CH_2SH above CH and O below C

Asparagine (Asn)

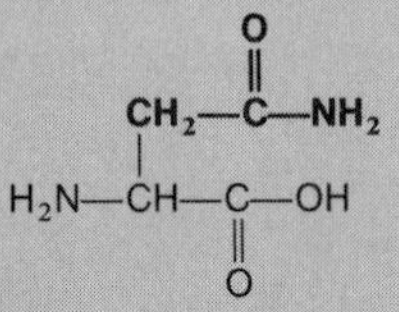

$$H_2N-CH-C-OH$$

with CH_2-C-NH_2 (O above C) above CH and O below C

Glutamine (Gln)

$$H_2N-CH-C-OH$$

with $CH_2-CH_2-C-NH_2$ (O above C) above CH and O below C

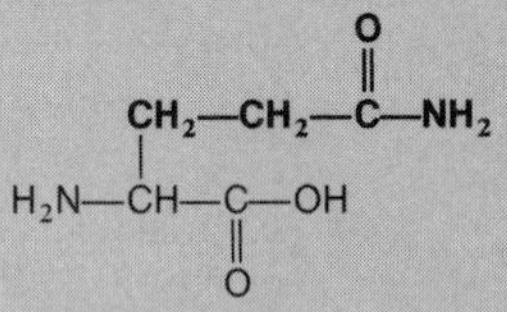

THOSE CONTAINING IONIC R GROUPS

Aspartic acid (Asp)

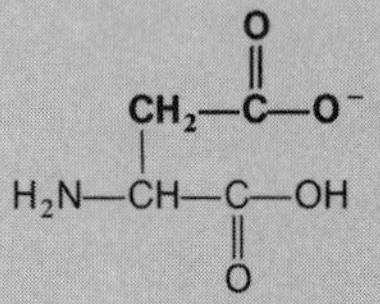

$$H_2N-CH-C-OH$$

with CH_2-C-O^- (O above C) above CH and O below C

Glutamic acid (Glu)

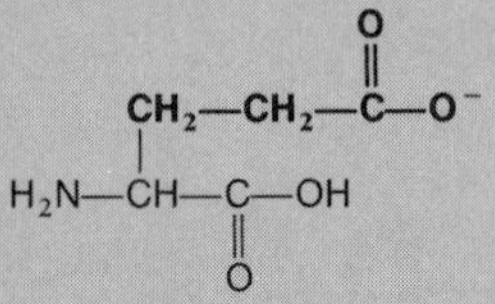

$$H_2N-CH-C-OH$$

with $CH_2-CH_2-C-O^-$ (O above C) above CH and O below C

Lysine (Lys)

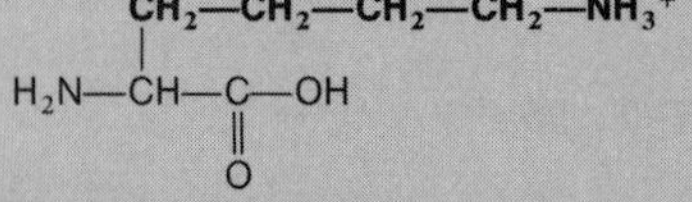

$$H_2N-CH-C-OH$$

with $CH_2-CH_2-CH_2-CH_2-NH_3^+$ above CH and O below C

Arginine (Arg)

$$H_2N-CH-C-OH$$

with $CH_2-CH_2-CH_2-NH-C-NH_2^+$ (NH_2 above C) above CH and O below C

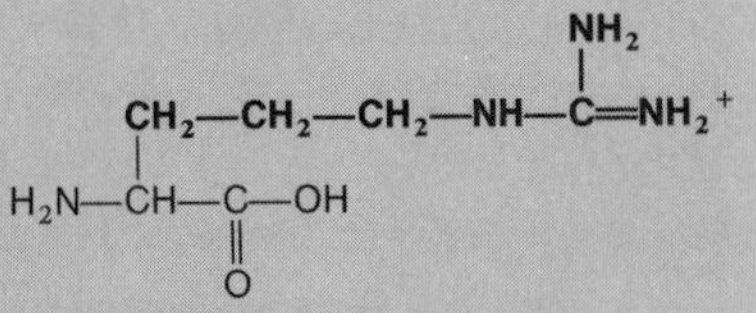

Histidine (His)

$$H_2N-CH-C-OH$$

with imidazole ring—CH_2 above CH and O below C

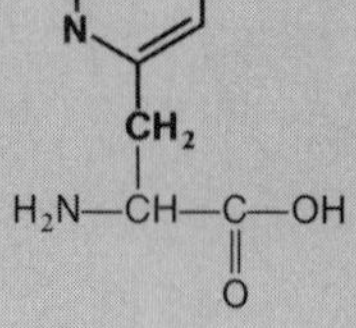

TABLE 15-6 *Essential amino acid requirements*

ESSENTIAL AMINO ACID	GRAMS/DAY (for a 150-pound adult)
Isoleucine	0.84
Leucine	1.12
Lysine	0.84
Methionine[a]	0.70
Phenylalanine[b]	1.12
Threonine	0.56
Tryptophan	0.21
Valine	0.98

[a] Includes cystine.
[b] Includes tyrosine.
Source: The National Research Council—National Academy of Sciences, 1974.

tryptophan, phenylalanine, methionine, threonine, leucine, isoleucine, and valine. Another amino acid, histidine, is sometimes listed with the essential amino acids because it is required by infants and children for proper growth and development—together with the other eight essential amino acids, of course (Table 15-6).

To stay healthy, we must eat foods containing the essential amino acids every day. *Complete proteins* contain all the amino acids essential for the maintenance of good health. Eggs, milk, lean meat, fish, poultry, and soybean are the best sources of complete protein in the diet. Cereal grain, nuts, and vegetables are not sources of complete protein, since they may be low in one or more of the amino acids. Proteins found in such foods are called *incomplete proteins*. Corn, for example, is missing the amino acids lysine and tryptophan. Rice is missing lysine and threonine. Wheat is missing lysine. People whose diets consist chiefly of one type of incomplete protein may suffer from malnutrition even though they are not going hungry. Mixtures of whole-grain cereals and vegetables, or mixtures of whole grains and milk, meat, or eggs, are good sources of complete protein. Generally speaking, proteins that come from vegetable sources are incomplete and those that come from animal sources are complete protein.

VITAMINS: THEY'RE VITAL FOR YOUR HEALTH

Vitamins are organic compounds that can't be produced by the body and must be taken in to keep our bodies working properly. What happens if you don't take vitamins either in your food or as a vitamin supplement? You will develop a deficiency disease. That's just what happens to people whose diets are lacking vitamins. Table 15-7 tells you all about the different vitamins and where to

TABLE 15-7   *Some important vitamins*

NAME	MINIMUM DAILY ADULT REQUIREMENT	FUNCTION OF VITAMIN	DEFICIENCY DISEASE OR SYMPTONS	SOURCE OF VITAMIN IN THE DIET
FAT SOLUBLE				
Vitamin A (retinol and dehydro-retinol)	5000 I.U.[a] (1.5 mg)	Helps to form visual pigments and maintains the normal epithelial structure	Night blindness	Fish liver oils, liver, egg yolk, green leafy or yellow vegetables
Vitamin D (ergocalciforal and cholecalciforal)	400 I.U.	Needed for good bone and teeth formation	Rickets, which results in defective bone formation	Fish liver oil, butter, egg yolk, [together with ultraviolet light (i.e., sunlight)]
Vitamin E (tocopherol)	30 mg as synthetic dl-α-tocopherol acetate	Stabilizes biological membranes and acts as an antioxidant	Red blood cells will have a greater tendency for hemolysis	Vegetable oil, wheat germ, leafy vegetables
Vitamin K (K_2-phylloquinone)	About 0.03 mg/kg of body weight	Needed for prothrombin formation and normal blood coagulation	Hemorrhage	Leafy vegetables, vegetable oils; also is produced by intestinal flora after the fourth day of life
WATER SOLUBLE				
Vitamin C (ascorbic acid)	60 mg	Needed for maintenance of connective tissues, vascular function, tissue respiration, and wound healing	Scurvy, which includes loose teeth and bleeding gums	Citrus fruits, tomatoes, potatoes
Folic acid (folacin)	0.1–0.5 mg, depending on source	Needed for the synthesis of purines and pyrimidines and for the maturation of red blood cells	Anemia	Fresh green leafy vegetables and fruit, organ meats, liver, dried yeast

NAME	MINIMUM DAILY ADULT REQUIREMENT	FUNCTION OF VITAMIN	DEFICIENCY DISEASE OR SYMPTONS	SOURCE OF VITAMIN IN THE DIET
WATER SOLUBLE (*cont.*)				
Niacin (nicotinic acid)	15–20 mg	Acts as a coenzyme in hydrogen transport; also used in carbohydrate and tryptophan metabolism	Pellagra	Dried yeast, liver, meat, fish, whole grains
Vitamin B_1 (thiamine)	1–1.5 mg	Needed for carbohydrate metabolism and for nerve cell function	Beriberi (includes acute cardiac symptoms and heart failure, also weight loss and nerve inflammation)	Whole grains, pork, liver, nuts, enriched cereal products
Vitamin B_2 (riboflavin)	1.0–1.7 mg	Coenzyme in energy transport; also involved in protein metabolism	Visual problems and skin fissures	Milk, cheese, liver, meat, eggs
Vitamin B_6 group (pyridoxine, pyridoxal, pyridoxamine)	2 mg	Necessary for cellular function and for the metabolism of certain amino and fatty acids	Convulsions in infants and skin disorders in adults	Dried yeast, liver, organ meats, fish, whole-grain cereals
Vitamin B_{12} (cyanocobalamin)	5 μg	Necessary for DNA synthesis related to folate coenzymes, also for maturation of red blood cells	Pernicious anemia and certain psychiatric syndromes	Liver, beef, pork, eggs, milk, and milk products
Vitamin H (biotin)	0.15–0.30 mg	Necessary for protein synthesis, transamination, and CO_2 fixation	Skin disorders	Liver, dried peas and lima beans, egg white (and by bacteria in the alimentary canal)
Pantothenic acid	10 mg	Forms part of coenzyme A	Fatigue, malaise, neuromotor, and digestive disorders	Dried yeast, liver, eggs

[a] I.U. stands for international unit. One international unit equals 0.3 μg of retinol.

get them. It also tells which are the water-soluble vitamins and which are the fat-soluble vitamins. Water-soluble vitamins are *not* stored in the body and must be replenished daily. The fat-soluble vitamins are absorbed from the intestine with the fats that we eat. They can be stored in the liver for a period of time.

MINERALS: SOME ARE YOURS AND SOME ARE MINE

The word mineral comes from the Latin word "minera," which means "mine." Minerals, as you might have guessed, come out of the ground. We get minerals from eating plants which grow in soil that contains minerals.

Minerals include about 24 of the chemical elements. Some of them are metals and some are nonmetals. They have various functions in the body (Table 15-8). Some of the most important minerals are calcium, which is necessary for proper growth of bones and teeth; iodine, for proper functioning of the thyroid gland; chlorine, for maintaining proper pressure in our cells; and iron, which helps to carry oxygen in the blood. (Note that these minerals

TABLE 15-8 *Some important minerals*

ELEMENT	IMPORTANCE IN NUTRITION
Calcium	Making bones and teeth
Chlorine (as chloride ion)	Maintaining proper fluid balance in the body
Chromium	Activates enzymes
Copper	Production of hemoglobin and the activation of certain enzymes
Fluorine (as fluoride ion)	Helps prevent tooth decay
Iodine (as iodide ion)	Proper functioning of the thyroid gland
Iron	Acts as oxygen carrier in hemoglobin
Magnesium	Activates enzymes and important in muscle function
Manganese	Activates enzymes and catalyzes synthesis of cholesterol
Molybdenum	Synthesis of enzymes
Phosphorus	Involved in energy production and conversion in the body
Sulfur	Synthesis of insulin
Vanadium	Involved in regulating circulation
Zinc	Synthesis of enzymes and hormones

are found as ions in the body, *not* as neutral elements. In other words, chlorine is found in the body as chloride ions, iodine as iodide ions, and calcium as calcium ions.)

ARE PROCESSED FOODS GOOD FOR US?

If you have a hamburger, french fries, and a cola drink for lunch, are you eating properly? That depends on how the food was prepared. "Convenience foods" and "fast foods" are highly processed. This means that the foods have been changed in some way either to preserve them for a longer time or just to make them more appealing to the consumer. In many cases, however, nutrients are replaced in the processing, so a "fast foods" lunch may supply you with a nutritious and delicious meal.

When we eat potatoes we usually discard the peel. Peeled potatoes look better than unpeeled potatoes. But many of the nutrients are found in the peel. It's the same with apples. The peel is sometimes removed to make the apple look better, but apple peels contain many nutrients.

How about white flour? Well, white flour comes from whole wheat, which provides a rich supply of vitamin B_1 and other nutrients. When white flour is produced, the wheat germ and the bran are removed from the grain. The proteins, vitamins, and minerals are removed. What's left is mainly starch. However, for just a penny or two a loaf, white flour can be enriched. This means that some vitamins and minerals can be re-added to the flour. This is usually done and it makes for a good advertisement and replaces certain nutrients that have been lost (Fig. 15-14).

What about rice? Processing has provided us with polished rice. This process makes appealing rice that is easily cooked. But the protein, vitamins,

Figure 15-14 Breads and other baked goods are usually made with enriched flour. (ITT Continental Bakery Company, Rye, New York)

and minerals are essentially removed when rice is polished. In countries where rice is a major portion of the diet, people came down with deficiency diseases after switching to polished rice. Nevertheless, processed foods are a great aid in helping people keep up in today's complex society, where saving time is a prime factor.

SOFT DRINKS: THE PAUSE THAT REFRESHES

Each year Americans consume more soft drinks than the year before. In the 1960's profits from wholesale soft drinks sales were $1.7 billion. Today, profits are over six times that amount. On the whole, each American drinks about 500 eight-ounce glasses of soft drink per year (Fig. 15-15).

Figure 15-15 Soft drinks are a very popular beverage today! (Courtesy of Goodyear, who produces these polyester bottles)

What's so bad about soft drinks? Research has implicated soft drinks in studies involving both diabetes and tooth decay. One-fourth of our total sucrose (table sugar) intake comes from soft drinks. They add empty calories to our diet since there are no nutrients to speak of in these drinks. And some contain caffeine, colors, flavors, and preservatives which are not particularly healthy either.

MILK: IT'S A NATURAL

In the early 1960's a general decline in milk consumption began in the United States. This goes hand in hand with the reduced proportion of children in the American population. Milk is one of the best foods around. It is filled with nutrients. Unfortunately, the calorie-filled, nutrient-empty soft drinks are taking the place of drinks such as milk.

There are several types of milk available today. Whole milk contains the most fat—4%. Low-fat milk contains 1 to 2% fat. Skim milk contains no fat (Fig. 15-16). Americans are increasing consumption of low-fat and no-fat milk, while whole milk consumption is dropping rather sharply. Milk is a good source of high-quality protein and ensures that calcium and riboflavin needs are met. Milk products are good as snack foods because they provide extra energy. Milk shakes, milk drinks, and cheeses are included in the milk group.

Figure 15-16   There are several types of milk available on the market today. (United Dairy Industry Association).

WHICH MILK IS BEST FOR A BABY?

At birth, the mother has a choice of breast feeding or bottle feeding her infant. There are many myths and social attitudes toward breast feeding which are being clarified in present times. The decision as to whether one should breast-feed or bottle-feed depends on knowledge of nutritional facts and on the feelings and life-style of the mother. Here are some of the nutritional facts.

Breast feeding is a simple, safe, and economical way in which to feed a baby. Studies have disproved myths about the difficulties of breast feeding. In fact, a recent study showed that 99% of all woman who desire to breast feed can do it successfully. All that is necessary is a positive attitude toward the experience. Human milk is tailormade for human infants. It provides all the nutrients necessary for normal growth and development of the child—proteins, carbohydrates, fats, vitamins, and minerals. The chief protein in breast milk is *lactalbumin*, which is easily digested by infants and causes less spitting up and less stomach discomfort. The sugar in breast milk is *lactose*. This sugar aids in the absorption of calcium, phosphorus, and magnesium. Breast milk contains higher levels of fat and cholesterol than are present in cow's milk. There is some recent evidence which shows that animals that consume breast milk are better able to break down cholesterol. It is thought that the metabolic pathways that will break down cholesterol throughout life are stimulated by the

Figure 15-17 Soy formulas are given to babies who can't digest milk. (Ross Laboratories, Columbus, Ohio)

higher cholesterol levels of breast milk. In addition, breast milk is low in sodium. Its vitamin and mineral content depend on the mother's diet, so a nursing mother should eat well and take a multi-vitamin/mineral supplement if it is deemed necessary by her physician. There are also protective elements which are passed to the breast-fed baby. They seem to lower the infant's chance of developing infections, gastrointestinal upsets, allergies, and diarrhea. It is also felt that children born to allergy-prone parents derive some additional protection from breast feeding.

Infant formulas are manufactured to be as close to breast milk as possible. Soy or coconut oil provide fat, lactose provides carbohydrates, and nonfat milk provides protein. The caloric content is the same as that of breast milk—20 kcal per ounce. Cow's milk-based formulas contain nutrients which are similar to those in cow's milk. The fat of cow's milk is removed and replaced with vegetable oil. Since the fat content of both cow's milk and breast milk are subject to environmental and industrial pollution, the substitution of vegetable oil reduces the level of agents such as pesticides which are frequently found in animal milk today. Milk-based formulas are also available in iron-fortified or nonfortified varieties. For babies who are allergic to cow's milk, soy protein formulas are available. Soy protein is used instead of nonfat milk, and lactose is replaced by sucrose and corn syrup. Some babies who are either sensitive to cow's milk protein or who are lactose intolerant do better on soy formula (Fig. 15-17).

FOOD ADDITIVES:
THEY ADD COLOR AND FLAVOR AND . . .

There are a variety of chemicals which are routinely added to foods to make them look better, taste better, last longer, and taste sweeter. Let's take a look at some of these chemicals.

Food Coloring

Each year we ingest many pounds of artificial colors, flavors, and preservatives. In fact, 95% of the coal-tar-based colors which the Food and Drug Administration (FDA) certifies for use each year go into various foods. The remaining 5% go into coloring drugs and cosmetics.

Sales of certified colors used as food-coloring additives averaged about 0.85 g (0.030 ounce) per person in 1940. Since then the use of certified colors has gone up eleven-fold, to about 9.7 g (0.34 ounce). In 1976, Red Number 2, a very popular food color, was removed from the FDA's approved list of food colors. This was because laboratory animals that ingested Red Number 2 dye developed cancer. Red Number 2 was then replaced by other colors.

Perhaps in the future artificial colors will be replaced by safer compounds, such as carotene (from carrots), beet juice extract, and caramel. This will be hard on companies who rely heavily on food coloring to replace a real food item that belongs in their product, such as eggs, fruit, or whole wheat flour.

Flavor Enhancers

There are many natural flavors added to foods to make them taste better. Many fruits and spices add natural flavor to the foods that you like to eat. Now we also have a whole variety of artificial or synthetic flavors as well. Let's take a look at some of them.

Number one on the list is monosodium glutamate (MSG). MSG is made from glutamic acid, which is an amino acid. MSG is the sodium salt of glutamic acid (Fig. 15-18). Have you ever eaten Chinese food and gotten a headache or felt weak after the meal? This problem has affected many people, and researchers learned that these symptoms were side effects of MSG, which is added to Chinese food in some restaurants. A tremendous variety of foods contain MSG. Canned soups are a good example. A trip down the soup aisle in a supermarket reveals that most canned soups contain MSG as a flavor enhancer. Mayonnaise and salad dressings may also contain MSG, although it doesn't have to be listed on the label. And seasoned salts which make various foods taste better have MSG as a prime ingredient. The trouble with MSG is that if you eat in a restaurant, you can't be sure that it hasn't been added to your food. MSG was tested on laboratory animals and found to cause numbness in a portion of the brain. It was even thought that it could cause severe defects in unborn babies. After this evidence, MSG was removed as an ingredient of baby foods. Now why in the world would baby foods ever need a flavor enhancer? Because parents taste foods before giving them to their babies and the MSG was added to please parents. Nowadays baby foods are as pure as possible—no sugar, salt, flavor enhancers, or artificial ingredients added in most cases. While they don't taste as good as they used to, they're healthier to eat (Fig. 15-19).

Another common flavor enhancer is "good old" sodium chloride—table salt. Potato chips, pretzels, soups, canned foods, breads, processed meats, and many, many other products have salt added. With the evidence of high

Figure 15-18 The formulas of MSG and the amino acid called glutamic acid, from which MSG is made.

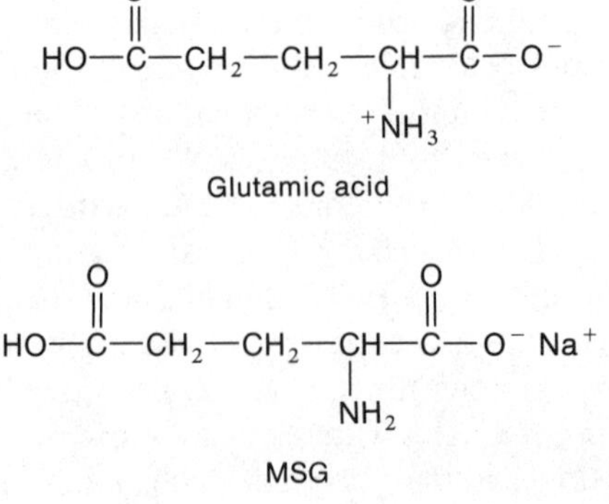

$$HO-\overset{\overset{\displaystyle O}{\|}}{C}-CH_2-CH_2-\underset{\underset{\displaystyle {}^+NH_3}{|}}{CH}-\overset{\overset{\displaystyle O}{\|}}{C}-O^-$$

Glutamic acid

$$HO-\overset{\overset{\displaystyle O}{\|}}{C}-CH_2-CH_2-\underset{\underset{\displaystyle NH_2}{|}}{CH}-\overset{\overset{\displaystyle O}{\|}}{C}-O^-\ Na^+$$

MSG

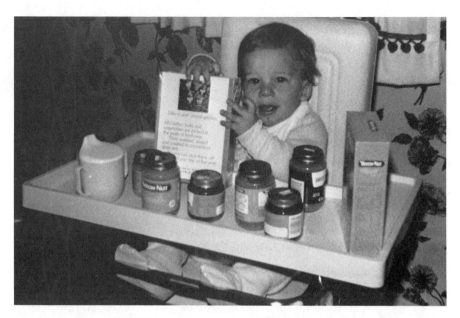

Figure 15-19 Baby foods don't taste as good as they used to, but they're healthier to eat! (Courtesy of Michael Gregory Sherman. Used with his permission)

blood pressure, related to high-sodium diets mounting, it's a good idea to limit your intake of sodium-containing foods.

The Inhibitors: Chemicals to Preserve Foods Longer

The newest additions to the list of suspected cancer-causing agents (carcinogens) in the diet are two chemicals added to processed meats to preserve them longer. Such chemicals are called *inhibitors*, since they retard or inhibit the growth of molds or bacteria that spoil food. The two chemicals under attack are sodium nitrate ($NaNO_3$) and sodium nitrite ($NaNO_2$). They keep foods such as salami and bologna pink and they also add the spicy flavor characteristic of hot dogs. Nitrites are added to meats to stop the growth of a type of bacterium called *Clostridium botulinum*, which causes botulism poisoning. But the amount of nitrites could be cut 10 times and still prevent botulism. So why do manufacturers add 10 times the amount needed? The answer is to keep the product looking the way we expect it to! How would you feel about eating a gray hot dog?

Let's see how nitrates and nitrites may contribute to cancer. In the stomach certain bacteria are present which convert nitrates to nitrites. Stomach acid contains hydrochloric acid (HCl), which converts nitrites to nitrous acid. And nitrous acid is capable of reacting with certain compounds called secondary amines, producing nitrosoamines. These nitrosoamines can cause cancer.

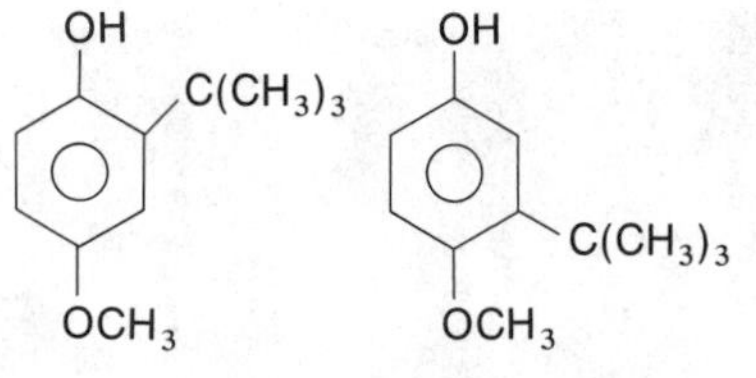

OH   C(CH$_3$)$_3$

OCH$_3$

OH

C(CH$_3$)$_3$

OCH$_3$

The two isomers of BHA

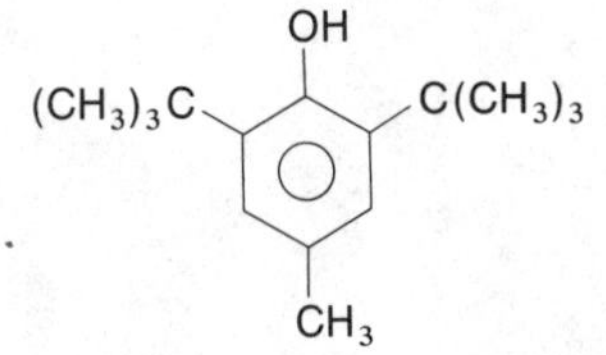

OH

(CH$_3$)$_3$C   C(CH$_3$)$_3$

CH$_3$

BHT

Figure 15-20 The formulas of BHA and BHT, two widely used food additives.

This is borne out by statistics which show that in countries where people don't eat processed meat products, the rate of stomach cancer is lower than in the more developed nations where such foods are consumed.

Another common chemical found in foods is sulfur dioxide (SO_2). When inhaled, it is a dangerous irritant and an air pollutant as well. So far it seems to be harmless when ingested. It stops food spoilage and acts as a bleach, an antioxidant (stops rancidity), a disinfectant, and a preservative. Foods that tend to turn brown, such as jellies and potatoes, are kept looking natural in the presence of sulfur dioxide. It also preserves raisins and apricots.

You probably know that foods such as peanut butter, packaged dry breakfast cereals, and bread turn rancid when kept for a long period of time. This is because they contain lipids—fats and oils, which become rancid after various amounts of time. Chemists have come up with chemicals called *antioxidants* to slow down this spoilage. Two common antioxidants are *butylated hydroxyanisole* (*BHA*) and *butylated hydroxytoluene* (*BHT*), whose structures are shown in Figure 15-20. These chemicals retard the chemical breakdown of food and retard spoilage. These chemicals have side effects— some good and some bad. Some people are allergic to them. Mice fed diets containing less than 1% BHA or BHT gave birth to offspring with some brain abnormalities. But rats fed large amounts of BHT lived 20 human-equivalent years longer than other rats. Perhaps BHT preserved life by slowing down the aging process. It's something to think about and study further.

Sweeteners for Our Powerful Sweet Tooth

Americans are eating more sweeteners than ever before (Fig. 15-21). We consume sucrose (table sugar), corn syrup, and saccharin in greater quantities than did our parents and grandparents. Some sweeteners contain calories and

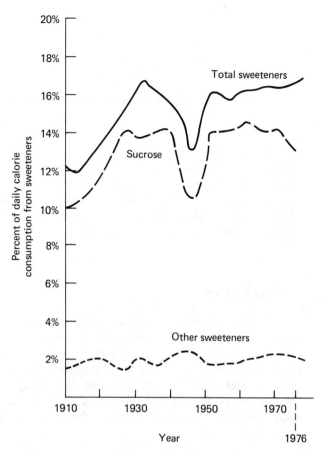

Figure 15-21 Americans are eating more sweeteners than ever before, especially in the form of sucrose (table sugar).

others don't. Today, we use saccharin (Fig. 15-22) to help control obesity since it is a noncaloric sweetener. Up until 1969, cyclamates were used (Fig. 15-23), but they were removed from the market because they were found to induce cancer in laboratory animals.

Now there is also some question as to the safety of saccharin. Rats fed large amounts of saccharin have developed bladder tumors. Future research will bring us new data concerning saccharin.

Figure 15-22 The structure of saccharin.

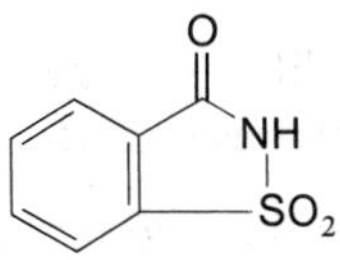

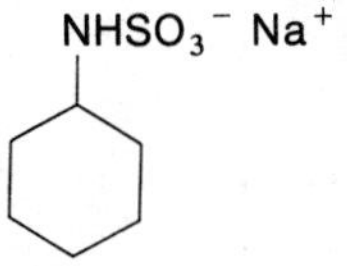

NHSO₃⁻ Na⁺ appears as: $NHSO_3{}^- \; Na^+$

Figure 15-23 The structure of sodium cyclamate.

What's so bad about a high-sugar diet? Sweeteners give us little more than empty calories. They are simple carbohydrates. They add no protein, fat, vitamins, minerals, or dietary fiber. We consume empty calories in place of nutritious foods. This leads to obesity, increased risk of heart disease, diabetes, and tooth decay. Perhaps some day we will control our intake of sweeteners and eat a more balanced diet.

SUMMARY

In this chapter we looked at what composes our food. First, we looked at our basic nutrients—carbohydrates, proteins, lipids, vitamins, and minerals. We took a brief look at food processing. We looked at two beverages, soft drinks and milk, and compared their values. One is prevalent in today's diet and the other was prevalent years ago. We surveyed two methods of feeding infants—breast feeding and bottle feeding. Finally, we took a look at some food additives—coloring agents, flavoring agents, preservatives, and sweeteners.

EXERCISES

1. Name five nutrients found in foods.

2. What are the three basic classes of carbohydrates? Give an example of each.

3. Why is it important to include starch in the diet?

4. If 500 g of carbohydrate is ingested, how much energy is produced?

5. What are some of the properties of lipids?

6. Distinguish among monoglycerides, diglycerides, and triglycerides. Use examples.

7. What is the difference between a saturated fat and an unsaturated fat? What is a polyunsaturated fat? Include examples.

8. What is accomplished by the process of hydrogenation?

9. If 50 g of fat is consumed, how much energy is produced?

10. Of what types of molecules are proteins composed?

11. Distinguish between a complete protein and an incomplete protein.

12. Name the fat-soluble vitamins.

13. What are the functions in the body of calcium, iodine, and chlorine?

14. Discuss some of the pros and cons of eating processed foods.

15. Discuss your feelings about the use of MSG in cooking.

16. What choices are involved in the nitrate–nitrite controversy?

17. What is the importance of limiting sugar intake?

18. State whether each of the following compounds are mono-, di-, or tri-glycerides:

(a)
$$CH_3-(CH_2)_{18}-\overset{\overset{\displaystyle O}{\|}}{C}-O-CH_2$$

$$CH_3-(CH_2)_{16}-\overset{\overset{\displaystyle O}{\|}}{C}-O-CH$$

$$HO-CH_2$$

(b)
$$CH_3-(CH_2)_{16}-\overset{\overset{\displaystyle O}{\|}}{C}-O-CH_2$$

$$CH_3-(CH_2)_{16}-\overset{\overset{\displaystyle O}{\|}}{C}-O-CH$$

$$CH_3-(CH_2)_{16}-\underset{\underset{\displaystyle O}{\|}}{C}-O-CH_2$$

(c)
$$CH_3-(CH_2)_{18}-\overset{\overset{\displaystyle O}{\|}}{C}-O-CH_2$$

$$HO-CH$$

$$HO-CH_2$$

19. For each of the following examples, state whether the fatty acid is saturated or unsaturated:

(a) $CH_3(CH_2)_6CH=CH_2CH=CHCH_2=CH(CH_2)_3-\overset{\overset{\displaystyle O}{\|}}{C}-OH$

(b) $CH_3-(CH_2)_{18}-\overset{\overset{\displaystyle O}{\|}}{C}-OH$

(c) $CH_3(CH_2)_7CH=CH(CH_2)_7-\overset{\overset{\displaystyle O}{\|}}{C}-OH$

20. Explain the difference between fat-soluble and water-soluble vitamins.

21. Discuss some of the positive and negative aspects of consuming soft drinks.

22. Explain the importance of milk in the diet.

23. Make a list showing the constituents of breast milk and infant feeding formulas. Discuss the advantages and disadvantages of using each method to feed a baby.

Chapter 16

Biochemistry: The Molecules of Life

Some Things You Should Know After Reading This Chapter

You should be able to:

1. Discuss the importance of proteins to life.
2. Explain how amino acids compose proteins.
3. State what an essential amino acid is.
4. Explain the difference between complete proteins and incomplete proteins.
5. Discuss the importance of the peptide bond.
6. Explain the difference between the primary, secondary, tertiary, and quaternary structure of a protein.
7. Explain the importance of enzymes in the body.
8. Explain the difference between a simple enzyme and a conjugated enzyme.
9. Discuss the theory of how enzymes work.
10. Discuss the importance of nucleic acids.
11. Name the three components of nucleotides.
12. Explain the importance of the double helix in DNA.
13. Define complementary nitrogen base pairing.
14. Name the nitrogen bases of DNA and RNA.
15. Discuss the importance of recombinant DNA technology to the future of humankind.
16. Explain how interferon may prevent viral infections.

INTRODUCTION

In Chapter 15 we studied some of the most important biochemical molecules, including carbohydrates, fats, and proteins. We look now at some other important molecules which are vital to life.

In the same way that chemicals produced in the laboratory influence us, the chemicals that make up our bodies affect us. The chemicals in our foods give us the energy and the materials that we need to grow, develop, and maintain our bodies in working order. Chemicals such as hormones and enzymes play important roles in controlling the chemical reactions that take place in our bodies.

In this chapter we take a closer look at proteins. We learn about amino acids; enzymes, which are a special group of proteins; and the all-important nucleic acids, including DNA and RNA. We'll survey recombinant DNA technology and look at a new hope for the future—human interferon—an antiviral protein.

ALL ABOUT PROTEINS

We discussed proteins in Chapter 15. Let's review what was said. Proteins are important molecules in the living cell. Without them the human being would not exist. Proteins are essential in sustaining normal life processes.

Figure 16-1 The structure of collagen, the major protein of connective tissue and bone. Three polypeptide chains form a triple helix in the basic unit of this protein, called tropocollagen.

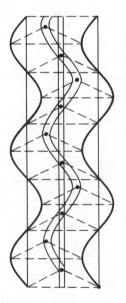
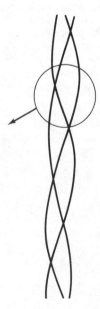

Some of the functions of proteins are: catalyzing reactions in the body, defending the body against disease, helping with the digestion of foods, transporting oxygen in the blood, aiding in the clotting of blood, regulating cellular activities, and participating in genetic functions. About one-third of all the proteins in the body are structural proteins which make up the membranes, musculature, and connective tissues of the body (Fig. 16-1).

Proteins are synthesized by living matter. Plants produce the greatest amount of protein. They synthesize carbohydrates and then combine these carbohydrates with nitrogen-containing compounds from the soil to produce proteins. Animals have very limited ability to produce proteins, so they must include them in the diet and transform them into other types of proteins in the body.

Proteins are composed of the elements carbon, hydrogen, oxygen, and nitrogen. Some proteins contain sulfur, iodine, phosphorus, or iron. Proteins are large molecules with weights ranging from 5000 to millions of grams per mole. They are such large molecules because they are polymers composed of simpler units called *amino acids*. In fact, it is the amino acids that are the key to understanding the functions and compositions of proteins.

AMINO ACIDS: THE BUILDING BLOCKS OF PROTEINS

Amino acids are the molecules that make up proteins. It is the union of amino acids in the proper sequence that gives each protein its special structure and function. Amino acids are held together to form proteins by special bonds called *peptide linkages*.

Amino acids are difunctional organic compounds that contain a carboxyl group ($-\overset{\parallel}{\underset{O}{C}}-OH$) on one end of the molecule and an amino group ($-NH_2$) on the other end of the molecule. The general structure of an amino acid is shown in Fig. 16-2.

Each amino acid has a different R group, which is what makes each amino acid unique. There are 20 major amino acids found in proteins. Table 15-5 lists these amino acids. Because amino acids are the building

Figure 16-2 The general structure of an amino acid.

The R group side chain

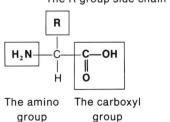

The amino   The carboxyl
group   group

TABLE 16-1 *Essential amino acid requirements*

ESSENTIAL AMINO ACID	GRAMS/DAY (for a 150-pound adult)
Isoleucine	0.84
Leucine	1.12
Lysine	0.84
Methionine[a]	0.70
Phenylalanine[b]	1.12
Threonine	0.56
Tryptophan	0.21
Valine	0.98

[a] Includes cystine.
[b] Includes tyrosine.
Source: The National Research Council—National Academy of Sciences, 1974.

blocks of proteins, they are very important to us biologically. Some amino acids can be synthesized by the body and others can be interconverted from closely related amino acids. There are also eight amino acids that can't be synthesized by the body in any way and must be included in the diet. These are called the *essential amino acids*: lysine, tryptophan, phenylalanine, methionine, threonine, leucine, isoleucine, and valine. Another amino acid, histidine, is sometimes listed with the essential amino acids because it is required by infants and children for proper growth and development (together with the other eight essential amino acids). Consumption of these essential amino acids is important for the maintenance of a healthy body (Table 16-1).

Complete proteins contain all the amino acids essential for the maintenance of good health. Eggs, milk, meat, fish, poultry, and soybean are the best sources of complete protein in the diet. Cereal grain, nuts, and vegetables are not sources of complete protein, since they may be low in one or more of the amino acids. Proteins found in such foods are called *incomplete proteins*. However, mixtures of whole-grain cereals and vegetables, or mixtures of whole grains and milk, meat, or eggs, are good sources of complete protein. Generally speaking, proteins that come from vegetable sources are incomplete and those that come from animal sources are complete.

PEPTIDE BONDS

Peptide bonds, or peptide linkages, hold amino acids together in a protein molecule. The carboxyl group of one amino acid bonds to the amino group

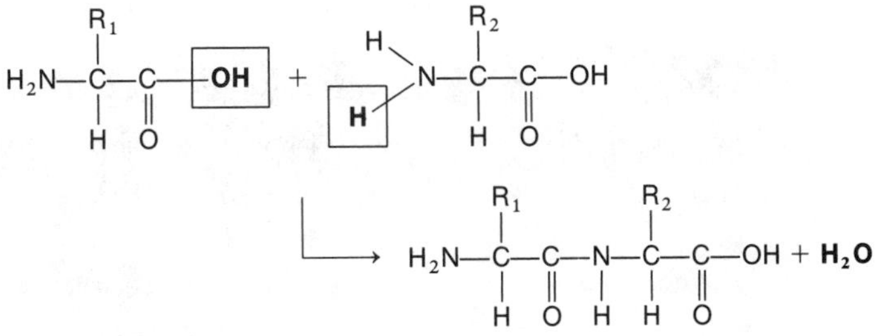

$$H_2N-\underset{\underset{H}{|}}{\overset{\overset{R_1}{|}}{C}}-\underset{\underset{O}{\|}}{C}-\boxed{OH} \quad + \quad \underset{\boxed{H}}{\overset{H}{\diagdown}}N-\underset{\underset{H}{|}}{\overset{\overset{R_2}{|}}{C}}-\underset{\underset{O}{\|}}{C}-OH$$

$$\longrightarrow \quad H_2N-\underset{\underset{H}{|}}{\overset{\overset{R_1}{|}}{C}}-\underset{\underset{O}{\|}}{C}-\underset{\underset{H}{|}}{N}-\underset{\underset{H}{|}}{\overset{\overset{R_2}{|}}{C}}-\underset{\underset{O}{\|}}{C}-OH + H_2O$$

Figure 16-3 When two amino acids combine to form a peptide bond, one molecule of water is eliminated in the reaction.

of another amino acid. When two amino acids combine to form a peptide bond, one molecule of water is eliminated in the reaction (Fig. 16-3).

Amino acids can form long protein chains because each amino acid has an amino end and a carboxyl end available for bonding. Let's look at the following situation to understand this better. Suppose that you were going to tie many pairs of sneakers together. Each pair of sneakers has a left lace and a right lace. You would tie the right lace of one pair of sneakers to the left lace of the next pair of sneakers. You would always end up with a free left lace at one end of the sneaker chain and a free right lace at the other end of the sneaker chain (Fig. 16-4). It's the same with protein chains. You bond the carboxyl group of one amino acid to the amino group of the next amino acid. You always end up with a free amino group at one end of the protein chain and a free carboxyl group at the other end of the protein chain.

Figure 16-4 The sneaker chain.

TABLE 16-2   *The classification of proteins by function*

CLASS	EXAMPLE
Blood proteins	Fibrinogen, to help blood clot
Catalytic proteins	Enzymes, to speed up and regulate cellular reactions
Contractile proteins	Myosin, which is found in muscle tissues
Digestive proteins	Pepsin, an *enzyme* that splits other proteins into smaller molecules
Hormonal proteins	Insulin, which regulates activities in cells
Natural-defense proteins	Antibodies, which protect the body from disease
Repressor proteins	Proteins that regulate the expression of genetic information
Respiratory proteins	The cytochromes, which help transport electrons to electron acceptors such as oxygen
Structural proteins	Collagen, which is found in connective tissue of vertebrates
Transport proteins	Hemoglobin, which moves oxygen through the bloodstream

THE FUNCTIONS OF PROTEINS

Proteins have a wide range of functions, which is why they are so important in nature. Proteins are divided into 10 categories according to their functions (Table 16-2). Since all proteins are important in sustaining life, no single functional classification system ranks above another. Proteins that function as enzymes, however, are of specific importance. This is because enzymes catalyze all the chemical reactions that take place in the cell. Enzymes speed up these reactions and determine how far each reaction will proceed. This enables the cell to adapt to its changing needs in an efficient and economical manner.

PROTEIN STRUCTURE: THE BIG PICTURE

Now that we have a basic understanding of proteins, let's take a look at a protein's total structure. You probably think that we've already done this. After all, we've learned that proteins are combinations of amino acids, arranged in a specific order. But this is only part of a protein's structure, called its *primary structure*. It is not unusual for a protein to have a secondary, tertiary, and quaternary structure as well. Let's see what all this means.

Val—His—Leu—Thr—Pro—Glu—Glu—Lys—Ser—Ala—Val—Thr—Ala—Leu—Trp—Gly—Lys—Val—
 1   2   3   4   5   6   7   8   9   10   11   12   13   14   15   16   17   18

Asp—Val—Asp—Glu—Val—Gly—Gly—Glu—Ala—Leu—Gly—Arg—Leu—Leu—Val—Val—Tyr—Pro—
19   20   21   22   23   24   25   26   27   28   29   30   31   32   33   34   35   36

Trp—Thr—Glu—Arg—Phe—Phe—Glu—Ser—Phe—Gly—Asp—Leu—Ser—Thr—Pro—Asp—Ala—Val-
37   38   39   40   41   42   43   44   45   46   47   48   49   50   51   52   53   54

Met—Gly—Asp—Pro—Lys—Val—Lys—Ala—His—Gly—Lys—Lys—Val—Leu—Gly—Ala—Phe—Ser—
55   56   57   58   59   60   61   62   63   64   65   66   67   68   69   70   71   72

Asp—Gly—Leu—Ala—His—Leu—Asp—Asp—Leu—Lys—Gly—Thr—Phe—Ala—Thr—Leu—Ser—Glu-
73   74   75   76   77   78   79   80   81   82   83   84   85   86   87   88   89   90

Leu—His—Cys—Asp—Lys—Leu—His—Val—Asp—Pro—Glu—Asp—Phe—Arg—Leu—Leu—Gly—Asp
91   92   93   94   95   96   97   98   99 100 101 102 103 104 105 106 107 108

Val—Leu—Val—Cys—Val—Leu—Ala—His—His—Phe—Gly—Lys—Glu—Phe—Thr—Pro—Pro—Val—
109 110 111 112 113 114 115 116 117 118 119 120 121 122 123 124 125 126

Glu—Ala—Ala—Tyr—Glu—Lys—Val—Val—Ala—Gly—Val—Ala—Asp—Ala—Leu—Ala—His—Lys—
127 128 129 130 131 132 133 134 135 136 137 138 139 140 141 142 143 144

Tyr—His
145 146

Figure 16-5 The primary structure of a beta chain of hemoglobin.

Primary Structure of Proteins

The primary structure of a protein is the basic sequence or order of amino acids that make up the protein (Fig. 16-5). We've already learned that there are 20 different amino acids that make up most of the known proteins. A survey of these known proteins shows that the average *chain length* (or number of amino acids bonded together) is somewhere between 100 and 150 amino acids per protein molecule. Needless to say, proteins are very complex molecules, and there are numerous methods employed by chemists for determining the sequence of amino acids in proteins. In 1956, Sanger determined the exact amino acid sequence of insulin. Once the amino acid sequence is known, chemists attempt to synthesize these molecules in the laboratory. This is not an easy task, but some peptide chains and some proteins have been synthesized. Vasopressin and oxytocin, which are hormones produced by the posterior pituitary gland, have been synthesized. In 1965, a Chinese group of scientists combined the research of others and produced the first all-synthetic insulin.

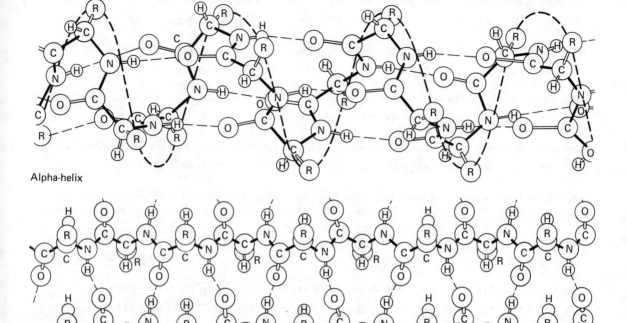

Alpha-helix

Beta configuration (pleated sheet)

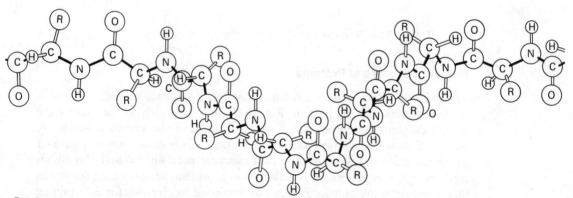

Random chain

Figure 16-6 Some secondary structures of proteins.

Secondary Structure of Proteins

The primary structure tells us the sequence of amino acids in a protein. But a protein is more than just a straight chain of amino acids. It has a special geometrical arrangement. The *secondary structure* of a protein tells us how the polypeptide chain is arranged in space. It tells us, for example, if the chain is in an ordered helix, which is a coiled spiral-like arrangement, or if it is in an orderly pleated sheet arrangement, or if it is in a disordered random arrangement (Fig. 16-6).

In the early 1950's, Linus Pauling and coworkers discovered that the polypeptide chains in the protein keratin are coiled in what is called an *alpha helix*. In an alpha helix there are 3.6 amino acids in every turn of the helix. The helix is held together by hydrogen bonds between a hydrogen atom (belonging to the amino part of a peptide bond) and an oxygen atom (belonging

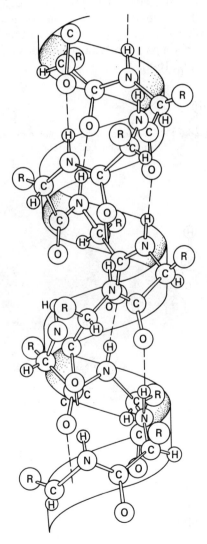

Figure 16-7 Hydrogen bonding in the alpha helix. (From M. M. Bloomfield, *Chemistry and the Living Organism*, 2nd ed., 1980. By permission of John Wiley & Sons, Inc.)

to the carboxyl part of a peptide bond). The hydrogen bonding takes place between atoms found along the turns of the helix, and since peptide bonds occur regularly along the helix, the entire helix is held together by these bonds (Fig. 16-7).

Tertiary Structure of Proteins

The *three-dimensional structure* of a protein is called its *tertiary structure*. It specifically pertains to the way the secondary structure is folded in space. Experiments with certain proteins, such as myoglobin, keratin, myosin, and albumin, have shown that the alpha helix does not remain as a long extended coil. In such molecules the helix folds upon itself several times to produce a compact molecule. The protein myoglobin, which is an oxygen-carrying protein found in the muscle tissue of mammals, if extended, would be an alpha helix whose length would be 20 times its width. But experiments show that it is a compact molecule which looks almost like a football because of the folding (Fig. 16-8).

Figure 16-8 A model of the myoglobin molecule, showing its tertiary structure. The disk represents the heme group.

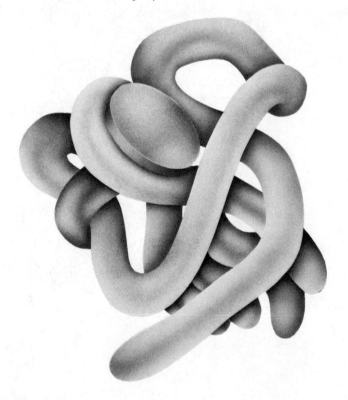

TABLE 16-3 *Some proteins with quaternary structures*

PROTEIN	NUMBER OF SUBUNITS	MOLECULAR WEIGHT
Insulin	2	11,466
Hemoglobin	4	64,500
Alcohol dehydrogenase	4	80,000
Apoferritin	20	480,000
Tobacco mosaic virus	2130	40,000,000

Quaternary Structure of Proteins

Some proteins have a *quaternary structure*. This means that they exist as groups of two or more polypeptide chains. The polypeptide chains are tertiary structures, so a quaternary structure is really made up of groups of two or more tertiary structures. Each tertiary structure is called a *subunit* (Table 16-3).

Hemoglobin is the classical example of a protein with a quaternary structure. Four peptide chains (which are tertiary structures, or subunits) make up the hemoglobin molecule. Each chain is able to combine with one oxygen molecule, so the hemoglobin molecule can combine with four oxygen molecules (Fig. 16-9). Hemoglobin is found in the blood and is the molecule that carries oxygen to all the cells of the body.

ENZYMES: A VERY SPECIAL GROUP OF PROTEINS

Enzymes are a group of proteins that regulate biological reactions in our bodies. They are biological *catalysts* which drive the chemical reactions that take place inside our cells. *Catalysts* are substances that affect the rate at which chemical reactions occur. They take part in chemical reactions, but are not changed chemically in the overall reaction. *Biological catalysts*, or enzymes, as they are called, are very specific in their actions. A particular enzyme will act only on certain substances, and will catalyze only certain types of reactions.

Over 1000 enzymes are known to us. Enzymes work in our bodies to help us digest our food, conduct nerve impulses throughout our bodies, and break down carbohydrates, lipids, and proteins into smaller molecules. If even one enzyme is missing in the body, the results can be disastrous. Enzymes are very important, indeed, and now we're going to take a look at this very special group of proteins.

ENZYME COMPOSITION

There are two types of enzymes: simple enzymes and conjugated enzymes. *Simple enzymes* are proteins that are composed entirely of amino acids. Examples of simple enzymes are pepsin, trypsin, and chymotrypsin. They are

entirely protein in composition. *Conjugated enzymes* are composed of a protein group and a nonprotein group. The nonprotein group is called a *cofactor*. Cofactors can be metal ions such as Fe^{+2}, Fe^{+3}, Mn^{+2}, Mg^{+2}, Zn^{+2}, or Cu^{+2}, or they can be complex organic molecules. When the cofactor of an enzyme is a complex organic molecule, it is called a *coenzyme*. The cofactor is essential for the biological activity of the conjugated enzyme. Some enzymes require both kinds of cofactors, a metal ion and a coenzyme, to be biologically active. Many of the vitamins are coenzymes.

THE DIFFERENT TYPES OF ENZYMES

Enzymes act on specific molecules. The *substrate* is the substance on which the enzyme acts. A *carbohydrase* is an enzyme that acts on a carbohydrate molecule and breaks it down into smaller units. The specific enzyme that breaks down the carbohydrate maltose is called malt*ase*.

There are enzymes that catalyze oxidation–reduction reactions in the body. These are called *oxidoreductases*. *Transferases* catalyze reactions in which a functional group is moved from one molecule to another. *Hydrolases* catalyze hydrolysis reactions to form smaller molecules by the addition of water. *Carbohydrases* break down carbohydrates into monosaccharides. *Proteases* break down proteins into amino acids and peptides. *Lipases* break down lipids into smaller molecules. *Amylases* break down starches into smaller molecules. And there are additional enzymes that function in the body.

A THEORY OF HOW ENZYMES WORK

Enzymes are biological catalysts that provide a reaction pathway that requires less energy than would otherwise be needed. But how do they do this?

An enzyme combines with a reactant called a *substrate*. Each enzyme binds with a specific substrate. The attraction of an enzyme to a substrate is called *substrate specificity*. You remember that proteins have a tertiary structure which gives them a three-dimensional shape. Since enzymes are proteins,

---

Figure 16-9 The quaternary structure of hemoglobin. A hemoglobin molecule, as deduced from x-ray diffraction studies, is shown from above (top) and from the side (bottom). The drawings follow the representation scheme used in three-dimensional models built by M. F. Perutz and his co-workers. The irregular blocks represent electron-density patterns at various levels in the hemoglobin molecule. The molecule is built up from four subunits: two identical alpha chains (white blocks) and two identical beta chains (dark gray blocks). The letter "N" in the top view identifies the amino ends of the two alpha chains; the letter "C" identifies the carboxyl ends. Each chain enfolds a heme group (light gray disk), the iron-containing structure that binds oxygen to the molecule. (From M. F. Perutz, The Hemoglobin Molecule, *Scientific American*, Nov. 1964, *211*, (5), pp. 64–76. Copyright © 1964 by Scientific American, Inc.)

Figure 16-10 The lock-and-key theory explains how enzymes and substrates react. The enzyme is like a lock, and the substrate is like a key that fits the lock.

they have a unique tertiary structure. Only a certain substrate can react with a particular enzyme because of the unique structure.

The *lock-and-key theory* explains how enzymes and substrates react. The enzyme is like a lock that has a special structure. The substrate is like a key which has a special structure that will fit the lock (Fig. 16-10). Some enzymes are very specific in their actions and others are more general. Some enzymes act on specific groups; for example, a *transaminase* transfers amino groups and a *methyltransferase* transfers methyl groups. A *decarboxylase* removes carboxyl groups. Some enzymes catalyze reactions that break specific types of bonds. An *esterase* cleaves ester linkages and a *protease* cleaves peptide bonds. Some enzymes catalyze very specific reactions. *Urease* has only one substrate: urea. *Maltase* works only on the maltose molecule. So we see that enzymes are very selective molecules.

Let's summarize what happens during an enzyme-catalyzed reaction, using the letters E to represent the enzyme, S to represent the substrate, and P to represent the products (Fig. 16-11).

1. E + S ⇌ ES. The substrate combines with the enzyme to form an enzyme–substrate complex (ES).

2. ES ⇌ EP. The active site is part of the enzyme where the reactants are converted to the products (EP).

3. EP ⇌ E + P. The products are released from the surface of the enzyme and the enzyme is now free to catalyze another reaction. The enzyme can be used over and over again.

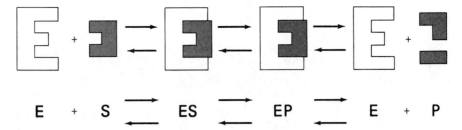

$$E + S \rightleftharpoons ES \rightleftharpoons EP \rightleftharpoons E + P$$

Figure 16-11 What happens during an enzyme-catalyzed reaction?

NUCLEIC ACIDS: THE BASIS OF LIFE

All human beings require the same basic nutrients which are used by the body to produce the correct set of proteins that make up each individual. The information needed for the synthesis of these proteins is found in the nucleus of each cell. The cell nucleus contains *nucleic acids*, which are responsible for linking amino acids together in specific sequences to make the various proteins found in the body. Nucleic acids are the major components of *chromosomes*. Chromosomes hold all the genetic information that we inherit from our parents. The nucleic acids found in our cells hold the information for the maintenance, growth, and reproduction of our cells. There are two types of nucleic acids: *deoxyribonucleic acid*, or DNA, and *ribonucleic acid*, or RNA. DNA is the genetic material found in the nucleus of each cell, which contains all the information needed for the development of an individual. It tells how a person will look and how that person's body will function. It determines what chemical reactions will take place in the body, and which proteins and enzymes must be produced in order to carry out these chemical reactions. RNA brings the information contained in the cell nucleus to another part of the cell, called the *ribosomes*, where the actual synthesis of protein occurs. It is the DNA and RNA, combined with some protein molecules, which make up our chromosomes. Each human cell contains 46 chromosomes, 23 from the mother and 23 from the father. (The sex cells contain only 23 chromosomes, so that when fertilization occurs the embryo has 46 chromosomes.) *Genes*, which carry the information for each specific characteristic that the person will have, are found on the chromosomes.

Let's now take a closer look at the composition of nucleic acids.

COMPONENTS OF NUCLEIC ACIDS

Nucleic acids are composed of many small units called *nucleotides*. Just as amino acids are the basic building blocks of proteins, and monosaccharides combine to form polysaccharides, *nucleotides* join together to form *nucleic acids* (Fig. 16-12).

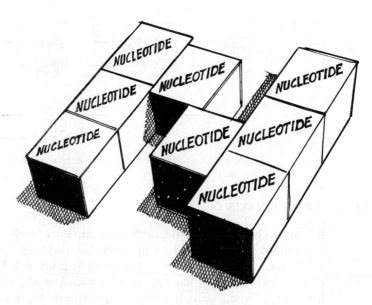

Figure 16-12 Nucleotides join together to form nucleic acids.

TABLE 16-4 *The composition of nucleotides*

SUBUNIT	SUBSTANCES FOUND IN RNA	SUBSTANCES FOUND IN DNA
Five-carbon sugar	Ribose	Deoxyribose
Phosphate group	HO—P—O— (with OH above and O below, double-bonded)	HO—P—O— (with OH above and O below, double-bonded)
Nitrogen-containing bases	Adenine, guanine, cytosine, and uracil	Adenine, guanine, cytosine, and thymine

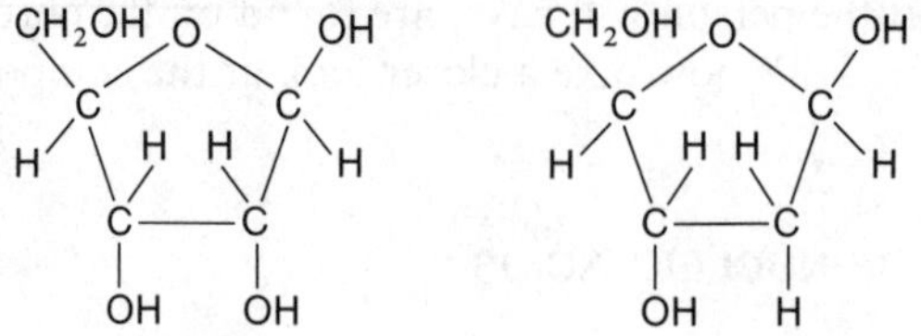

Ribose   Deoxyribose

Figure 16-13 The five-carbon sugars found in nucleotides. Ribose is found in the nucleotides of ribonucleic acid, and deoxyribose is found in the nucleotides of deoxyribonucleic acid.

Nucleotides are composed of three subgroups:

1. A five-carbon sugar

2. A phosphate group

3. A nitrogen-containing base (Table 16-4)

The *five-carbon sugar* found in nucleotides is either *ribose* or *deoxyribose* (Fig. 16-13). Ribose is present in ribonucleic acid and deoxyribose is present in deoxyribonucleic acid.

The *phosphate group* bonds nucleotides together. The structure of the phosphate group is as follows:

$$HO-\overset{\overset{\displaystyle OH}{|}}{\underset{\underset{\displaystyle O}{\|}}{P}}-O-$$

A *nitrogen-containing base* is a cyclic compound containing both nitrogen and carbon in the ring. The ring can have either one or two cyclic parts. There are five different nitrogen-containing bases which can be found in DNA or RNA. They fall into two classes: the purines and pyrimidines. The *purines* are double-ring structures and the *pyrimidines* are single-ring structures. The

Figure 16-14 The nitrogen-containing bases.

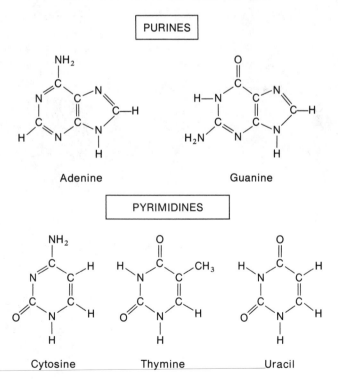

PURINES

Adenine   Guanine

PYRIMIDINES

Cytosine   Thymine   Uracil

TABLE 16-5 *The nitrogen-containing bases of DNA and RNA*

NUCLEIC ACID	NITROGEN-CONTAINING BASES
DNA	Adenine, guanine, cytosine, and *thymine*[a]
RNA	Adenine, guanine, cytosine, and *uracil*[a]

[a] Both DNA and RNA contain the bases adenine, guanine, and cytosine. However, only DNA contains thymine and only RNA contains uracil.

purines are *adenine* and *guanine* and the pyrimidines are *cytosine*, *thymine*, and *uracil* (Fig. 16-14).

The *DNA nucleotides* are made of the nitrogen-containing bases adenine, guanine, cytosine, and thymine. The *RNA nucleotides* are made of the nitrogen-containing bases adenine, guanine, cytosine, and uracil. In other words, DNA can't contain uracil and RNA can't contain thymine (Table 16-5).

THE STRUCTURE OF NUCLEIC ACIDS

DNA

In 1953, J. Watson and F. Crick proposed a structure for the DNA molecule, for which they received the Noble Prize in Physiology and Medicine. The *Watson–Crick model of DNA* shows the nucleotides arranged as a double-

Figure 16-15 The double helix of DNA. (a) A drawing of a double helix. (b) A photograph of an actual model of DNA (The Ealing Corporation).

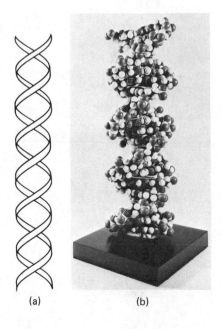

(a)   (b)

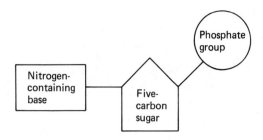

Figure 16-16 A nucleotide is composed of a phosphate group bonded to a five-carbon sugar, which in turn is bonded to a nitrogen-containing base.

stranded coil, or what is better known as a *double helix* (Fig. 16-15). Let's look at the nucleotides of DNA.

We said earlier that a nucleotide is made up of a nitrogen-containing base, a five-carbon sugar, and a phosphate group. The bonding sequence in the nucleotide has the phosphate group bonded to the five-carbon sugar, which in turn is bonded to the nitrogen-containing base (Fig. 16-16). A nucleic acid is a long chain of nucleotides bonded together (Fig. 16-17). The phosphate

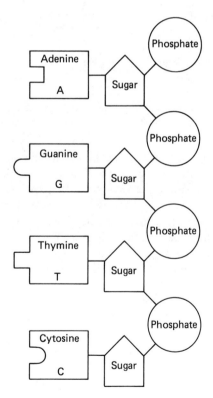

Figure 16-17 A nucleic acid is a long chain of nucleotides bonded together. The phosphate group of one nucleotide bonds to a sugar group on the next nucleotide.

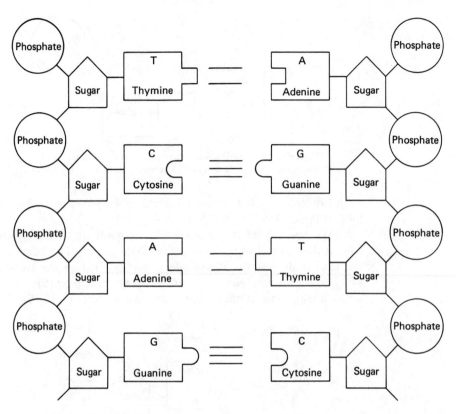

Figure 16-18 Hydrogen bonding between complementary base pairs holds together the two nucleotide chains that compose DNA.

group of one nucleotide bonds to the sugar group on the next nucleotide. This is how we explain the bonding in each of the nucleotide chains of DNA. But what holds the two chains together to make the double helix? The two chains are held together by hydrogen bonding between the nitrogen-containing bases. A nitrogen-containing base from one chain bonds with a nitrogen-containing base from the other chain. The bonding between the nitrogen-containing bases in DNA is very specific. Adenine will bond only with thymine, and cytosine will bond only with guanine. This is known as *complementary nitrogen base pairing*. It means that the two strands of DNA are not identical, but are complementary. In other words, where adenine appears on one strand of DNA, thymine appears on the other, and where guanine appears on one strand of DNA, cytosine appears on the other. The adenine and thymine are held together by two hydrogen bonds and the cytosine and guanine are held together by three hydrogen bonds. Figure 16-18 shows what this hydrogen bonding would look like if we could uncoil the two nucleotide chains. But remember, the two chains actually exist in a helical arrangement (Fig. 16-19).

Another question we might ask is: What makes the DNA molecules that produce one type of cell in the body different from the DNA molecules

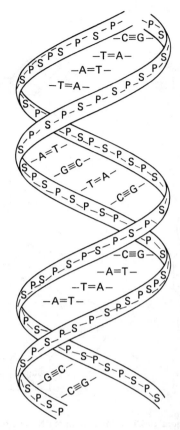

Figure 16-19 The helical arrangement of DNA in the Watson-Crick model. The S and P are the sugar and phosphate groups, respectively.

that produce another type of cell? Each DNA molecule must contain phosphate groups, five-carbon sugars, and nitrogen-containing bases. But the number of base pairs and the sequence of base pairs can vary from molecule to molecule. Also, the number of nucleotides can vary from molecule to molecule. The range is from 5000 to 5 million nucleotides per DNA molecule. It is these variations in DNA molecules that allow one type of cell in the body to be different from another type. The sequence of bases along the chain of the DNA molecule determines what proteins will be made, which in turn determines what characteristics a person will have.

RNA

RNA molecules are single chains of nucleotides, unlike DNA molecules, which are double-coiled chains of nucleotides. RNA nucleotides are made up of a nitrogen-containing base: a five-carbon sugar, which is ribose; and a phosphate group. When the phosphate group of one RNA nucleotide bonds

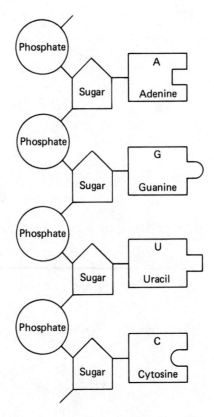

Figure 16-20   A long chain of RNA nucleotides makes up a molecule of RNA.

to the sugar group on the next RNA nucleotide, a long chain of RNA nucleotides is formed. A long chain of RNA nucleotides make up a molecule of RNA (Fig. 16-20). The nitrogen-containing bases of RNA are adenine, guanine, cytosine, and uracil. The complementary base pairs of RNA are adenine and uracil bonding with each other, and guanine and cytosine bonding with each other.

THE REPLICATION OF DNA

When our cells are damaged and when we grow, we must produce new cells. The information needed to produce a new cell, which is exactly like the original cell, is contained in the DNA molecules that make up our chromosomes. The process of reproducing a new cell is called *mitosis*. In this process, a cell divides, forming two new cells. The new cells are called the *daughter cells* and the original cell is called the *parent cell*. The daughter cells will undergo mitosis to form more identical cells. This is how damaged cells can be replaced and how growth can occur.

Each time a cell undergoes mitosis and reproduces itself, the DNA in the cell nucleus must also reproduce itself. The DNA in a cell directs the reproduction or replication of another identical DNA molecule.

You might remember that a DNA molecule is a double-helical chain of DNA nucleotides which is held together by hydrogen bonds between nitrogen base pairs. When DNA replication occurs, an enzyme present in the cell causes the double-helical chain to uncoil, breaking the hydrogen bonds and separating the nitrogen-containing bases. When the helix is completely un-coiled, two separate DNA nucleotide chains remain. In the presence of the

Figure 16-21 The replication of DNA. The parent DNA uncoils into two new strands. New complementary strands are built on each of the parent strands, forming two new molecules of DNA, which are exactly like the original.

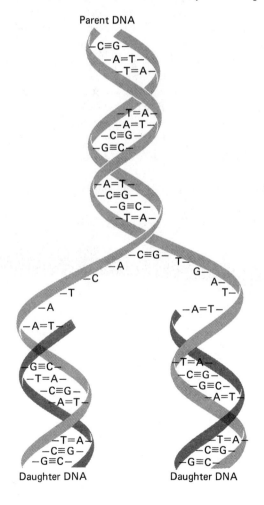

Parent DNA

Daughter DNA   Daughter DNA

proper enzymes and the necessary nucleotides, two new DNA nucleotide chains which are complementary to each of the existing DNA nucleotide chains are formed. Each newly formed chain remains united with its complementary chain, and two new molecules of DNA which are exactly like the original DNA molecule are formed (Fig. 16-21). The process of DNA replication is completed.

RECOMBINANT DNA: THE PROCESS OF GENE SPLICING

In laboratories all over the world, biologists are taking genes from one organism and planting them into another organism. This process is known as *gene splicing* and is part of the new technology known as recombinant DNA. Here's how it works (Fig. 16-22).

A ring of DNA, called a *plasmid*, is isolated from a bacterium. *Escherichia coli* (*E. coli*, for short) is a bacterium that normally flourishes in the intestinal tract of human beings. *E. coli* contains rings of DNA called plasmids. Researchers remove a plasmid from a bacterium of *E. coli*. Then an enzyme is added to the plasmid, which cuts the ring of DNA at a specific site. Next, a gene for protein, such as human insulin or growth hormone, is taken from another cell. It is cut with the same enzyme that was used on the plasmid. The new gene is then inserted into the opened plasmid. It fits exactly because the same enzyme was used in cutting both the plasmid and the new gene. The ring is then closed with another enzyme and the plasmid is put into the bacterium. The plasmid now contains the recombinant DNA. When the *E. coli* divides, it passes the new gene along to the next generation. Within hours the protein that was inserted into the bacterium is reproduced. The desired product is now available for use and has been produced at a faster rate and cheaper cost than ever before.

RECOMBINANT DNA AND THE FUTURE

Human insulin will be one of the first products to be manufactured in large quantities by recombinant DNA technology. The insulin now used by diabetics comes from pigs or cattle and contains impurities that can cause allergic reactions.

Interferon, a natural virus fighter, is another important human protein being produced. Interferon may help prevent flu, hepatitis, and other viral infections. It is now being tested against cancer. Large quantities of interferon may soon be produced using recombinant DNA technology.

Someday, bacteria will probably be turned into living factories using recombinant DNA. The bacteria will produce large amounts of vital medical substances, such as serums and vaccines, to fight anything from the common cold to cancer.

Recombinant DNA technology will also help us to understand and perhaps treat diseases that cause birth defects. It will also help us to understand our bodies better.

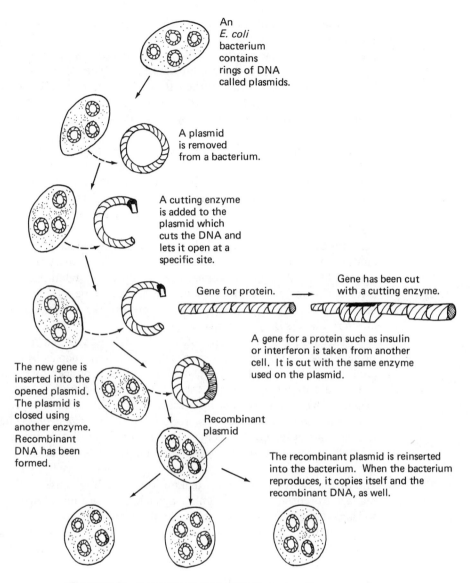

An *E. coli* bacterium contains rings of DNA called plasmids.

A plasmid is removed from a bacterium.

A cutting enzyme is added to the plasmid which cuts the DNA and lets it open at a specific site.

Gene for protein. → Gene has been cut with a cutting enzyme.

A gene for a protein such as insulin or interferon is taken from another cell. It is cut with the same enzyme used on the plasmid.

The new gene is inserted into the opened plasmid. The plasmid is closed using another enzyme. Recombinant DNA has been formed.

Recombinant plasmid

The recombinant plasmid is reinserted into the bacterium. When the bacterium reproduces, it copies itself and the recombinant DNA, as well.

Figure 16-22 Recombinant DNA technology—how the process of gene splicing works.

INTERFERON: THE NEW DRUG FOR CANCER

Interferon was discovered in 1957 by virologists Alick Isaacs and Jean Lindenmann at London's National Institute for Medical Research. The two scientists were fascinated with a biological phenomenon known as viral interference. Doctors had observed that a victim of one kind of virus-caused illness just about never came down with another viral disease at the same time. It seems that when one kind of virus is present, infection by any other virus is inhibited.

Isaacs and Lindenmann wanted to know why this was so, and they experimented. Here's what they did. Inside a chicken eggshell is a thin membrane which lines the eggshell. The researchers took several pieces of chicken membranes and grew them in a nutrient broth. Then they exposed the membranes to influenza viruses. When they added other viruses to the culture, they found that the cells resisted further infection. The first set of viruses seemed to be warding off an attack by the second set of viruses, as the researchers expected. Next, they removed all traces of the viruses and the chicken membranes, and left only the culture brew. To this solution they added healthy cells and tried to infect them with a new virus. These new cells remained uninfected. It seemed that the initial virus infection had stimulated the chicken cells to produce a substance that interfered with further viral assaults. This substance stayed behind in the solution when the original cells and viruses were removed. Lindenmann called this substance interferon.

Further research showed that interferon is produced only in very small amounts in living cells. All vertebrate animals produce interferon, but it works only in the type of animal that produces it. Unlike insulin, which is extracted from cattle and pig glands, and used by people, human interferon must be used by human beings. Pig interferon works only in pigs, and monkey interferon works only in monkeys.

Research on interferon continued. Researchers learned that it is a protein produced by cells in response to stimulation, such as by a virus. There are three kinds of interferon known. One kind is produced by the white blood cells, and another by the cells that form connective tissue in skin and other organs. This second type of interferon is called *fibroblast interferon*. A third kind, called *immune interferon*, is produced by the *T lymphocytes*, which are cells that are part of the body's immune system. Once a cell is destroyed by a virus, the cell's nucleus sends out signals for interferon production. This triggers the production of antiviral proteins in neighboring cells and prevents the virus from reproducing and attacking other healthy cells. If the virus does manage to reproduce in a neighboring cell, the reproduced viruses are unable to leave the cell. This is how the cycle of infection is broken.

Interferon has been successfully used to treat chicken pox and shingles. It has prevented the recurrence of a chronic viral disease called CMV which sometimes endangers kidney-transplant patients and newborn babies. Israeli doctors have used interferon eyedrops to combat pink eye. Soviet doctors claim to have warded off respiratory infections with weak sprays of interferon. Researchers at Standford University together with a group of British researchers have exposed volunteers to common cold viruses both before and shortly after receiving interferon nasal spray. None of the volunteers caught cold, whereas members of a control group that had no interferon treatment did come down with colds.

How about cancer? An American scientist named Ion Gresser, working in France, tested interferon on a group of mice injected with a virus that causes leukemia, a blood cancer. After a month, the interferon-treated mice were in good health. An untreated control group contained mice afflicted with leukemia.

Inspired by Gresser's work, Hans Strander of Stockholm's Karolinska Institute injected interferon into children suffering from a rare and deadly bone cancer called *osteogenic sarcoma*. Conventional treatment is amputation of the affected limb. This disease kills 80% of its victims in 2 years. Of the patients treated with interferon, 50% are alive after 5 years. Of the untreated group, less than 25% are alive.

Jordon Gutterman of the M.D. Anderson Hospital and Tumor Institute in Houston was encouraged by Strander's work. He tried interferon on 38 patients with advanced breast cancer, multiple myeloma, and lymphoma. Once again, the results were encouraging. Although some patients responded much better than others did, on the whole tumors decreased in size and people felt better.

The big problem with interferon is getting enough of it. Scientists all over the world are using recombinant DNA technology to try and get *E. coli* to churn out human interferon. Only when it can be extensively tested can its potential be realized. The future holds great promise.

SUMMARY

In this chapter we surveyed the the topic of biochemistry and looked at some of the molecules important to life.

We began our study by discussing proteins and amino acids. We learned the difference between complete and incomplete proteins, as well as the various functions of proteins in the body. Next, we examined the structure of proteins and discovered that a protein could have a primary, secondary, tertiary, and quaternary structure.

Next, we looked at enzymes, a very special group of proteins which act as biological catalysts. We examined the composition of enzymes, as well as the theory of how they work. We also reviewed the different types of enzymes and their functions.

In this chapter we also studied nucleic acids. We learned that nucleic acids are composed of units called nucleotides and we examined the three subgroups that make up the nucleotides. We also looked at some of the major nucleic acids, including RNA and DNA, and discussed the importance of the nucleic acids. We also learned about DNA replication and how our bodies grow new cells.

In the final section we looked at recombinant DNA technology. We discussed the process of gene splicing and learned how this new technology could affect our lives in the future. We also looked at the antiviral substance interferon, which could become readily available as a result of recombinant DNA technology. Interferon research may result in a cure for various forms of human cancer.

EXERCISES

1. Discuss why proteins are necessary for life as it exists on our planet.

2. Write the general formula for an amino acid.

3. Define an essential amino acid.

4. Define a complete protein and an incomplete protein. Give an example of each type.

5. Say whether each of the following statements describes a protein's primary, secondary, tertiary, or quaternary structure.
   (a) In sickle cell anemia, the protein hemoglobin is defective because out of about 300 amino acids, glutamic acid is replaced in one position by the amino acid valine.
   (b) In bovine ribonuclease the disulfide linkages between cysteine groups cause the protein to fold.
   (c) The protein collagen exists in the form of a triple helix.
   (d) The four subunits of hemoglobin must be present for this protein to carry out its function as an oxygen carrier.

6. Explain the importance of the peptide bond. Show how this bond forms between two amino acids.

7. What is the difference between a simple enzyme and a conjugated enzyme?

8. Why are enzymes important to the body?

9. For each of the following substrates, write the name of the enzyme.
   (a) sucrose   (b) maltose   (c) lactose
   (d) lipids   (e) proteins

10. For the following pairs, state which is the enzyme and which is the substrate.
    (a) urea, urease   (b) protease, protein
    (c) maltose, maltase   (d) carbohydrate, carbohydrase

11. Explain how a catalyst works.

12. Explain the lock-and-key theory of enzymes.

13. Complete the following equation for an enzyme reaction.

$$E + S \rightleftharpoons \; ? \; \rightleftharpoons \; ? \; \rightleftharpoons \; E + P$$

14. Discuss the importance of nucleic acids.

15. What are the components of nucleotides?

16. Name the two five-carbon sugars and state the nucleic acids with which they are associated.

17. List the two types of nitrogen-containing bases, and state whether they

are single-ring or double-ring compounds. Also, name the five nitrogen-containing bases.

18. Which nitrogen-containing bases are found in DNA and which are found in RNA?

19. Write the complementary base pairs for the following nitrogen bases of DNA.
(a) adenine   (b) guanine   (c) cytosine   (d) thymine

20. Write the complementary base pairs for the following nitrogen bases of RNA.
(a) adenine   (b) guanine   (c) cytosine   (d) uracil

21. What is the importance of the double helix in DNA?

22. Match the word on the left with its definition on the right.
(1) amino acids   (a) basic building blocks of nucleic acids
(2) nucleotides   (b) basic building blocks of proteins
(3) monosaccharides   (c) basic building blocks of carbohydrates

23. Name the two types of nucleic acids.

24. The nucleic acid whose structure has a double helix is _____ (RNA or DNA)?

25. Define the following terms.
(a) mitosis   (b) parent cell   (c) daughter cell

26. Discuss the implications of recombinant DNA technology.

27. Explain how recombinant DNA technology works.

28. Summarize the current developments dealing with interferon research.

Chapter 17

Pharmaceuticals and Drugs

Some Things You Should Know After Reading This Chapter

You should be able to:

1. Discuss the use of drugs in earlier societies.
2. Differentiate among the brand name, chemical name, and generic name of a drug.
3. Define an analgesic as a pain-relieving substance and an antipyretic as a fever-reducing substance.
4. Discuss the properties of salicylic acid, sodium salicylate, phenyl salicylate, and acetylsalicylic acid (aspirin).
5. Discuss the properties of methyl salicylate (oil of wintergreen).
6. Compare the effectiveness of aspirin and other pain-relieving products.
7. Explain how aspirin works in the body.
8. Discuss the various antacid products and explain how they work.
9. Write a balanced chemical equation to show how an antacid neutralizes stomach acid.
10. Define chemotherapy and discuss its beginnings.
11. Explain how sulfanilamide works.
12. Discuss the discovery of penicillin and ex-

plain why researchers continued looking for new antibiotics.
13. Define hormones and name the major male and female sex hormones.
14. Discuss the development of the birth control pill from a chemical point of view.
15. Explain the differences among stimulants, depressants, and narcotics. Cite examples for each group.
16. Discuss the properties of ethyl alcohol as a depressant and explain a chemical mechanism for alcohol addiction.
17. Explain what is meant by a synergistic effect and tell how this relates to barbiturates.
18. Discuss the properties of morphine, heroin, and codeine.
19. Explain the action of a tranquilizer.
20. Discuss the mind-altering drugs and cite some examples.
21. Explain the differences in the various types of marijuana.

INTRODUCTION

Did school give you a headache today? Take a pain pill. Perhaps you're nervous about an upcoming test? Take a tranquilizer and calm down. Or perhaps you have to stay up late tonight? Take an amphetamine. Yes, our society consumes a lot of drugs and other medications. We take drugs to relieve pain, put us to sleep, settle an upset stomach, and fight disease germs in our bodies. We use synthetic hormones for birth control, stimulants to make us feel good, and tranquilizers to calm us down. There are those who inject narcotics and mind-altering drugs into their bodies in order to escape into an *unreal* world.

In this chapter we are going to examine the various kinds of pharmaceuticals and drugs that have overwhelmed us today. We look at both the positive and negative aspects of these substances and gauge their impact on our society. But before we do, let's take a look at a day in the life of a married couple whom we call George and Mary.

SCENARIO

> ### A DAY IN THE LIFE OF GEORGE AND MARY
>
> The digital alarm clock is buzzing. George, a man of 41, chief accountant for a large oil company, is awakening from a night's sleep. It is Monday, February 25, 1985, time to get ready for another week of work. The first thing that George does is take his Hydro-Diuril, a pill for high blood pressure.
>
> George's wife, Mary, age 38, is also awakening. The first thing that she does is to take her birth control pill. George and Mary have three children and don't want to have any more. At breakfast George and Mary each have two cups of coffee. The caffeine in the coffee helps to wake them up so they will be more alert.
>
> George is off to work. When he arrives, the first thing he does is to have another cup of coffee. Next, be begins to work on an accounting project. But working with all those numbers gives George a headache. George pops a couple of aspirin. The morning progresses and it's getting close to the time when George has to present his report, so he rushes lunch. Sure enough, he gets an upset stomach, so he takes some Maalox, an antacid medication.
>
> The hours pass quickly and it's time for George's presentation to the directors of the company. George is a little nervous, so he takes a tranquilizer to help calm his nerves. The presentation goes fine.
>
> On the way home from work George stops at the dentist to have a cavity filled. The dentist administers novocaine as a pain killer. George leaves the dentist and begins his long ride home. Before dinner George has a scotch and soda to help him unwind. Later that night he takes his sleeping pill so that he can get a good night's sleep.
>
> And what has Mary been doing all day? She started her day off by taking a birth control pill and drinking two cups of coffee. But it wasn't long before she too required aspirin for a headache and a tranquilizer to calm her down. After all, it's not easy getting your husband off to work, the children off to school, and yourself ready for your job as a real estate agent. Lunch with some clients includes a few alcoholic drinks. When Mary has some time home alone she

likes to smoke a marijuana cigarette, and sometimes she takes a tranquilizer before going to bed.

What a day!

MEDICINE, DRUGS, AND SOCIETY

Ours is not the first drugged society. The ancient Chinese knew of substances that could be obtained from plants that could be used to stop pain and put people into euphoric states. Around the year 300 B.C., the ancient Greeks learned about opium. The Indians of the southwest United States and the Indians of Mexico knew how to extract mind-altering drugs from certain mushrooms. And the use of marijuana can be traced back thousands of years.

TABLE 17-1 *Various types of prescription and over-the-counter drugs*

TYPE OF DRUG	EXAMPLE
Analgesics (pain relivers) and/or antipyretics (fever reducers)	Aspirin, acetaminophen, codeine, Darvon[a]
Antibacterials/antiseptics	Tincture of iodine, mercurochrome, bacitracin, hexachlorophene
Antibiotics	Penicillin, tetracyclines
Antiinflammation drugs	Cortisone
Appetite suppressants	Amphetamines
Cough and cold medications	Antihistamines, cough suppressants, decongestants
Diabetic drugs	Insulin
Diuretics (increase urine production)	Hydro-Diuril[a]
Epilepsy and other nervous system drugs	Anticonvulsants such as phenobarbital
Heart and blood vessel drugs	Digitalis and nitroglycerin
Hormones	Birth control pills
Sedatives	Barbituates, Seconal sodium[a]
Stomach and intestinal drugs	Anatacids and laxatives
Tranquilizers	Librium,[a] Valium[a]
Vitamins and minerals	One-A-Day[a], etc.

[a]Brand name.

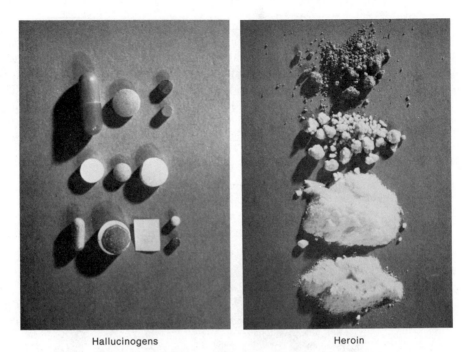

Hallucinogens   Heroin

Figure 17-1 Besides the billions spent on legal drugs and medicines, there
are countless billions spent on illegal drugs each year. (Drug Enforcement
Administration, U.S. Department of Justice)

Also, many cultures knew how to ferment grapes and other fruits to make
alcoholic beverages.

So drugs have been with us a long time. But what makes our society
different from earlier ones? The answer is the number of drugs available to
the public today. Drugs are big business. Annual sales by pharmaceutical
companies run about $11 billion a year. Drug prescriptions alone number
about 2 billion a year. That averages out to about 10 prescriptions a year for
every man, woman, and child in the United States. Table 17-1 lists the various
types of prescription and over-the-counter (OTC) drugs available to the
American public. Besides the billions spent on legal drugs and medicines,
there are countless billions spent on illegal drugs each year (Fig. 17-1).

ALL ABOUT PRESCRIPTION DRUGS

There are numerous prescription drugs on the market, and many of them are
expensive. In some cases they're more expensive than they have to be. That's
because when a pharmaceutical company develops a drug they usually give
it a *brand name*. The drug also has a very specific *chemical name*, which is usually
too difficult or too long to pronounce. And the drug also has a *generic name*,

which is easier to pronounce. The generic name is chosen jointly by the Food and Drug Administration (FDA) and the pharmaceutical company that discovered the drug. It usually has some relationship to the chemical name. For example, the pharmaceutical company Hoffman–La Roche discovered a tranquilizer which they named *Valium* (their *brand name*). The *chemical name* for this compound is 7-chloro-1,3-dihydro-1-methyl-5-phenyl-2H-1,4-benzodiazepin-2-one. The *generic name* for this compound is *diazepam*. Soon after Roche began marketing Valium, other companies began developing and

TABLE 17-2 *The 12 most frequently dispensed prescription drugs in the United States, 1975*

FUNCTION	EXAMPLES
Analgesics (pain relievers) and antipyretics (fever reducers)	Aspirin, codeine, Darvon,[a] morphine
Antibacterials and antiseptics (fight bacteria)	Hexachlorophene
Antibiotics (substances produced by microorganisms to fight other microorganisms)	Ampicillin, penicillin, tetracyclines
Antidepressants	Alertness elevators, mood elevators
Antiinfectives (fight amoebae, parasites, viruses, cancer)	Sulfa drugs
Antiinflammation drugs	Antiarthritics (arthritis drugs)
Appetite suppressants	Amphetamines
Biologicals	Antigens, serums, vaccines
Cough and cold drugs	Antihistamines, cough suppressants, decongestants
Diabetic drugs	Insulin
Diuretics (increase urine production)	Thiazides (Diuril)[a]
Epilepsy and other nervous system drugs	Anticonvulsants
Heart and blood vessel drugs	Digitalis, nitroglycerin
Hormones and steroids	Birth control pills, cortisone, thyroxin
Sedatives and sleeping aids	Barbiturates
Stomach and intestinal drugs	Antispasmodics, antacids, laxatives
Tranquilizers and muscle relaxants	Equanil,[a] Librium,[a] Valium[a]
Vitamins, minerals, and nutrients	Many combinations

[a]Brand name. All other listings are names of drug families within each general function category.
Source: Robert L. Wolke, *Chemistry Explained,* Prentice-Hall, Inc., Englewood Cliffs, N.J., 1980. Used by permission.

selling their own brands of diazepam, usually at a price much cheaper than Roche's Valium. What does this mean to the consumer?

A doctor who prescribes Valium for a patient forces the pharmacist to dispense the Roche product—even though he or she may have an identical and less expensive generic brand sitting on the shelf. Only if the doctor writes the generic name of the prescription on the prescription form may the pharmacist dispense the less expensive drug. (Some states, for example New Jersey, have passed laws to allow pharmacists to dispense generic brands of drugs even if the M.D. writes a brand name.)

Many doctors are not familiar with generic names of prescriptions. That's because the big pharmaceutical companies spend millions of dollars in promoting and advertising their brand name products. It's only fair to point out that these major pharmaceutical companies spend large sums of money developing and researching a new drug, so of course they want the doctors to prescribe the drug by its brand name.

Why does the American public consume so many pharmaceutical products? Is it to fight and prevent disease? Table 17-2 lists the 12 most frequently dispensed drugs in the United States in 1975. A quick glance at this table shows that two out of the top four drugs are tranquilizers, *not* antibiotics. The best-selling drug, Valium, had 60 million prescriptions written for it in 1977. The third-best-selling drug, Darvon-65, is a pain reliever that may not be any more effective than aspirin. It seems that many people are taking drugs for many reasons other than curing or preventing disease.

We'll spend the remainder of this chapter surveying some of the major classes of prescription and over-the-counter drugs. We also take a brief look at some illegal drugs.

THE PAIN-RELIEVING DRUGS

A variety of medications are available for pain relief. Some of these medications are available only by prescription, whereas others are available on an *over-the-counter basis*. Pain-relieving medications are sometimes called *analgesics*.

Around 1860 an important analgesic was isolated from willow bark. It was called *salicylic acid* (Fig. 17-2). Besides being an analgesic (pain reliever), this substance also had good fever-reducing properties (such substances are known as *antipyretics*). Unfortunately, although salicylic acid had good

Figure 17-2 An important analgesic is salicylic acid.

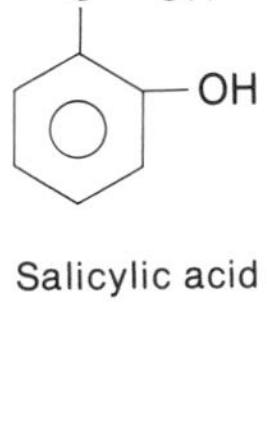

Salicylic acid

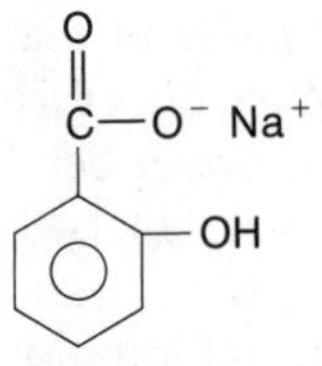

Sodium salicylate

Figure 17-3 The compound sodium salicylate was an improvement over salicylic acid.

fever-reducing properties, it was irritating to the stomach and it had a very unpleasant sour taste.

It took only 15 years, until 1875, for chemists to come up with an improved form of salicylic acid. The compound was *sodium salicylate*, the sodium salt of salicylic acid (Fig. 17-3). Although this substance had a better taste, it was still very irritating to the stomach.

Another attempt at producing a derivative of salicylic acid gave rise to a substance called *phenyl salicylate*, but better known as *salol* (Fig. 17-4). Salol was used in the 1880's and 1890's. The advantage of salol over salicylic acid and sodium salicylate was that it passed through the stomach unchanged—therefore, no stomach upset! In the small intestine the salol was converted to salicylic acid, but also a substance known as phenol was formed. In great enough concentrations, phenol can be toxic to human beings. This could occur if someone were to take a large overdose of salol.

Finally, in 1899 another derivative of salicylic acid was produced— *acetylsalicylic acid* (Fig. 17-5). Commonly known as *aspirin*, this substance had

Figure 17-4 Phenyl salicylate, better known as salol, was used as an analgesic during the 1880's and 1890's.

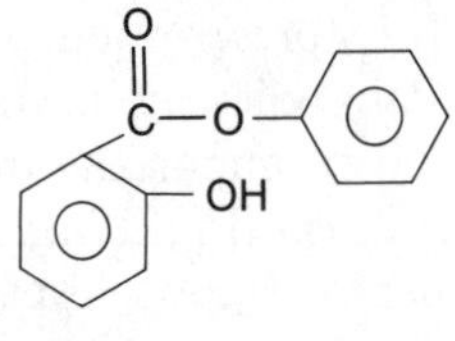

Figure 17-5 After its discovery in 1899, aspirin became one of the best-known and most widely used over-the-counter drugs.

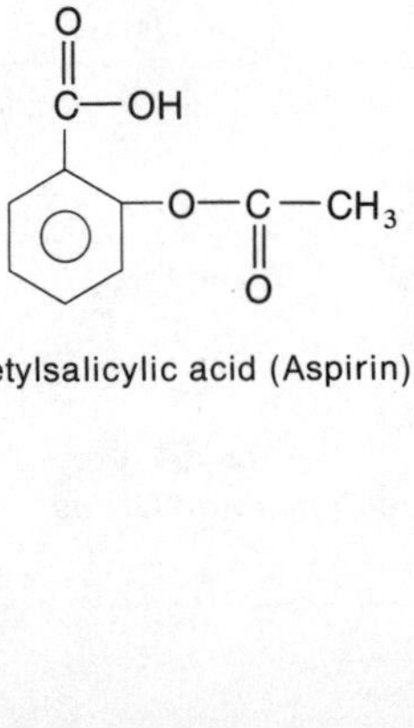

Acetylsalicylic acid (Aspirin)

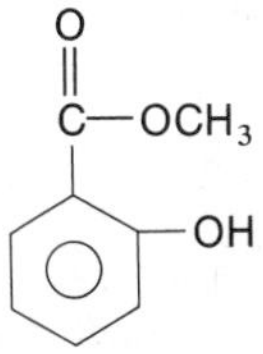

Methyl salicylate (oil of wintergreen)

Figure 17-6 Methyl salicylate, also known as oil of wintergreen, is used as a flavoring agent and as an ingredient in rubbing compounds.

both analgesic and antipyretic properties. It was also safe to take. Most individuals who used aspirin in the early 1900's showed very little stomach upset when compared to the people who used other derivatives of salicylic acid. Aspirin quickly became one of the best known and most widely used over-the-counter drugs in the world. It also proved to be one of the most effective medications for general pain relief, and an excellent substance for reducing fever and swelling.

We should point out, however, that approximately 2 people in 1000 are allergic to aspirin. This includes from 6 to 20% of asthmatics. Also, because aspirin tends to increase bleeding, individuals who have a history of ulcers should not take aspirin. Hemophiliacs should not take aspirin for the same reason. And people who are facing surgery, or women in their last three months of pregnancy, should not take aspirin because of its anticlotting properties.

Also, about 100 people a year die from aspirin overdose in the United States. Most of these people are children accidentally given an overdose by their parents. These deaths could easily be avoided. The first sign of aspirin overdose is a ringing in the ears. This means that you should stop taking the aspirin immediately! A doctor should be consulted if there is a chance that too much aspirin has been ingested.

By the way, another derivative of salicyclic acid was also discovered. This compound was called *methyl salicylate*, but is better known as *oil of wintergreen* (Fig. 17-6). In small concentrations oil of wintergreen is used as a flavoring agent. In larger concentrations it is used as an ingredient in rubbing compounds. When applied to the skin the oil of wintergreen causes a burning sensation which makes sore muscles feel good.

ASPIRIN VERSUS BUFFERIN
AND THOSE OTHER PAIN RELIEVERS

Americans spend nearly $1 billion a year on pain relievers, most of it for aspirin. Yet you probably wouldn't know it by a casual walk through your neighborhood pharmacy, or listening to the advertisements on radio and television, or reading the newspaper.

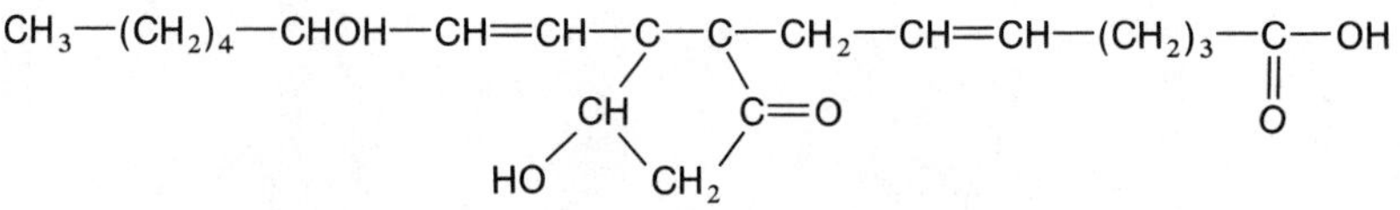

$$CH_3-(CH_2)_4-CHOH-CH=CH-\underset{\underset{\underset{HO}{\diagdown}\ CH_2}{\overset{\diagup}{CH}}}{C}-\underset{\underset{\overset{\|}{O}}{C=O}}{C}-CH_2-CH=CH-(CH_2)_3-\underset{\underset{\overset{\|}{O}}{}}{C}-OH$$

Figure 17-7 The formula of a typical prostaglandin. The name of this compound is prostaglandin E_2.

Anacin promises "fast pain relief," and Bufferin says that it's "twice as fast as aspirin." The makers of Alka-Seltzer claim that their product has "the sound of fast relief." Yet the pain-relieving ingredient in each of these products is aspirin, and the only thing *significantly different* about each is their cost!

Also, aspirin is aspirin, regardless of who makes it. The typical tablet usually contains about 325 mg of acetylsalicylic acid, together with a binder to hold the tablet together. The binder is usually made of starch or some other inert material. Most of you are probably familiar with Bayer Aspirin. Yet independent studies have shown that all brands of aspirin are essentially the same, and that the only major difference is the cost per tablet. You can pay about 3 cents a tablet for the 24-tablet size of Bayer Aspirin, or less than $\frac{1}{2}$ cent a tablet for the large size of a non-brand name tablet. Yet both tablets contain the exact same thing—325 mg of aspirin.

Aspirin works to relieve pain, reduce fever, and suppress inflammation by inhibiting the synthesis of *prostaglandins*. These compounds, produced in the body, are involved in the inflammation process. They are also involved in regulating blood pressure and in the contraction of smooth muscle. We'll have more to say about prostaglandins later in this chapter when we discuss birth control pills (Fig. 17-7).

Now, what about some of these other high-priced pain and fever reducers? Anacin contains aspirin plus the stimulant caffeine. The caffeine amounts to that found in a quarter of a cup of coffee. Anacin has slightly more aspirin per tablet than regular aspirin—about 400 mg per tablet instead of 325 mg. But you take fewer tablets, so it all evens out (Fig. 17-8).

Bufferin contains aspirin plus two buffering agents. The buffering agents are supposed to reduce the acidity of the stomach caused by the acidic properties of the aspirin. But the amount of buffering may be insufficient to completely overcome aspirin's acidity. And some doctors recommend taking aspirin with

Figure 17-8 An Anacin tablet contains aspirin plus caffeine.

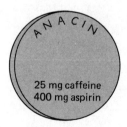

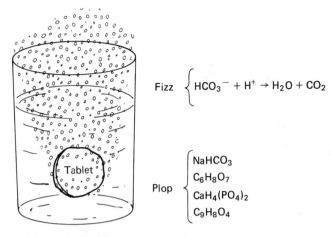

Figure 17-9 is caused
by bicarbonate ions from the $NaHCO_3$ reacting with hydrogen ions from the
acidic components of the tablet.

Figure 17-9 An Alka-Seltzer tablet has many ingredients. The fizz is caused
by bicarbonate ions from the $NaHCO_3$ reacting with hydrogen ions from the
acidic components of the tablet.

food, milk, or an antacid such as Maalox, Di-Gel, or Mylanta to achieve the
same or better results than taking Bufferin.

Alka-Seltzer contains aspirin plus large amounts of buffers, adequate to
overcome aspirin's acidity. But Alka-Seltzer also has a high sodium content,
so may not be advisable for people who must be on low-salt diets (Fig. 17-9).

A number of products, including Tylenol, Datril, Bayer Non-Aspirin
Pain Reliever, and Anacin-3, contain acetaminophen, the leading aspirin
substitute (Fig. 17-10). Acetaminophen has the same pain- and fever-reducing
properties as aspirin but has no effect on inflammation. So acetaminophen is
of no use for the pain of arthritis or other types of swelling. However, acetaminophen causes less stomach upset than aspirin.

There are also a number of *combination drugs* available, such as Excedrin,
Vanquish, and Empirin Compound. These products contain aspirin or acetaminophen in combination with caffeine, buffers, or antihistamines (which
promote drowsiness). Advertisements for these substances claim relief superior
to that of plain aspirin or acetaminophen, but there is no good scientific proof
for that claim. Some of these products even contain phenacetin (Fig. 17-11),
a pain reliever that the FDA concluded was not safe, owing to the possibility
of kidney damage with repeated use. Also phenacetin is metabolized in the

Figure 17-10 Acetaminophen is the leading aspirin substitute.

$$CH_3-\underset{\underset{O}{\|}}{C}-NH-\bigcirc-OH$$

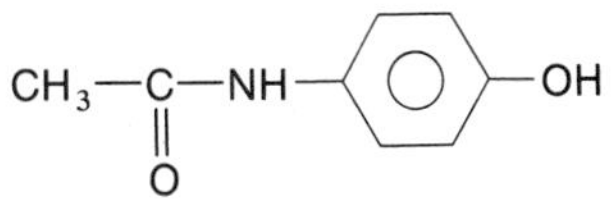

Acetaminophen

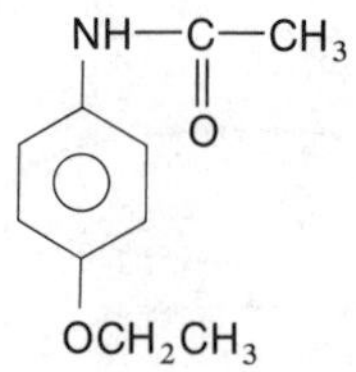

$$\underset{\text{Phenacetin}}{\overset{\displaystyle NH{-}C{-}CH_3}{\overset{\displaystyle \|}{\overset{\displaystyle O}{\bigcirc}}}\atop OCH_2CH_3}$$

Phenacetin

Figure 17-11 Phenacetin, a pain reliever, may cause kidney damage with chronic use.

TABLE 17-3 *Toxicity of some drugs found in pain-relieving medications*

SUBSTANCE	LD$_{50}^{a}$
Acetaminophen	3.7 g/kg body weight in rabbits
Aspirin	1.75 g/kg body weight in rats
Caffeine	0.2 g/kg body weight in rats
Phenacetin	0.00165 g/kg body weight in rats
Salicylic acid	1.3 g/kg body weight in rabbits

Source: The Merck Index, 9th ed., Merck & Co., Rahway, N.J.

[a] The term LD$_{50}$ means the dose that is lethal to 50% of the animals tested. The data given here are for oral administration of the substance.

body to acetaminophen, so a dose of acetaminophen would have the same effect with less danger.

In summary, it seems that plain, old aspirin is the safest and cheapest pain reliever and fever reducer for most people. It is also one of the safest medications in terms of an accidental overdose (Table 17-3).

THE ANTACIDS: DRUGS FOR STOMACH UPSET

The process of digestion involves the production of hydrochloric acid in the stomach. Sometimes the stomach produces too much HCl. This may be due to emotional stress or overeating. Stomach upset and heartburn are the result. The cure for such a problem may be any number of antacid medications (Table 17-4).

The various antacid medications work by neutralizing the excess acid. The reactions that take place may be represented as follows:

(For antacids that contain hydroxide ions) $OH^- + H^+ \longrightarrow H_2O$

TABLE 17-4 *Some commercial antacid medications*

NAME	ACTIVE INGREDIENTS
Alka-Seltzer	Monocalcium phosphate, $CaH_4(PO_4)_2$
	Sodium bicarbonate, $NaHCO_3$
	Citric acid, $C_6H_8O_7$
	Aspirin (for headache pain)
Maalox	Magnesium hydroxide, $Mg(OH)_2$
	Aluminum hydroxide, $Al(OH)_3$
Rolaids	Aluminum sodium dihydroxy carbonate, $AlNa(OH)_2CO_3$
Tums	Calcium carbonate, $CaCO_3$
	Magnesium carbonate, $MgCO_3$
	Magnesium trisilicate, $Mg_2Si_3O_8$

(For antacids that contain carbonate ions)

$$CO_3^{-2} + 2\,H^+ \longrightarrow H_2O + CO_2$$

In other words, the consumption of an antacid medication causes an acid–base reaction to occur in the stomach. The overall reaction for an antacid that contains magnesium hydroxide would look like this:

$$Mg(OH)_2 + 2\,HCl \longrightarrow MgCl_2 + 2\,H_2O$$

It's also interesting to point out that certain antacids, such as magnesium carbonate ($MgCO_3$) and magnesium hydroxide, may act as laxatives. In other words, in small doses these substances act as antacids and in large doses they act as laxatives.

THE ANTIBIOTICS: LIFESAVING DRUGS

At the beginning of the twentieth century, infectious diseases were still killing millions of people each year as they had done throughout history. But by the middle of the twentieth century most of these diseases were brought under control. How was this accomplished? It was accomplished by the discovery and wide-scale use of a class of drugs known as antibiotics (Fig. 17-12).

It all began in 1909 when the German bacteriologist Paul Ehrlich used the synthetic substance *arsphenamine* (Fig. 17-13) as a therapeutic agent against *syphilis*. This event is taken by many to mark the beginnings of *chemotherapy*— the treatment of a disease by the use of a specific chemical agent or agents.

Figure 17-12 Penicillin mold. This symmetrical colony of green mold is *Penicillium chrysogenum*, a mutant form of which now produces almost all of the world's penicillin. (Pfizer, Inc.)

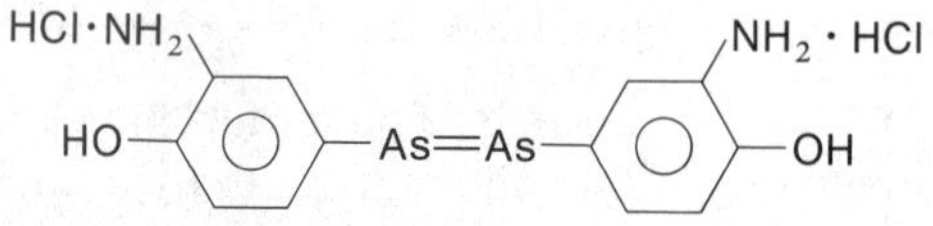

Arsphenamine

Figure 17-13 The substance arsphenamine was used in the early 1900's as a therapeutic agent against syphilis.

Figure 17-14 The compound sulfanilamide was the first of the sulpha drugs.

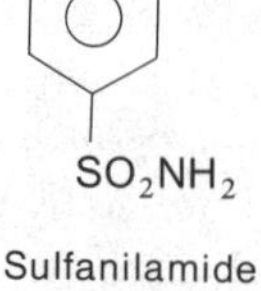

Sulfanilamide

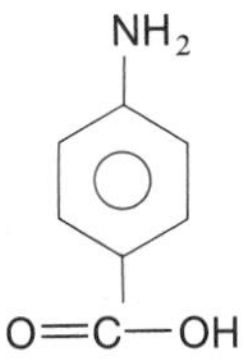

Para-aminobenzoic acid

Figure 17-15   The structure of *para*-aminobenzoic acid.

In 1908, the compound *sulfanilamide* was synthesized (Fig. 17-14), but it wasn't until 1932 that the research of the German chemist Gerhad Domagk showed that sulfanilamide and certain related compounds, which came to be known as sulfa drugs, could be used to fight a variety of infectious diseases.

Sulfanilamide works by stopping the growth of bacteria. It's able to do this because its chemical structure is similar to the structure of a substance known as para-aminobenzoic acid, PABA for short (Fig. 17-15). PABA is required by bacteria to synthesize the compound folic acid (Fig. 17-16). The folic acid is needed for bacterial growth. Bacterial ezymes can't differentiate between PABA and sulfanilamide. If sulfanilamide is incorporated into the bacteria's folic acid synthesis a nonfunctioning folic acid type compound is produced. The bacteria stop growing.

In practice, a doctor administers sulfanilamide to a patient suffering from a bacterial disease. The sulfanilamide finds its way to the bacteria and begins to get incorporated into the bacteria's production of folic acid. The pseudo folic acid that forms is not capable of supporting bacterial growth, so the bacteria die and the patient is cured.

The sulfa drugs, which are synthetic compounds, proved to be a great boon to medicine in the 1930's. However, it wasn't long before natural products caught up and surpassed the synthetics.

The first of the natural antibiotics was penicillin (Fig. 17-17), whose existence was discovered accidentally by the Scottish bacteriologist Alexander Fleming in 1928. Fleming had left a culture of staphylococcus germs uncovered for a number of days and then found that it had become moldy. Around

Figure 17-16   Folic acid is needed for bacterial growth.

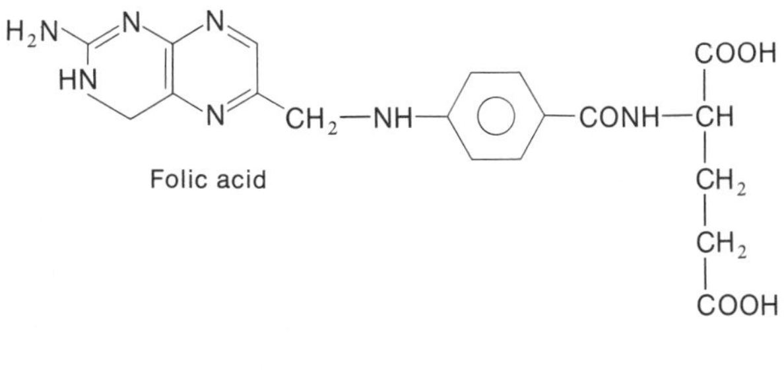

Folic acid

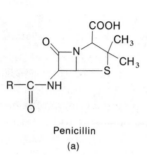

Penicillin

(a)

(b)

Figure 17-17   (a) A general structure for penicillin. Different R groups lead to the different types of penicillin.
(b) Dr. Alexander Fleming, noted British scientist and discoverer of penicilllin, at The Squibb Institute for Medical Research during a trip to the United States. Penicillin was discovered in 1929 by Dr. Fleming at St. Mary's Hospital, London. E. R. Squibb & Sons was one of three pharmaceutical companies which first produced penicillin in the United States. The first culture of *Penicillin notatum* from Dr. Fleming's laboratory in England went to the Squibb Institute for Medical Research. (c) Antibiotic manufacturing begins in the spore laboratory where organisms that produce penicillin and streptomycin are carefully grown in sterile flasks. The organisms may be transferred to larger jars, such as those shown here, containing penicillin culture. This penicillin culture will then be used to inoculate germinator tanks containing several hundred gallons of nutrient. (Photographs courtesy of E. R. Squibb and Sons, Inc., Princeton, N.J.)

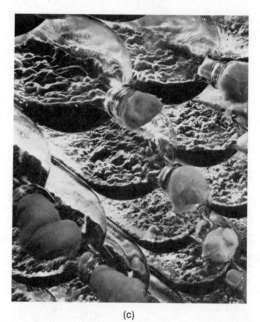

(c)

every speck of mold spore there was a clear area where the bacterial culture did not grow. Fleming suspected that the mold spore contained some antibacterial substance, but he had difficulty in isolating the specific compound.

It wasn't until the onset of World War II that the Australian-English pathologist Howard Florey and the German-English biochemist Ernst Chain

TABLE 17-5 *The penicillins*

NAME	COMMENT
Penicillin G	Usually given by injection; used for acute and chronic gonorrhea; side effects include allergic reactions and diarrhea
Penicillin V	Usually given orally—in other words, an oral penicillin; similar in effectiveness to penicillin G, and with similar side effects
Phenethicillin	Similar to penicillin V; used for mild or moderate infections due to streptococci and pneumonococci
Ampicillin	Given by injection or taken orally; very effective in urinary tract infections
Methicillin	Given by injection; extremely effective in certain types of staphylococci infections
Oxacillin	An oral form of methicillin; more active against pneumococci and streptococci
Nafcillin	Similar to oxacillin, but given by injection
Dicloxacillin	Similar activity as oxacillin; can be given orally and is absorbed readily from the intestines
Carbenicillin	A semisynthetic penicillin that must be given by injection; excreted slowly from the body, thus maintaining high serum levels

Source: *The Merck Manual*, 12th ed., Merck & Co., Rahway, N.J.

isolated and determined the structure of penicillin. By 1945, a process for producing about $\frac{1}{2}$ ton of penicillin a month was developed.

Continued research into the structure of penicillin soon had chemists altering the basic structure of the molecule. The results were a number of synthetic penicillins, which in some instances had superior germ-fighting properties to those of the original (Table 17-5).

During the 1940's and 1950's other antibiotics, such as streptomycin and the tetracyclines, terramycin and aureomycin, were isolated (Fig. 17-18). The tetracyclines are known as *broad-spectrum antibiotics*. That's because they are effective against some viruses as well as a large number of bacteria (Fig. 17-19).

Pharmaceutical companies are constantly looking for new antibiotics or modifications of existing ones. One of the reasons for this is because of the development of resistant microorganism strains. For example, certain microorganisms that were easily controlled with penicillin in the late 1940's and early 1950's developed penicillin-resistant strains in the late 1960's and early 1970's. One of the most widely publicized examples of this was the discovery of a penicillin-resistant strain of gonococcus—the bacterial strain that produces gonorrhea. An epidemic of this type of gonorrhea was prevalent among our servicemen during the Vietnam war. The epidemic was controlled by the use of various antibiotics. However, because bacterial organisms can develop resistant strains fairly rapidly, scientists are always searching for new drugs.

415 *Pharmaceuticals and Drugs*

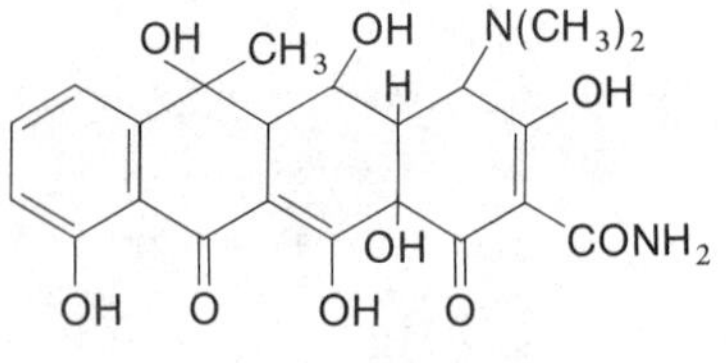

Terramycin

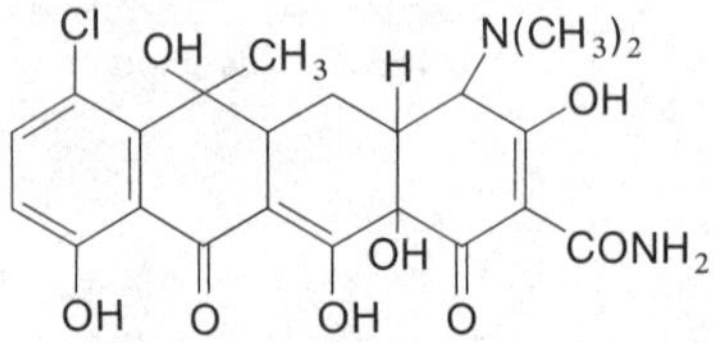

Aureomycin

Figure 17-18 During the 1940's antibiotics such as Terramycin and Aureomycin were developed.

Figure 17-19 Mighty Mold: Shown is a colony of *Streptomyces rimosus*, the microorganism used in the production of the antibiotic Terramycin. This antibiotic was discovered by a team of Pfizer scientists in 1949 and is effective in the treatment of more than 100 different diseases. (Photo courtesy Pfizer, Inc.)

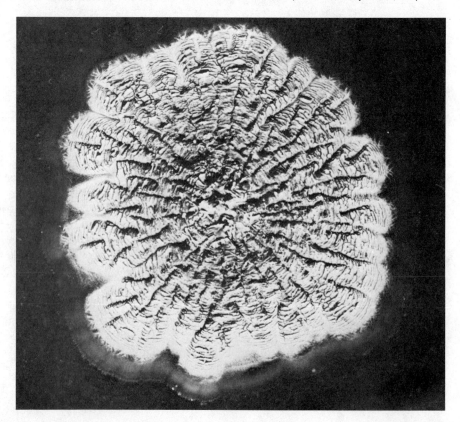

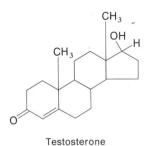

Testosterone

Figure 17-20   The most important male hormone is testosterone.

BIRTH CONTROL PILLS AND SYNTHETIC HORMONES

In the 1930's and 1940's chemists were busy determining the structure of various hormones in the body. It wasn't long before some of these hormones were synthesized in the laboratory.

At first many hormones that were synthesized were used for therapeutic reasons. They would be given by injection to patients whose bodies did not produce enough of a particular hormone.

Of the various types of hormones in the body, the sex hormones are extremely important. The male sex hormones are known as *androgens.* They are produced in the testes and are responsible for the development of the male secondary sex characteristics such as the appearance of facial hair and the deepening of the voice. The most important male hormone is *testosterone* (Fig. 17-20).

The two categories of female sex hormones are known as *estrogens* and *gestagens.* The estrogens are produced primarily in the ovaries and are responsible for the development of the female secondary sex characteristics, including breast development, skeletal size, hair distribution, and the control of the menstrual cycle. The gestagens are also produced in the ovaries and are responsible for preparing the uterus for pregnancy.

There are three very important female hormones: estradiol, estrone, and progesterone (Fig. 17-21). The hormones estradiol and estrone are estrogens. Besides being important in the development of secondary sexual characteristics, they play an important part in the first half of the female menstrual cycle. They are responsible for egg maturation. The hormone progesterone, a gestagen, plays an important role in the second half of the menstrual cycle, preparing the uterus for pregnancy and if pregnancy occurs, halting the release of other eggs from the ovaries.

Armed with the knowledge of how the female reproductive cycle operates, scientists in the 1930's and 1940's began looking for ways to develop a safe and easily administered *contraceptive.* It was known that progesterone was the key to the puzzle, because it could trick the body into thinking that a pregnant condition was present. A large injection of progesterone could serve

417   *Pharmaceuticals and Drugs*

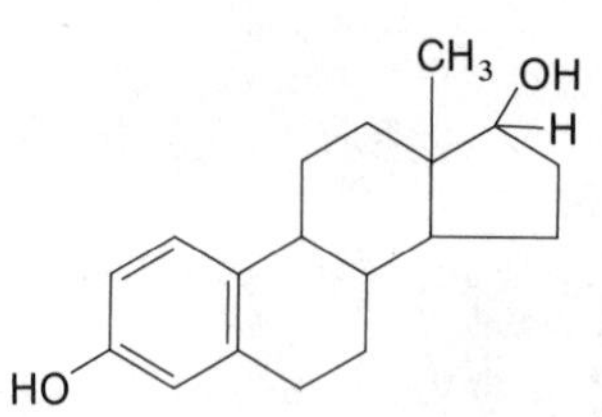

Estradiol

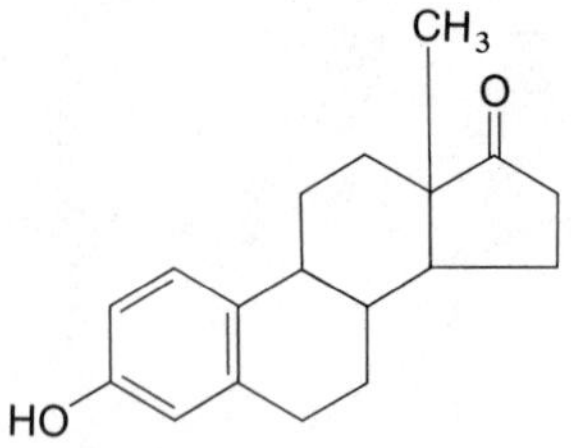

Estrone

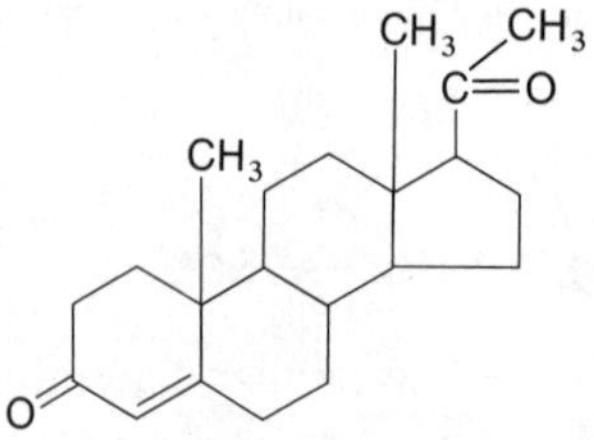

Progesterone

Figure 17-21 Three very important female hormones.

as an effective birth control drug. However, this did not prove to be an ideal solution. A major breakthrough occurred in 1951 when Carl Djerassi discovered that removal of a methyl group from progesterone increased its activity fourfold. In other words, this substance was very effective in simulating pregnancy, making it a very effective birth control drug (Fig. 17-22). However, this demethylated progesterone, which was named 19-norprogesterone, still had to be administered by injection.

Djerassi's next breakthrough was the discovery that the replacement of the $-\overset{\displaystyle ||}{\underset{\displaystyle O}{C}}-CH_3$ group by the $-C\equiv CH$ group would allow the compound to

be administered orally. The result was the substance Norlutin, chemically

Figure 17-22 The compound 19-norprogesterone had the potential of being an effective birth control substance, but it had to be administered by injection.

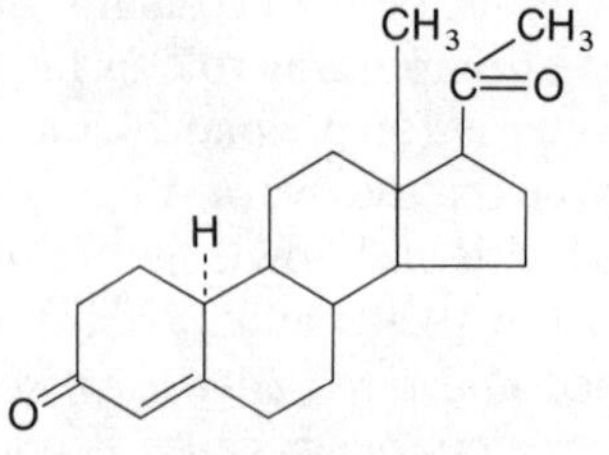

19-Norprogesterone

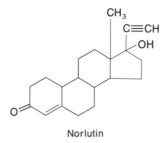

Norlutin

Figure 17-23 Norlutin proved to be an effective oral birth control pill even when administered in small doses.

known as 17α-ethynyl-19-nortestosterone (Fig. 17-23). Notice how this substance differs from progesterone. A methyl group is gone. This makes the drug more potent than progesterone. And the $-\overset{\text{O}}{\underset{\|}{\text{C}}}-CH_3$ group from pro-

gesterone has been replaced by a $-C\equiv CH$ group and an OH group. This allows the Norlutin to be administered orally. Norlutin proved to be an effective birth control substance even when administered in small doses.

In an actual birth control pill, the Norlutin, which falls into the gestagen category, is mixed with an estrogen. For example, in the brand name product known as Ortho-Novum, the Norlutin is compounded with the estrogen mestranol. The mestranol controls the menstrual cycle and the Norlutin acts to establish a state of false pregnancy. Because there is no ovulation, a real pregnancy cannot occur.

Numerous brand names of birth control pills are available today. Some of them use different estrogens and gestagens. But they all operate in pretty much the same way (Fig. 17-24).

Figure 17-24 Examples of birth control pills available on the market today. (Ortho Pharmaceutical Corp., Raritan, New Jersey, and Searle & Co., San Juan, Puerto Rico)

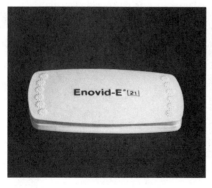
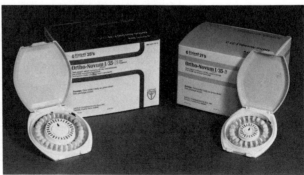

BIRTH CONTROL PILLS AND SAFETY

Are birth control pills safe to take? This question has received a lot of attention over the years. There does indeed seem to be side effects for *some* women who take the pill. These side effects include high blood pressure and blood clotting—to name the most serious ones. Such side effects could be lethal if they cause a heart attack or stroke. However, the death rate associated with the pill is low; about 3 per 100,000 users. (That's low, of course, if you're not one of the three!) But keep in mind that the death rate for women during pregnancy is about 30 per 100,000. So on a competitive basis, taking the pill is safer than getting pregnant. Still, women who are known to have high blood pressure or blood-clotting problem, or who come from families where these conditions are prevalent, should not take the "pill."

Also keep in mind that what we just discussed refers only to the short-term effects of the pill. The long-range effects are yet to be seen. The decades of the 1980's and 1990's may yield some answers to the question of long-range effects.

There is of course continuing research on new types of birth control pills. Some of the prostaglandins that we discussed earlier in this chapter have the ability to abort a fertilized ovum in the early weeks of pregnancy. In the 1970's when these compounds were being tested in Europe, the press called these prostaglandins "morning-after pills." However, they were anything but that, since they had to be given intravenously over a long period of time.

HOW ABOUT A BIRTH CONTROL PILL FOR MALES?

Many of the same pharmaceutical companies that brought us the "pill" for women are actively engaged in supporting research programs that will hopefully lead to a safe and effective *male* birth control pill. Up to now the compounds that have been tested work on the principle of stopping sperm production. But they also have many negative side effects.

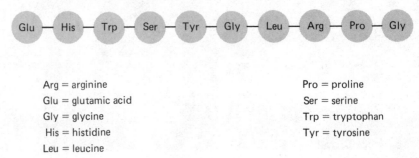

Figure 17-25 The substance leutinizing releasing factor helps to control the development of secondary sex characteristics and certain aspects of sexual behavior, such as libido.

Arg = arginine

Glu = glutamic acid

Gly = glycine

His = histidine

Leu = leucine

Pro = proline

Ser = serine

Trp = tryptophan

Tyr = tyrosine

In February 1980 several scientists working under National Institute of Health grants announced that preliminary studies on a group of compounds known as *peptide hormones* showed that they had certain contraceptive properties. These substances, one of which is known as *leutinizing releasing factor*, helps to control the development of secondary sex characteristics and also certain aspects of sexual behavior, such as libido (Fig. 17-25). The researchers are trying to develop analogs of these compounds that interfere with conception but not with other functions. According to Wylie Vale and his colleagues at the Salk Institute in San Diego, it may be possible to develop a peptide hormone that will interfere selectively with sperm development without affecting other traits. The male birth control pill may not be far away!

STIMULANTS, DEPRESSANTS, AND NARCOTICS

The ancient Greeks knew that the extracts of various plants and herbs could make a person more alert, put him or her into a stupor, or knock the person out completely. They also knew that some of these substances had addicting properties. Today we classify these substances as *stimulants*, *depressants*, and *narcotics*.

Stimulants

Probably the most widely used stimulant in the world is *caffeine* (Fig. 17-26). We all ingest caffeine when we drink coffee, tea, or cola. The effect of this substance on our bodies is to make us more awake and alert. Some of us are very dependent on this compound. Just think of all of the people you know who *must* have that morning cup of coffee or tea in order for them to begin functioning.

Is all of this caffeine that we ingest harmless? In early 1980 the Food and Drug Administration (FDA) reported that caffeine caused birth defects in rats. The amount of caffeine given the rats was more than a human being would normally consume, but an FDA spokesman said the gap between the two amounts was "not a comfortable difference." Earlier studies had linked

Figure 17-26 Probably the most widely used stimulant in the world today is caffeine.

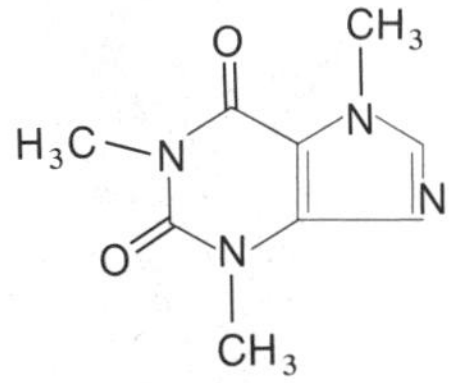

Caffeine

caffeine to birth defects in human beings, but a final decision on what to do about the problem has not been made. It appears at this point that excessive caffeine should be avoided by expectant mothers. Perhaps in the near future we will see warning labels on packages of coffee, tea, and cola: "Caution, this substance may be harmful to your offspring."

The "Uppers"

Caffeine is a mild stimulant; however, there are many other more potent stimulants available today. The so-called "uppers," chemically known as *amphetamines*, are probably the most abused drugs on the market. Originally used to treat patients on weight reduction programs, amphetamines found their way into the illegal drug market during the 1960's and 1970's, where people bought them for their "high" effect. Amphetamine use was most noticeable among students, truck drivers, and army personnel, who coined the term "bennies" and "copilots" to describe these compounds. Table 17-6 lists the names and formulas of some of these "uppers."

TABLE 17-6 *Examples of some stimulants*

NAME	FORMULA	COMMENT
Nicotine		Can be highly toxic in its pure form; found in tobacco and may be addicting
Cocaine		Very powerful stimulant; also acts as a local anesthetic; can cause heart attack and death if an overdose is taken; usually sniffed by addicts
Amphetamine	CH_2CH-NH_2 with CH_3	Used for weight reduction, but causes high blood pressure, restlessness, and insomnia—a feeling of uneasiness; usually not recommended for weight reduction anymore because of wide-scale abuse by public as an "upper"
Methamphetamine	$CH_2CH-NH-CH_3$ with CH_3	A very potent stimulant, sometimes called "speed"; can be addicting and toxic in low concentrations (the LD_{50} in mice is 50 mg/kg body weight)

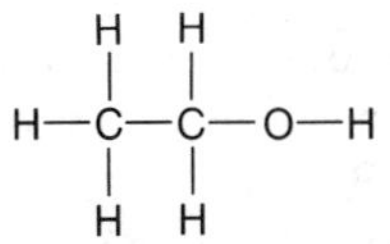

$$H-\overset{\underset{|}{H}}{\underset{|}{C}}-\overset{\underset{|}{H}}{\underset{|}{C}}-O-H$$

Ethyl alcohol (Ethanol)

Figure 17-27 Perhaps the most abused depressant in the world is ethyl alcohol.

Depressants

The *stimulants* are known as "uppers." The *depressants* are known as "downers." Perhaps the most abused depressant in the world is ethyl alcohol (Fig. 17-27). In the United States alone it is estimated that there are about 10 million alcoholics.

When we drink an alcoholic beverage, the alcohol is quickly absorbed into the blood. The initial effect of drinking results in a relaxing of tensions. In fact, many people have a drink after work just to unwind. As the amount of alcohol builds up in the blood a tranquilizing effect occurs, followed by a dulling of the senses and finally sedation (Table 17-7).

Most people who drink, and that includes something like two-thirds to three-fourths of the adult population, do so responsibly. But then there are the approximately 10 million alcoholics—people who seem to be *addicted* to alcohol. This addiction may be physiological as well as psychological.

In 1971, a group of researchers at the University of Michigan isolated a liver enzyme called cytochrome P-450. This enzyme, together with other substances manufactured by the liver is important in the detoxification process of alcohol, as well as other drugs and other substances. The research group headed by Minor Coon showed that large amounts of cytochrome P-450 built up in the liver of heavy alcoholic users and drug users. Is this good? At first glance it seems to be, because only in this way can the body detoxify the large

TABLE 17-7 *Alcohol blood levels and their effect on an individual*

AMOUNT (mg alcohol/100 ml blood)	EFFECT
50–150	Lack of coordination[a]
150–200	Causes intoxication
300–400	Causes unconsciousness
500 mg or more	May be fatal

[a] The legal driving level is 50 mg per 100 ml of blood or less. This level could be achieved by drinking about two "shots" of 90 proof whiskey, one right after another. A "shot" is about 30 ml.

423 *Pharmaceuticals and Drugs*

amounts of alcohol or drugs that the alcoholic or drug addict has consumed. But there's another side to the coin. Once begun, the continuous buildup of cytochrome P-450 could be responsible for causing an individual to crave alcohol. In other words, with the cytochrome P-450 already produced, the body calls for alcohol so that the enzyme can do its job. This series of events could be a partial explanation for alcohol addiction, and perhaps even drug addiction.

TABLE 17-8   *Some barbiturates and their uses*

NAME	FORMULA[a]	COMMENT
Phenobarbital	$A = C_2H_5-$	Used as an anticonvulsant for epilepsy; also acts as a sedative and hypnotic
Amobarbital	$A = C_2H_5$ $B = (CH_3)_2CHCH_2CH_2-$	Acts as a sedative and hypnotic; not as strong as phenobarbital
Secobarbital	$A = CH_2{=}CHCH_2-$ $B = CH_3CH_2CH_2\underset{\underset{CH_3}{\vert}}{CH}-$	A short-acting sedative and hypnotic
Pentobarbital	$A = C_2H_5$ $B = CH_3CH_2CH_2\underset{\underset{CH_3}{\vert}}{CH}-$	A short-acting hypnotic
Butabarbital	$A = C_2H_5-$ $B = CH_3CH_2\underset{\underset{CH_3}{\vert}}{CH}-$	Acts as a sedative
Thiopental (penthiobarbital)		A short-acting anesthetic

[a] The general formula for barbiturates is

Source: *The Merck Index*, 9th ed., and *The Merck Manual*, 12th ed., Merck & Co., Rahway, N.J.

Other Depressants

There are, of course, many depressant drugs besides alcohol. A whole class of compounds known as *barbiturates* can produce effects from mild sedation, to deep sleep, to *permanent* sleep if an overdose of the drug is taken.

The parent compound of this class of drugs is known as *barbituric acid*. It was first synthesized in 1864 by the German chemist Johann Friedrich Wilhelm Adolf von Baeyer! This compound was produced by Baeyer when he reacted urea with malonic acid.

$$H_2N-\underset{\underset{O}{\|}}{C}-NH_2 + HO-\underset{\underset{O}{\|}}{C}-CH_2-\underset{\underset{O}{\|}}{C}-OH \longrightarrow$$

Urea   Malonic acid

Barbituric acid

But it wan't until 1903 that derivatives of barbituric acid were found to be useful as depressant drugs. Table 17-8 lists a number of barbiturates in use today. Many of the barbiturates are used by individuals as sleeping pills. However, there are many serious problems that arise when they are used for this reason. For example, many patients build up tolerances to these drugs and require more as time goes on to achieve the same effect. This, of course, could lead to overdose and death.

Also, many people who take barbiturates also consume alcoholic beverages. This can also lead to death because alcohol has an *enhancing* effect on the action of barbiturates. This enhancing action is known as a *synergistic effect*. For example, an individual who takes the prescribed dose of a sleeping pill called Seconal sodium, and has also been "drinking" that evening, can suffer overdose symptoms. His body will react as if he had taken a dose many times that of one tablet. This could lead to respiratory arrest, brain damage, and death. Unfortunately, many such occurrences have taken place over the past 20 years.

Narcotics

Depressant drugs which historically have been known to be addictive are called *narcotics*. Their medicinal use is regulated by federal law. The narcotic drugs act on the body to produce both a pain-relieving effect and sedative effect. However, the initial effect of taking certain narcotics is one of exhilaration.

Many narcotics are obtained from plants (Fig. 17-28). For example, *opium* is obtained from an oriental poppy and has been used for centuries in China. The opium is obtained by squeezing the juice from the unripe seeds of the poppy. The dried juice is then smoked, eaten, or injected. In ancient China the opium was usually smoked. Because it relieves pain and causes

Marijuana plants

Psilocybe mushroom

Coca plant

Figure 17-28 Many narcotics are obtained from plants. (U.S. Department of Justice)

sedation, opium was used in many cough preparations in this country in the early 1900's.

The active ingredient in opium is an alkaloid compound called *morphine* (from Morpheus, the Greek god of dreams) (Fig. 17-29). The *alkaloids* are a family of nitrogen-containing compounds found in plants. Morphine was first isolated in 1805 by a German pharmacist, Frederich Sertürner. It had extremely effective pain-relieving properties. In the 1850's the hypodermic syringe was invented and it became possible to inject morphine solutions

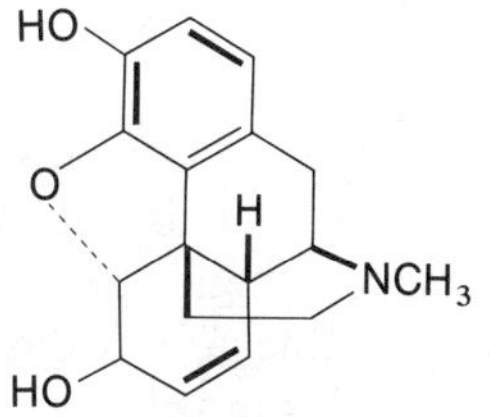

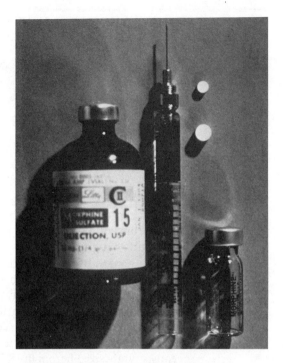

Morphine

Figure 17-29 The active ingredient in opium is the alkaloid compound morphine.

directly into the bloodstream. The drug had a pain-relieving factor about 50 times that of aspirin.

Morphine was used quite extensively during the American Civil War to help relieve the pain of soldiers wounded in battle. Unfortunately, many soldiers became addicted to the drug. Morphine had other undesirable properties including constipation, drowsiness, and confusion.

A less potent form of morphine can be synthesized by converting one of morphine's hydroxyl groups to a methyl ether. The result is codeine (Fig. 17-30), a substance that has narcotic properties but in a milder form. Codeine

Figure 17-30 The narcotic codeine can be synthesized from morphine by converting one of morphine's hydroxyl groups to a methyl ether.

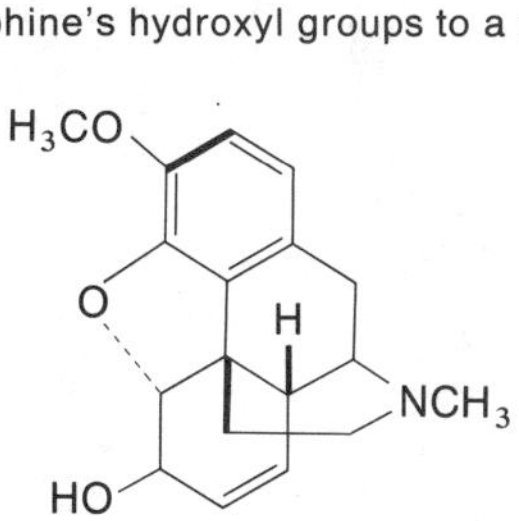

Codeine

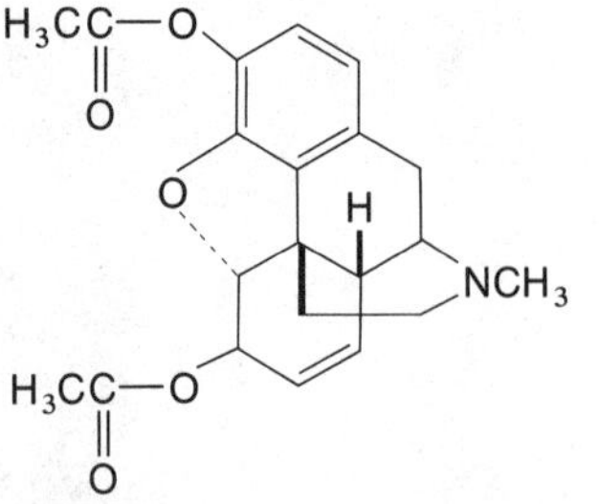

Figure 17-31 The narcotic heroin can be synthesized by converting both of morphine's hydroxyl groups into acetate groups.

can be taken orally, but it's about six times less potent than morphine. In fact, in concentrations of less than 2.2 g/ml, codeine may be used in non-prescription cough preparations.

Conversion of both of morphine's hydroxyl groups into acetate groups produces the compound heroin (Fig. 17-31). This substance is much more addictive than morphine and gives a bigger "high" to the user. However, its very addictive properties make it undesirable for medical reasons and it is illegal to use this substance for such purposes in this country.

It's interesting to note that the German-based Bayer Company, whose chemists synthesized heroin in 1874, were using it in cough syrups around 1900, because of its sedative properties.

Other narcotics, both natural and synthetic, are also known (Table 17-9). The synthetic narcotics have excellent pain-relieving properties and are a bit less addictive than morphine. One of the best known of these compounds is

TABLE 17-9 *Some synthetic narcotics*

NAME	FORMULA	COMMENT
Methadone	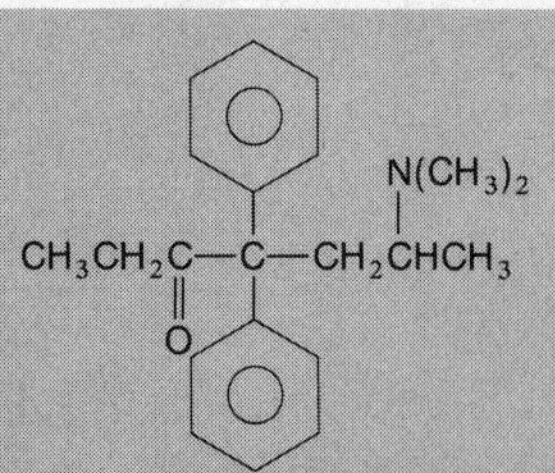	Used in the treatment of heroin addiction, but is also addictive; may be taken orally; does not put the individual into a stupor
Meperidine (Demerol)	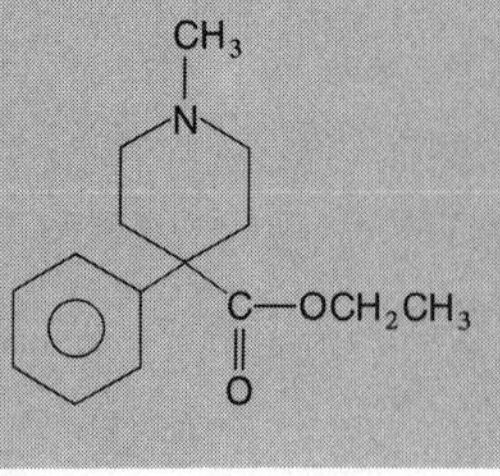	Very effective pain reliever; less powerful than morphine, but does not cause nausea; continuous use can lead to addiction

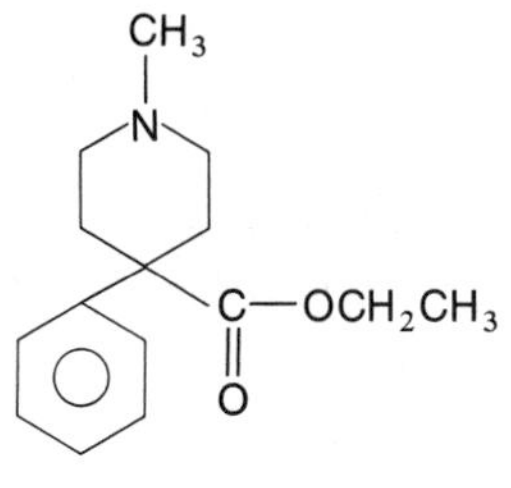

CH₃ ... (Meperidine structure)

Meperidine (Demerol)

Figure 17-32 One of the best known synthetic narcotics is meperidine.

meperidine, better known as demerol (Fig. 17-32). It is usually given to a patient to relieve pain after surgery.

Tranquilizers

Tranquilizers can also be thought of as depressant drugs. However, their intended effect is to calm a patient without making the person tired. Three of the best known tranquilizers are Librium, Miltown, and Valium (all brand names) (Fig. 17-33).

Figure 17-33 Three of the best known tranquilizers are Librium, Miltown, and Valium.

$$H_2N-\underset{O}{\underset{\|}{C}}-O-CH_2-\underset{\underset{CH_2CH_2CH_3}{|}}{\overset{\overset{CH_3}{|}}{C}}-CH_2-O-\underset{O}{\underset{\|}{C}}-NH_2$$

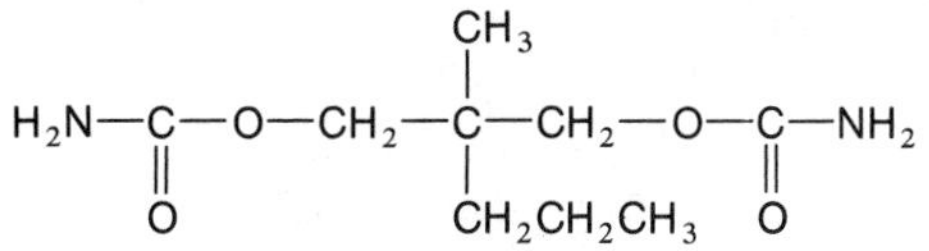

Miltown (meprobamate)

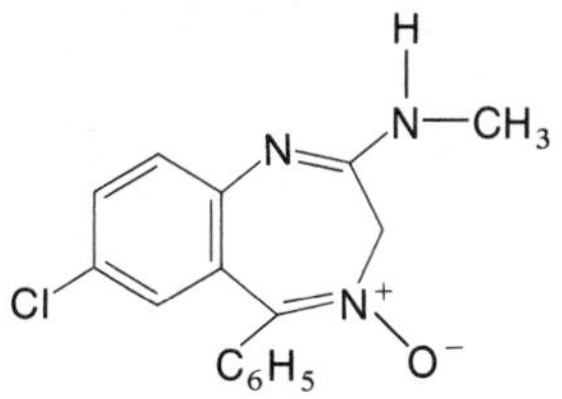

Librium (chlorodiazepoxide)

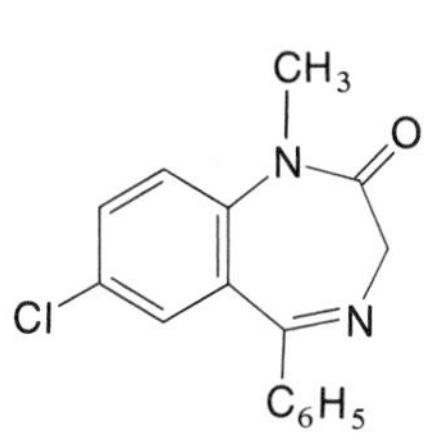

Valium (diazepam)

How effective these drugs actually are is still a question of controversy. Studies have been conducted that yield contradictory results. Some of these studies have shown that patients who thought that they were being given a tranquilizer, but in reality were being given a sugar pill, behaved similarly to patients who were actually given the tranquilizer. Yet millions of people take these tranquilizers daily.

The Mind-Altering Drugs

In the 1960's drug enthusiasts found a new product on the market. This drug wasn't a "downer" or an "upper." In other words, it didn't depress or stimulate the mind when ingested. What it did do was change the way in which things were perceived. The name of this drug was LSD (lysergic acid diethylamide) (Fig. 17-34).

During the 1960's certain government agencies were experimenting with LSD as a mind control drug. But these tests ended in failure. However, when LSD hit the illegal drug market, many young people were eager to try it.

LSD is an extremely potent substance. Only a few micrograms (not *milli*grams but *micro*grams) are needed to produce hallucinogenic effects. An individual under the influence of LSD is said to be on a "trip." The problem is that the trip may be a bad one. A person under the influence of LSD might be happy, depressed, or even psychotic. The individual may not be able to differentiate between fantasy and reality. In fact, there have been reports of LSD users jumping out of windows in high buildings because they thought they could fly.

Another major problem that afflicts LSD users is that without even taking another dose of the drug it is possible to have a spontaneous "trip." Also, there seems to be some evidence that LSD may affect genes and certain hormone levels in the blood.

In the late 1970's another—perhaps even more powerful—mind-altering drug came on the street scene. Its name is PCP (chemically known as phencyclidine) or "angel dust" (Fig. 17-35). The effects of this drug on an individual are completely unpredictable. A state of depression, drunkenness, elation, or hallucination is possible. So too, are rage, violence, amnesia, or

Figure 17-34 Lysergic acid diethylamide (LSD), one of the mindbenders.

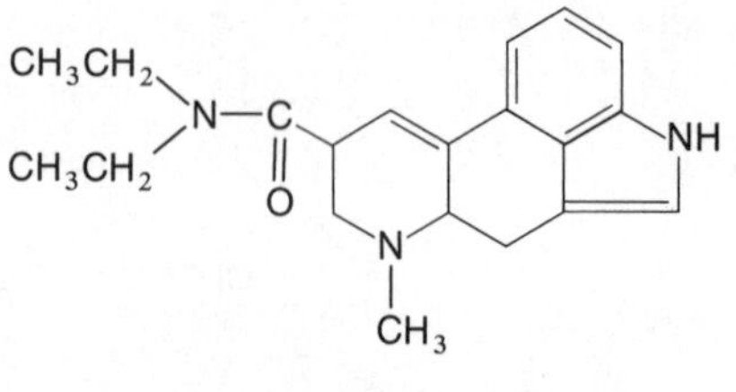

Lysergic acid diethylamide (LSD)

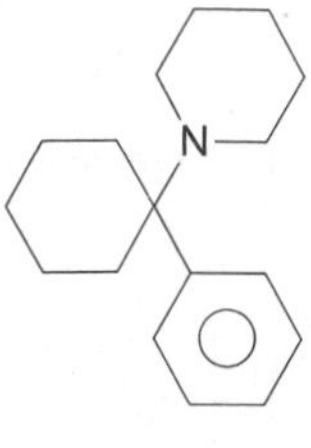

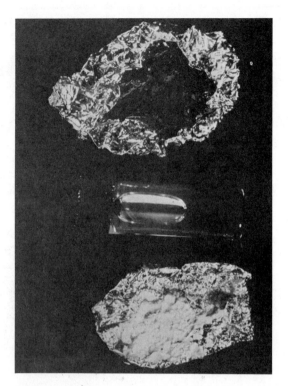

PCP

Figure 17-35 Phencyclidine (PCP), also known as angel dust, is a very powerful mind-altering drug.

paranoia. Law enforcement agencies claim that this drug has been responsible for murders, suicides, and drownings.

Ganga, bhang, hashish, or what we call marijuana, comes from the Indian hemp plant *Cannabis sativa*. The active ingredient in marijuana is tetrahydrocannabinol (Fig. 17-36). Various strains of marijuana have different amounts of this active ingredient tetrahydrocannabinol, sometimes called THC, for short (Table 17-10). Also, different parts of the plant produce a product of varying potency.

Figure 17-36 The active ingredient in marijuana is tetrahydrocannabinol (THC).

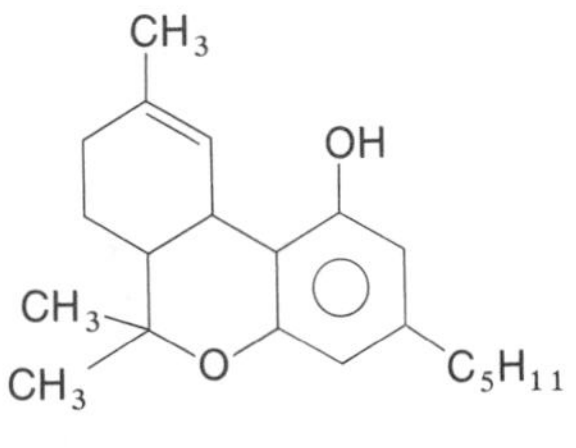

THC

TABLE 17-10 *The* THC *content of various types of marijuana*

TYPE	THC CONTENT (APPROX. %)
Mexican	1
Southeast Asian	2–4
American	Less than 0.2
Jamaican ganja	4–8
Indian hash	5–12

Source: Bureau of Narcotics and Dangerous Drugs.

TABLE 17-11 *Some effects of marijuana smoking*

1. Increased pulse rate and blood pressure
2. Time discrimination affected
3. Mental sluggishness
4. Possible brain damage on prolonged use
5. Possible hallucinatory effect
6. Has been shown to cause brain lesions in rats

People have been debating whether marijuana should be legalized. There have also been many debates on the short- and long-term effects of this substance. Results of many of the tests done on marijuana have been contradictory because the THC content varies so much. People who use marijuana on a continuous basis are never sure that they're getting the same product. There's just no quality control for this substance! However, the experts do seem to agree that although marijuana does not seem to be physiologically addicting like heroin and morphine, there are other negative aspects to smoking this substance (Table 17-11).

SUMMARY

In this chapter we surveyed various kinds of pharmaceuticals and drugs. We began by tracing the use of medicines and drugs from ancient times to modern times. Next we learned about prescription drugs and the differences between generic names, brand names, and chemical names.

We also studied about the various types of drugs. We began by looking at the pain-relieving drugs. We compared aspirin with other pain relievers. Next we surveyed some of the antacid medications, and learned how they work. Our study of antibiotics began with the use of the substance arsphenamine and the beginnings of chemotherapy. We also learned about sulfanilamide and its mode of action. And we learned about penicillin and other broad-spectrum antibiotics.

Our next topic of survey was the birth control pills and synthetic hormones. We discovered how the chemical structures of these substances affected their properties. We also glimpsed at the future by looking at birth control pills for males.

The final section of the chapter dealt with stimulants, depressants, and narcotics. We studied the properties and structures of members from each class of drugs. We looked at their chemical and psychological effects on people. We also reviewed the effects and properties of some of the mind-altering drugs.

Perhaps the best way to end this chapter is to remind you, if you do take prescription drugs or other drug-related substances, that not all drugs are compatible. There can be serious consequences for an individual who mixes incompatible drugs. Because all of us use pharmaceutical substances at one time or another, we have arranged a partial list of incompatible drugs (Table 17-12). We have also arranged a list of foods that are incompatible with certain drugs (Table 17-13). You might want to keep this list handy for future reference. Many pharmacies today keep a record of all prescription items given to a patient.

Our science and technology has given us an abundance of chemical substances that can be used to help us achieve good health. But there's also the other side of the coin. The choice is ours to make.

TABLE 17-12 *Drugs that don't mix*

DRUG	INCOMPATIBLE DRUG	POSSIBLE INTERACTION ON MIXING
Antibiotics such as tetracyclines	Antacids, iron compounds, milk, kaolin-containing substances such as Kaopectate	Decreases drug activity or drug effect
Antidepressants such as Norpramin and Tofranil	Anticoagulants	Causes internal bleeding
	Tranquilizers	Increases sedative power
	Monoamine oxidase inhibitors such as Nardil, Marplan, and Eutron	Causes wide blood pressure fluctuations, fever, and hyperexcitability
Oral anticoagulants such as warfarin, Dicumarol, and Coumadin	Aspirin, Chloromycetin, cholesterol-lowering drugs, thyroid preparations, and barbiturates	Increases bleeding and possibility of hemorrhage
	Birth control pills	Decreases drug effect
Aspirin-containing substances	Anticoagulants	Increases bleeding and possibility of hemorrhage
	Oral antidiabetic drugs	Increases effect of antidiabetic drug
Cold and cough preparations containing antihistamines	Sedatives	Causes dizziness and drowsiness
	Codeine	Extreme dizziness and drowsiness
Birth control pills	Oral anticoagulants	Decreases anticoagulant effect
	Antidiabetics	Decreases effect of diabetic drug
	Rifampin (Rifadin)	Decreases contraceptive effect
Sedatives, sleeping pills, and certain tranquilizers such as Equanil, Librium, Miltown, and Valium	Antidepressant drugs and antihistamines	Exaggerates effect of the sedative
Stimulants and anti-obesity drugs such as amphetamines and catecholamines	Tricyclic antidepressants and monoamine oxidase inhibitors	Causes severe headaches and rapid blood pressure rise
	Darvon (high doses)	Causes convulsions

TABLE 17-13 *Drugs and foods that don't mix*

FOOD	INCOMPATIBLE DRUG	POSSIBLE INTERACTION ON MIXING
Alcohol	Aspirin	Gastrointestinal bleeding
	Anticoagulants	Decrease or increase of blood-thinning effect
	Antidepressants	Deep sedation, lowering of body temperature, and possibility of death
	Antihistamines	Possible drowsiness and loss of consciousness
	Antihypertensives	Causes dizziness and fainting
	Darvon	Causes dizziness and drowsiness; large amounts can be fatal
	Sedatives, sleeping pills, and certain tranquilizers such as Valium, Librium, and Miltown	Severe central nervous system effects; large amounts can be fatal
Green leafy vegetables	Anticoagulants such as Coumadin and Dicumarol	Cancels blood-thinning effects
Dairy products	Tetracycline antibiotics	Reduces therapeutic effect
Pickled herring, ripened or aged cheese, Chianti wine, chicken liver, excessive chocolate, and other tyramine-containing foods	Monamine oxidase inhibitors such as Nardil, Marplan, or Parnate for depression, and Eutonyl or Eutron for blood pressure	Causes rapid rise in blood pressure, bad headaches, vomiting, and possibly death
Monosodium glutamate (MSG; for example, the major ingredient of seasoning salts)	Diuretics	May cause elimination of excessive amounts of sodium
Fruit juices	Ampicillin, cloxacillin, certain types of penicillin, and erythromycin	Decreases drug effect

EXERCISES

1. Discuss the use of drugs in today's society and compare it to the use of drugs in earlier societies. What are the major similarities and differences?

2. A pain-relieving drug has the following names: Propoxyphene, Darvon, (S)-α-[2-(Dimethylamino)-1-methylethyl]-α-phenylbenzeneethanol propanoate. Which name is (a) the brand name; (b) the chemical name; (c) the generic name?

3. Discuss the brand name–generic name controversy. Why do consumers want generic names? Why do pharmaceutical companies want brand names?

4. Define what is meant by an analgesic and cite an example. Define what is meant by an antipyretic and cite an example. Name a substance that has both analgesic and antipyretic properties.

5. What is an OTC drug?

6. Compare and contrast the properties of salicylic acid, sodium salicylate, phenyl salicylate (salol), and acetylsalicylic acid (aspirin).

7. Certain individuals should not take aspirin. Why?

8. How can aspirin overdose be avoided? In other words, what symptoms should you look for?

9. Discuss the properties and uses of methyl salicylate (oil of wintergreen).

10. Compare the effectiveness of aspirin and the other pain-relieving products, such as Bufferin, Anacin, and Alka-Seltzer.

11. Explain how aspirin works in the body.

12. Discuss some of the problems with combination-type pain relievers.

13. Write a chemical equation to explain how an antacid that contains hydroxide ion works. Then write an equation to explain how an antacid that contains carbonate ion works.

14. Define chemotherapy and discuss its beginnings.

15. Explain how sulfanilamide works in the body to kill bacteria.

16. How was penicillin discovered? Trace the development of penicillin from its discovery to the synthesis of various forms of this substance.

17. What is a broad-spectrum antibiotic?

18. Why is it so important for researchers to continue searching for new antibiotics?

19. Name the major male and female sex hormones and explain the functions of each.

20. Trace the development of the first oral contraceptive, norlutin.

21. Discuss the safety of birth control pills.

22. How do you feel about birth control pills such as the prostaglandins—"morning-after pills"?

23. How do you feel about the development of male birth control pills?

24. Discuss the use of caffeine as a stimulant. Cite some food products that contain caffeine. What might be some problems of high caffeine ingestion?

25. Match the drug with its description.
    (1) stimulant   (a) ethyl alcohol
    (2) depressant   (b) amphetamines
    (3) narcotic   (c) Librium
    (4) tranquilizer   (d) morphine

26. Explain a possible chemical mechanism for alcohol addiction.

27. What is a barbiturate? What is the parent compound of these drugs?

28. Explain what is meant by a synergistic effect and explain how this relates to barbiturates.

29. Define narcotic. Discuss the properties of opium, morphine, and codeine, and state some of their uses.

30. What is a synthetic narcotic? Give an example of such a compound.

31. What is a tranquilizer suppose to do? How does a tranquilizer differ from a depressant?

32. Discuss the properties of the mind-altering drugs. Cite some examples.

33. Explain the differences among the various types of marijuana.

Chapter 18

The Chemistry of Home Care and Personal Products

Some Things You Should Know After Reading This Chapter

You should be able to:

1. Discuss how early societies used saponins to clean their clothes.
2. Explain how soap was made by early societies.
3. Explain how soap works to clean people and clothes.
4. Discuss the conditions under which soap doesn't clean well.
5. Explain what a syndet is and give some examples.
6. Discuss the environmental impact of synthetic detergents.
7. Discuss the differences between LAS and ABS detergents.
8. Define the terms surfactant and builder, and give some examples.
9. Explain how phosphate detergents have an environmental impact on bodies of water.
10. Explain the chemical differences among anionic, cationic, and nonionic surfactants.
11. Explain how an automatic dishwasher detergent works.
12. Tell how a chlorine bleach works.
13. Discuss the differences between chlorine bleaches and all-fabric bleaches.
14. Explain how brighteners work on fabrics.
15. Define the term cosmetic.
16. Discuss the major constituents of lipsticks, lotions, and creams.
17. Explain two ways in which acne can be treated.
18. Discuss the major constituents of perfumes, colognes, and aftershave lotions, and state the purpose of each.
19. State the differences between a deodorant and an antiperspirant.
20. Discuss the major constitutents of toothpastes, and state the purpose of each.
21. Explain how fluoride prevents tooth decay.
22. Tell how mouthwashes work.
23. List the basic ingredients of shampoos.
24. Explain the differences between temporary and permanent dyes.
25. Discuss the process used to straighten or curl hair.

IT'S WASH DAY AT THE JONES'S HOUSE

Today is Saturday and it's wash day at the Jones's house. Mr. and Mrs. Jones take a trip to the supermarket to purchase some products to clean the house, do the laundry, and wash the dishes. They spend well over $20 on cleaning products.

When they get home and read the labels on their cleaning products, the Joneses feel like chemists. Here's a list of the chemicals they purchased: sodium carbonate, sodium tripolyphosphate, sodium silicate, sodium citrate, potassium pyrophosphate, sodium sesquicarbonate, trisodium phosphate, sodium perborate, aluminosilicates, sodium sulfate, perchloroethylene, petroleum distillates, ammonia, anionic surfactants, nonionic surfactants, artificial color, perfume, ethyl alcohol, and finally, water (Fig. 18-1).

Because they are wise consumers, the Joneses read all of the labels and follow the directions carefully. Many of the products are eye irritants and they are careful not to get them on their skin or in their eyes. Many of the products have pleasant odors, but there is one that has an awful smell and should be used in a well-ventilated area. This product is a laundry stain remover. There are detergents that soften clothes, remove static cling, whiten, brighten, and also remove dirt from clothes. There are appliance cleaners which will do a good job on many surfaces and remove several types of dirt and stains.

Certainly, these products have made wash day at the Jones's house a little easier—but more expensive than ever before. As long as these products are used according to the directions on the cans, boxes, and containers, the Joneses can spend a part of their day using their newly bought chemicals, and end up with a clean house, whiter, brighter clothes, and dishes so clean that they can see their reflections in them!

Figure 18-1 There are many cleaning products available on the market today. (A. Sherman)

OLD-FASHIONED CLEANERS

Years ago, people were not as concerned with cleaning as we are today. Part of the reason was lack of knowledge. They didn't know that microorganisms were the causes of disease, and they did little to keep themselves clean.

In primitive societies people used stones and rocks to beat their clothes and then washed them in a stream or lake. The first detergents probably came from plants such as the *soapberries* or *soapworts*, which produce a lather. These plants contain chemical compounds called *saponins*, which produce the soapy lather. Two other chemicals, sodium carbonate (Na_2CO_3) and potassium carbonate (K_2CO_3), are formed in plant ashes. When the carbonate ion reacts with water, an alkaline solution is formed. This solution has detergent properties and served as a cleaning solution for many people. In fact, *washing soda*, or sodium carbonate, is still in use today (Fig. 18-2).

People have known how to make soap for at least 2000 years. The Romans cooked potash (obtained from the ashes of wood) and fat together to make soap. The Spanish used olive oil to make castile soap, and the English used whale oil. The pioneers in this country saved all the fats from cooking and combined these with wood ashes to make soap.

Figure 18-2 Sodium carbonate, also known as washing soda, is still used today. (Church & Dwight Co., Inc., Piscataway, New Jersey)

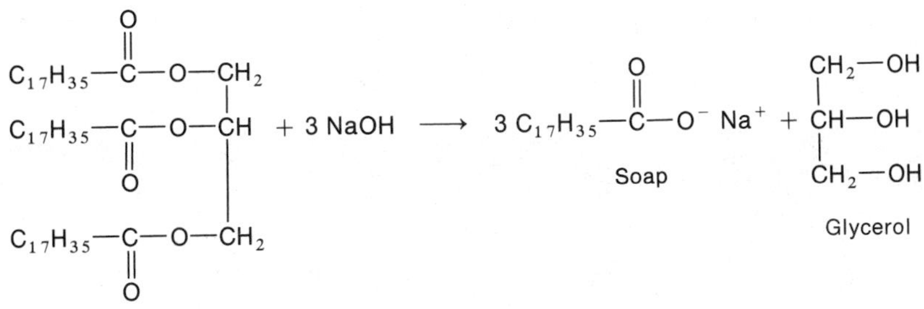

$$C_{17}H_{35}-\overset{\overset{\displaystyle O}{\|}}{C}-O-CH_2$$

$$C_{17}H_{35}-\overset{\overset{\displaystyle }{\|}}{\underset{\underset{\displaystyle O}{\|}}{C}}-O-CH \quad +\ 3\ NaOH \longrightarrow 3\ C_{17}H_{35}-\overset{\overset{\displaystyle O}{\|}}{C}-O^-\ Na^+ + \begin{array}{l} CH_2-OH \\ CH-OH \\ CH_2-OH \end{array}$$

$$C_{17}H_{35}-\overset{\overset{\displaystyle }{\|}}{\underset{\underset{\displaystyle O}{\|}}{C}}-O-CH_2$$

Soap   Glycerol

Tristearin (a fat)

Figure 18-3 To make soap, fat or oil from an animal or vegetable source is reacted with sodium hydroxide. The products of the reaction are soap and glycerol.

$$CH_3(CH_2)_{12}-\overset{\overset{\displaystyle O}{\|}}{C}-O^-\ K^+$$

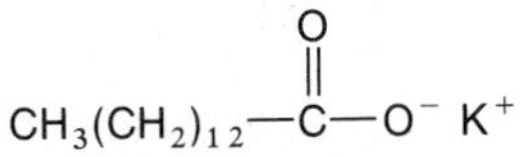

Potassium palmitate, a potassium soap

Figure 18-4 Soaps made with potassium ions instead of sodium ions produce a finer lather.

To make soap, fat or oil from either an animal or vegetable source must be reacted with sodium hydroxide (lye). The mixture is heated and the products of the reaction are soap and glycerol (Fig. 18-3).

SOAP TODAY

Soap today is not what it used to be. It is a much more refined product. Fats and oils are often hydrolyzed using superheated steam and then fatty acids are neutralized. A variety of chemicals are then added to make the soap suit its purpose. Soaps used for scouring contain silica or pumice, which are *abrasives*. Soaps used for the body contain perfume, cream, oil, and dyes to color the product. Deodorant soaps contain perfume to cover up odor. Floating soaps are made by blowing air into the soap to lower the density. (If you've forgotten what density is, see Appendix B.) Soaps made with potassium ions instead of sodium ions produce a finer lather (Fig. 18-4).

HOW SOAP WORKS

Dirt combines with grease and oil and sticks to such surfaces as skin, clothing, floors, walls, and automobiles. Water alone does not remove greasy, oily dirt because grease and oil are not soluble in water.

    Soap molecules have two ends which have different properties. One end is an ionic end, which allows soap to dissolve in water. The other end is

(a) $2\ CH_3-(CH_2)_{18}-\overset{\overset{O}{\parallel}}{C}-O^- + Mg^{+2} \longrightarrow \left[CH_3-(CH_2)_{18}-\overset{\overset{O}{\parallel}}{C}-O^- \right]_2 Mg^{+2}$

Soap   Soap scum

(b) $CH_3-(CH_2)_{18}-\overset{\overset{O}{\parallel}}{C}-O^-\ Na^+ + H^+ \longrightarrow CH_3-(CH_2)_{18}-\overset{\overset{O}{\parallel}}{C}-OH + Na^+$

Soap   Acid   Free fatty acid

Figure 18-5   (a) When hard water that contains metal ions such as magnesium, calcium, or iron reacts with soap, a scum forms instead of lather. This is due to the reaction between the metal ions and the soap molecules. (b) In acidic solutions, soap molecules are converted to free fatty acids that separate from the water as a greasy scum.

nonpolar and dissolves nonpolar grease and oil. So the nonpolar greases and dirt get attached to the nonpolar end of the soap molecule and get carried away with water.

There are two conditions in which soap doesn't work very well. The first condition arises when hard water is present. This makes the soap form a scum instead of lather. Hard water is water that contains metal ions, such as magnesium, calcium, and iron. The ionic end of the soap molecule reacts with these ions to form what we call soap scum or "bathtub ring" (Fig. 18-5a).

The second condition arises when soap is placed in acid solution. Soap doesn't work well in acidic solutions, because in such solutions soap molecules are converted to free fatty acids. These free fatty acids have a nonpolar end but no ionic end, so they can't dissolve in water. What happens is that the fatty acids separate as a greasy scum (Fig. 18-5b).

The advantages of soap are many. First, in soft water, it works well. It is a relatively nontoxic substance. It comes from natural sources—animal and vegetable fats and oils—and the sources can be renewed. It is also biodegradable.

SYNTHETIC DETERGENTS: SYNDETS

To overcome the inability of soaps to work in hard water and in acidic solutions, *synthetic detergents* were developed. *Sodium lauryl sulfate* was one of the first of these products developed. It is derived from the fat *trilaurin*, which is reduced with hydrogen in the presence of a catalyst. This is followed by reaction with sulfuric acid and finally by neutralization (Fig. 18-6). The only problem with sodium lauryl sulfate is its cost, which is high because of the many steps required for its production.

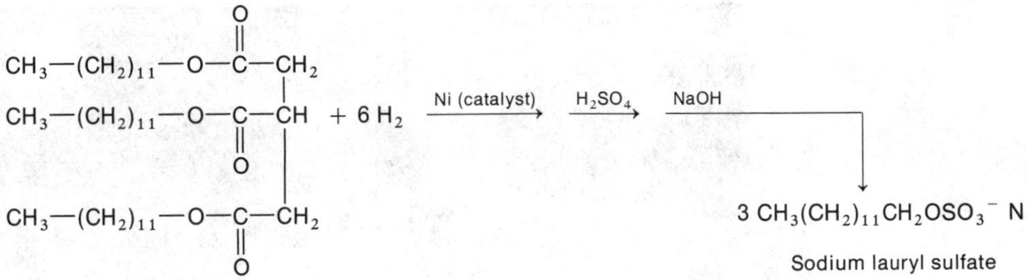

$$CH_3-(CH_2)_{11}-O-\overset{\overset{\displaystyle O}{\|}}{C}-CH_2$$

$$CH_3-(CH_2)_{11}-O-\underset{\underset{\displaystyle O}{\|}}{C}-CH + 6\,H_2 \xrightarrow{\text{Ni (catalyst)}} \xrightarrow{H_2SO_4} \xrightarrow{\text{NaOH}}$$

$$CH_3-(CH_2)_{11}-O-\underset{\underset{\displaystyle O}{\|}}{C}-CH_2$$

Trilaurin

$$3\ CH_3(CH_2)_{11}CH_2OSO_3^-\ N$$

Sodium lauryl sulfate

Figure 18-6 Sodium lauryl sulfate, one of the first synthetic detergents, is produced by reacting trilaurin with hydrogen in the presence of a catalyst. This is followed by reaction with sulfuric acid and finally by neutralization with sodium hydroxide.

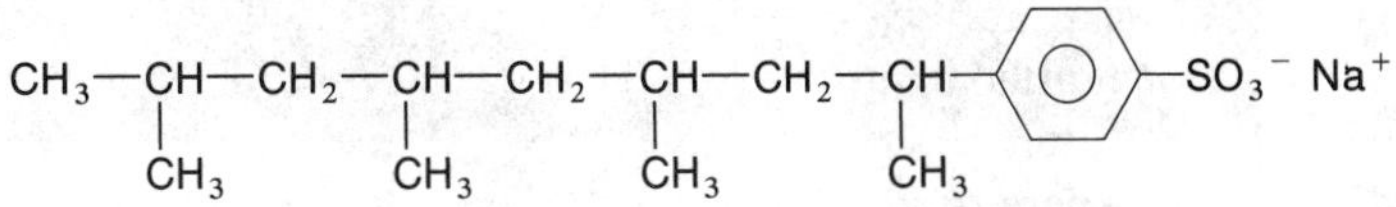

$$CH_3-\underset{\underset{\displaystyle CH_3}{|}}{CH}-CH_2-\underset{\underset{\displaystyle CH_3}{|}}{CH}-CH_2-\underset{\underset{\displaystyle CH_3}{|}}{CH}-CH_2-\underset{\underset{\displaystyle CH_3}{|}}{CH}-\bigcirc-SO_3^-\ Na^+$$

Figure 18-7 An ABS detergent.

Soon after the development of the first synthetic detergents came less costly ones. The alkene *propylene*, $CH_3-CH{=}CH_2$, which is a petroleum product, served as the basis for the production of *alkyl benzene sulfonates* (ABS). To produce ABS, chemists combined propylene, benzene, sulfuric acid, and a base such as sodium carbonate (Na_2CO_3) (Fig. 18-7).

ABS detergents were widely used for over 10 years until their negative environmental impact was realized. The molecules could not be broken down by the microorganisms in sewage treatment plants. These were nonbiodegradable products which endangered the whole ground water supply. Suds appeared in sewage treatment plants, in streams and rivers, and some people even had foam in their drinking water (Fig. 18-8). Legislation soon appeared forcing industries to put biodegradable products on the market.

The next set of synthetic detergents to appear on the market were the *linear alkylsulfonates*, or LAS for short. These detergents are made of long chains of carbon atoms (Fig. 18-9) and can be broken down by microorganisms. Unlike the ABS detergents, which consist of branched chains of carbon atoms, the LAS detergents can be degraded, two carbon atoms at a time, so they don't produce foam and suds in streams and rivers. However, some environmentalists feel that the LAS detergents form *phenol*, a toxic chemical, when they are degraded, and this chemical will kill fish. The detergent industry feels that this is not the case. Only time will tell who is right and who is wrong (Fig. 18-10).

Figure 18-8 The ABS detergents produced suds in streams, rivers, and lakes. They even contaminated our drinking water! (Robert S. Sherman)

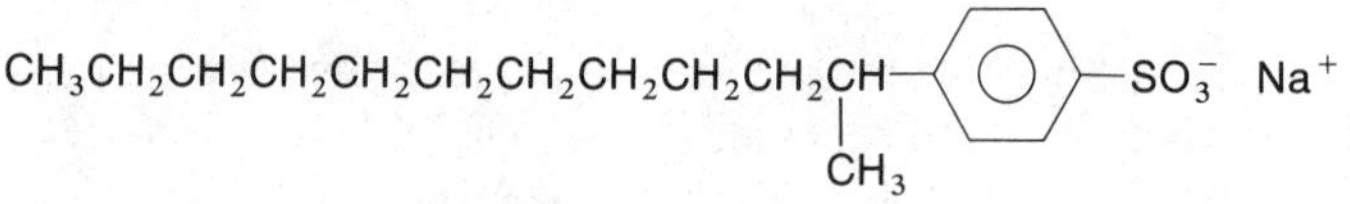

$$CH_3CH_2CH_2CH_2CH_2CH_2CH_2CH_2CH_2CH_2CH\!-\!\!\langle\bigcirc\rangle\!-\!SO_3^-\ Na^+$$
$$| \atop CH_3$$

Figure 18-9 An LAS detergent.

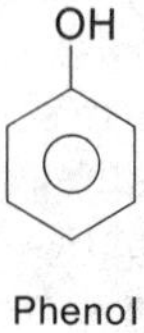

OH

Phenol

Figure 18-10 Some environmentalists feel that LAS detergents form phenol, a toxic chemical, when they are degraded.

SURFACTANTS AND BUILDERS:
THEY MAKE YOUR CLOTHES WHITER AND BRIGHTER

Any substance that stabilizes (or holds together) the suspension of nonpolar substances in water is called a *surfactant*. For example, soap is a substance that stabilizes a suspension of grease and oil (which are nonpolar substances) in water and so it is a *surface-active agent* or *surfactant*. Synthetic detergents are also classified as surfactants.

Chemicals are often added to a surfactant to make it clean better. Any substance that increases the detergency of a surfactant when added to it is called a *builder*. Some chemicals containing phosphate, such as *sodium tripolyphosphate* ($Na_5P_3O_{10}$), are builders. This particular chemical reacts with Ca^{+2} or Mg^{+2} ions in water and softens the water. It also raises the pH of the water a bit and makes the detergent work better.

More than a decade ago *phosphate detergents* were widely used until their negative environmental impact was noted. When the phosphates in detergents reached a body of water, they caused a major problem. Phosphate ions (PO_4^{-3}) serve as nutrients, which leads to the overgrowth of algae. When the algae die, they add to the organic waste matter in the body of water. Oxygen is needed to break down this dead organic matter. Since the body of water contains only a certain amount of dissolved oxygen (see Chap. 9 for a discussion of dissolved oxygen in water), this additional organic matter depletes the oxygen supply, which leads to the death of the lake, stream, or river. Some bacteria that don't require oxygen (*anaerobic* bacteria) begin to thrive and produce foul-smelling chemicals such as hydrogen sulfide (H_2S), methanethiol (CH_3SH), and various amines. Phosphate detergents contributed to the eutrophication of bodies of water.

Other builders used today include sodium perborate ($NaBO_2 \cdot H_2O_2$) and sodium metasilicate (Na_2SiO_3). These are not as harmful to the environment as were phosphates, but they do have other effects. They form caustic solutions which are irritating to the skin. There have been incidences where children were injured by exposure to these phosphate substitutes.

When sodium perborate reacts with water, it forms a solution that is strongly alkaline.

$$NaBO_2 \cdot H_2O_2 + H_2O \longrightarrow NaOH + HBO_2 + H_2O_2$$

Sodium
perborate

Sodium
hydroxide
(forms a strongly
alkaline solution)

Hydrogen peroxide is formed, which serves as a bleach and decomposes to liberate oxygen.

$$2 H_2O_2 \longrightarrow 2 H_2O + O_2$$

Hydrogen
peroxide

When sodium metasilicate reacts with water, it, too, forms a strongly alkaline solution.

$$Na_2SiO_3 + 2 H_2O \longrightarrow 2 NaOH + H_2SiO_3$$

Sodium
metasilicate

Sodium
hydroxide
(forms strongly
alkaline solution)

Today some detergents contain small amounts of phosphates as builders instead of the more caustic chemicals. Research is still continuing to find better cleaning agents.

ANIONIC, CATIONIC, AND NONIONIC SURFACTANTS: CLEANERS FOR EVERY CLEANING CHORE

The surfactants that we've discussed so far are *anionic surfactants*. These are the soaps and synthetic detergents that have molecules with two ends—a *nonpolar* end and an *ionic* end. These molecules all contain an anion (negative ion) end and so they're classified as *anionic surfactants* (Fig. 18-11).

Liquid laundry detergents contain two types of surfactants—anionic and nonionic. LAS is usually the detergent used, which is an *anionic surfactant*. Sometimes *alcohol ether sulfates* are used, which are good anionic surfactants but are quite costly (Fig. 18-12).

The *nonionic surfactants*, sometimes found in liquid laundry detergents, are the *alcohol ethoxylates* and alkylphenol ethoxylates (Fig. 18-13). They are soluble in water because they contain several oxygen atoms. They are great for removing oil and grease and are more soluble in cold water than in hot water.

Dishwashing liquids generally contain the sodium or triethanolamine salts of LAS, together with perfumes and coloring agents. Some contain

Figure 18-11 An anionic surfactant has a nonpolar end on one side of the molecule and a negative ion (anion) end on the other side of the molecule.

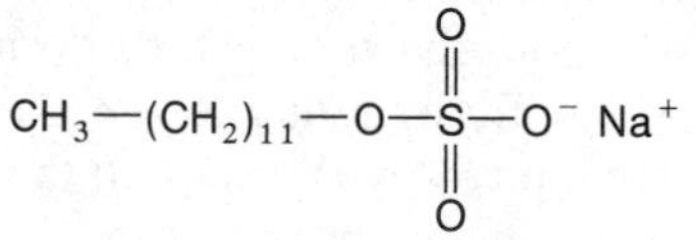

$$CH_3-(CH_2)_{11}-O-\overset{\displaystyle O}{\underset{\displaystyle O}{\overset{\|}{\underset{\|}{S}}}}-O^-\ Na^+$$

Sodium lauryl sulfate

Figure 18-12 Alcohol and ether sulfates are good anionic surfactants.

$$CH_3-(CH_2)_m-O-(CH_2CH_2O)_n-SO_3^-\ Na^+$$

$$m = 7\ to\ 13$$
$$n = 7\ to\ 13$$

Figure 18-13 Examples of nonionic surfactants.

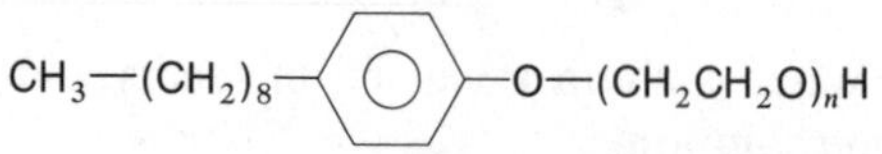

$$CH_3-(CH_2)_8-\!\!\left\langle\!\bigcirc\!\right\rangle\!\!-O-(CH_2CH_2O)_nH$$

An alkylphenol ethoxylate

$$CH_3-(CH_2)_m-CH_2-O-(CH_2CH_2O)_nH$$

An alcohol ethoxylate

$$m = 7\ to\ 13$$
$$n = 6\ to\ 13$$

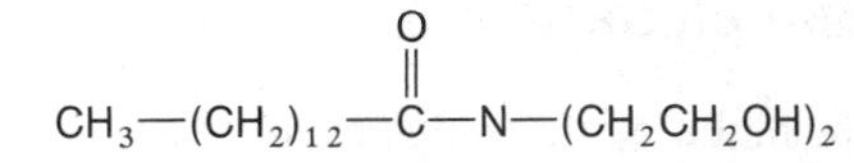

$$CH_3-(CH_2)_{12}-\overset{\overset{\displaystyle O}{\|}}{C}-N-(CH_2CH_2OH)_2$$

Cocamido DEA

Figure 18-14   The compound cocamido DEA is a nonionic surfactant used in dishwashing detergents.

$$CH_3-(CH_2)_{14}-CH_2-\overset{\overset{\displaystyle CH_3}{|}}{\underset{\underset{\displaystyle CH_3}{|}}{N^+}}-CH_3 \; Cl^-$$

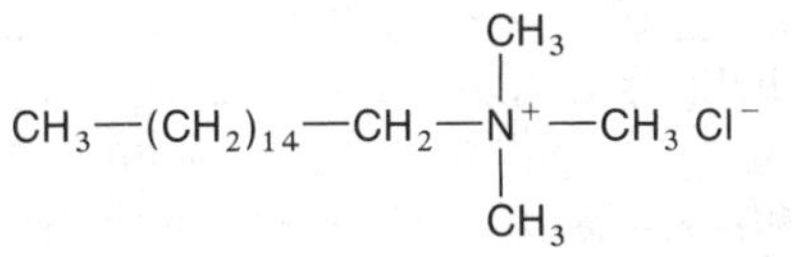

Hexadecyltrimethylammonium chloride

Figure 18-15   The quaternary ammonium salts are cationic surfactants which contain four groups attached to a nitrogen atom, which carries a positive charge.

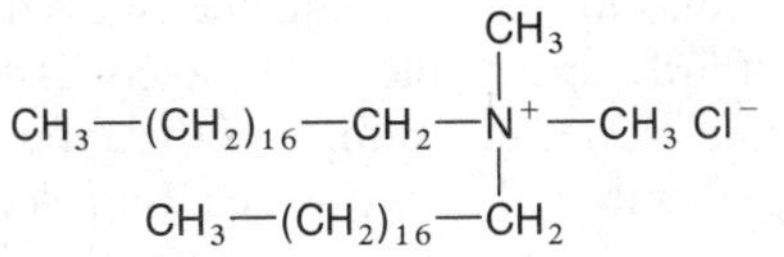

$$CH_3-(CH_2)_{16}-CH_2-\overset{\overset{\displaystyle CH_3}{|}}{\underset{\underset{\displaystyle CH_3-(CH_2)_{16}-CH_2}{|}}{N^+}}-CH_3 \; Cl^-$$

Dioctadecyldimethylammonium chloride
(a fabric softener)

Figure 18-16   Some cationic surfactants act as fabric softeners.

nonionic surfactants, such as the alcohol ethoxylates or amides made from diethanolamine, and fatty acids, such as cocamido DEA (Fig. 18-14).

Dishwashing detergents for use in automatic dishwashers are another story. They contain chemicals which are caustic, such as sodium tripolyphosphate and sodium metasilicate, chlorine bleach, and little surfactant. They work because of the very hot water in the dishwasher and the strong alkaline solution formed by the caustic chemicals. Dishwashing detergents should never be used on the skin.

Fabric softeners and disinfectants sometimes contain *cationic surfactants* in which the surfactant part of the molecule is a positive ion. The *quaternary ammonium salts* are cationic surfactants which contain four groups attached to a nitrogen atom which carries a positive charge (Fig. 18-15). *Cationic surfactants* and *anionic surfactants* can't be used together because the oppositely charged ions would react and cancel each other out.

Some cationic surfactants are used as disinfectants because of their germicidal action. Others, such as dioctadecyldimethylammonium chloride, are fabric softeners (Fig. 18-16). Such compounds form a film on the fabric's surface and lubricate the fibers, making a softer, more flexible fabric.

Bleaches make white things even whiter. There are chlorine bleaches and all fabric bleaches. Chlorine bleaches contain 4.25% sodium hypochlorite solution (NaOCl). In these bleaches the chlorine hits the fabric all at once and can damage some fabrics if direct contact is made. The active ingredient in chlorine bleaches is the hypochlorite ion (ClO^{-1}). These bleaches work because they remove loosely bound electrons which color fabrics. These loosely bound electrons can color or dull fabrics because they have the ability to absorb visible light as they move to higher energy levels. The absorption of visible light causes the dulling effect. By immobilizing or removing these electrons, the chlorine bleach makes whites look whiter. The equation for this reaction may be represented as

$$ClO^{-1} + H_2O + 2\,e^{-1} \longrightarrow Cl^{-1} + 2\,OH^{-1}$$

Chlorine bleach works well on cotton and linen fabrics.

Another type of bleaching agent is composed of the *hydantoin* and *cyanurate* bleaches (Fig. 18-17). These bleaches release chlorine slowly in water. A lower concentration of chlorine is less damaging to the fabric.

A third type of bleaching agent is the oxygen-releasing bleach. We've already mentioned an example of this type of bleach when we discussed *builders*. You'll remember that we stated that sodium perborate, a builder, releases hydrogen peroxide, which acts as a bleach. Oxygen bleaches aren't as active as chlorine bleaches but they are better for bleaching synthetic fibers. You need to use more bleach, hotter water, and a more alkaline solution with these than with chlorine bleach, but they do work well when used properly.

Borax, also known as sodium pyroborate, $Na_2B_4O_7$, is often added to oxygen-releasing bleaches because of its ability to clean. Its pH is about 9.5, so it produces an alkaline solution which makes the all-fabric bleaches work better.

Brighteners are very interesting compounds. Many of them contain *optical brighteners*, which are colorless dyes. They absorb radiation which comes from sunlight and is invisible to us. Then these dyes reemit this radiation to us as visible light. The reemitted light is at the blue end of the spectrum, so

Figure 18-17 Hydantoin and cyanurate bleaches release chlorine slowly in water and so there is less chance of damage to certain fabrics.

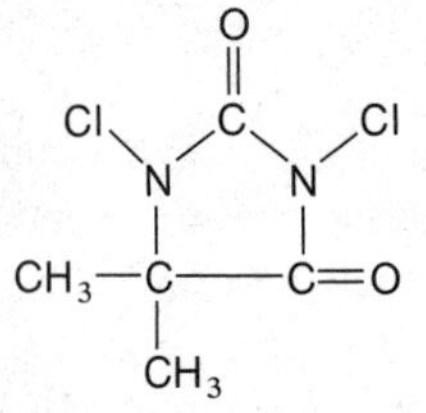

A hydantoin-type bleach 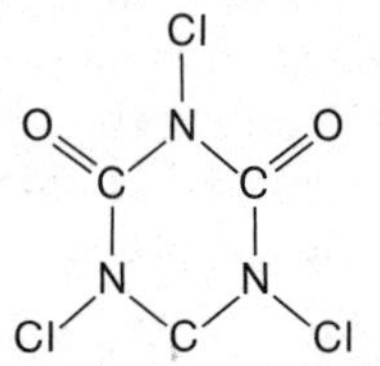   A cyanurate-type bleach

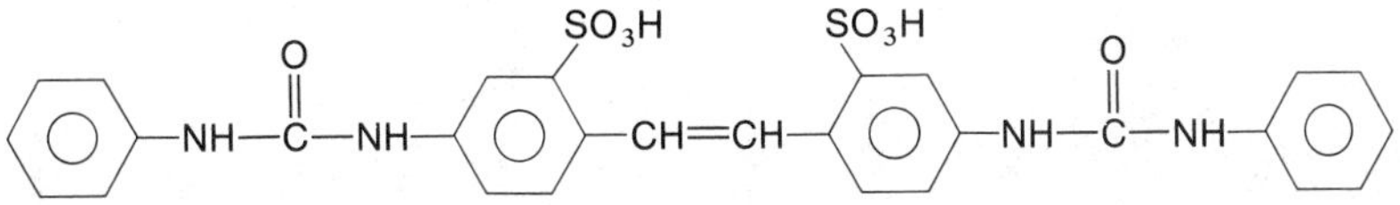

Blancophor R

Figure 18-18 Blancophor R, an optical brightener, is able to convert ultraviolet light into blue light and therefore to hide yellowing in fabrics.

the clothes take on a bluish tint, which masks any yellowing and makes the fabric look brighter. *Blancophors* are the chemical names of these compounds (Fig. 18-18). The safety of these chemicals is not yet known with certainty. They don't appear to harm human beings, but they have caused minor mutations in microorganisms. However, they are known to cause skin rashes on some people.

A WORD FOR THE CONSUMER

The cleaning products on the market today are suited for a variety of cleaning chores. They are easy to use, but in many cases are expensive. They must also be used with caution and you must always read the label and follow the directions. Never mix cleansers or household products. Seven fatalities in the United States between the years 1974 and 1977 were caused by chloramines, which are formed when cleansers containing ammonia are mixed with chlorine bleach. By following the manufacturer's directions, you'll get the best and the safest results.

Part 2 The Chemistry of Cosmetics

THE COSMETICS OF YESTERDAY

It seems that people have always tried to alter their appearance to make some change for the better. They've used products to cleanse the skin and makeup to appear more attractive to others. Cold cream, perfume, eye shadow, and face powder have been in use for years and years. Cosmetics were not always as safe as they are today. In eighteenth-century Europe many women covered their faces with lead(II) carbonate ($PbCO_3$) to make their skin appear whiter. The only problem with this is that many lead compounds are poisonous and many people died from lead poisoning.

Figure 18-19 In the world of cosmetics there are thousands of products. There are preparations that will reduce or stop us from perspiring, perfumes and colognes to make us more alluring, and all kinds of creams and lotions to keep us looking young. (Mary Kay Cosmetics, Dallas, Texas).

Today we spend billions of dollars a year on cosmetics. Every few months it seems, people are afflicted with another condition that can be remedied by a cosmetic preparation. Commercial television reminds us of the current "epidemic" and, of course, suggests a cure. For example, in the past we were plagued with an "epidemic" of greasy hair, followed by an outbreak of bad breath and then yellow teeth. There are preparations that will reduce or stop us from perspiring, perfumes and colognes to make us more alluring, and all kinds of creams and lotions to keep us looking young (Fig. 18-19). Seriously, though, these products can make people look and feel better and the psychological benefit of such preparations is hard to measure. Now let's take a look at some of these products.

WHAT ARE COSMETICS?

In 1938, the *United States Food, Drug and Cosmetic Act* defined cosmetics as "articles intended to be rubbed, poured, sprinkled or sprayed on, introduced into, or otherwise applied to the human body or any part thereof, for cleansing, beautifying, promoting attractiveness or altering the appearance . . . ." Over-the-counter pills and preparations that affect the body's functioning are not classified as cosmetics. Neither are soaps, antidandruff shampoos, or antiperspirants.

When a drug is put on the market it requires extensive testing to be sure that it is safe and that it works. Cosmetics need not be tested before they go on the market. If you do have an adverse reaction to a cosmetic, you should certainly report it to the company that manufactured the product as well as to the FDA (Food and Drug Administration).

CHEMICALS FOR THE SKIN:
LIPSTICKS, LOTIONS, AND CREAMS

The skin is the largest organ of the body. It has many different functions. It protects the internal organs from invasion by bacteria and against mechanical injury. It also serves as an organ which helps regulate body temperature by preventing excess loss of fluids from the tissues and allowing waste products to be excreted. The skin is an excretory organ as well as a sensory organ.

The skin is constantly exposed to nature's elements—sun, wind, rain— and conditions brought about by people—pollution, dirt, and smog. Exposure to these elements can leave the skin dry. Frequent washing also leaves the skin dry. So cosmetic companies have come up with products to help remedy this condition. Most creams and lotions contain *petroleum jelly* and *mineral oil*, which form a protective film over the skin. These products also soften the skin. They keep the skin moist by cutting down on the loss of water by evaporation because of the presence of the protective film.

There are a wide variety of creams and lotions on the market which are supposed to have different functions. *Emollients* are skin softeners. They contain petroleum jelly as a major ingredient. *Moisturizers* keep the skin soft and prevent loss of moisture from the skin. They contain mixtures of petroleum jelly and mineral oil or other fatty or oily substances. *Night creams* are generally thicker than products that you would use during the day. They generally contain petroleum jelly, mineral oil, beeswax, and paraffin as hardeners, which thicken the product. *Cleansing creams* are used to remove dirt from the skin without the drying effects of soap. They contain mineral oil, water, and paraffin as the main ingredients, together with petroleum jelly, beeswax, and glycerin. The dirt and grime are carried away when you wipe off the product. Many skin products also contain *lanolin*, which is a fat obtained from the wool of sheep.

The question often asked is: Are all these products worth anything? There are hundreds and hundreds of skin care products on the market. They all perform the same basic functions: they soften the skin and prevent loss of moisture from the skin. Together with the ingredients already mentioned are water, perfumes, emulsifiers to keep the oily substances in solution with the water, and coloring agents. You might prefer one product to another because of the way in which it is formulated. You might prefer the color or the odor of one product to another. The cost of these products varies greatly and you might choose the product that best fits your budget. Packaging and advertising might sway you to one particular product. But basically, they're all very similar.

Lipsticks are closely related to creams and lotions. They're made with oil and wax. Unlike creams, they contain more wax than oil, to produce a harder product. The waxes used include beeswax, candelilla wax, and carnauba wax. Sometimes castor oil or mineral oil is used. Then coloring agents such as dyes or pigments are added. Chemicals called *antioxidants* are added which slow down the process of rancidity or spoiling of the fatty substances. Perfumes and flavorings are added to make the product have a pleasant taste and odor. Then comes the packaging, marketing, and advertising. Comparing lipsticks and lip glosses by their chemical formulation will show that they're basically all the same. You, as the consumer, may favor one over another for purely personal reasons.

PREPARATIONS FOR ACNE

Almost everybody either had or has acne as they go through adolescence. This ailment usually begins at puberty, although it can affect post-adolescents. It

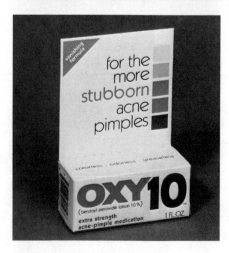

Figure 18-20 A number of products are available for treating acne. (Norcliff Thayer Inc., Yonkers, New York)

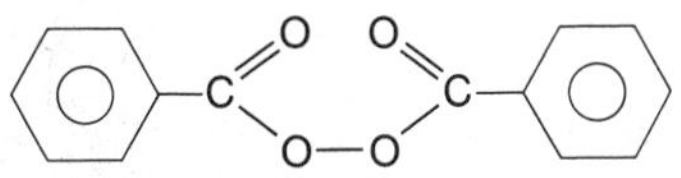

Benzoyl peroxide

Figure 18-21 Benzoyl peroxide seems to be an effective acne preparation.

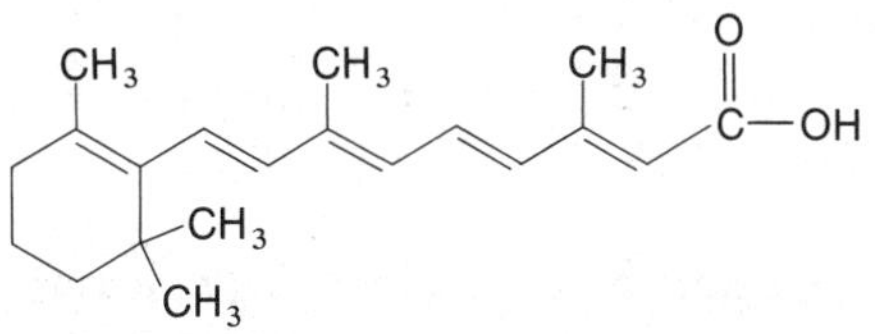

Figure 18-22 Retinoic acid is the tissue-active form of vitamin A.

is caused by internal as well as external factors. Hormonal changes prior to a menstrual period may cause an eruption in women. Stress, also hormonal, via the adrenal glands, could cause skin breakouts. Makeup or simple mechanical or physical irritation by clothing touching the skin or leaning on a hand can irritate the skin and hair follicles in which the disease develops.

Chemists have developed some products which are effective in treating acne (Fig. 18-20). First, blackheads can't be washed off, so washcloths, sandpaper, scrubbing creams, and buffers won't do much good. They only irritate the skin further. Since acne is caused by neither dirt nor infection, antibacterial scrubs are useless. Sunlight irritates the skin and could lead to skin cancer, so it's best to stay away from too much ultraviolet light. The newest acne treatment can be purchased without a prescription and is called benzoyl peroxide (Fig. 18-21). It is applied directly to the skin and helps dry up existing pimples. Another recent treatment is the use of the tissue-active form of vitamin A, called retinoic acid (Fig. 18-22). This is applied to the skin and provides benefits that would require dangerously large doses of vitamin A by oral administration. This treatment must be prescribed by a doctor.

CHEMICALS TO MAKE US SMELL BETTER

Perfumes, colognes, and after-shave lotions are products that make us smell better. Our bodies produce their own characteristic odors through sweating, and many people choose to mask this odor by smelling like a particular flower or fruit.

Perfumes, colognes, and *after-shave lotions* contain odorous compounds and ethyl alcohol. The ethyl alcohol acts as the solvent for the perfume and any other additives, such as the coloring agents that may be added to these products. Sometimes substances known as *fixatives* are added to these products. *Fixatives* are substances that tone down the odor of very strong smelling perfumes. We'll have more to say about fixatives shortly.

453 *The Chemistry of Home Care and Personal Products*

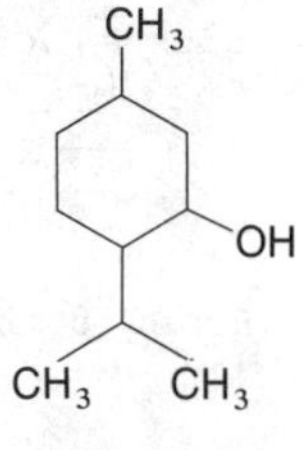

Menthol

Figure 18-23 Many after-shave lotions contain menthol, which cools the skin and makes it tingle.

Perfumes contain 10 to 25% odorous compounds, whereas colognes and after-shave lotions contain only 1 to 2% odorous compounds, so they are more dilute and less expensive than perfumes. After-shave lotions also contain *menthol* (Fig. 18-23), which cools the skin and makes it tingle.

The components of a perfume are called *notes*. There are three notes that make up a perfume: the top note, the middle note, and the end note. The *top note* contains the compounds that vaporize most easily and are therefore the most volatile compounds. You smell these just as soon as you apply the perfume. The *middle note* compounds are not as volatile as the top notes, but are more volatile than the end notes. These compounds cause the lingering odor of the perfume. The *end-note* compounds are the least volatile. An expensive

TABLE 18-1   *Compounds that produce particular odors in perfumes*

NAME	STRUCTURE	ODOR
Citral	$CH_3C{=}CHCH_2CH_2C{=}CHC{<}^O_H$ with CH_3 groups	Lemon
Geraniol	$CH_3C{=}CHCH_2CH_2C{=}CHCH_2OH$ with CH_3 groups	Rose
Irone	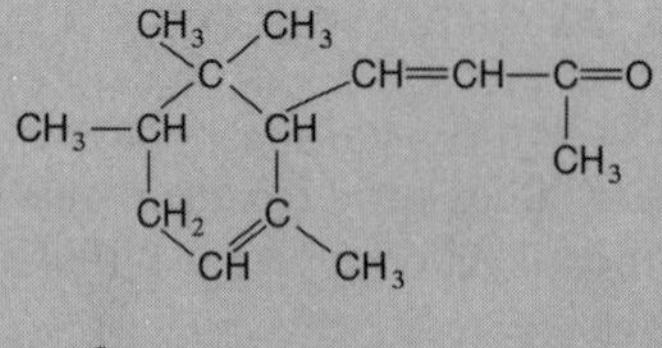	Violet
Jasmone	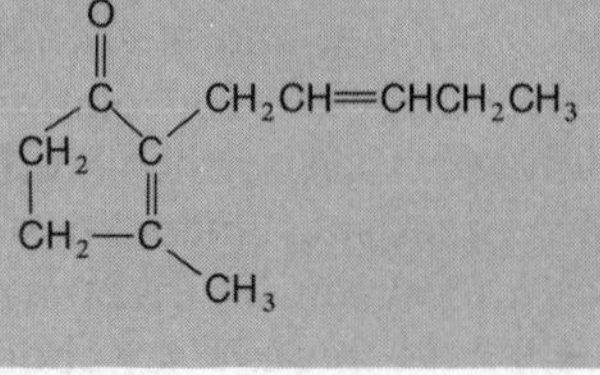	Jasmine

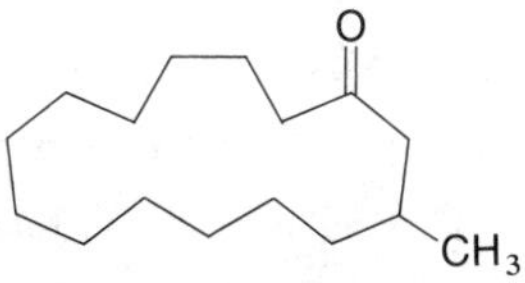

Muscone

Figure 18-24 Muscone is used as a fixitive in perfumes.

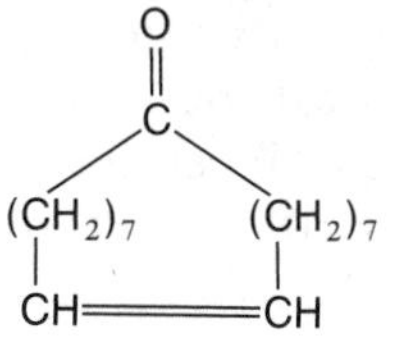

Civetone

Figure 18-25 Civetone is also used as a fixitive in strong-smelling perfumes.

perfume may have numerous compounds composing one note. Originally, these compounds came from natural sources, but chemists are now able to synthesize many of them in the laboratory. Table 18-1 lists some compounds that produce particular odors in perfumes.

A very strong perfume solution made of a large amount of odorous compounds and a small amount of solvent is not really pleasant to smell. These solutions must be diluted several times to produce a nice-smelling product. Sometimes musk compounds (Fig. 18-24) must be added to tone down the very sweet smells of some perfumes. Compounds such as civetone (Fig. 18-25) from the civit cat and muscone from the musk deer are added as *fixitives* to tone down the overpowering odors of some perfumes.

CHEMICALS FOR THE UNDERARM AREA

Deodorants and *antiperspirants* cover up the natural body odor. Because antiperspirants actually constrict the opening of sweat glands to decrease the amount of sweat released, they are classified as drugs. The chemical aluminum chlorohydrate (Fig. 18-26) is an *astringent* which is usually dissolved in alcohol and acts as the active ingredient in antiperspirants. (An *astringent* is a substance that contracts the tissues or canals of the body to stop the flow of a liquid, such as sweat). Antiperspirants come in many forms: roll-ons, sprays, and creams. Most antiperspirants have perfume added to them so that they

Figure 18-26 Aluminum chlorohydrate is an astringent which is usually the active ingredient found in antiperspirants.

$$Al_2(OH)_5Cl \cdot 2 H_2O$$

Aluminum chlorohydrate

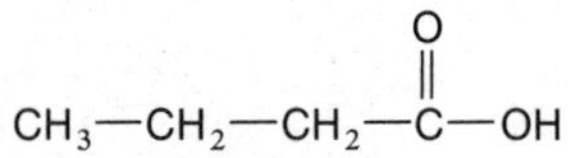

$$CH_3—CH_2—CH_2—\overset{\displaystyle O}{\overset{\displaystyle \|}{C}}—OH$$

Butyric acid

Figure 18-27 The compound butyric acid is one of the substances that causes body odor.

also function as deodorants. (A substance that acts just as a deodorant has only perfume in it, no astringent.)

When we sweat, our bodies secrete oil from the sebaceous glands (oil glands) in the skin. When this oil breaks down, the chemicals produced are foul smelling (Fig. 18-27). People try to mask this odor with deodorants or antiperspirants. Actually, a daily bath or shower and some talcum powder and clean clothes are probably just as effective as using a deodorant or antiperspirant. Perhaps many people have been brainwashed to believe that their natural body odors will offend others unless they cover up these odors.

TOOTHPASTES AND MOUTHWASH: WHITER TEETH AND CLEANER BREATH

Toothpastes are cosmetics that clean our teeth. They contain a *detergent* and an *abrasive*. The detergent sodium lauryl sulfate is often used (Fig. 18-28). Abrasives such as calcium carbonate or titantium dioxide are often used (Fig. 18-29). Together with these ingredients are sweeteners, preservatives, thickeners, and flavoring agents (Table 18-2).

In addition to these ingredients, stannous fluoride (SnF_2) is also added to some toothpastes. Research has shown that in areas of the country where drinking water has fluoride levels of 1 part per million or above, children have 60% fewer cavities than do children living in areas with lesser fluoride con-

Figure 18-28 The compound sodium lauryl sulfate is used as the detergent in some toothpastes.

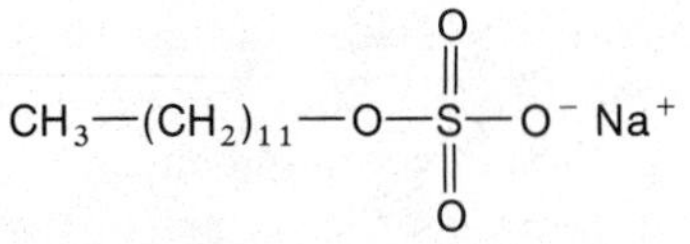

$$CH_3—(CH_2)_{11}—O—\overset{\displaystyle O}{\underset{\displaystyle O}{\overset{\displaystyle \|}{\underset{\displaystyle \|}{S}}}}—O^- \ Na^+$$

Sodium lauryl sulfate

Figure 18-29 The compounds calcium carbonate and titanium dioxide are often used as abrasives in toothpastes.

$CaCO_3$	TiO_2
Calcium carbonate	Titanium dioxide

TABLE 18-2   *Toothpaste ingredients*

Detergent	Sodium lauryl sulfate (sodium dodecyl sulfate)
Abrasives	Calcium pyrophosphate, $Ca_2P_2O_7$
	Calcium hydrogen phosphate, $CaHPO_4$
	Hydrated alumina, $Al_2O_3 \cdot nH_2O$
	Hydrated silica, $SiO_2 \cdot nH_2O$
	Insoluble sodium metaphosphate, $(NaPO_3)_n$
	Precipitated calcium carbonate, $CaCO_3$
	Titanium dioxide, TiO_2
	Tricalcium phosphate, $Ca_3(PO_4)_2$
Sweeteners	Sorbitol
	Glycerol
	Saccharin
Thickeners	Cellulose gum
	Polyethylene glycols
Preservative	Sodium benzoate
Flavoring agents	Mint
	Peppermint oil
Fluoride	Stannous fluoride, SnF_2

centrations. Fluoride can be ingested by taking fluoride-containing vitamins, a fluoride supplement, or by drinking fluoridated water. Fluoride treatments given by a dentist or using toothpaste with fluoride also help. The enamel on teeth and the bones of the body are made of a compound called *hydroxyapatite* $[Ca_{10}(PO_4)_6(OH)_2]$. Fluoride present in the body produces *fluorapatite* $[Ca_{10}(PO_4)_6F_2]$, which is more resistant to decay.

Mouthwash is a product that is good to use between brushings. Actually, mouth odor comes from other sources as well as from the mouth. Stomach gases that pass upward from the esophagus cause some mouth odors. Food stuck between teeth which are acted upon by mouth bacteria, as well as tooth decay, gum decay, and smoking, all contribute to mouth odor. Mouthwashes may inhibit or prevent certain types of bacterial growth, but they shouldn't kill all microorganisms in the mouth, since some of these microorganisms take part in the digestion of food. Mouthwash temporarily freshens the mouth. Some just contain a chemical to sweeten the breath and others actually kill odor-producing microorganisms temporarily (Table 18-3).

TABLE 18-3 *Some germ-killing agents used in mouthwashes*

Boric acid,	H_3BO_3
Cetylpyridinium chloride	
Domiphen bromide	
Methyl salicylate (oil of wintergreen)	
Phenol	

CHEMICALS FOR THE HAIR: SHAMPOOS, CREAM RINSES, DYES, PERMANENT WAVES, AND STRAIGHTENERS

Today's shampoos contain a synthetic detergent to clean the dirt and oil from the hair. The detergent is generally an anionic surfactant such as sodium lauryl sulfate (which we discussed earlier). Baby shampoos contain a surfactant which is amphoteric and will react with both acids and bases. This is not as irritating to the eyes as a conventional shampoo.

Other ingredients in shampoos are fragrances, coloring agents, and thickeners. Shampoos come in all colors and in flavors and fragrances such as strawberry, green apple, lemon, honey, and herbal, to name a few. The pH-balanced shampoos have pH values between 5 and 8. This pH range is compatible with the pH of hair and scalp and is less damaging to hair than highy acidic or highly alkaline shampoos.

Cream rinses leave hair soft, fluffy, and silky smooth. They leave a protein film on the hair which will collect dust and natural oil, so you may have to shampoo more often.

Texturizers or thickeners leave a film on the hair. Some contain white glues which are proteins, and these can glue the split ends of the hair together and make it feel thicker.

Dyes for hair come in two types: temporary and permanent. *Permanent* dyes penetrate the hair shaft and remain there. *Temporary* dyes wash out as

Para-phenylenediamine

Figure 18-30 Permanent hair dyes are made from derivatives of the compound *para*-phenylenediamine.

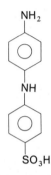

Para-aminodiphenylaminesulfonic acid

Figure 18-31 The compound *para*-aminodiphenylaminesulfonic acid is used in hair dyes to make hair blonde.

the hair is shampooed. The permanent dyes are made from derivatives of a compound called *para*-phenylenediamine (Fig. 18-30). Various colors are formed by placing different substituents on the molecule. For example, a derivative of *para*-phenylenediamine is *para*-aminodiphenylaminesulfonic acid (Fig. 18-31). This compound is used in hair dyes to make hair blonde.

Hair can be bleached by using a product such as hydrogen peroxide. The colored pigments in the hair (those which give your hair its natural color) are oxidized to colorless products. Then a lighter-colored dye can be applied to the hair to change its color.

As you may know, hair is a protein. The protein molecules that make up hair are held together by disulfide linkages (Fig. 18-32). Hair can be made either curly or straight by breaking the disulfide linkages and changing their positions. The first step in either waving or straightening hair is to treat the hair with a reducing agent such as thioglycolic acid (Fig. 18-33). When the disulfide linkages are broken, the chains can be pulled apart. If hair is to be made wavy, small curlers are used and a mild oxidizing agent is added which allows the disulfide linkages to re-form, making the hair take on the shape of the curlers. If hair is to be straightened, large rollers can be used to take out some of the curl and a mild oxidizing agent such as hydrogen peroxide is used

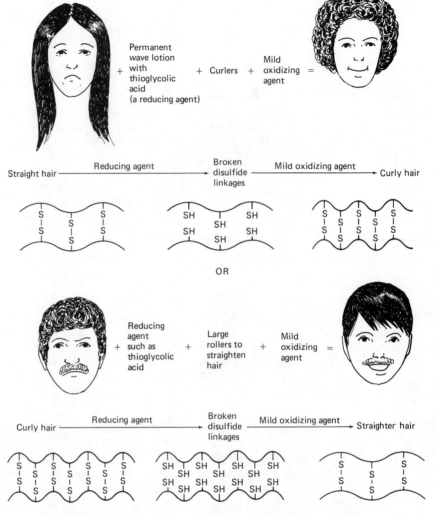

Figure 18-32 Protein molecules that make up hair are held together by disulfide linkages. Hair can be made either curly or straight by breaking the disulfide linkages and changing their positions.

Figure 18-33 The compound thioglycolic acid is used as a reducing agent to break the disulfide linkages that hold hair together.

$$H-S-CH_2-\overset{\overset{\displaystyle O}{\|}}{C}-OH$$

Thioglycolic acid

to re-form the new disulfide linkages. When the hair grows, the disulfide linkages are just where Mother Nature put them, and the waving or straightening process must be repeated on the new hair.

SUMMARY

In this chapter we surveyed some home care and personal products available to consumers today. We began our survey by examining some old-fashioned cleaners, and discovered that certain plants produce lathers because they contain compounds called saponins. We also learned that people have produced soap for over 2000 years by cooking potash and fat together. Next, we looked at modern soaps and discussed the mechanism by which soap cleans.

Because there are disadvantages to soap in certain types of water, synthetic detergents were developed. In our survey we examined ABS and LAS detergents. We also looked at surfactants and phosphate detergents, as well as "builders" used to make the detergents work better. Next, we examined anionic, cationic, and nonionic surfactants, and discussed the uses of each. We also examined the uses of bleach, borax, and brighteners and learned how they work on fabrics.

In the second part of the chapter we surveyed some cosmetics and products used for personal hygiene. We began by looking at lipsticks, lotions, and creams and we examined the major ingredients in these products. Next, we discussed acne and acne preparations and treatments, and we learned some ways to deal with this problems, as well as some things to avoid.

We continued our survey by examining perfumes, colognes, and after-shave lotions. We discussed the differences between each of these products, as well as the basic ingredients of each. Next, we learned about deodorants and antiperspirants and discovered how they work. We also discussed the differences between a deodorant and an antiperspirant.

Our discussion continued with the topic of toothpastes and mouthwashes. Again, we looked at the basic ingredients of these materials and discussed how the products work. We completed our survey by examining various hair care products, such as shampoos, cream rinses, dyes, permanent wave products, and hair straighteners.

We hope that this chapter has given you sufficient information to be a better consumer.

EXERCISES

1. Name two plants that were used to produce lather in primitive societies. What chemical compounds were responsible for the lather?

2. Explain how soap was made by early societies.

3. Explain how soap works to clean people and clothes.

4. List two conditions when soap doesn't clean well.

5. Explain what a syndet is and give an example.

6. Discuss the environmental problems caused by the ABS detergents.

7. Discuss the differences between ABS and LAS detergents.

8. (a) Explain what a surfactant does.
   (b) Explain what a builder does.

9. Discuss the problems caused by phosphate detergents to the environment.

10. Sodium perborate is a builder; however, it also acts as a bleach. Explain why this occurs.

11. Explain the chemical differences among anionic, cationic, and nonionic surfactants.

12. Explain how an automatic dishwasher detergent works.

13. Explain why cationic surfactants and anionic surfactants can't be used together.

14. Explain how a chlorine bleach works.

15. Discuss the differences between chlorine bleaches and all-fabric bleaches.

16. What is the advantage of hydantoin and cyanurate bleaches over chlorine bleach?

17. Explain how brighteners work on fabrics.

18. List the major constituents of perfumes, colognes, and after-shave lotions, and state the purpose of each one.

19. List the major constituents of lipsticks, lotions, and creams, and state the purpose of each one.

20. Name two treatments for acne and explain how they work.

21. What is the difference between a deodorant and an antiperspirant?

22. What is the difference between the top note, middle note, and end note of a perfume?

23. List the major constituents of toothpaste and state the purpose of each.

24. Explain how fluoride prevents tooth decay.

25. How do mouthwashes work?

26. List the basic ingredients of shampoos and state the purpose of each one.

27. What is the difference between a temporary and a permanent hair dye?

28. How do hair texturizers work?

29. What is the chemical mechanism behind straightening and curling hair?

30. Name an advantage and a disadvantage of using a cream rinse on your hair.

Chapter 19

Chemistry and Outer Space

Some Things You Should Know After Reading This Chapter

You should be able to:

1. Discuss a theory of how the universe was formed.
2. Write an equation for some nuclear reactions taking place in stars.
3. Define the terms galaxy, photosphere, chromosphere, corona, and sun's core.
4. Discuss the major features of the sun and define the terms sunspots, prominences, flares, and coronal holes.
5. State the relationship between the solar wind and the auroras seen on Earth.
6. Name the nine planets of our solar system and state some of the major features of each.
7. Discuss what we mean by "giant gas planets."
8. Explain what is meant by a galaxy and a nebula.
9. Discuss the evolution of our sun, from its birth to its death.
10. Explain what will happen to a star more massive than our sun when it reaches the red giant stage.
11. Explain what is meant by a pulsar and a black hole.
12. Discuss why many scientists believe that life exists on other planets outside our own solar system.
13. Explain how the space program has helped improve the human condition on Earth.
14. Give an example of a spin-off from space and explain how it has affected our lives.

THE UNIVERSE*

The study of the universe spans almost inconceivable extremes of size and distance and time—from the vast island of stars we call a *galaxy* (Fig. 19-1) to the tiny atom and the particles that comprise it; from cosmic events that occurred billions of years in the past to microcosmic events in the present that endure for only billionths of a second. To explore the universe at these extremes,

Figure 19-1　The Andromeda galaxy. (NASA)

* We would like to acknowledge the Public Affairs Office of NASA and Brian Duff, its director, for permission to use the script from the film, *The Universe*, and for allowing us to obtain and use the photographs that you will see in this chapter. We would especially like to thank Robert Schulman, Graphics Coordinator for NASA, and Les Gaver, Audio-Visual Chief for NASA, for their help in developing this chapter. We dedicate this chapter to the people of NASA and to their mission of exploring space—the final frontier. For in performing their mission, they are helping to improve the human condition here on Earth.

Figure 19-2 NASA's infrared telescope, atop 14,000-foot Mauna Kea in Hawaii. (NASA)

scientists build instruments that extend their reach and vision. The great eye of the telescope has the light-gathering power of a million human eyes. It peers, not only into the depths of space, but far back in time, since the light it *now* observes may have left its source when dinosaurs inhabited the Earth (Fig. 19-2).

Our scientific ear is tuned to the invisible radio sky (Fig. 19-3). It detects not objects, but the radio regions associated with them, and at distances far beyond the range of the largest optical telescope. But the radio waves and the visible light that pass through the Earth's atmosphere to these ground-based telescopes are only part of a broad spectrum of radiation, most of which is blocked by the atmosphere. So electronic instruments are lifted above this murky and turbulent layer. Airborne by rockets, by balloons, in unmanned astronomical observatories, in manned laboratories, and in spacecraft orbiting the planets, instruments probe the near and distant environments of space and open new windows on the universe (Fig. 19-4).

In the beginning, it is believed, all the matter of the universe was contained in a primordial atom—an unimaginable concentration of elementary particles. In one gigantic detonation the contents of this cosmic fireball were hurled outward in all directions. After 1 million years of expansion, the universe was an intense blaze of light.

Then the radiation cooled and after hundreds of millions of years, great clouds of hydrogen gas began contracting, and, in time, evolved into the galaxies we now observe.

Figure 19-3 A 210-foot tracking dish for NASA's Deep Space Network. (NASA)

Figure 19-4 A view of Skylab taken from the Apollo command module. There was much information obtained about our universe from project Skylab. (NASA)

Inside the galactic whirlpools, smaller fields of gravity condensed hydrogen into stars. Stars are inconceivably hot. They are so hot that they sustain thermonuclear reactions that transform hydrogen into heavier elements. For example,

$$\,_{1}^{2}H + \,_{1}^{2}H \xrightarrow[\text{millions of degrees}]{\text{temperatures in the}} \,_{2}^{3}He + \,_{0}^{1}n + \text{Energy}$$

$$\,_{1}^{2}H + \,_{1}^{3}H \xrightarrow[\text{millions of degrees}]{\text{temperatures in the}} \,_{2}^{4}He + \,_{0}^{1}n + \text{Energy}$$

Sometimes their hydrogen fuel burns so fast that they flare out in violent explosions, hurling new elements across space. Like a great wind, the radiant energy of starlight drives these clouds of dust and gas throughout the galaxy. Out of these clouds evolve new generations of stars.

More than half the stars in our galaxy travel in groups of two or more, orbiting around a common center of gravity. Like galactic comets, immense clusters of stars swing in and out of the galaxy in vast eccentric paths. Some small stars do not travel in the company of other stars. Our own sun is one of these.

To the astronomer, the sun is a vast laboratory for the detailed study of a star's structure and energy. *Solar observatories* track the sun and gather its rays for analysis. In one such observatory a vertical tower supports a *heliostat mirror*, which tracks the sun, gathers its rays, and reflects them down a light shaft that extends 300 feet below ground. At the end of the shaft, the rays are cast back to an observing room, where minute-by-minute changes across the face of the sun are observed (Fig. 19-5). Another mirror projects a light beam to a *spectroscope*, an instrument that splits the light into its component colors to produce a visible spectrum (Fig. 19-6).

The dark lines that cut across the spectrum band are produced by the radiation from the sun's interior shining through its atmosphere. Each line is the signature of a chemical element, such as sodium, iron, or calcium. It is this array of lines that forms the code that describes the properties and motions of a star. By narrowing the view of the sun to a single line of the spectrum, each level of the solar atmosphere can be photographed, and each reveals a remarkably different aspect. And with the addition of computer mapping and color processing—which distinguishes levels of brightness—a detailed and multi-dimensional picture is obtained of a sun undergoing dramatic and turbulent change.

The sun is a sphere of hot seething gases and surges of radiation. Most of the light we get from the sun comes from the thin, bright layer that defines its visible edge—the *photosphere*. Above it, the *chromosphere*, a region of flaming outbursts of gas, extends through a transition zone to the thin outer atmosphere of the *corona*.

Once thought to be a quiet layer of the solar atmosphere, the corona is

Figure 19-5 An enormous solar eruption in which cool gases at 50,000°K were shot 300,000 km (500,000 miles) into the hot rarified gas of the solar corona, which is at a temperature of about 2,000,000°K. (NASA)

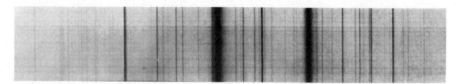

Figure 19-6 A portion of the yellow region (5880 Å) of the solar spectrum. The two strong absorption lines (Fraunhofer D-lines) are due to sodium vapor.

now revealed to be a region of dramatic large-scale changes and unexpected turbulence, with temperatures reaching millions of degrees.

Deep beneath the sun's atmospheric shell is the *core*, a violent nuclear furnace. Here hydrogen is fused into helium, and in the process some of the matter is converted into energy. Radiating outward as a gas, it convects like a boiling liquid beneath the surface. The turbulent bubbling motion is visible in the granular cells of the photosphere.

Sunspots, regions of intense magnetic fields, appear on the surface, disappear in a few hours, or grow and persist for months in a mysterious 11-year cycle (Fig. 19-7).

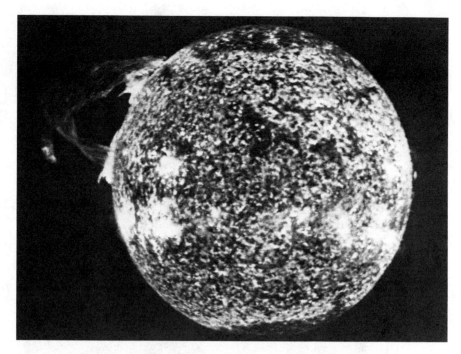

Figure 19-7 Sun spots and solar flares as seen from Skylab. (NASA)

The sun rotates once in 27 days. Because its equatorial regions rotate faster than the polar caps, the shearing action in the gas contorts the magnetic field into tangled structures. Shaped by these magnetic fields are the spectacular *prominences*, titanic streamers of gas, reaching heights of more than half a million miles above the surface.

The greatest explosions in the solar system are *flares*—intense bursts of light, erupting with the force of billions of hydrogen bombs. They move at hundreds of miles a second, then, after minutes or hours, they fade away.

The dark areas across the solar disk are *coronal holes*, which may provide new clues to the sun's interior, and may be the source of the *solar wind* that blows outward to the farthest planets. On Earth, the effects of these solar events are visible when *auroras* light up the dark arctic sky, and radio communication is disrupted (Fig. 19-8).

The sun is an average, middle-aged star. Yet it will generate heat and light for billions of years to come, as it has for 5 billion years past. It dominates the motions of all bodies in the *solar system*.

Nearest the sun and obscured by its intense glare is *Mercury*, a cratered planet much like our moon (Fig. 19-9). Temperatures rise to about 187°C (369°F), but no clouds or atmosphere protect its ancient surface from the searing heat.

Figure 19-8   Events like these on the sun cause auroras on Earth. (NASA)

Figure 19-9   A mosaic of over 200 high-resolution photographs from Mariner 10 of the south pole of Mercury. (NASA)

Figure 19-10 Artist's concept of the Pioneer probe, which descended through the atmosphere of Venus in December 1978. (NASA)

Moving outward from Mercury, we encounter *Venus*. Its perpetual cloud cover traps the radiant energy of the sun within an atmosphere of incredible pressures. From the surface, only a reddish glow reveals the presence of the sun (Fig. 19-10).

Beyond Venus, 93 million miles from the sun, is *Earth* (Fig. 19-11). Its great oceans, forming clouds and air currents that warm and irrigate the planet, shape its continents and nourish life.

Earth's satellite, the *moon*, airless, waterless, and scarred by meteors that have bombarded it since the time of its formation, now bears the imprints of our astronauts.

A probe of the planet *Mars* has discovered a dynamic and evolving planet with unexpected geological features (Fig. 19-12): a volcanic mountain, many times larger than the largest volcano on Earth; a vast and deep canyon, extending for 2500 miles; and dry, riverlike channels that may have been carved by running water (Fig. 19-13).

Beyond the orbit of Mars is the belt of *asteroids*, craggy chunks of rock and metal, some as small as boulders, others hundreds of miles in diameter.

About 500 million miles from the sun we encounter the first of the giant gas planets, *Jupiter*, the colossus of the solar system. It is more massive than all of the other planets combined. Deep beneath the maelstrom of clouds that band its surface is a primordial atmosphere much like that in which life awakened on Earth millions of years ago. And drifting on its surface is the mysterious *red spot*—an immense cyclonic storm that has raged for hundreds of years

Figure 19-11 The earth as seen by the Apollo 16 astronauts. (NASA)

Figure 19-12 An intricate network of canyons on Mars, taken by Mariner 9. This photo covers an area 542 km wide and 426 km high (336 miles by 264 miles). (NASA)

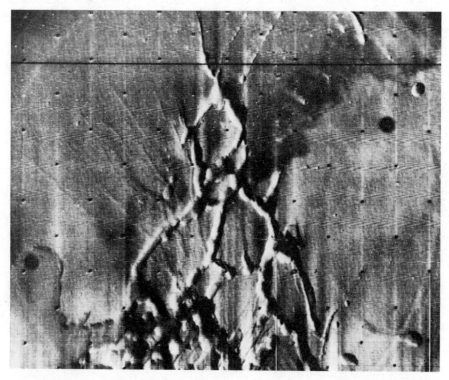

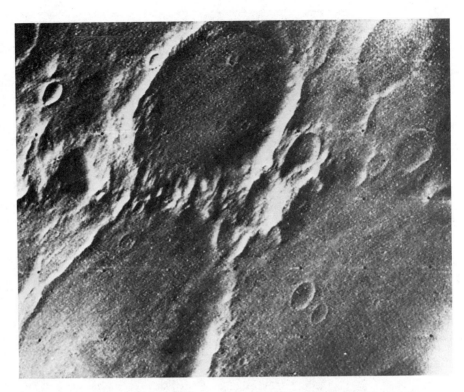

Figure 19-13 The surface of Mars. Called the "Giant's Footprints," these two adjacent craters measure 85 by 200 miles. This photo was taken by Mariner 7. (NASA)

and continues unabated. Radiating more energy that it receives from the sun, and circled by 14 moons, Jupiter is like a miniature solar system (Fig. 19-14).

The next largest of the gas planets is *Saturn* (Fig. 19-15), girded by rings which, as we approach them, resolve into countless particles of frozen debris and ice. Each is a tiny moon orbiting the massive planet.

And as we continue past the frozen worlds of *Uranus* and *Neptune*, we arrive at the outermost planet in the solar system, *Pluto*. It moves in a dim twilight of unimaginable cold. The sun 4 billion miles away, is only a brilliant light in the night sky.

To travel beyond the solar system to the nearest star would require a journey of more than 5 trillion miles. Yet our sun is only one of 100 billion stars widely separated from one another in time and space, but all bound by gravity, and all revolving around the central core of our galaxy, the *Milky Way*.

Drifting between the stars are vast clouds of gas and dust—the *nebulae*, made luminous by the radiation of stars within or near them, or darkly obscuring the light of whatever lies behind them. Here, new stars are being born.

About half a century ago our galaxy was thought to be alone in the universe. We now know it to be one of a local group of about 20 galaxies.

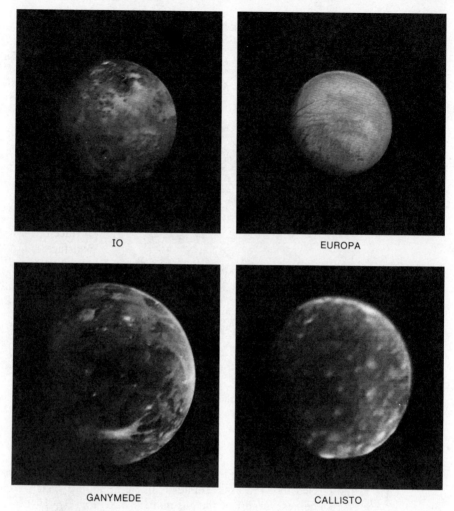

IO   EUROPA

GANYMEDE   CALLISTO

Figure 19-14   The four largest moons of Jupiter as seen by Voyager 1. (NASA)

Figure 19-15   The planet Saturn. (NASA)

And strewn throughout the vast reaches of space are more than 10 billion galaxies grouped in clusters as far as our most sensitive instruments can reach. Little is known about the evolution of galaxies and why some are formless or irregular, others elliptical, and still others spiral-shaped (Fig. 19-16). And we know as little about the galactic core and its role in the galaxy's evolution and structure.

The problem has become more perplexing by the discovery that some galaxies are in a state of extreme disarray, exploding, ejecting gaseous matter, or interacting with other galaxies. Even more puzzling are *quasars*, starlike objects that emit as much energy every second as the sun radiates in some 10 million years. They appear to be among the most remote objects in space.

Stars are born, live out their life spans, and die. The life history of a star is marked by an opposition of two kinds of pressure. One is created by the energy in the core of the star, pushing the surface *outward*. The other is the crushing force of gravity pulling the star's surface *inward*. When these are balanced, a star becomes stable and shines steadily. As hydrogen fuel is depleted, the release of energy is insufficient to withstand the gravitational pressure, and the core collapses.

But compression by gravity *raises* the temperature in the core and helium ash rekindles the nuclear fires. Vast amounts of energy are released which lift the outer zones against the forces of gravity. The star is now a red giant. In the final stage of its evolution, it is the *mass* of a star that determines its fate.

The sun, a medium-size star, will remain stable for approximately 10 billion years. Then it will expand to 400 times its present diameter. As it

Figure 19-16 The great spiral galaxy in Andromeda. (NASA)

Figure 19-17 A black hole results when a very massive star collapses. The presence of a black hole can be deduced only by its influence on a visible companion star, distorted out of shape by the black hole's gravitational attraction. Gas pulled off the visible star circulates about the black hole and is pulled into it.

expands it will engulf the inner planets, Mercury, Venus, Earth, and Mars. And it will create a nebula extending past the outer planets. After millions of years, its reserves of nuclear fuel will be exhausted, its outer layers will have dissipated, and only a white dwarf star will remain, no larger than the Earth. Slowly cooling to zero temperature, it will end its life as a black stellar corpse.

When a star more massive than the sun reaches the red giant stage, the collapse of its core raises its temperature billions of degrees and triggers a spectacular detonation—a supernova explosion. At the center of the explosion a residue of the star is crushed by gravity to a neutron core, only a few miles across but so dense that 10 billion tons of its matter would fill only a tablespoon. It spins rapidly, generating radio signals in its strong magnetic field. A radiation beam sweeping past the Earth is observed as a pulse. The star is known as a *pulsar*.

An even stranger end is predicted for *very* massive stars. According to the laws of gravity as presently understood, nothing can stop its collapse. The star disappears from our universe, leaving a *black hole* in space (Fig. 19-17). Its presence can be deduced only by its influence on a visible companion star, distorted out of shape by the black hole's gravitational attraction. Gas, pulled off the visible star, circulates about the black hole. And in the dizzying plunge it emits x rays which can be detected in space. No light or matter can ever leave the intense gravitational field of this cosmic abyss. The physical laws that govern the conditions within this bizzare object are totally unknown to us.

The evolving universe must come to an end. If it continues to expand indefinitely, the light of every star will, in time, be extinguished. The galaxies

Figure 19-18 The Pioneer F spacecraft before launch at the Kennedy Space Center in Florida. (NASA)

will disappear into infinite darkness. But if gravity halts the expansion, the universe will fall back on itself. Galaxies will lose their separate identities, stars will explode, and the sky will again be ablaze with light. Finally, all matter will be engulfed in a fireball like that from which it emerged.

All things on Earth, living and inert, are formed from the elements forged in some distant and unknown star. On Earth, atoms, joined together in definite numbers and patterns, compose the organic molecules that form living cells. Since the discovery of complex molecules in the chill vacuum of interstellar space, there is reason to believe that among the countless galaxies in the universe, there are stars orbited by planets favorable for the evolution of intelligent life. Is space travel to these planets possible? Time and distance may be insurmountable barriers.

The spacecraft Pioneer, now speeding toward the outer planets and beyond, traveling at 35,000 miles an hour, would take almost 80,000 years to reach the nearest star, *Alpha Centauri* (Fig. 19-18). A spacecraft traveling 2500 times faster than Pioneer, at 10% the speed of light, would require so great an expenditure of energy that until new sources have been tapped, it must remain an invention of science fiction.

Figure 19-19 The Space Telescope will shed new light on our universe.

A more practical strategy in the search for extraterrestrial life is to tune in on radio signals traveling at the speed of light, beamed perhaps by creatures on the planet of a distant star.

Some day, an array of telescopes, Earth bound, or lifted to the far side of the moon, may hear faint but unmistakably meaningful sounds amidst the din of cosmic radio chatter (Fig. 19-19). That moment will signal a change in the human condition that we cannot foresee or imagine. "For man," wrote H. G. Wells, "there is no rest and no ending. He must go on, conquest beyond conquest; and when he has conquered all the deeps of space and all the mysteries of time, still he will be beginning."

Part 2   Spin-offs from Space

In Part 1 of the chapter we surveyed the evolution of our universe. We reviewed some of the latest facts and theories about how our universe operates. You might say that we looked at the total cosmic picture. But how does all of this relate to us on our small planet, in a practical everyday sense? Sure, human beings are curious creatures who want to understand how the universe works. And yes, there are many of us who would like to travel into space and visit unknown worlds. But are there any immediate benefits that can be derived from space research for the vast majority of people on Earth?

The answer to this question is yes. The NASA Space Program has greatly benefited humankind with many technological advances over the past 20 years. In medicine, engineering, and science, the NASA people have found ways to use the ideas and information from the space program to benefit human beings on Earth. In this section of the chapter we survey some of these ideas.

INSULATING THE ALASKA PIPELINE

You probably remember the 800-mile-long Trans-Alaska Pipeline, which was built to transport crude oil in the cold Arctic weather of Alaska. Have you ever thought that the oil might get too cold to flow in the Arctic? Well, General Electric Company's Space Division in Valley Forge, Pennsylvania, has used chemistry to deal with the problem. They've developed a product called *Therm-O-Trol* to insulate the Alaska Pipeline and keep the oil heated to about 180°F (what's this in Celsius?) which maintains the fluidity of the oil. Therm-O-Trol is a metal-bonded polyurethane foam especially formulated for the Arctic weather. This product is a spin-off of the Apollo and Gemini space missions.

SAVING ENERGY AND INSULATING WINDOWS

Over 25 years ago a company called National Metallizing, a division of Standard Packaging Corporation, located in Cranbury, New Jersey, helped to develop a reflective insulating film for use on the Echo communications satellite. This film helped protect NASA spacecraft from intense solar radiation and is now available commercially for use as a window insulating product. The film reflects the sun's heat by filtering out 75% of the sun's infrared heat rays. This stops heat from penetrating through the glass and heating the interior space of a room. This in turn cuts down on energy costs for cooling. In the winter, the film blocks the escape of interior heat through the glass. The product is called *Nunsun* and it is a thin metallized film which is bonded to windows with adhesive. National Metallizing has also developed several more advanced reflective films for use on Apollo, Skylab, and the Space Shuttle.

VACUUM DRYING TECHNIQUE

General Electric Company's Space Division has built a large environmental chamber which is used for simulating the operating conditions of an orbiting

Figure 19-20 The General Electric Company's environmental chamber being used to dry water-damaged books. (General Electric Company's Space Division, Valley Forge, Pennsylvania)

spacecraft. The environmental chamber has also proven to be useful in restoring water-damaged objects such as books, blueprints, paper products, and textiles.

In 1972, the Temple University Klein Law Library caught fire and 60,000 books were damaged by water. Some of the books were irreplaceable. The normal procedure for drying water-soaked books involves blotting each page by hand and then air-drying them. This costs about $100 per book. The environmental chamber was used instead (Fig. 19-20). The books were sealed in the environmental chamber and the pressure in the chamber was reduced to 0.01 atm of pressure. This promoted the evaporation of water. As the water evaporated, it froze because of the *near-vaccum* conditions in the chamber. After 24 hours, the pressure in the chamber was increased to eliminate the vacuum. Next, hot *freon gas* entered the chamber to melt the ice. The water was then drained off. The process continued until the water-soaked materials were completely dry. The cost of this operation was about $2 per book.

A NEW SYSTEM FOR ANALYZING MICROORGANISMS IN THE HUMAN BODY

Years of NASA-sponsored research by the McDonnell Douglas Corporation led to the development of a fully automated system for detecting and identifying microorganisms for the space program. This device is now in use at various hospitals throughout the country.

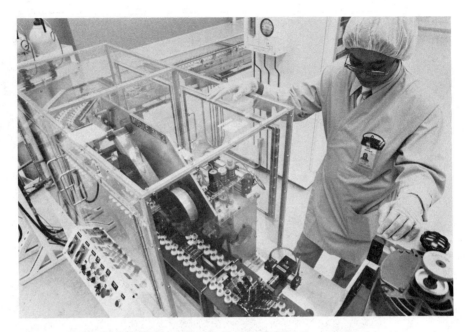

Figure 19-21 The AutoMicrobic System (AMS) being used in a laboratory. (Vitek Systems, Inc., Division of McDonnell Douglas Corporation, St. Louis, Missouri)

Analyzing a specimen for harmful microorganisms traditionally involved growing cultures containing various body fluids. These cultures were incubated for two or three days and then the disease-producing organisms that were present were identified.

The AutoMicrobic System (AMS) does the same job in a shorter time. Specimens of body fluid are exposed to microbe nutrients for the nine most common disease-producing organisms. The cultures are incubated for cycles ranging from 4 to 13 hours. During this time an electrooptical scanner checks each specimen once every hour to monitor changes in cell growth. When cell growth reaches a certain predetermined level, the presence of disease-producing organisms is confirmed. Pushing a button gives a visual display of the number and type of organisms present and reports the information on a printout. There is even a more advanced AMS model available which not only detects which microorganisms are present but also subjects these organisms to different drugs to obtain the most effective drug for treatment. The time savings of AMS over conventional methods is 50 to 80% (Fig. 19-21).

STRAIGHTENING CROOKED TEETH MORE EASILY

A new type of wire, called *Nitinol*, was developed by the Naval Ordnance Laboratory in White Oaks, Maryland. The wire is an alloy of the metals nickel

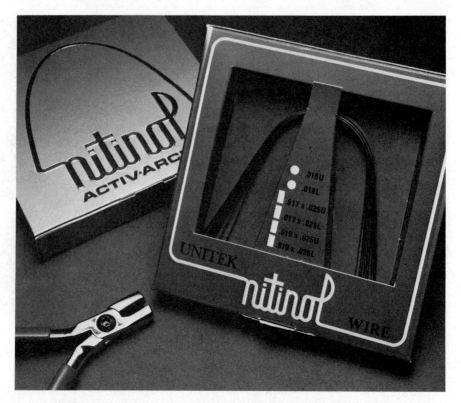

Figure 19-22 Nitinol, an alloy of the metals nickel and titanium, being used in dental research. (Unitek Corporation, Monrovia, California)

and titanium. NASA conducted further research on this alloy and together with Unitek Corporation of Monrovia, California, an orthodontic application surfaced.

George Andreasen, an orthodontist at the University of Iowa, used this wire to help straighten teeth. Traditionally, stainless steel wire is bent and attached to the crooked teeth. The stainless steel wire loses its ability to exert sufficient pressure after being bent, so must be adjusted or changed frequently. Nitinol does not lose its ability to exert pressure, even after being bent. It continues to pull on the teeth and maintains its original shape even after being bent 100 times. This means fewer vists to the orthodontist, and in some cases, a shorter overall treatment time (Fig. 19-22).

SMOKE DETECTORS

A common item now on the market is a smoke detector which sounds a loud ear-piercing alarm when smoke is sensed by the detector. Did you know that the development of smoke and fire detectors incorporates technology developed for very sophisticated smoke and fire detection systems for NASA's Skylab? Although there are various types of smoke and fire detectors on the market, some battery powered and some which run on house current, they all sound a loud alarm to warn the occupants of the area of danger.

There are two types of smoke detectors: ionization and photoelectric units. The ionization detector contains a radioactive source, usually americium-241, which emits mainly alpha particles which ionize air particles. The ionized air particles are attracted by a voltage charge on the collector electrode. This results in a current flow of less than 1 μA.

When smoke particles ranging in size from 0.01 to 1 micrometer (μm; micron) diameter enter the chamber, they unite with the ionized air particles, forming heavier particles which are slow moving. Since fewer of the heavier particles reach the electrode, current flow decreases and this change is signaled by an alarm sounding.

A photoelectric detector features a beam of light that is projected at almost a right angle to the face of a photocell in a darkened chamber. When smoke particles ranging in size from 0.5 to 1000 μm in diameter enter the chamber, the light from the particles is scattered and reflected onto the photocell. The photocell current is increased and amplified and an alarm sounds. Humankind has certainly been advantageously affected by this space-based technology.

LIQUID-COOLED GARMENTS

Astronauts who reached the moon had to deal with temperatures up to 121°C (250°F) on the surface of the moon. NASA's Ames Research Center developed liquid-cooled garments which were worn under the astronauts' space suits to maintain their body temperatures. By adaptation of this portable cooling system, two teenagers afflicted with rare diseases were helped.

A young woman afflicted with "burning limb syndrome" was treated with ice baths and confined to bed. This disease causes severe pain in the limbs and only cooling relieves the pain. Bill Williams of NASA's Ames Research Center designed a garment to wrap around her thighs, and circulate cool water which relieved pain and prevented further tissue deterioration for the young woman.

A young man suffering from a rare disease causing scaling of the outer layer of skin was helped by a cooling unit and a garment to help his condition and enable him to spend time outdoors instead of in an air-cooled environment.

CANCER THERAPY

Conventional treatment of cancerous tumors include surgery, x-ray therapy, cobalt therapy, or chemotherapy. One alternative to conventional x-ray therapy is *fast neutron radiation*, which is used when a patient has a cancerous tumor that has not spread and can't be treated by conventional methods. Fast neutron therapy is an experimental technique and research is being carried out under an agreement between NASA's Lewis Research Center in Cleveland, Ohio, and the Cleveland Clinic Foundation. The Lewis Research Center houses a 50 million-volt cyclotron which generates neutron radiation by bombarding a beryllium target. This causes a nuclear reaction which produces fast neutron particles. This research is being funded by a grant from the National Cancer Institute.

SUMMARY

In this chapter we learned about our universe. We began by studying the "big-bang" theory of how our universe was born. We learned how galaxies were formed and how stars came to be. We also looked at the chemical processes that take place in a star which allow it to generate tremendous quantities of heat, light, and energy.

Next, we learned about our sun and the reactions taking place in its various layers (or parts). Then we read about the planets in our solar system and learned a little about the features of each one. We also studied our galaxy, the Milky Way, and learned that it is just one of billions of galaxies strewn through space.

We concluded the first part of the chapter by reviewing the fate of stars, including our own sun, and we studied the possibilities that await our universe in the distant future.

In the second part of the chapter we looked at how the NASA space program has produced advances in science, technology, and engineering here on planet Earth. These *spin-offs from space* are examples of how the space program has produced an abundance of information to make life better for all of us.

Our Earth is just a small planet revolving around an average middle-aged star, which is a member of a not-so-unusual galaxy. But it is our only home, and it is important for us to understand our place in the universe.

EXERCISES

1. Discuss the theory of how the universe was formed.

2. Write two equations for some nuclear reactions taking place in stars.

3. Define the following terms.
   (a) galaxy   (b) photosphere   (c) chromosphere
   (d) corona   (e) sun's core

4. Explain each of the following.
   (a) sunspots   (b) prominances   (c) flares   (d) coronal holes

5. What is the solar wind? What is the effect of the solar wind on the Earth?

6. Match the planet with its features.

(1) Mercury   (a) the largest planet in our solar system

(2) Venus   (b) the only planet in our solar system known to contain great oceans of water

(3) Earth

(4) Mars   (c) the closest planet to the sun

(5) Jupiter   (d) a large gas planet with rings

(6) Saturn   (e) Earth's sister planet, shrouded by clouds

(7) Uranus   (f) the planet that is usually farthest from the sun

(8) Pluto   (g) the red planet that has canals and volcanic mountains

   (h) a planet that has recently been discovered to have rings around it

7. What is a galaxy? What is a nebula?

8. Discuss the evolution of our sun, from its birth to its death.

9. Explain what happens to a star more massive than our sun when it reaches the red giant stage.

10. Explain what a pulsar is.

11. How does a black hole form?

12. Discuss the possibilities of life on other planets outside our solar system.

13. How has the space program helped to improve the human condition on Earth?

14. Give an example of a spin-off from space research and explain how it has affected our lives.

15. Discuss the H. G. Well's statement: "For man, there is no rest and no ending. He must go on, conquest beyond conquest; and when he has conquered all the deeps of space and all the mysteries of time, still he will be beginning."

Appendix A

Basic Mathematics

Some Things You Should Know After Reading This Appendix

You should be able to:

1. Add numbers algebraically.
2. Subtract numbers algebraically.
3. Multiply exponential numbers.
4. Divide exponential numbers.
5. Write numbers in scientific notation.
6. Solve problems using the factor-unit method.
7. Solve an algebraic equation for an unknown.
8. Solve a word equation for an unknown letter.

This appendix is taken from Alan Sherman and Sharon J. Sherman, *The Elements of Life: An Approach to Chemistry for the Health Sciences*, Prentice-Hall, Inc., Englewood Cliffs, N.J., 1980.

INTRODUCTION

You say that you're rusty in mathematics and that you need some review. Well, we've got just what you need; a brief review of the math that you'll need for this course. Just peruse the material and try the sample problems. See if it doesn't refresh your memory.

ADDING AND SUBTRACTING ALGEBRAICALLY

In the study of chemistry you will come into contact with positive and negative numbers. You may have to add and subtract these numbers, so remember the following rules.

ADDITION RULE

1. *When the signs are alike*   When the signs are the same, add the numbers and keep the same sign. For example,

$$+5 + 2 = +7$$
$$-5 - 2 = -7$$

2. *When the signs are different*   When the signs are not the same, subtract the numbers and keep the sign of the larger number. For example.

$$-5 + 2 = -3$$
$$+5 - 2 = +3$$

SUBTRACTION RULE

Change the sign of the *number being subtracted* and follow the rules for addition. (*Note*: Plus signs before positive numbers are usually left out, but minus signs before negative numbers are always written.) For example.

$$3 - (-2) \text{ becomes } \quad 3 + 2 = 5$$
$$3 - (+2) \text{ becomes } \quad 3 - 2 = 1$$
$$-3 - (+2) \text{ becomes } -3 - 2 = -5$$
$$-3 - (-2) \text{ becomes } -3 + 2 = -1$$

Example A-1   Add the following numbers algebraically.

(a) $25 - 60$ (b) $-25 - 60$

(c) $-25 + 60$ (d) $25 + 60$

Solution   Follow the rules for algebraic addition.

(a) $25 - 60 = -35$ (b) $-25 - 60 = -85$

(c) $-25 + 60 = 35$ (d) $25 + 60 = 85$

Example A-2 Subtract the following numbers algebraically.

(a) $30 - (+50)$ (b) $30 - (-50)$

(c) $-30 - (+50)$ (d) $-30 - (-50)$

Solution Follow the rules for algebraic subtraction.

(a) $30 - (+50)$ becomes $30 - 50 = -20$

(b) $30 - (-50)$ becomes $30 + 50 = 80$

(c) $-30 - (+50)$ becomes $-30 - 50 = -80$

(d) $-30 - (-50)$ becomes $-30 + 50 = 20$

EXPONENTS: THOSE NUMBERS ABOVE

Exponents are numbers written to the right of another number (called the *base number*) and above it. They tell you to perform an operation on the base number. For example, $(3)^2$ means to multiply 3 by itself: 3×3. Three is the base number and 2 is the exponent.

$$(3)^2 \longleftarrow \text{Exponent}$$
$$ \longleftarrow \text{Base number}$$

Sometimes exponents are seen as negative numbers. For example, $(3)^{-2}$ actually means $\dfrac{1}{(3)^2}$, or if the number is $\dfrac{1}{(3)^{-2}}$, this actually means $\dfrac{(3)^2}{1}$.

Let's solve some problems using exponents.

Example A-3 Perform the indicated operation.

(a) $(4)^2$ (b) $(5)^3$

(c) $(5)^1$ (d) $(2)^{-4}$

(e) $\dfrac{1}{(2)^5}$ (f) $\dfrac{1}{(3)^{-4}}$

Solution (a) $(4)^2 = 4 \times 4 = 16$

(b) $(5)^3 = 5 \times 5 \times 5 = 125$

(c) $(5)^1 = 5$

(d) $(2)^{-4} = \dfrac{1}{(2)^4} = \dfrac{1}{2 \times 2 \times 2 \times 2} = \dfrac{1}{16}$

489 *Basic Mathematics*

(e) $\dfrac{1}{(2)^5} = \dfrac{1}{2 \times 2 \times 2 \times 2 \times 2} = \dfrac{1}{32}$

(f) $\dfrac{1}{(3)^{-4}} = \dfrac{(3)^4}{1} = 3 \times 3 \times 3 \times 3 = 81$

There will be times when we have to multiply exponential numbers that have the same base. The rule for doing this is to *keep the base number and add the exponents algebraically*. Let's see how this is done.

Example A-4   Perform the indicated operation.

(a) $(3)^3(3)^1$ (b) $(3)^3(3)^{-1}$ (c) $(3)^{-3}(3)^1$

(d) $(3)^{-3}(3)^{-1}$ (e) $(2)^4(2)^1$ (f) $(2)^4(2)^{-1}$

Solution   We will keep the same base and add the exponents algebraically.

(a) $(3)^3(3)^1 = (3)^4$ (b) $(3)^3(3)^{-1} = (3)^2$

(c) $(3)^{-3}(3)^1 = (3)^{-2}$ (d) $(3)^{-3}(3)^{-1} = (3)^{-4}$

(e) $(2)^4(2)^1 = (2)^5$ (f) $(2)^4(2)^{-1} = (2)^3$

There will be times when we have to divide exponential numbers that have the same base. The rule for doing this is to *keep the base number and subtract the exponents algebraically*. Let's see how this is done.

Example A-5   Perform the indicated operation.

(a) $\dfrac{(3)^7}{(3)^2}$ (b) $\dfrac{(3)^7}{(3)^{-2}}$ (c) $\dfrac{(3)^{-7}}{(3)^2}$ (d) $\dfrac{(3)^{-7}}{(3)^{-2}}$

Solution   We will keep the same base and subtract the exponents algebraically.

(a) $\dfrac{(3)^7}{(3)^2} = (3)^{7-2} = (3)^5$

(b) $\dfrac{(3)^7}{(3)^{-2}} = (3)^{7-(-2)} = (3)^{7+2} = (3)^9$

(c) $\dfrac{(3)^{-7}}{(3)^2} = (3)^{-7-2} = (3)^{-9}$

(d) $\dfrac{(3)^{-7}}{(3)^{-2}} = (3)^{-7-(-2)} = (3)^{-7+2} = (3)^{-5}$

There will be times when we have to multiply or divide more than two exponential numbers. Let's try some of these examples.

Example A-6   Perform the indicated operation.

        (a) $(2)^3(2)^7(2)^4$ (b) $(2)^3(2)^7(2)^{-4}$

        (c) $\dfrac{(3)^2(3)^5}{(3)^3}$ (d) $\dfrac{(a)^8(a)^{-3}}{(a)^2(a)^7}$

Solution   (a) $(2)^3(2)^7(2)^4 = (2)^{14}$

        (b) $(2)^3(2)^7(2)^{-4} = (2)^6$

        (c) $\dfrac{(3)^2(3)^5}{(3)^3} = \dfrac{(3)^7}{(3)^3} = (3)^{7-3} = (3)^4$

        (d) $\dfrac{(a)^8(a)^{-3}}{(a)^2(a)^7} = \dfrac{(a)^5}{(a)^9} = (a)^{5-9} = (a)^{-4}$

WRITING NUMBERS AS POWERS OF 10: SCIENTIFIC NOTATION

In the physical sciences we often deal with numbers that are very large and other numbers that are very small: for example, 100,000,000,000 (a hundred billion) or 0.0000008 (eight ten-millionths). The number 100,000,000,000 can be written as 1×10^{11}, and 0.0000008 as 8×10^{-7}. In the first example we move the decimal point 11 places to the left.

$$1\underbrace{00,000,000,000.}_{} = 1 \times 10^{11}$$

This corresponds to multiplying by a positive power of 10. In the second example we moved the decimal point seven places to the right.

$$0.\underbrace{0000008}_{} = 8 \times 10^{-7}$$

This corresponds to multiplying by a negative power of 10. The exponent indicates the number of places that we moved the decimal point. The exponent is written as a positive number if the decimal point is moved to the left and a negative number if the decimal point is moved to the right. When you write numbers in scientific notation it is only necessary to write the nonzero numbers, since the zeros are incorporated into the exponent. (Of course, a zero appearing between two nonzero numbers must be written. For example, the number 805,000 is written as 8.05×10^5.) Also, the usual rule for writing numbers in scientific notation is to move the decimal point so that only one number remains to the left of the decimal point. Review the following examples and see if you get the idea.

Example A-7   Write the following numbers in scientific notation.

        (a) 350,000,000,000   (b) 807,000,000

        (c) 4,205,000,000   (d) 0.00000765

        (e) 0.00204   (f) 0.001008

Solution   (a) $350,000,000,000. = 3.5 \times 10^{11}$

        (b) $807,000,000. = 8.07 \times 10^{8}$

        (c) $4,205,000,000. = 4.205 \times 10^{9}$

        (d) $0.00000765 = 7.65 \times 10^{-6}$

        (e) $0.00204 = 2.04 \times 10^{-3}$

        (f) $0.001008 = 1.008 \times 10^{-3}$

THE FACTOR-UNIT METHOD, OR HOW TO WORK WITH UNITS

When you work with numbers in the physical sciences they are almost always accompanied by units. You should learn how to work with these units as well as with their numbers. For example, let's say that we wanted to change 4 feet into inches. You could probably do this in your head, but we're going to learn how to use a process called the factor-unit method. This method involves setting up the problem in the following manner.

What I want to know = (quantity given)(factor unit)

The idea is that when you multiply the "quantity given" by the "factor unit," the proper units cancel to give the desired quantity. Let's see how this works.

Example A-8   Change 4 feet into inches by the factor-unit method.

Solution   In this problem the factor unit is 12 inches = 1 foot, which can be written in two ways:

$$\frac{12 \text{ inches}}{1 \text{ foot}} \quad \text{or} \quad \frac{1 \text{ foot}}{12 \text{ inches}}$$

Our job is to choose the proper factor unit so that our answer comes out in feet. We proceed as follows:

What I want to know = (quantity given)(factor unit)

$$? \text{ inches} = (4 \text{ feet})\left(\frac{12 \text{ inches}}{1 \text{ foot}}\right)$$

$$= 48 \text{ inches}$$

Notice how the term "feet" cancels to give inches. If we had chosen the other factor unit, the terms wouldn't have canceled. Let's do it so you can see for yourself.

$$? \text{ inches} = (4 \text{ feet})\left(\frac{1 \text{ foot}}{12 \text{ inches}}\right)$$

$$= 0.33 \frac{\text{feet}^2}{\text{inches}}$$

The units don't cancel and we don't get the answer *inches*.

In doing problems by the factor-unit method, always write the factor unit so that the term you want is in the numerator, and the term you want to cancel is in the denominator of the factor unit. See what we do in the next example.

Example A-9   Change 60 inches into feet by the factor-unit method.

Solution   Again, the factor unit is 12 inches = 1 foot.

What I want to know = (quantity given)(factor unit)

$$? \text{ feet} = (60 \text{ inches})\left(\frac{1 \text{ foot}}{12 \text{ inches}}\right)$$

$$= 5 \text{ feet}$$

Example A-10   Solve each of the following conversions by the factor-unit method.

(a) 27 feet = ? yards
(b) 0.5 yard = ? feet

(c) $\dfrac{60 \text{ miles}}{1 \text{ hour}} = \dfrac{? \text{ feet}}{1 \text{ second}}$

Solution   (a) The factor unit we must know is 3 feet = 1 yard.

What I want to know = (quantity given)(factor unit)

$$? \text{ yards} = (27 \text{ feet})\left(\frac{1 \text{ yard}}{3 \text{ feet}}\right)$$

$$? \text{ yards} = 9 \text{ yards}$$

(b) Again, the factor unit is 3 feet = 1 yard.

What I want to know = (quantity given)(factor unit)

$$? \text{ feet} = (0.5 \text{ yard})\left(\frac{3 \text{ feet}}{1 \text{ yard}}\right)$$

$$? \text{ feet} = 1.5 \text{ feet}$$

(c) In this problem we need two factor units: one to change miles to feet, and one to change hours to seconds. These factor units are

$$1 \text{ mile} = 5280 \text{ feet}$$

$$1 \text{ hour} = 3600 \text{ seconds}$$

We set the conversion up as follows:

What I want to know = (quantity given)(factor units)

$$\frac{? \text{ feet}}{1 \text{ second}} = \left(\frac{60 \text{ miles}}{1 \text{ hour}}\right)\left(\frac{5280 \text{ feet}}{1 \text{ mile}}\right)\left(\frac{1 \text{ hour}}{3600 \text{ seconds}}\right)$$

$$= \frac{88 \text{ feet}}{1 \text{ second}}$$

SOLVING FOR THE UNKNOWN IN AN ALGEBRAIC EQUATION

In your study of chemistry you will have to know how to solve simple algebraic equations. The following examples will help you to refresh your memory.

Example A-11   Solve each of the following equations for the unknown quantity.

(a) $4a = 8$ (b) $6x + 5 = 53$

(c) $8y - 3 = 69$ (d) $\dfrac{5b}{3} = \dfrac{10}{4}$

Solution   (a) $4a = 8$ Isolate the unknown by dividing each side of the equation by 4.

$$\frac{4a}{4} = \frac{8}{4}$$

$$a = 2$$

(b) $6x + 5 = 53$ Remove the 5 from the left-hand side of the equation by subtracting five from both sides of the equation.

$$6x + 5 - 5 = 53 - 5$$

$$6x = 48$$

Now divide each side of the equation by 6.

$$\frac{\cancel{6}x}{\cancel{6}} = \frac{48}{6}$$

$$x = 8$$

(c) $8y - 3 = 69$ Remove the 3 from the left-hand side of the equation by adding three to both sides of the equation.

$$8y - 3 + 3 = 69 + 3$$
$$8y = 72$$

Now divide both sides of the equation by 8.

$$\frac{\cancel{8}y}{\cancel{8}} = \frac{72}{8}$$

$$y = 9$$

(d) $\dfrac{5b}{3} = \dfrac{10}{4}$ Remove the 3 from the left-hand side of the equation by multiplying both sides of the equation by 3.

$$(\cancel{3})\left(\frac{5b}{\cancel{3}}\right) = \left(\frac{10}{4}\right)(3)$$

$$5b = \frac{30}{4}$$

Remove the 5 from the left-hand side of the equation by dividing both sides of the equation by 5.

$$\frac{\cancel{5}b}{\cancel{5}} = \frac{30}{(4)(5)}$$

$$b = \frac{30}{20} = 1.5$$

Many of the mathematical equations we use in chemistry are word equations, such as

$$\text{density} = \frac{\text{mass}}{\text{volume}}$$

which is usually abbreviated as

$$D = \frac{m}{V}$$

As written, the equation is set up so that we can solve for the density of a substance if we are given its mass and volume. However, in some problems we are given the density and volume of a substance and asked to solve for its mass. In other problems we may be given the density and mass of a substance and asked to solve for its volume. What do we do then?

Example A-12   Given $D = \dfrac{m}{V}$ solve this equation for m.

Solution $D = \dfrac{m}{V}$ To isolate the m, multiply both sides of the equation by V.

$$(D)(V) = \left(\frac{m}{\cancel{V}}\right)(\cancel{V})$$

$$(D)(V) = m$$

Example A-13   Given $D = \dfrac{m}{V}$ solve this equation for V.

Solution   The V is in the denominator, so we must get it into the numerator. We can do this by multiplying each side of the equation by V. We've already done this in Example A-12 and obtained

$$(D)(V) = m$$

To isolate the V, we must divide each side of the equation by D.

$$\frac{(\cancel{D})(V)}{(\cancel{D})} = \frac{m}{D}$$

$$V = \frac{m}{D}$$

SUMMARY

In these few pages we've reviewed the basic mathematics that you will need in your chemistry course. Our purpose here was to refresh your memory and give you a place to look up some basic math rules in case you get stuck. If you want to be sure that you've mastered the topics in this appendix, try the exercises that follow.

EXERCISES

1. Add the following numbers algebraically.
   (a) $18 + 35$ (b) $-64 - 13$ (c) $-20 + 59$
   (d) $30 - 19$ (e) $19 - 30$

2. Subtract the following numbers algebraically.
   (a) $18 - (+35)$ (b) $-64 - (-13)$ (c) $-20 - (+59)$
   (d) $30 - (-19)$ (e) $19 - (-30)$

3. Perform the indicated operation.
   (a) $(3)^2$ (b) $(4)^3$ (c) $(6)^1$

   (d) $(3)^{-4}$ (e) $\dfrac{1}{(2)^4}$ (f) $\dfrac{1}{(2)^{-4}}$

4. Perform the indicated operation.
   (a) $(2)^3(2)^1$ (b) $(2)^3(2)^{-1}$ (c) $(2)^{-3}(2)^1$ (d) $(2)^{-3}(2)^{-1}$

5. Perform the indicated operation.

   (a) $\dfrac{(2)^5}{(2)^8}$ (b) $\dfrac{(2)^5}{(2)^{-8}}$ (c) $\dfrac{(2)^{-5}}{(2)^8}$ (d) $\dfrac{(2)^{-5}}{(2)^{-8}}$

6. Perform the indicated operation.

   (a) $\dfrac{(3)^5(3)^7}{(3)^2}$ (b) $(4)^8(4)^{-5}$

   (c) $\dfrac{(2)^{14}(2)^{-10}}{(2)^3(2)^4}$ (d) $\dfrac{(y)^{-5}(y)^3}{(y)^{-2}(y)^5}$

7. Write the following numbers in scientific notation.
   (a) 789,000,000,000,000   (b) 0.0000001003
   (c) 3,050,000   (d) 0.0021

8. Perform the indicated operation and leave the answer in scientific notation.
   (a) $(3 \times 10^5)(2 \times 10^3)$ (b) $(3 \times 10^5)(2 \times 10^{-3})$

   (c) $(3 \times 10^{-5})(2 \times 10^3)$ (d) $(3 \times 10^{-5})(2 \times 10^{-3})$

   (e) $\dfrac{8 \times 10^9}{2 \times 10^5}$ (f) $\dfrac{8 \times 10^9}{2 \times 10^{-5}}$

   (g) $\dfrac{8 \times 10^{-9}}{2 \times 10^5}$ (h) $\dfrac{8 \times 10^{-9}}{2 \times 10^{-5}}$

9. Solve each of the following problems by the factor-unit method.
   (a) 32 pints = ? quarts (*Hint*: 2 pints = 1 quart.)
   (b) 21,120 feet = ? miles (*Hint*: 1 mile = 5280 feet.)
   (c) 25 yards = ? feet = ? inches (*Hint*: 1 yard = 3 feet,
       1 foot = 12 inches.)
   (d) 500 cm = ? m (*Hint*: This is a metric conversion. We are going to
       discuss the metric system in Appendix B. However, see if you can
       solve this problem with the following hint: 1 m = 100 cm.)

10. Solve each of the following equations for the unknown quantity.
    (a) $3x = 21$ (b) $5a + 15 = 45$

    (c) $2c + 3 = 24 - 5c$ (d) $\dfrac{8m}{5} = \dfrac{160}{2}$

11. Given the formula $PV = nRT$, solve for each of the following in terms of the other letters.
    (a) $P =$ (b) $V =$
    (c) $n =$ (d) $R =$
    (e) $T =$

Appendix B

The Metric System and Measurement

Some Things You Should Know After Reading This Appendix

You should be able to:

1. Convert between the units of mm, cm, dm, m, and km.
2. Convert between the units of mg, cg, dg, g, and kg.
3. Convert between the units of ml, cl, dl, liters, and kl.
4. Convert from a metric unit to a corresponding English unit, and vice versa.
5. Use the density formula, $D = \dfrac{m}{V}$ to calculate the density, mass, or volume of a substance when given the other two values.
6. Convert Celsius temperatures to Fahrenheit, and vice versa.
7. Define exothermic and endothermic reactions.
8. Define enthalpy and calorie.
9. Calculate the calories absorbed by a sample of water when given its mass and temperature change.
10. Calculate the ΔH of a substance when given the experimental data.
11. Convert the ΔH of a substance from kcal/mole to kcal/g, and vice versa.
12. Know how to use significant figures.

This appendix is taken from Alan Sherman and Sharon J. Sherman, *The Elements of Life: An Approach to Chemistry for the Health Sciences*, Prentice-Hall, Inc., Englewood Cliffs, N.J., 1980.

INTRODUCTION

The *metric system* was developed by the French in the late 1700's. The metric system is currently used in almost all parts of the world and in just about every branch of science. It is a very easy system to use since it is based on units of 10, much like our American money system. This means, for example, that conversion between one metric unit of length and another is simply a move of the decimal point (in other words, a multiplication or division of a power of 10). This is much simpler than the English system of measurement, where if you want to convert from one unit of length to another you must *memorize* all the "odd-ball" conversion factors (for example, 12 inches = 1 foot, 3 feet = 1 yard, 5280 feet = 1 mile). As we review the metric system over the next few pages you will see just how easy it is.

THE METRIC SYSTEM: A SHORT COURSE

The metric system—like the English system—has units of length, mass, and volume. These are the *meter, gram,* and *liter*, respectively. For a mass that is smaller than a gram, we simply use a prefix (Tables B-1 and B-2). A milligram (mg) is equal to one one-thousandth of a gram (1/1000 or 0.001 g). A kilogram (kg) is a thousand times as large as a gram and is therefore equal to 1000 g. The following examples show how easy it is to convert from one metric unit to another.

TABLE B-1   *Prefixes and abbreviations used in the metric system*

nano- = 0.000000001	nanometer = nm
micro- = 0.000001	micrometer = μm (Greek lowercase mu)
milli- = 0.001	millimeter = mm
centi- = 0.01	milliliter = ml
deci- = 0.1	milligram = mg
deca- = 10	centimeter = cm
kilo- = 1000	centigram = cg
	decimeter = dm
	decigram = dg
	kilometer = km
	kilogram = kg

TABLE B-2 *The metric system*

LENGTH

1 millimeter = 0.001 meter = 1/1000 meter	1 meter = 1000 millimeters
1 centimeter = 0.01 meter = 1/100 meter	1 meter = 100 centimeters
1 decimeter = 0.1 meter = 1/10 meter	1 meter = 10 decimeters
1 kilometer = 1000 meters	1 meter = 0.001 kilometer

MASS

1 microgram = 0.000001 gram	1 gram = 1,000,000 micrograms
1 milligram = 0.001 gram	1 gram = 1000 milligrams
1 centigram = 0.01 gram	1 gram = 100 centigrams
1 decigram = 0.1 gram	1 gram = 10 decigrams
1 kilogram = 1000 grams	1 gram = 0.001 kilogram

VOLUME

1 milliliter = 0.001 liter	1 liter = 1000 milliliters
1 milliliter = 1 cubic centimeter (cubic centimeter is abbreviated cc or cm^3)	1 liter = 1000 cubic centimeters
1 centiliter = 0.01 liter	1 liter = 100 centiliters
1 deciliter = 0.1 liter	1 liter = 10 deciliters
1 kiloliter = 1000 liters	1 liter = 0.001 kiloliter

Example B-1 Change 8 grams to milligrams.

Solution We will use the factor-unit method that we discussed in Appendix A. The factor unit that we *must* know is

$$1 \text{ gram} = 100 \text{ milligrams} \quad \text{(Table B-2)}$$

$$? \text{ milligrams} = (8 \cancel{\text{ grams}})\left(\frac{1000 \text{ milligrams}}{1 \cancel{\text{ gram}}}\right)$$

$$= 8000 \text{ milligrams}$$

Example B-2 Change 3500 centimeters to meters.

Solution We'll use the same method as before, only this time we'll use abbreviations for the units (in other words, m for meters and cm for centimeters). The factor unit we *must* know is

$$1 \text{ m} = 100 \text{ cm} \quad \text{(Table B-2)}$$

$$? \text{ m} = (3500 \text{ cm})\left(\frac{1 \text{ m}}{100 \text{ cm}}\right)$$

$$= 35 \text{ m}$$

As you can see, this isn't really too difficult. Let's try some additional examples.

Example B-3 Change 35,000 ml into

(a) liters   (b) centiliters   (c) deciliters   (d) kiloliters

Solution First, we'll change the 35,000 ml into liters. Then we'll change the liters into cl, dl, and kl. The reason that we'll do it in this order is because we have the necessary conversion factors in Table B-2.

(a) The factor unit we *must* know is

$$1 \text{ liter} = 1,000 \text{ ml}$$

$$? \text{ liters} = (35,000 \text{ ml})\left(\frac{1 \text{ liter}}{1000 \text{ ml}}\right)$$

$$= 35 \text{ liters}$$

(b) The factor unit we *must* know is

$$1 \text{ liter} = 100 \text{ cl}$$

$$? \text{ cl} = (35 \text{ liters})\left(\frac{100 \text{ cl}}{1 \text{ liter}}\right)$$

$$= 3500 \text{ cl}$$

(c) The factor unit we *must* know is

$$1 \text{ liter} = 10 \text{ dl}$$

$$? \text{ dl} = (35 \text{ liters})\left(\frac{10 \text{ dl}}{1 \text{ liter}}\right)$$

$$= 350 \text{ dl}$$

(d) The factor unit we *must* know is

$$1 \text{ kl} = 1000 \text{ liters}$$

$$? \text{ kl} = (35 \text{ liters})\left(\frac{1 \text{ kl}}{1000 \text{ liters}}\right)$$

$$= 0.035 \text{ kl}$$

TABLE B-3 *Conversion factors for English-metric and metric-English conversions*

LENGTH	MASS	VOLUME
1 inch = 2.54	1 ounce = 28.35 g	1 liter = 1.06 quarts
1 foot = 0.30 m	1 pound = 454 g	

Note: The conversions in this table can be used to go from English to metric or metric to English simply by writing the conversions as factor units, for example, $\dfrac{1 \text{ inch}}{2.54 \text{ cm}}$ or $\dfrac{2.54 \text{ cm}}{1 \text{ inch}}$.

Example B-4   Change 255 mg to dg.

Solution   First, we'll change the 255 mg to grams. Then we'll change the grams into dg. (Of course, you may go directly from mg to dg if you know that there are 100 mg in 1 dg.)

$$? \text{ g} = (255 \text{ mg})\left(\frac{1 \text{ g}}{1000 \text{ mg}}\right)$$

$$= 0.255 \text{ g}$$

$$? \text{ dg} = (0.255 \text{ g})\left(\frac{10 \text{ dg}}{1 \text{ g}}\right)$$

$$= 2.55 \text{ dg}$$

We can now see that conversion between units of the metric system is easy. But what if you want to convert metric into English units, or vice versa? Table B-3 should help you in this task. Let's use this table and the factor-unit method to do some English-metric and metric-English conversions.

Example B-5   Convert 12 inches into centimeters.

Solution   Table B-3 tells us that 1 inch = 2.54 cm.

$$? \text{ cm} = (12 \text{ inches})\left(\frac{2.54 \text{ cm}}{1 \text{ inch}}\right)$$

$$= 30.5 \text{ cm}$$

Example B-6   Convert 2270 grams into pounds.

Solution   Table B-3 tells us that 1 pound = 454 grams.

$$? \text{ pounds} = (2270 \text{ g})\left(\frac{1 \text{ pound}}{454 \text{ g}}\right)$$

$$= 5 \text{ pounds}$$

503   *The Metric System and Measurement*

Although it is always handy to be able to convert between systems, it is more important for you to be able to associate metric measurements with objects in the world around you. You've got to begin to *think metric*. Only in this way will the metric system mean something to you. Here are some things to think about to help you begin to think metric.

A meter stick is slightly longer than a yard stick (about $3\frac{3}{8}$ inches longer).

A liter is just a little more than a quart. (Think of a quart milk container; a liter of milk would be almost the same size.)

A standard-size paper clip weighs about 1 gram.

DENSITY: AN INTERESTING PROPERTY OF MATTER

Which is heavier: glass, iron, or wood from an oak tree? Naturally, it depends on the size of each piece. However, what if all three were the same size? In other words, what if we got hold of cubes of glass, iron, and oak wood, each with a volume of 1 cm³? (Remember, a cube with a volume of 1 cm³ would be 1 cm long, 1 cm wide, and 1 cm high; see Fig. B-1.) Let's say that we weigh each cube to determine which is the heaviest. We find that 1 cm³ of iron weighs 7.9 g, 1 cm³ of glass weighs 2.4 g, and 1 cm³ of oak wood weighs 0.6 g. We conclude that, for a *particular volume*, the iron has the greatest mass. We have now developed the idea of *density*.

Density is the mass per unit volume of a substance. Mathematically, we define density as

$$D = \frac{m}{V}$$

where D is density, m is mass, and V is volume. If we find the mass of our substance in grams and the volume in cubic centimeters, then the units of density become

$$\frac{\text{grams}}{\text{cubic centimeter}} \quad \text{or} \quad \frac{\text{g}}{\text{cm}^3}$$

Let's see how we would calculate the density of the object in the following example.

Figure B-1   Three cubes of equal volume.

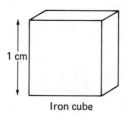

Iron cube

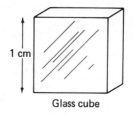

Glass cube

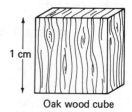

Oak wood cube

Example B-7 A block of the element barium is 5 cm long, 3 cm wide, and 1 cm high, and it weighs 52.5 g. What is the density of the element barium?

Solution First, calculate the volume of the block.

$$\text{Volume} = \text{length} \times \text{width} \times \text{height}$$

$$V = 5 \text{ cm} \times 3 \text{ cm} \times 1 \text{ cm} = 15 \text{ cm}^3$$

We now know that $V = 15 \text{ cm}^3$ and $m = 52.5$ g, so all we have to do is solve the density formula for D.

$$D = \frac{m}{V} = \frac{52.5 \text{ g}}{15 \text{ cm}^3} = 3.5 \frac{\text{g}}{\text{cm}^3}$$

(which also can be thought of as $\dfrac{3.5 \text{ g}}{1 \text{ cm}^3}$, which reads that there is 3.5 g for every cubic centimeter of barium.)

Example B-8 Suppose you are told that 400 g of alcohol occupies a volume of 500 ml. What is the density of the alcohol?

Solution In this problem we are given the mass and volume of the substance whose density we are asked to find. But note that the volume of the alcohol is given in milliliters (ml). Remember that in the metric system 1 *milliliter is equal to* 1 *cubic centimeter*. That is, 1 ml = 1 cm^3 (look back at Table B-2). So it's okay to interchange the terms ml and cm^3. Now let's solve this problem.

$$D = \frac{m}{V}$$

$$= \frac{400 \text{ g}}{500 \text{ cm}^3}$$

$$= 0.8 \frac{\text{g}}{\text{cm}^3}$$

Sometimes you may know the volume and density of an object and you may be asked to calculate its mass. Let's see how we use the density formula to do this.

Example B-9 A piece of aluminum in the shape of a perfect cube measures 10 cm on each side. The density of aluminum is 2.7 g/cm^3. What is the mass of this block of aluminum?

Solution First calculate the volume of the block.

$$\text{Volume} = \text{length} \times \text{width} \times \text{height}$$

$$V = (10 \text{ cm})(10 \text{ cm})(10 \text{ cm}) = 1000 \text{ cm}^3$$

We now know that the volume is 1000 cm³ and the density is 2.7 g/cm³. We may now solve the density formula for m, plug in the numbers, and obtain the mass of the aluminum.

$$D = \frac{m}{V}; \quad \text{therefore, } (D)(V) = m$$

$$m = (D)(V)$$

$$= \left(2.7 \frac{g}{\cancel{cm^3}}\right)(1000 \ \cancel{cm^3})$$

$$= 2700 \text{ g}$$

Sometimes you may be given the mass and density of an object and be asked to solve for the volume. Let's see how we do this.

Example B-10 The density of blood serum is about 1.014 g/cm³. What would be the volume of 5070 g of blood serum (which is about the mass of blood in an adult)?

Solution Let's solve the density formula for V.

$$D = \frac{m}{V}; \quad \text{therefore, } (D)(V) = m \text{ and finally } V = \frac{m}{D}$$

$$V = \frac{m}{D}$$

$$= \frac{5070 \ \cancel{g}}{1.014 \ \cancel{g}/cm^3}$$

$$= 5000 \text{ cm}^3 \quad (\text{or 5 liters})$$

TEMPERATURE: THE INTENSITY OF HOTNESS OR COLDNESS

When we light an oven and the gas starts to burn, we know that heat energy is being generated. How can we find out how intense this heat is? Well, naturally, when we want to know how hot something is, we measure its temperature. What we find out by measuring the temperature is the *intensity* of the object's hotness or coldness. Temperature tells us nothing about the *quantity* of heat that has entered into a substance or a place; we will learn about that later in this supplement.

In 1724, a German scientist by the name of Gabriel Daniel Fahrenheit devised a way of measuring the heat intensity of various substances. He used a sealed tube containing mercury (a thermometer). When he placed the thermometer in a substance, the mercury in the thermometer would move up or down the tube. To give some meaning to the height of the mercury column,

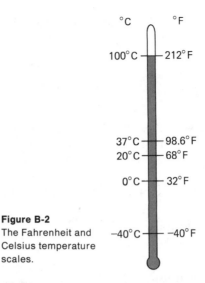

Figure B-2
The Fahrenheit and
Celsius temperature
scales.

Fahrenheit set up a temperature scale. On his scale, the temperature of a mixture of ice and salt water is zero degrees. The temperature of a mixture of ice and water is 32 degrees (the freezing point of water), and the temperature of a water–steam mixture (at normal atmospheric pressure) is 212 degrees (which is called the normal boiling point of water). Notice that on the Fahrenheit scale there are 180 equal divisions between the freezing point and boiling point of water. Each division corresponds to a degree Fahrenheit. But the Fahrenheit scale, though very useful, was soon replaced by a new temperature scale devised by a Swedish astronomer named Anders Celsius.

Celsius devised his Celsius or centigrade temperature scale in 1742. Scientists liked it better than Fahrenheit's, since the Celsius scale uses zero degrees as the freezing point of water and 100 degrees as the boiling point of water. There are 100 equal divisions between these two points; each division is called 1 degree Celsius (or centigrade) (Fig. B-2).

Scientists throughout the world use the Celsius thermometer to measure temperatures, mainly because of the convenience of the 100-degree scale. It fits right in with the metric system of measurement, as you can see. Just travel outside the United States and ask any health practitioner, "What's normal body temperature?" Don't be surprised when he or she tells you that it's 37! That's 37 degrees Celsius (37°C), of course.

Converting Celsius and Fahrenheit Degrees

As we were saying, on the Celsius scale there are 100 divisions between the boiling point and the freezing point of water. Therefore, 100 Celsius degrees are equal to 180 Fahrenheit degrees, or $1°C = 1.8°F$. The formula you may use to convert from one system to another is

$$9°C = 5°F - 160$$

[Some instructors prefer to use the following two formulas for temperature conversions.

$$°F = (1.8 \times °C) + 32$$

and
$$°C = \frac{°F - 32}{1.8}$$

You may use these two formulas if you wish, but we'll use our formula in the following examples.] That looks pretty odd, but let's use this formula to convert from Fahrenheit to Celsius and Celsius to Fahrenheit degrees.

Example B-11 Convert 68°F (average room temperature) into °C.

Solution Substitute 68°F into the conversion formula and solve for degrees Celsius.

$$9°C = 5°F - 160$$
$$= (5)(68°F) - 160$$
$$°C = \frac{(5)(68) - 160}{9}$$
$$= \frac{340 - 160}{9}$$
$$= \frac{180}{9}$$
$$= 20$$

Therefore, 68°F = 20°C.

Example B-12 Convert 37°C into °F.

Solution Substitute 37°C into the conversion formula and solve for degrees Fahrenheit.

$$9°C = 5°F - 160$$
$$(9)(37°C) = 5°F - 160$$
$$(9)(37) + 160 = 5°F$$
$$\frac{(9)(37) + 160}{5} = °F$$
$$\frac{333 + 160}{5} = °F$$
$$\frac{493}{5} = °F$$
$$98.6 = °F$$

Therefore, 37°C = 98.6°F (normal body temperature).

HEAT AND CHEMICAL REACTIONS

Every substance—foods, minerals, and even people—has a specific amount of energy associated with it: a sort of chemical potential energy. When a substance reacts to form a new substance, it can either lose energy to the surroundings or gain energy from the surroundings. This is because, in chemical reactions, old bonds between atoms are broken and new bonds are formed. Breaking a bond usually requires additional energy, while forming a bond usually releases energy. Depending on the number of bonds broken and the number of bonds formed, as well as the strength of the bonds, energy is either released to the surroundings or absorbed from them.

Reactions that release energy to the surroundings are called *exothermic reactions.* For example, the explosion of a stick of dynamite is a chemical reaction that releases a tremendous amount of heat energy to the environment. Reactions that absorb energy from the surroundings are called *endothermic reactions.* For example, plants live and grow by a process called photosynthesis. In this process the plants use carbon dioxide, water, and *energy from the sun* (sunlight) to make glucose and oxygen.

We can usually calculate the energy changes that result from reactions by measuring the *heat* released or absorbed during the reaction. This is because each chemical substance has a certain heat content associated with it. The heat content is called *enthalpy,* and has the symbol *H.* One mole of a chemical substance has a definite heat content (enthalpy). (If you've forgotten what a mole is, see Chap. 2.) Although we can't measure the heat content of specific substances, we can measure the *change* in heat content during chemical reactions. The change in heat content during a chemical reaction is called the *heat of reaction.* We measure this heat in units of *calories. One calorie is the amount of heat needed to raise the temperature of one gram of water 1°C.* Large heat changes are measured in *kilocalories* (1000 cal = 1 kcal).

Example B-13 You have a liter of water (1000 g) that you heat from 20°C to 80°C. How many calories of heat energy did the water absorb?

Solution Based on our definition, every gram of water that we heat 1°C gains a calorie of heat energy. Therefore, we can derive an equation that says:

(for water) calories = (grams of water)(change in °C) or in abbreviated form, cal = $(m)(\Delta t)$. The symbol Δ means "change in."

$$? \text{ cal} = (m)(\Delta t)$$
$$= (1000 \text{ g})(80°C - 20°C)$$
$$= (1000 \text{ g})(60°C)$$
$$= 60,000 \text{ cal} \quad \text{or} \quad 60 \text{ kcal}$$

As you can see, it's easy to measure the calories of heat gained by a sample of water. But how can we measure the change in heat content of other substances

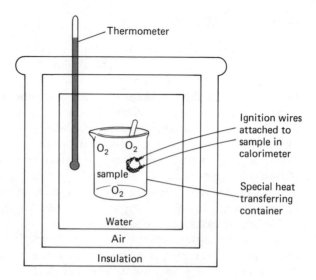

Figure B-3   A calorimeter.

when they undergo chemical reactions? We can, by using a device called a *calorimeter* (Fig. B-3). The calorimeter is frequently used to measure changes in the heat content of substances (for example, foods) as they react with oxygen. The way such a determination is carried out is as follows:

We want to measure the *heat of reaction* of glucose, as it reacts with oxygen. A sample of glucose is obtained, for example 1.8 g. It is placed inside the special heat-transferring container that fits into the calorimeter. The container is positioned in the calorimeter and is attached to an ignition switch which will be used to cause a spark that will allow the sugar and oxygen to react. The calorimeter is filled with a known amount of water, for example 1000 g. The temperature of the water is measured and is found to be 25°C. The ignition switch is pressed and a great amount of heat is liberated by the reaction between glucose and oxygen.

$$C_6H_{12}O_6 + 6O_2 \longrightarrow 6CO_2 + 6H_2O + \text{energy}$$

The heat energy passes through the special heat-transferring container to the water, whose temperature increases to 31.7°C. With this information we can now calculate the calories of heat liberated by the 1.8-g sample of glucose. We proceed as follows:

$$? \text{ cal (absorbed by water)} = (m_{\text{water}})(\Delta t_{\text{water}})$$
$$= (1000 \text{ g})(31.7°C - 25°C)$$
$$= (1000 \text{ g})(6.7°C)$$
$$= 6700 \text{ cal} \quad \text{or} \quad 6.7 \text{ kcal}$$

Therefore, the 1.8-g sample of glucose liberated 6.7 kcal of heat energy to the water. Therefore, the 1.8-g sample of glucose lost 6.7 kcal of energy. We usually like to express *heats of reaction* in kcal per mole. The molecular weight of glucose is 180. So we can obtain the *heat of reaction* in kcal/mole as follows:

$$? \frac{\text{kcal}}{\text{mole}} = \left(\frac{6.7 \text{ kcal}}{1.8 \text{ g}} \right) \left(\frac{180 \text{ g}}{1 \text{ mole}} \right)$$

$$= 670 \frac{\text{kcal}}{\text{mole}}$$

The symbol for heat content (enthalpy) is H. We represent the *change* in heat content by the symbol ΔH. For exothermic reactions the ΔH value is always preceded by a *minus* sign. For endothermic reactions the ΔH has a *positive* value. In our example of glucose we express the heat of reaction as follows:

$$\Delta H = -670 \text{ kcal/mole}$$

indicating it was an exothermic reaction.

The nutritional value of many food substances are expressed in kcal/mole. However, sometimes they are also expressed in kcal per gram (kcal/g).

Example B-14   The nutritional value of glucose is 670 kcal/mole. Express this in kcal/g.

Solution   We will use the factor-unit method. We must also know the molecular weight of glucose, which is 180. In other words, 1 mole of glucose = 180 g

$$? \frac{\text{kcal}}{\text{g}} = \left(\frac{670 \text{ kcal}}{1 \text{ mole}} \right) \left(\frac{1 \text{ mole}}{180 \text{ g}} \right) = 3.7 \frac{\text{kcal}}{\text{g}}$$

SIGNIFICANT FIGURES

When we weigh an object on a balance, we make a measurement. We measure the mass of the object and get a numerical or quantitative description of its mass. But how accurate *is* this number that we get? No measuring instrument is perfect, so the accuracy of our number depends on the accuracy of our measuring device.

When chemists perform an experiment, they repeat it as many times as possible. If you were told to do an experiment five times, you might get five slightly different results. If the results were numerical, you would have to evaluate them in terms of *precision* and *accuracy*. You may think that these two terms mean the same thing, but they don't.

To understand the difference between precision and accuracy, think of a game of darts. Accuracy depends on how close the darts are to the bullseye. Precision depends on how close the darts are to each other. You may throw five darts that land very close together, but are in an outer ring. This would show precision, but very little accuracy (with respect to hitting the bullseye). If one dart landed right on the bullseye and the rest of the darts landed in various places on the dart board, then you would have one accurate result, but little precision. Of course, if all five darts hit the bullseye, your score would be excellent in both precision *and* accuracy.

Suppose you are doing an experiment measuring the temperature of a liquid, and suppose the thermometer you are using is calibrated, or marked off, only in whole degrees. Imagine that the mercury in the thermometer is halfway between 34 and 35 degrees (Fig. B-4(a)). You can then guess that the temperature of the liquid is 34.5 degrees.

Now you pick up another thermometer, which is calibrated in *tenths* of degrees (Fig. B-4(b)). When you read this thermometer, it looks as if the temperature of the liquid is 34.55 degrees. It seems that there is a difference in the sensitivity of the two thermometers, but which measurement is correct?

To answer this question, we look at *significant figures*, which are numbers that give the greatest amount of correct, measured information and no information that is in doubt. The temperature 34.5 degrees given by the first thermometer is correct to three significant figures. The temperature 34.55 degrees given by the second thermometer is correct to four significant figures. In other words, both measurements are correct, but one is more precise than the other.

Let's do another experiment to see how significant figures play an important role in measurement. Suppose that we have to measure a certain piece of glass to find its perimeter (we do this by measuring the four edges of the glass and adding the lengths). We have two rulers, one calibrated in centimeters and the other in millimeters. We set out to measure each edge of the glass with the

Figure B-4 (a) We can measure the temperature of a liquid using a thermometer calibrated only in degrees. (b) We can also measure the temperature of a liquid using a thermometer calibrated in tenths of degrees.

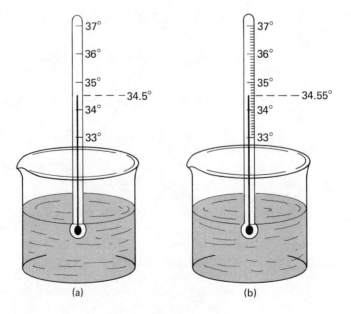

(a)   (b)

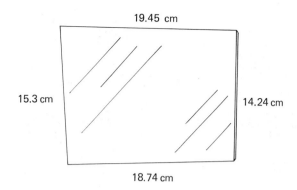

Figure B-5 Measurement of a piece of glass.

more precise ruler (the one calibrated in millimeters). However, when we get to the fourth side of the glass, we absent-mindedly pick up the less precise ruler (the one calibrated in centimeters). Looking at the glass, shown in Figure B-5, we can see that three edges are measured to two decimal places and the fourth is measured to one decimal place. What is the perimeter of the glass? Is it 67.73 cm or 67.7 cm? The answer is 67.7 cm. We can report only what we know for sure, and that is the answer to one decimal place. *Our least precise measurement determines the number of decimal places in our answer.*

Whenever we *add* or *subtract* measured quantities, we report answers in terms of the least precise measurement. We do this by rounding off to the least number of significant figures. This means that our answer must have no more decimal places than our least precise quantity.

Example B-15 Add 18.7444 and 13, and use the appropriate number of significant figures in your answer.

Solution The 13 is the less precise measurement; therefore we add and round to whole numbers.

$$
\begin{array}{r}
18.7444 \\
+\,13 \\
\hline
31.7444
\end{array}
$$ Round off to 32 (two significant figures).

Example B-16 Subtract 0.12 from 48.743, and use the appropriate number of significant figures in your answer.

Solution The 0.12 is the less precise number, since we know it to only two decimal places. Therefore we subtract and round off to two decimal places.

$$
\begin{array}{r}
48.743 \\
-\,0.12 \\
\hline
48.623
\end{array}
$$ Round off to 48.62 (four significant figures).

Here are some rules for significant figures which are useful in *multiplying and dividing* measured quantities.

RULE 1. Count the number of significant figures in each of the quantities to be multiplied or divided.

RULE 2. Report the answer to the least number of significant figures as determined by following rule 1.

Example B-17 What is the area of a square, if the length of a side is measured as 1.5 cm?

Solution The area of a square is equal to the length of its side squared, or

$$A = s \times s = 1.5 \text{ cm} \times 1.5 \text{ cm} = 2.25 \text{ cm}^2 \qquad \text{Round off to } 2.3 \text{ cm}^2.$$

(We can report only to two significant figures, since the measurement had only two significant figures.)

Example B-18 Using significant figures, divide 20.8 by 4.

Solution Divide and round off your answer to one significant figure.

$$\frac{20.8}{4} = 5.2 \qquad \text{Round off to } 5.$$

Example B-19 Using significant figures, multiply 20.8 by 4.1.

Solution Multiply and round off your answer to two significant figures.

$$20.8 \times 4.1 = 85.28 \qquad \text{Round off to } 85.$$

Another problem arises when we use significant figures: What do we do about zeros? Are they significant figures, or not? Here are some helpful rules. Read them very carefully, and then try to do Example B-20.

1. Zeros *between* nonzero digits are significant. (*Digit* is another word for figure.) 4.004 has four significant figures.

2. Zeros to the *left* of nonzero digits are *not* significant, because these zeros are only showing the position of the decimal point.

   0.00254 has three significant figures.

   0.0146 has three significant figures.

   0.06 has one significant figure.

3. Zeros that fall at the *end* of a number are not significant, unless they are marked as significant. If a zero does indicate the number's precision, we can mark it as significant by *placing a line over it*.

84,000 has two significant figures.

84,0̄00 has three significant figures.

84,0̄0̄0 has four significant figures.

84,0̄0̄0̄ has five significant figures.

84,0̄0̄0̄.0̄ has six significant figures.

In the last example, it really isn't necessary to put a line over each zero, because the number shows that you have measured to one decimal place and found it to be zero. Therefore all the zeros must be significant.

Example B-20   Here's a problem that will give you a chance to see whether you can find the number of significant figures in different kinds of numbers. (Cover the answers with your hand, then look later, after you've answered.)

Solution   (a) 0.00087 has two significant figures.

(b) 1.004 has four significant figures.

(c) 873.005 has six significant figures.

(d) 9.00000 has six significant figures.

(e) 320,000 has two significant figures.

(f) 180,0̄00 has four significant figures.

(g) 180,000.0 has seven significant figures.

(h) 2,0̄0̄0 has four significant figures.

SUMMARY

The purpose of this supplement was to give you some review of measurement in the sciences. We began with a discussion of the metric system and looked at the basic components of this system. We then looked at the concept of density and learned how you would go about measuring the density of an object in the laboratory. Next, we discussed the Fahrenheit and Celsius temperature scales and practiced switching from one scale to the other. Finally, we discussed the concept of heat and learned how to measure heats of reactions in calories. See if you've mastered the major points of this supplement by trying the exercises that follow.

EXERCISES

1. Convert 855 mm into
   (a) cm   (b) dm   (c) m   (d) km

2. Convert 4.38 liters into
   (a) ml   (b) cl   (c) dl   (d) kl

3. Find the area of a rug that measures 5 m by 2 m. Report the answer in
   (a) m^2 and (b) cm^2.

4. Find the volume of a cube that measures 5 cm on each side. Report the
   answer in (a) cm^3 and (b) mm^3.

5. Express 6 ft in
   (a) m   (b) cm   (c) mm

6. Express 2 quarts in (a) ml (b) liters

7. Express 2 liters in
   (a) quarts   (b) fluid ounces

8. A block of gold in the shape of a cube measures 20 cm on each side. The
   density of gold is 19.3 g/cm^3. Determine the mass of this block of gold in
   (a) grams and (b) pounds.

9. A 71-g sample of diethyl ether has a volume of 100 ml. What is the density
   of this compound?

10. The density of isopropyl alcohol is 0.79 g/ml. What is the volume of
    158 g of this compound?

11. Convert the following temperatures to °C.
    (a) 50°F   (b) 185°F   (c) −40°F

12. Convert the following temperatures to °F.
    (a) 230°C   (b) 490°C   (c) −148°C

13. A 500-g sample of water is heated from 20°C to 50°C. How many calories
    of heat did the water absorb?

14. A 1-g sample of sucrose-raises the temperature of 500 g of water from
    25°C to 33.2°C. Determine the enthalpy change for sucrose in kcal/mole.
    (*Hint*: The molecular weight of sucrose is 342.)

15. The caloric value of vanillin is 914 kcal/mole. Determine its caloric
    value in kcal/g. (*Hint*: The molecular weight of vanillin is 152.)

16. The caloric value of ethyl alcohol is 7.12 kcal/g. Determine its caloric
    value in kcal/mole. (*Hint*: The molecular weight of ethyl alcohol is 46.)

Appendix C

Expanded Rules
for Writing and Naming
Chemical Compounds

COMMON NAMES

There are no rules behind the common names of compounds. Most common names have been derived from common usage, or just handed down through history. For example, table salt is sodium chloride, and lye is sodium hydroxide. Table C-1 lists some additional examples of common names.

NAMES FOR BINARY COMPOUNDS

Binary compounds are derived from the names of the ions or elements that compose them. Let's first look at the rules for positive ions, and then we'll look at the rules for negative ions.

Positive Ions

The name of the element is the name of the ion. For example, Na^{+1} is the sodium ion, Mg^{+2} is the magnesium ion, and Al^{+3} is the aluminum ion. For those elements that can exist as either one of two positive ions, the Stock System,

This appendix is taken from Alan Sherman, Sharon Sherman, and Leonard Russikoff, *Basic Concepts of Chemistry*, 2nd ed., Houghton Mifflin Company, Boston, 1980. Used by permission.

TABLE C-1 *Common names of some chemical compounds*

COMMON NAME	FORMULA	CHEMICAL NAME
Baking soda	$NaHCO_3$	Sodium hydrogen carbonate
Borax	$Na_2B_4O_7 \cdot 10\,H_2O$	Sodium tetraborate decahydrate
Epsom salts	$MgSO_4 \cdot 7\,H_2O$	Magnesium sulfate heptahydrate
Gypsum	$CaSO_4 \cdot 2\,H_2O$	Calcium sulfate dihydrate
Laughing gas	N_2O	Nitrous oxide
Quicklime	CaO	Calcium oxide
Marble	$CaCO_3$	Calcium carbonate
Milk of magnesia	$Mg(OH)_2$	Magnesium hydroxide
Muriatic acid	HCl	Hydrochloric acid
Oil of vitriol	H_2SO_4	Sulfuric acid
Saltpeter	$NaNO_3$	Sodium nitrate

which is recommended by the IUPAC, is used to name the ion. In this system, a Roman numeral is used to designate the ion. For example, Cu^{+1} is called copper(I), and Cu^{+2} is called copper(II). The ion Fe^{+2} is called iron(II), and Fe^{+3} is called iron(III). (Note that the roman numeral refers to the *charge on the ion*.)

Although the Stock System is the system of choice, the old system of naming positive ions that can exist with more than one charge is still used today. Therefore, it is important to review the rules of this system. The ion with the lower charge is given the ending *-ous*, and the ion with the higher charge is given the ending *-ic*. Also, the Latin form of the element is used. For example, Fe^{+2} is called ferr*ous*, and Fe^{+3} is called ferr*ic*. Note the use of the suffixes *-ous* and *-ic*, and also the use of *ferr-*, not iron, as the main stem of the element. Another example is copper. The Cu^{+1} is called cupr*ous* and Cu^{+2} is called cup*ric*.

Table C-2 lists some additional examples of positive ions and their names, using the rules of both systems.

Negative Ions

The name of the negative ion is the name of the element with the ending *-ide*. For example, Cl^{-1} is called chlor*ide* ion, and O^{-2} is called ox*ide* ion. The ion S^{-2} is called sulf*ide* and P^{-3} is called phosph*ide*.

If the negative ion is a polyatomic ion (one consisting of two or more elements), the name is *best learned* by memorizing it along with its formula. For example, $(PO_4)^{-3}$ is called the phosphate ion and $(SO_4)^{-2}$ is called the

TABLE C-2 *Some positive ions used in chemistry*

+1			+2		
Hydrogen	H^{+1}		Calcium	Ca^{+2}	
Lithium	Li^{+1}		Magnesium	Mg^{+2}	
Sodium	Na^{+1}		Barium	Ba^{+2}	
Potassium	K^{+1}		Zinc	Zn^{+2}	
Mercury(I)[a]	Hg^{+1} (also called mercurous)		Mercury(II)	Hg^{+2} (also called mercuric)	
Copper(I)	Cu^{+1} (also called cuprous)		Tin(II)	Sn^{+2} (also called stannous)	
Silver	Ag^{+1}		Iron(II)	Fe^{+2} (also called ferrous)	
Ammonium	$(NH_4)^{+1}$		Lead(II)	Pb^{+2} (also called plumbous)	
Rubidium	Rb^{+1}		Copper(II)	Cu^{+2} (also called cupric)	
Cesium	Cs^{+1}		Strontium	Sr^{+2}	
			Nickel(II)	Ni^{+2}	
			Chromium(II)	Cr^{+2} (also called chromous)	
			Cobalt(II)	Co^{+2} (also called cobaltous)	
			Manganese(II)	Mn^{+2} (also called manganous)	

+3			+4		
Aluminum	Al^{+2}		Tin(IV)	Sn^{+4} (also called stannic)	
Iron(III)	Fe^{+3} (also called ferric)		Lead(IV)	Pb^{+4}	
Bismuth(III)	Bi^{+3}		Manganese(IV)	Mn^{+4}	
Chromium(III)	Cr^{+3} (also called chromic)				
Cobalt(III)	Co^{+3} (also called cobaltic)				

[a] Note that the mercury (I) ion is a diatomic ion. In other words, you never find Hg^{+1} ion alone, but always as Hg^{+1}—Hg^{+1}. The two Hg^{+1} ions are bonded to each other.

sulfate ion. The polyatomic ion $(NO_2)^{-1}$ is the nitrite ion, and $(NO_3)^{-1}$ is the nitrate ion. Table C-3 gives a summary of negative ions, including polyatomic ions.

Writing the Names and Formulas of Binary Compounds

Using Tables C-2 and C-3, we can now write the formulas of binary compounds if we are given their names, or write the names of binary compounds if we are given their formulas. Here's how we do it.

TABLE C-3 *Some negative ions used in chemistry*

−1		−2		−3	
Fluoride	F^{-1}	Oxide	O^{-2}	Nitride	N^{-3}
Chloride	Cl^{-1}	Sulfide	S^{-2}	Phosphide	P^{-3}
Hydroxide	$(OH)^{-1}$	Sulfite	$(SO_3)^{-2}$	Phosphate	$(PO_4)^{-3}$
Nitrite	$(NO_2)^{-1}$	Sulfate	$(SO_4)^{-2}$	Arsenate	$(AsO_4)^{-3}$
Nitrate	$(NO_3)^{-1}$	Carbonate	$(CO_3)^{-2}$	Borate	$(BO_3)^{-3}$
Acetate	$(C_2H_3O_2)^{-1}$	Chromate	$(CrO_4)^{-2}$		
Chlorate	$(ClO_3)^{-1}$	Dichromate	$(Cr_2O_7)^{-2}$		
Bromide	Br^{-1}	Oxalate	$(C_2O_4)^{-2}$		
Iodide	I^{-1}				
Hypochlorite	$(ClO)^{-1}$				
Chlorite	$(ClO_2)^{-1}$				
Chlorate	$(ClO_3)^{-1}$				
Perchlorate	$(ClO_4)^{-1}$				
Permanganate	$(MnO_4)^{-1}$				
Cyanide	$(CN)^{-1}$				
Hydrogen sulfite	$(HSO_3)^{-1}$				
Hydrogen sulfate	$(HSO_4)^{-1}$				
Hydrogen carbonate	$(HCO_3)^{-1}$				

Example C-1 Write the chemical formula for sodium chloride.

Solution Since the sodium ion has a charge of $+1$ (Na^{+1}), and the chloride ion has a charge of -1 (Cl^{-1}), the formula for sodium chloride is

$$Na_1^{+1}Cl_1^{-1}$$

The subscripts show that we need one ion of each element to obtain a neutral compound. The subscript 1 is usually not written. Similarly, the charges $+1$ and -1 are left out, giving us the chemical formula: NaCl.

Let's study the formation of chemical compounds in more detail. We assume that any compound has a positive part and a negative part. The oxidation number tells us *how* positive or negative a specific ion is. It's our job to choose the right subscripts to balance the charges, so that the compound is electrically neutral.

Here are suggestions to help you choose the right subscripts.

1. Write the symbol of each element of the compound, together with its oxidation number. For example, in the compound zinc chloride, write

$$Zn^{+2}Cl^{-1}$$

2. You want the positive side of the compound to balance the negative side. You can do this by crisscrossing the numbers.

$$Zn^{+2}Cl^{-1} \quad \text{or} \quad Zn_1^{+2}Cl_2^{-1} \quad \text{or} \quad ZnCl_2$$

3. Note that the subscript numbers are written without charge (that is, without a plus or minus sign). For example, the formula for aluminum oxide is

$$Al^{+3}O^{-2} \quad \text{or} \quad Al_2^{+3}O_3^{-2} \quad \text{or} \quad Al_2O_3$$

4. Also note that subscript numbers should be in least-common-denominator form. For example, the formula for barium oxide is

$$Ba^{+2}O^{-2} \quad \text{or} \quad Ba_2^{+2}O_2^{-2} \quad \text{or} \quad Ba_1^{+2}O_1^{-2} \quad \text{or} \quad BaO$$

Example C-2   Write the formula for barium chloride.

Solution   Tables C-2 and C-3 list the oxidation numbers of the barium ion and chloride ion.

$$Ba^{+2}Cl^{-1}$$

Now choose the right subscripts to balance the charges.

$$Ba_1^{+2}Cl_2^{-1} \quad \text{or} \quad BaCl_2$$

Example C-3   Write the formula for aluminum sulfide.

$$Al^{+3}S^{-2}$$

Solution   Now choose the right subscripts to balance the charges.

$$Al_2^{+3}S_3^{-2} \quad \text{or} \quad Al_2S_3$$

You can use the same procedure when you're dealing with poly-atomic ions. If two or more polyatomic ion groups are needed to balance the charges, remember to put parentheses around the polyatomic ion.

Example C-4   Write the chemical formula of sodium carbonate.

$$Na^{+1}(CO_3)^{-2}$$

Solution   Now choose the right subscripts to balance the charges.

$$Na_2^{+1}(CO_3)_1^{-2} \quad \text{or} \quad Na_2CO_3$$

Example C-5 Write the chemical formula of aluminum sulfate.

$$Al^{+3}(SO_4)^{-2}$$

Solution Now choose the right subscripts to balance the charges.

$$Al_2^{+3}(SO_4)_3^{-2} \quad \text{or} \quad Al_2(SO_4)_3$$

Example C-6 Write the chemical formula for copper chloride.

Solution Which copper compound do we mean? We can have copper(I) chloride or copper(II) chloride. The Roman numeral tells us the oxidation number of the copper compound that is formed.

If we mean copper(I) chloride, we find the formula in the following way. We start with

$$Cu^{+1}Cl^{-1}$$

Now we choose the right subscripts to balance the charges.

$$Cu_1^{+1}Cl_1^{-1} \quad \text{or} \quad CuCl$$

If we mean copper(II) chloride, we find the formula in the following way. We start with

$$Cu^{+2}Cl^{-1}$$

Now we choose the right subscripts to balance the charges.

$$Cu_1^{+2}Cl_2^{-1} \quad \text{or} \quad CuCl_2$$

Each copper chloride is a distinctly different compound, with its own properties.

Example C-7 Write the chemical formula for iron(III) oxide.

Solution The (III) tells us that the iron is the Fe^{+3} ion.

$$Fe^{+3}O^{-2}$$

Now we choose the right subscripts to balance the charges.

$$Fe_2^{+3}O_3^{-2} \quad \text{or} \quad Fe_2O_3$$

Example C-8 Write the chemical formula of copper(I) phosphate.

Solution The (I) tells us that the copper is the Cu^{+1} ion.

$$Cu^{+1}(PO_4)^{-3}$$

Now we choose the right subscripts to balance the charges.

$$Cu_3^{+1}(PO_4)_1^{-3} \quad \text{or} \quad Cu_3PO_4$$

We may write the names of chemical compounds from their formulas by following the rules we just studied. Remember that if the positive-ion part of the compound can exist as more than one positive ion, we must give the charge of the ion in that particular compound.

Example C-9   Name the following compounds.

(a) $CuCl_2$	(b) $HgSO_4$	(c) $Cu(NO_3)_2$	(d) $Fe_2(SO_3)_3$
(e) Ag_2CrO_4	(f) $Ba_3(PO_4)_2$	(g) $MnCl_2$	(h) MnO_2
(i) Cu_2S	(j) SnO_2	(k) Cs_2CO_3	(l) UBr_4

Solution   (a) $CuCl_2$ is called copper(II) chloride or cupr*ic* chloride.

(b) $HgSO_4$ is called mercury(II) sulfate or mercur*ic* sulfate.

(c) $Cu(NO_3)_2$ is called copper(II) nitrate or cupr*ic* nitrate.

(d) $Fe_2(SO_3)_3$ is called iron(III) sulfite or ferr*ic* sulfite.

(e) Ag_2CrO_4 is called silver chromate.

(f) $Ba_3(PO_4)_2$ is called barium phosphate.

(g) $MnCl_2$ is called manganese(II) chloride or mangan*ous* chloride.

(h) MnO_2 is manganese(IV) oxide (commonly called manganese dioxide).

(i) Cu_2S is called copper(I) sulfide or cupr*ous* sulfide.

(j) SnO_2 is called tin(IV) oxide or stann*ic* oxide (commonly called tin dioxide).

(k) Cs_2CO_3 is called cesium carbonate.

(l) UBr_4 is called uranium(IV) bromide (commonly called uranium tetrabromide).

NAMES FOR TERNARY AND HIGHER COMPOUNDS (THREE OR MORE ELEMENTS)

For compounds that are ternary or higher, we follow the rules for binary compounds. This is because the negative part of ternary compounds is usually composed of polyatomic ions. That's what makes them ternary. For example, Na_3PO_4 is sodium phosphate, and $Fe(NO_3)_3$ is iron(III) nitrate.

Example C-10   Name the following ternary compounds.

(a) $Ca(NO_3)_2$	(b) $Zn_3(PO_4)_2$
(c) $Cu(C_2H_3O_2)_2$	(d) K_2SO_4

Solution (a) calcium nitrate   (b) zinc phosphate

(c) copper(II) acetate   (d) potassium sulfate

NAMES FOR INORGANIC ACIDS

According to Arrhenius, an acid is a substance that releases hydrogen ions, H^{+1}, in aqueous solutions. Two major classes of inorganic acids exist, and the naming rules differ for each group. First we'll examine the rules for naming the non-oxygen-containing acids. Then we'll examine the rules for naming the oxygen-containing acids.

Non-Oxygen-Containing Acids

These acids usually consist of hydrogen plus a nonmetal ion. For example, HCl, which is hydrogen chloride gas in its pure form, becomes hydrochloric acid when dissolved in water. The covalently bonded HCl molecules ionize in water to form hydrogen ions and chloride ions. We derive the names of acids like HCl by adding the prefix *hydro-* to the name of the nonmetal, which is given an *-ic* suffix. For example,

hydro— chlor— ic— acid
(Prefix)   (Name of   (Suffix)
              nonmetal)

Example C-11 Name the following non-oxygen-containing acids.

(a) HF   (b) HBr   (c) H_2S (d) HCN

Solution We'll first give the name of each substance as a covalent compound and then as an acid in aqueous solution.

Name as covalent compound	*Name as acid in aqueous solution*
(a) HF Hydrogen fluoride	Hydrofluoric acid
(b) HBr Hydrogen bromide	Hydrobromic acid
(c) H_2S Hydrogen sulfide	Hydrosulfuric acid
(d) HCN Hydrogen cyanide	Hydrocyanic acid

Oxygen-Containing Acids

Oxygen-containing acids formed from hydrogen and polyatomic ions (containing oxygen) are named as follows.

1. If the polyatomic ion has an *-ate* suffix, you name the acid by replacing the *-ate* suffix with an *-ic* suffix and adding the word *acid*. For example,

HNO_3, composed of a hydrogen ion and nitr*ate* ion, is called nitr*ic* acid in aqueous solution. Example C-12 gives some additional examples.

Example C-12 Name the following oxygen-containing acids.

(a) H_2SO_4 (b) $HC_2H_3O_2$ (c) $HBrO_3$

Solution These three acids are composed of hydrogen plus a polyatomic ion that has the suffix -*ate*.

(a) H_2SO_4 is called sufur*ic* acid. (SO_4^{-2} is the sulf*ate* ion.)

(b) $HC_2H_3O_2$ is called acet*ic* acid. ($C_2H_3O_2^{-1}$ is the acet*ate* ion.)

(c) $HBrO_3$ is called brom*ic* acid. (BrO_3^{-1} is the brom*ate* ion.)

(Note that the H_2SO_4 is an exception to our rule, in that the whole name of the element is used rather than the name of the sulfate group.)

2. If the polyatomic ion has an -*ite* suffix, you name the acid by replacing the -*ite* suffix with an -*ous* suffix and adding the word *acid*. For example, HNO_2, composed of a hydrogen ion and a nitr*ite* ion, is called nitr*ous* acid. Example C-13 provides some additional examples.

Example C-13 Name the following oxygen-containing acids.

(a) H_2SO_3 (b) H_3PO_3 (c) $HBrO_2$

Solution These three acids are composed of hydrogen plus a polyatomic ion that has the suffix -*ite*.

(a) H_2SO_3 is called sulfur*ous* acid. (SO_3^{-2} is the sulf*ite* ion.)

(b) H_3PO_3 is called phosphor*ous* acid. (PO_3^{-2} is the phosph*ite* ion.)

(c) $HBrO_2$ is called brom*ous* acid. (BrO_2^{-1} is the brom*ite* ion.)

(Note that the H_2SO_3 is an exception to our rule, in that the whole name of the element is used rather than the name of the sulfite group.)

3. Some elements form more than two oxygen-containing acids. Most notable are the acids formed by the elements chlorine, bromine, and iodine. This occurs because these elements can form four polyatomic ions. For example, bromine forms the following ions:

BrO_4^{-1} Perbrom*ate* ion BrO_3^{-1} Brom*ate* ion

BrO_2^{-1} Brom*ite* ion BrO^{-1} Hypobrom*ite* ion

Note that the first two ions listed have the suffix -*ate* and the last two ions listed have the suffix -*ite*. Therefore, we use the same rules just discussed.

Keep the name of the polyatomic ion, but change the *-ate* suffix to *-ic* or the *-ite* suffix to *-ous*. Example C-14 reviews these rules.

Example C-14   Name the following oxygen-containing acids.

(a) $HBrO_4$ (b) $HBrO_3$ (c) $HBrO_2$ (d) $HBrO$

Solution   (a) $HBrO_4$ is called perbrom*ic* acid. (BrO_4^{-1} is the perbrom*ate* ion.)

(b) $HBrO_3$ is called brom*ic* acid. (BrO_3^{-1} is the brom*ate* ion.)

(c) $HBrO_2$ is called brom*ous* acid. (BrO_2^{-1} is the brom*ite* ion.)

(d) $HBrO$ is called hypobrom*ous* acid. (BrO^{-1} is the hypobrom*ite* ion.)

Important Tables

TABLE 1 Prefixes and abbreviations used in the metric system

nano- = 0.000000001	nanometer = nm	centigram = cg
micro- = 0.000001	micrometer = μm (Greek lowercase mu)	decimeter = dm
milli- = 0.001		decigram = dg
centi- = 0.01	millimeter = mm	kilometer = km
deci- = 0.1	milliliter = ml	kilogram = kg
deca- = 10	milligram = mg	
kilo- = 1000	centimeter = cm	

TABLE 2 The metric system

LENGTH

1 millimeter = 0.001 meter = 1/1000 meter	1 meter = 1000 millimeters
1 centimeter = 0.01 meter = 1/100 meter	1 meter = 100 centimeters
1 decimeter = 0.1 meter = 1/10 meter	1 meter = 10 decimeters
1 kilometer = 1000 meters	1 meter = 0.001 kilometer

MASS

1 microgram = 0.000001 gram	1 gram = 1,000,000 micrograms
1 milligram = 0.001 gram	1 gram = 1000 milligrams
1 centigram = 0.01 gram	1 gram = 100 centigrams
1 decigram = 0.1 gram	1 gram = 10 decigrams
1 kilogram = 1000 grams	1 gram = 0.001 kilogram

VOLUME

1 milliliter = 0.001 liter	1 liter = 1000 milliliters
1 milliliter = 1 cubic centimeter (cubic centimeter is abbreviated cc or cm^3)	1 liter = 1000 cubic centimeters
1 centiliter = 0.01 liter	1 liter = 100 centiliters
1 deciliter = 0.1 liter	1 liter = 10 deciliters
1 kiloliter = 1000 liters	1 liter = 0.001 kiloliter

TABLE 3 Conversion factors for English-metric and metric-English conversions

LENGTH	MASS	VOLUME
1 inch = 2.54	1 ounce = 28.35 g	1 liter = 1.06 quarts
1 foot = 0.30 m	1 pound = 454 g	

Note: The conversions in this table can be used to go from English to metric or metric to English simply by writing the conversions as factor units, for example, $\dfrac{1 \text{ inch}}{2.54 \text{ cm}}$ or $\dfrac{2.54 \text{ cm}}{1 \text{ inch}}$.

TABLE 4 *Naturally occurring isotopes of the first 15 elements*

NAME	SYMBOL	ATOMIC NUMBER	MASS NUMBER	PERCENTAGE NATURAL ABUNDANCE
Hydrogen-1	$_1^1H$	1	1	99.985
Hydrogen-2	$_1^2H$	1	2	0.015
Hydrogen-3	$_1^3H$	1	3	Negligible
Helium-3	$_2^3He$	2	3	0.00013
Helium-4	$_2^4He$	2	4	99.99987
Lithium-6	$_3^6Li$	3	6	7.42
Lithium-7	$_3^7Li$	3	7	92.58
Beryllium-9	$_4^9Be$	4	9	100
Boron-10	$_5^{10}B$	5	10	19.6
Boron-11	$_5^{11}B$	5	11	80.4
Carbon-12	$_6^{12}C$	6	12	98.89
Carbon-13	$_6^{13}C$	6	13	1.11
Nitrogen-14	$_7^{14}N$	7	14	99.63
Nitrogen-15	$_7^{15}N$	7	15	0.37
Oxygen-16	$_8^{16}O$	8	16	99.759
Oxygen-17	$_8^{17}O$	8	17	0.037
Oxygen-18	$_8^{18}O$	8	18	0.204
Fluorine-19	$_9^{19}F$	9	19	100
Neon-20	$_{10}^{20}Ne$	10	20	90.92
Neon-21	$_{10}^{21}Ne$	10	21	0.257
Neon-22	$_{10}^{22}Ne$	10	22	8.82
Sodium-23	$_{11}^{23}Na$	11	23	100
Magnesium-24	$_{12}^{24}Mg$	12	24	78.70
Magnesium-25	$_{12}^{25}Mg$	12	25	10.13
Magnesium-26	$_{12}^{26}Mg$	12	26	11.17
Aluminium-27	$_{13}^{27}Al$	13	27	100
Silicon-28	$_{14}^{28}Si$	14	28	92.21
Silicon-29	$_{14}^{29}Si$	14	29	4.70
Silicon-30	$_{14}^{30}Si$	14	30	3.09
Phosphorus-31	$_{15}^{31}P$	15	31	100

Source: Alan Sherman, Sharon Sherman, and Leonard Russikoff, *Basic Concepts of Chemistry*, Houghton Mifflin Company, Boston, 1976. By permission of the publisher.

TABLE 5 *Electron configurations of the elements*

ELEMENT	ATOMIC NUMBER	ELECTRON CONFIGURATION						
		K	L	M	N	O	P	Q
H	1	1						
He	2	2						
Li	3	2	1					
Be	4	2	2					
B	5	2	3					
C	6	2	4					
N	7	2	5					
O	8	2	6					
F	9	2	7					
Ne	10	2	8					
Na	11	2	8	1				
Mg	12	2	8	2				
Al	13	2	8	3				
Si	14	2	8	4				
P	15	2	8	5				
S	16	2	8	6				
Cl	17	2	8	7				
Ar	18	2	8	8				
K	19	2	8	8	1			
Ca	20	2	8	8	2			
Sc	21	2	8	9	2			
Ti	22	2	8	10	2			
V	23	2	8	11	2			
Cr	24	2	8	13	1			
Mn	25	2	8	13	2			
Fe	26	2	8	14	2			
Co	27	2	8	15	2			
Ni	28	2	8	16	2			
Cu	29	2	8	18	1			
Zn	30	2	8	18	2			

TABLE 5 *Electron configurations of the elements (cont.)*

ELEMENT	ATOMIC NUMBER	ELECTRON CONFIGURATION K	L	M	N	O	P	Q
Ga	31	2	8	18	3			
Ge	32	2	8	18	4			
As	33	2	8	18	5			
Se	34	2	8	18	6			
Br	35	2	8	18	7			
Kr	36	2	8	18	8			
Rb	37	2	8	18	8	1		
Sr	38	2	8	18	8	2		
Y	39	2	8	18	9	2		
Zr	40	2	8	18	10	2		
Nb	41	2	8	18	12	1		
Mo	42	2	8	18	13	1		
Tc	43	2	8	18	14	1		
Ru	44	2	8	18	15	1		
Rh	45	2	8	18	16	1		
Pd	46	2	8	18	18			
Ag	47	2	8	18	18	1		
Cd	48	2	8	18	18	2		
In	49	2	8	18	18	3		
Sn	50	2	8	18	18	4		
Sb	51	2	8	18	18	5		
Te	52	2	8	18	18	6		
I	53	2	8	18	18	7		
Xe	54	2	8	18	18	8		
Cs	55	2	8	18	18	8	1	
Ba	56	2	8	18	18	8	2	
La	57	2	8	18	18	9	2	
Ce	58	2	8	18	20	8	2	
Pr	59	2	8	18	21	8	2	
Nd	60	2	8	18	22	8	2	
Pm	61	2	8	18	23	8	2	

TABLE 5 *Electron configurations of the elements (cont.)*

ELEMENT	ATOMIC NUMBER	ELECTRON CONFIGURATION						
		K	L	M	N	O	P	Q
Sm	62	2	8	18	24	8	2	
Eu	63	2	8	18	25	8	2	
Gd	64	2	8	18	25	9	2	
Tb	65	2	8	18	27	8	2	
Dy	66	2	8	18	28	8	2	
Ho	67	2	8	18	29	8	2	
Er	68	2	8	18	30	8	2	
Tm	69	2	8	18	31	8	2	
Yb	70	2	8	18	32	8	2	
Lu	71	2	8	18	32	9	2	
Hf	72	2	8	18	32	10	2	
Ta	73	2	8	18	32	11	2	
W	74	2	8	18	32	12	2	
Re	75	2	8	18	32	13	2	
Os	76	2	8	18	32	14	2	
Ir	77	2	8	18	32	15	2	
Pt	78	2	8	18	32	17	1	
Au	79	2	8	18	32	18	1	
Hg	80	2	8	18	32	18	2	
Tl	81	2	8	18	32	18	3	
Pb	82	2	8	18	32	18	4	
Bi	83	2	8	18	32	18	5	
Po	84	2	8	18	32	18	6	
At	85	2	8	18	32	18	7	
Rn	86	2	8	18	32	18	8	
Fr	87	2	8	18	32	18	8	1
Ra	88	2	8	18	32	18	8	2
Ac	89	2	8	18	32	18	9	2
Th	90	2	8	18	32	18	10	2
Pa	91	2	8	18	32	20	9	2
U	92	2	8	18	32	21	9	2

TABLE 5 *Electron configurations of the elements (cont.)*

ELEMENT	ATOMIC NUMBER	ELECTRON CONFIGURATION						
		K	L	M	N	O	P	Q
Np	93	2	8	18	32	22	9	2
Pu	94	2	8	18	32	24	8	2
Am	95	2	8	18	32	25	8	2
Cm	96	2	8	18	32	25	9	2
Bk	97	2	8	18	32	26	9	2
Cf	98	2	8	18	32	28	8	2
Es	99	2	8	18	32	29	8	2
Fm	100	2	8	18	32	30	8	2
Md	101	2	8	18	32	31	8	2
No	102	2	8	18	32	32	8	2
Lr	103	2	8	18	32	32	9	2
Rf	104	2	8	18	32	32	10	2
Ha	105	2	8	18	32	32	11	2

TABLE 6 *Table of ions frequently used in chemistry*

+1		+2		+3	
Hydrogen	H^{+1}	Calcium	Ca^{+2}	Iron(III)	Fe^{+3}
Lithium	Li^{+1}	Magnesium	Mg^{+2}	Aluminium	Al^{+3}
Sodium	Na^{+1}	Barium	Ba^{+2}		
Potassium	K^{+1}	Zinc	Zn^{+2}		
Mercury(I)	Hg^{+1}	Mercury(II)	Hg^{+2}		
Copper(I)	Cu^{+1}	Tin(II)	Sn^{+2}		
Ammonium	$(NH_4)^{+1}$	Iron(II)	Fe^{+2}		
Silver	Ag^{+1}	Lead(II)	Pb^{+2}		
		Copper(II)	Cu^{+2}		

−1		−2		−3	
Fluoride	F^{-1}	Oxide	O^{-2}	Nitride	N^{-3}
Chloride	Cl^{-1}	Sulfide	S^{-2}	Phosphate	$(PO_4)^{-3}$
Hydroxide	$(OH)^{-1}$	Sulfite	$(SO_3)^{-2}$	Arsenate	$(AsO_4)^{-3}$
Nitrite	$(NO_2)^{-1}$	Sulfate	$(SO_4)^{-2}$		
Nitrate	$(NO_3)^{-1}$	Carbonate	$(CO_3)^{-2}$		
Acetate	$(C_2H_3O_2)^{-1}$	Chromate	$(CrO_4)^{-2}$		

TABLE 7   *The activity series*

METALS	NONMETALS
Lithium	Fluorine
Potassium	Chlorine
Calcium	Bromine
Sodium	Iodine
Magnesium	
Aluminum	
Zinc	
Chromium	
Iron	
Nickel	
Tin	
Lead	
Hydrogen[a]	
Copper	
Mercury	
Silver	
Platinum	
Gold	

[a] Hydrogen is in italic type because the activities of the other elements were calculated relative to hydrogen.

TABLE 8 *Solubilities*[a]

	Acetate	Arsenate	Bromide	Carbonate	Chlorate	Chloride	Chromate	Hydroxide	Iodide	Nitrate	Oxide	Phosphate	Sulfate	Sulfide
Aluminum	W	a	W	—	W	W	—	A	W	W	a	A	W	d
Ammonium	W	W	W	W	W	W	W	W	W	W..	—	W..	W..	W
Barium	W	w	W	w	W	W	A	W	W	W	W	A	a	d
Cadmium	W	A	W	A	W	W	A	A	W	W	A	A	W	A
Calcium	W	w	W	w	W	W	W	W	W	W	w	w	w..	w
Chromium	W	—	W[b]	W	—	I	—	A	W	W	a	w	W[c]	d
Cobalt	W	A	W	A	W	W	A	A	W	W	A	A	W	A
Copper(II)	W	A	W	—	W	W	—	A	a	W	A	A	W	A
Hydrogen	W	W	W	—	W	W	—	—	W	W	W	W	W	W
Iron(II)	W	A	W	w	W	W	—	A	W	W	A	A	W	A
Iron(III)	W	A	W	—	W	W	A	A	W	W	A	w	w	d
Lead(II)	W	A	W	A	W	W	A	w	w	W	w	A	w	A
Magnesium	W	A	W	w	W	W	W	A	W	W	A	w	W	d
Mercury(I)	w	A	A	A	W	a	w	—	A	W	A	A	w	I
Mercury(II)	W	w	W	—	W	W	w	A	w	W	w	A	d	I
Nickel	W	A	W	w	W	W	A	w	W	W	A	A	W	A
Potassium	W	W	W	W	W	W	W	W	W	W	W	W	W	W
Silver	w	A	a	A	W	a	w	—	I	W	w	A	w	A
Sodium	W	W	W	W	W	W	W	W	W	W	d	W	W	W
Strontium	W	w	W	w	W	W	w	W	W	W	W	A	w	W
Tin(II)	d	—	W	—	W	W	A	A	W	d	A	A	W	A
Tin(IV)	W	—	W	—	—	W	W	w	d	—	A	—	W	A
Zinc	W	A	W	w	W	W	w	A	W	W	w	A	W	A

[a] W = soluble in water; A = insoluble in water, but soluble in acids; w = only slightly soluble in water, but soluble in acids; a = insoluble in water, and only slightly soluble in acids; I = insoluble in both water and acids; d = decomposes in water.
[b] $CrBr_3$.
[c] $Cr_2(SO_4)_3$.
Source: Reprinted in part from *The Handbook of Chemistry and Physics,* Chemical Rubber Company, Cleveland, Ohio, 1977–78.

TABLE 9   *Pressure of water vapor, P_{H_2O}, at various temperatures*

TEMPERATURE (°C)	PRESSURE (torr)	TEMPERATURE (°C)	PRESSURE (torr)
0	4.58	32	35.66
5	6.54	33	37.73
10	9.21	34	39.90
15	12.79	35	42.18
16	13.63	36	44.56
17	14.53	37	47.07
18	15.48	38	49.69
19	16.48	39	52.44
20	17.54	40	55.32
21	18.65	45	71.88
22	19.83	50	92.51
23	21.07	55	118.04
24	22.38	60	149.38
25	23.76	65	187.54
26	25.21	70	233.7
27	26.74	75	289.1
28	28.35	80	355.1
29	30.04	85	433.6
30	31.82	90	525.8
31	33.70	95	633.9
		100	760.0

Glossary

The number in parentheses that follows each entry indicates the chapter in the text in which the subject is introduced.

ABSOLUTE ZERO (7)   The coldest possible temperature that matter can reach, which is $0°K$ or $-273°C$.

ACID (8)   A substance that releases hydrogen ions in solution.

ACTIVATED SLUDGE PROCESS (9)   A method of secondary wastewater treatment.

ACTIVITY SERIES (5)   A list of elements grouped according to their reactivities.

ADDITION POLYMER (13)   A polymer in which a small molecule adds to itself, over and over, to form the polymer.

ADVANCED WASTEWATER TREATMENT (9)   A method of wastewater treatment which involves chemical processes. It usually removes nitrate ions and phosphate ions from the water.

AEROBIC MICROORGANISMS (9)   Microorganisms that require free oxygen in the water to survive.

ALCOHOLS (12)   A class of organic compounds with the general formula R—OH.

ALDEHYDE (12)   A class of organic compounds with the general formula

$$R-\underset{\underset{O}{\|}}{C}-H$$

ALKALOIDS (17)   A class of basic nitrogenous organic compounds occurring in plants.

ALKANES (12)   A group of organic hydrocarbons containing only single carbon-carbon bonds.

ALKENES (12)   A group of organic hydrocarbons each of which contains a carbon-carbon double bond in its structure.

ALKYL BENZENE SULFONATE (ABS) DETERGENTS (18)   The first low-cost synthetic detergents. These detergents proved to be non-biodegradable.

ALKYL GROUP (12)   An organic group such as a CH_3—(methyl) or CH_3CH_2—(ethyl) group, which is attached to the functional group of an organic compound.

ALKYNE (12)   A group of organic hydrocarbons each of which contains a carbon-carbon triple bond in its structure.

ALPHA HELIX (6)   A coiled protein chain in which there are 3.6 amino acids per turn of the helix.

ALPHA RAYS (3)   A type of nuclear radiation that seems like helium atoms with their electrons removed.

ALVEOLI (7)   Small sacs at the end of the lung which receive the oxygen.

AMIDES (12)   A class of organic compounds with the general formula

$$R-\underset{\underset{O}{\|}}{C}-NH_2$$

AMINES (12)   A class of organic compounds with the general formula $R-NH_2$.

AMINO ACIDS (15)   Difunctional organic compounds that contain a carboxyl group at one end of the molecule and an amino group at the other end of the molecule.

AMORPHOUS SOLID (7)   Solids that do not have a well-defined crystalline structure.

AMPHETAMINES (17)   A group of compounds that act as potent central nervous system stimulants.

ANAEROBIC MICROORGANISMS (9)   Microorganisms that do not need free oxygen to survive. They are able to take oxygen out of organic compounds.

ANALGESIC (17)   A medication that acts as a pain reliever.

ANDROGENS (17)   The male sex hormones.

ANODE (3)   A positively charged electrode.

ANTACIDS (17) Medications used to neutralize excess stomach acid. They usually contain hydroxide or carbonate ions.

ANTHRACITE COAL (6) The hardest form of coal. It also has the highest carbon content of the various types.

ANTIBIOTICS (17) A class of pharmaceuticals developed over the last 50 years which are used to treat a wide variety of bacterial and viral infections.

ANTIPYRETIC (17) A medication that acts as a fever reducer.

AROMATIC HYDROCARBONS (12) A series of cyclic hydrocarbons that contain alternating single and double bonds.

ASBESTOS (10) A mineral substance used for making fireproof articles.

ASTRINGENT (18) A compound that contracts the tissues or canals of the body to stop the flow of a liquid, such as sweat.

ATMOSPHERE (7) A unit of pressure. One atmosphere equals 760 torr.

ATOM (2) The smallest part of an element that can enter into chemical combination.

ATOMIC MASS UNIT (amu) (3) The unit used to describe the mass of an atom. (1 amu $= 1.66053 \times 10^{-24}$ g.)

ATOMIC NUMBER (3) The number of electrons or protons in a neutral atom.

ATOMIC WEIGHT (or atomic mass) (2) The mass of an element in relation to the mass of an atom of carbon-12, which has a mass of 12 amu.

BARBITURATES (17) A class of depressant drugs that can produce mild sedation or very deep sedation and death.

BASE (8) A substance that releases hydroxide ions in solution.

BETA RAYS (3) High-speed "electron-like" particles that come from the nuclei of radioactive atoms.

BHA (15) Butylated hydroxyanisole—a food additive that prevents spoilage.

BHT (15) Butylated hydroxytoluene—a food additive that prevents spoilage.

BITUMINOUS COAL (6) A type of soft coal that has impurities in it such as pitch and tar.

BLACK HOLES (19) The phenomenon that results when a very massive star undergoes collapse. The star disappears from sight, because it pulls all matter, including light, into its infinitely dense center.

BLANCOPHORS (18) Chemical compounds that act as clothes brighteners in detergent products.

BOILING POINT (7) The temperature at which the equilibrium vapor pressure of a liquid equals the atmospheric pressure.

BRACYTHERAPY (11) A method of radiation therapy in which a radioactive isotope is placed into the area to be treated.

BREEDER REACTOR (11)   A type of nuclear reactor that uses plutonium-239 and uranium-238. The plutonium-239 releases neutrons and these convert uranium-238 to plutonium-239. In other words, this reactor actually breeds fuel.

BRONCHITIS (10)   A respiratory disease in which the bronchia in the lungs become swollen, secrete mucous, and become obstructed.

BUILDER (18)   Any substance that increases the detergency of a surfactant.

CALORIE (6)   A measure of heat energy. The amount of heat needed to raise the temperature of one gram of water one degree Celsius.

CARBOHYDRATES (15)   Organic compounds which are generally composed of the elements C, H, and O in the ratio of $1:2:1$.

CARBOXYLIC ACIDS (12)   A class of organic acids with the general formula

$$R-\overset{\displaystyle O}{\underset{\displaystyle \|}{C}}-OH$$

CATALYSTS (16)   Substances that affect the rate at which chemical reactions occur.

CATHODE (3)   A negatively charged electrode.

CELLULOID (13)   The first synthetic plastic. Originally, it was used to make billiard balls.

CELSIUS TEMPERATURE SCALE (Appendix B)   A temperature scale on which the freezing point of water is $0°C$ and the normal boiling point of water is $100°C$.

CHAIN REACTION (11)   A self-sustaining nuclear reaction.

CHEMOTHERAPY (17)   The treatment of a disease by chemical agents.

CHLORINATED HYDROCARBON(14)   A substance composed of the elements carbon, hydrogen, and chlorine, for example the insecticide DDT.

CHROMOSOMES (16)   Substances in the cells of the body that hold the genetic information that we inherit from our parents.

COEFFICIENT (5)   A number placed before an element or compound in a chemical equation so that the equation can be balanced.

COLLOIDAL DISPERSION (8)   A type of solution in which the solute particles don't dissolve to the point they do in a true solution.

COMBINATION REACTIONS (5)   Reactions in which two or more substances combine to form a more complex substance.

COMPLETE PROTEINS (16)   Those that contain all the amino acids essential for the maintenance of good health.

COMPOUND (2)   A chemical combination of two or more elements.

CONDENSATION POLYMER (13)   A polymer formed when two or more monomer units react with each other and split out water.

CONTINUOUS SPECTRUM (3)   When one color of light merges into the next, for example, the solar spectrum.

CONTRACEPTIVE (17)   A device or drug used to prevent pregnancy.

COVALENT BOND (4)   A bond formed by the sharing of electrons between two atoms.

CRYSTAL (7)   A solid that has geometrically arranged plane surfaces and a symmetrical structure inside.

CRYSTAL LATTICE (7)   The symmetrical structure formed by the particles throughout a crystal.

CURIE (11)   A measure of activity of a radioactive source. (1 Ci $= 3.7 \times 10^{10}$ disintegrations per second.)

CYCLIC HYDROCARBONS (12)   A class of organic compounds that form ring-like chains.

DECOMPOSITION REACTIONS (5)   Reactions that involve the breakdown of a complex substance into simpler substances.

DEFOLIANT (14)   A chemical that causes plants to lose their leaves.

DENSITY (Appendix B)   The mass per unit volume of a substance.

DEOXYRIBONUCLEIC ACID (DNA) (16)   The genetic material found in the nucleus of each cell which contains all the information needed for the development of that individual.

DEPRESSANTS (17)   A class of drugs that lowers the vital activities of the central nervous system. Many depressants have sedative properties.

DIATOMIC ELEMENTS (4)   Elements that are found naturally as molecules with two atoms apiece, for example H_2, O_2, and N_2.

DIGLYCERIDES (15)   Fats in which two of the three glycerol groups have undergone esterification.

DISACCHARIDE (15)   A carbohydrate containing two sugar groups in the molecule.

DOSIMETER (11)   A device used to measure the amount of radiation an individual receives.

DOUBLE BOND (4)   A covalent bond in which each atom donates two electrons to form the bond.

DOUBLE REPLACEMENT REACTIONS (5)   Reactions in which two compounds exchange ions with each other.

ELASTOMERS (13)   Polymers that have elastic properties; in other words, when stretched they tend to return to their original form.

ELECTROLYSIS (3)   The decomposition of a compound into simpler substances by the use of electricity.

ELECTROLYTE (3)   A substance that when placed in solution conducts electricity.

ELECTROMAGNETIC SPECTRUM (3)   The range of electromagnetic radiation. All the types of this radiation travel at the same velocity, but vary in wavelength. They include radio waves, infrared waves (these have longer wavelengths), visible light, and x-rays (these have shorter wavelengths).

ELECTRON (3)   A particle with a relative negative charge of one unit and a mass of 0.0005486 amu.

ELECTRONEGATIVITY (4)   The attraction that an atom of an element has for shared electrons in a molecule.

ELECTROSTATIC PRECIPITATOR (10)   An air pollution device in which dust particles that have been electrically charged are collected between highly charged electrical plates.

ELEMENTS (2)   The basic building blocks of matter.

EMOLLIENTS (18)   Compounds that act as skin softeners.

EMPHYSEMA (10)   A chronic lung disease in which the air sacs in the lungs become enlarged and may be destroyed. This causes shortness of breath and eventual death.

ENDOTHERMIC REACTION (6)   A reaction that absorbs energy from the surroundings.

ENERGY (2)   The ability to do work. Energy appears in many forms—for example, heat, chemical, electrical, mechanical, and radiant (light) energy.

ENERGY LEVELS (3)   Various regions, outside the nucleus of an atom, in which the electrons move.

ENTHALPY (6)   The heat content of a chemical substance, represented by the symbol H.

ENZYMES (16)   A group of proteins that regulate biological reactions in our bodies.

EQUILIBRIUM VAPOR PRESSURE (7)   The pressure exerted by a vapor when it is in equilibrium with its liquid at any given temperature.

ESSENTIAL AMINO ACIDS (16)   The eight amino acids that can't be synthesized by the body and must be included in the diet.

ESTERS (12)   A class of organic compounds with the general formula

$$R-\overset{\displaystyle O}{\underset{\displaystyle \|}{C}}-O-R$$

ESTROGENS (17)   A class of female sex hormones.

ETHERS (12)   A class of organic compounds with the general formula R—O—R'.

EUTROPHICATION (9)   The filling in of a lake or river with plant life due to excessive nutrients (such as nitrates and phosphates) being dumped into the water.

EVAPORATION (7)   The change in state of molecules from a liquid to a gas.

EXOTHERMIC REACTION (6)   A reaction that releases energy to the surroundings.

FAHRENHEIT TEMPERATURE SCALE (Appendix B)   A temperature scale on which the freezing point of water is 32 degrees and the normal boiling point of water is 212 degrees.

FAMILY (of elements) (3)   A vertical column of elements in the periodic table.

FATTY ACIDS (15)   Organic compounds composed of a long chain of carbon atoms connected to a carboxyl group.

FILM BADGE (11)   A device used to detect exposure to radiation.

FIRST LAW OF THERMODYNAMICS (6)   (see Law of Conservation of Mass and Energy)

FISSION (nuclear) (5)   The splitting of an atom into two or more different atoms when bombarded by neutrons.

FIXATIVES (18)   Substances that tone down the odor of very strong smelling perfumes.

FORMULA (chemical) (2)   A notation that shows the number of atoms of each element that makes up a molecule of a compound.

FORMULA WEIGHT (2)   The sum of the atomic weights of the elements that make up a compound.

FOSSIL FUELS (6)   Fuels formed by the breakdown, over millions of years, of plant and animal microorganisms. Coal, oil, and natural gas are the three fossil fuels.

FREE RADICALS (11)   Highly reactive uncharged species. They can be atoms or groups of atoms having an odd (unpaired) number of electrons.

FREEZING POINT (Appendix B)   The temperature at which the solid and liquid phase of a substance exist together.

FUNCTIONAL GROUP (12)   The reactive part of an organic molecule.

FUSION (nuclear) (5)   The combination of two atoms to form a new heavier atom. The process releases huge amounts of energy.

GALAXY (19)   A group or cluster of millions or even billions of stars.

GAMMA RAYS (3)   A high-energy form of electromagnetic radiation given off by many radioactive substances.

GAS (7)   One of the states of matter. Gases have no definite shape or volume.

GEIGER–MÜLLER COUNTER (11)   A device used to detect and count nuclear radiation.

GENE SPLICING (16)   The process by which scientists take genes from one organism and plant them into another organism.

GENERIC NAME (17)   The name of a drug that has been agreed on by the pharmaceutical company that developed the drug and the FDA. It usually has some relationship to the chemical name of the drug. It is *not* the *brand name* of the drug!

GENES (16)   Substances found on the chromosomes which carry the information for each specific characteristic that a person has.

GRAM (Appendix B)   The metric unit of mass.

GRAM ATOMIC WEIGHT (2)   The atomic weight of an element expressed in grams.

GREEN REVOLUTION (14)   The term applied to modern agriculture, which uses chemical fertilizers and hybrid plants to grow enormous quantities of food per acre of land. This type of agriculture is said to be energy intensive because large amounts of energy are needed to produce the fertilizers and equipment needed to practice this type of farming.

GROUP (of elements) (3)   A vertical column of elements in the periodic table.

HABER PROCESS (14)   The production of ammonia from atmospheric nitrogen and hydrogen.

HALF-LIFE (11)   The time required for half the atoms originally present in a radioactive sample to decay.

HEAT OF FORMATION (6)   The heat released or absorbed when one mole of a compound is formed from its elements.

HEAT OF REACTION (6)   The change in heat content of a substance during a chemical reaction.

HERBICIDES (14)   Chemicals that kill weeds but don't harm plants.

HERPES VIRUS (1)   A group of viruses that cause cold sores, genital sores, and have been implicated in certain cancers.

HETEROGENEOUS MATTER (2)   Matter that has different parts with different properties.

HOMOGENEOUS MATTER (2)   Matter that has similar properties throughout.

HOMOLOGOUS SERIES (12)   A series of compounds of the same chemical type which differ only by fixed increments of the constituent elements.

HYDRATES (9)   Compounds that contain chemically combined water in definite proportions.

HYDROCARBONS (16)   Organic compounds composed only of carbon and hydrogen.

HYDROGEN BOND (as applied to water molecules) (9)   A chemical bond formed between the hydrogen atom of one water molecule and the oxygen atom of another water molecule.

HYDRONIUM ION (8)   The ion formed by the reaction of a hydrogen ion with a water molecule. It can be written as H_3O^{+1}.

HYDROXIDE ION (8)   The OH^{-1} ion.

INSECTICIDE (14)   A chemical that kills insects.

INSOLUBLE (5)   The description of a substance that does not dissolve in solution.

INTERFERON (16)   A protein produced in the human body to fight viral infections.

IONIC BOND (4)   A bond formed by the transfer of electrons from one atom to another. The atoms are always of different elements.

IONIZATION (3)   The breakup into ions of some substances when they dissolve in solution.

IONS (3)   Atoms that have gained or lost electrons and therefore have a positive or negative charge.

ISOMERS (12)   Compounds with the same molecular formula, but different structural formulas.

ISOTOPES (3)   Atoms that have the same number of electrons and protons, but different numbers of neutrons.

KELVIN TEMPERATURE SCALE (7)   Degrees Celsius $+ 273$.

KETONES (12)   A class of organic compounds with the general formula

$$R-\underset{\underset{O}{\|}}{C}-R'$$

KHEMIA (1)   A mystical religious form of "chemistry" that was practiced by the ancient Egyptians.

KINETIC ENERGY (6)   The energy of motion.

LAW OF CONSERVATION OF MASS AND ENERGY (6)   Matter and energy are neither created nor destroyed, but they can change from one form into another.

LD_{50}^{30} (11)   The term used to express the lethal dose for 50% of the population within 30 days.

LEGUMES (14)   A group of plants whose roots contain bacteria and algae capable of fixing atmospheric nitrogen.

LEWIS DOT STRUCTURE (4)   A notation that shows the symbol of the element and the number of outer electrons in an atom of the element.

LIGNITE (6)   A soft kind of coal with a woody appearance.

LINE SPECTRA (3)   Spectra produced by elements which appear as a series of bright lines at specific wavelengths, separated by dark bands.

LINEAR ALKYLSULFONATE (LAS) DETERGENTS (18)   A class of synthetic detergents that are biodegradable.

LIPIDS (15)   Biochemical substances found in plant and animal tissues. They include fats and other compounds that resemble fats in physical properties.

LIQUIDS (7)   One of the states of matter. Liquids have definite volume but indefinite shape.

LITER (Appendix B)   A unit of volume in the metric system.

MACROMOLECULES (13)   Large molecules containing many atoms per molecule.

MASS NUMBER (3)   The relative weight of an isotope of an element equal to the sum of its protons and neutrons.

MATTER (2)   Anything that occupies space and has mass.

MESOTHELIOMA (10)   A form of cancer found among industrial workers involved in the processing of asbestos fibers.

METABOLISM (14)   The process by which the food we eat is transformed in the body by a series of complex chemical reactions, catalyzed by enzymes, to provide material for cellular functions and energy to power the body.

METALLOIDS (2)   A class of elements that have some of the properties of metals and some of the properties of nonmetals.

METALS (2)   A class of elements that conduct electricity and heat, have luster, and take on a positive oxidation number when they bond.

METER (Appendix B)   The basic units of length in the metric system.

MINERALS (15)   A group of elements that have important nutritional value.

MOISTURIZERS (18)   Substances that keep the skin soft and prevent loss of moisture.

MOLARITY (8)   A concentration unit for solutions: moles of solute per liter of solution.

MOLE (2) 6.023×10^{23} items.

MOLECULAR FORMULA (12)   A chemical formula showing the number of atoms of each element in a molecule of a compound.

MOLECULE (2)   The smallest part of a compound that can enter into chemical combination.

MONOGLYCERIDES (15)   Fats in which only one of the three glycerol groups has undergone esterification.

MONOSACCHARIDE (15)   A carbohydrate containing one sugar group.

NARCOTICS (17)   A class of depressant drugs which are also addictive.

NEBULAE (19)   A cloud of interstellar gas or dust. Sometimes a galaxy.

NEUTRON (3) A particle with a mass of 1.008665 amu and zero charge.

NITROGEN BASES (of DNA and RNA) (16) Cyclic compounds containing both nitrogen and carbon in the ring. There are five nitrogen-containing bases found in DNA and RNA: adenine, guanine, cytosine, thymine, and uracil.

NOMENCLATURE (4) A system of names (for example, the system of names for chemical compounds).

NONELECTROLYTE (8) A substance that when placed in solution does not conduct electric current.

NONMETALS (2) A class of elements that are not good conductors of heat and electricity, and that usually take on a negative oxidation number when they bond.

NONPOLAR BOND (4) A covalent bond in which there is no electronegativity difference between the elements forming the bond.

NUCLEIC ACIDS (16) The major components of chromosomes.

NUCLEOTIDES (16) The subunits of nucleic acids.

NUCLEUS (3) The center of the atom where protons and neutrons reside.

OCTET RULE (3) When there are eight electrons in the outermost energy level of an atom, the atom tends to be nonreactive.

OIL SHALE (6) Rocks that contain oil. These oil shale deposits represent an enormous storehouse of energy in the United States.

ORBITAL (3) A region in space in which an electron can be found with a 95% probability.

OSMOSIS (13) The passage of water through a semipermeable membrane.

OXIDATION (5) Loss of electrons.

OXIDATION NUMBER (4) A number that expresses the charge of an element or a polyatomic ion in a compound.

OXIDE (8) A compound containing oxygen and another element.

OXIDIZING AGENT (5) A substance that causes something else to be oxidized while it, itself, gets reduced.

PARTICULATE MATTER (10) Particles of smoke, dust, fumes, and aerosols in the atmosphere.

PASSIVE IMMUNIZATION (1) Immunity to a disease acquired by injection of an antiserum or antitoxin which gives immediate immunity to the individual.

PEAT (6) A dark, wet, spongy material that is nature's first step in the production of coal.

PEPTIDE BOND (16) The bond that holds amino acids together in proteins.

PERIOD (3) A horizontal row of elements in the periodic table.

PERIODIC TABLE OF ELEMENTS (3)   The chemical table which lists all the known elements and places those with similar chemical properties in the same vertical column.

pH SCALE (8)   A scale that runs from zero to 14 and that expresses the acid–base strength of a solution. Zero is very acidic, 7 is neutral, and 14 is very basic (alkaline).

PHEROMONES (1)   Chemical sex attractants used by many insects for mating purposes. Some insect pheromones have been synthesized in the laboratory and are used for pest control.

PHLOGISTON (1)   A substance which all eighteenth-century chemists thought all materials capable of burning contained.

PHOTOCHEMICAL SMOG (10)   The result of reactions of chemicals in the atmosphere; in these reactions, sunlight is a catalyst.

PHOTOSYNTHESIS (6)   A chemical reaction involving carbon dioxide (from air) and water (from soil) in the presence of sunlight, to produce a simple sugar (a carbohydrate) and oxygen.

PLASTICS (13)   Polymeric substances that can be molded into various shapes or forms.

PLASTICIZERS (13)   Substances added to plastics to make them more pliable.

POLAR BOND (4)   A covalent bond in which there is unequal sharing of electrons.

POLYATOMIC ION (4)   A charged group of covalently bonded atoms.

POLYMER (13)   A large organic molecule made up of repeating units of a smaller molecule (or molecules).

POLYSACCHARIDES (15)   Carbohydrates containing many sugar groups per molecule.

POTABLE WATER (9)   Drinking water. To be considered potable, the water must meet certain requirements of the U.S. Public Health Service.

POTENTIAL ENERGY (6)   Stored energy.

PRIMARY WASTEWATER TREATMENT (9)   A system of wastewater treatment which employs physical methods to remove settleable and some suspended and floating solids.

PRODUCT (of a chemical reaction) (5)   That which results from a chemical reaction.

PROSTAGLANDINS (17)   Cyclic fatty acids containing 20 carbon atoms that form a five-membered ring in the center of the molecule.

PROTEINS (15)   Important biological molecules composed of amino acids linked together.

PROTON (3)   A particle with a relative charge of one unit and a mass of 1.0072766 amu.

PULSARS (19)   Pulsating radio sources that are probably rotating neutron stars.

PURINES (16)   The double-ring nitrogen-containing bases adenine and guanine.

PYRIMIDINES (16)   The single-ring nitrogen-containing bases cytosine, thymine, and uracil.

QUASARS (19)   Starlike objects that emit immense quantities of energy and that appear to be extremely distant from the earth.

R GROUP (12)   Any number of carbon atoms attached to the reactive part of an organic molecule.

RAD (radiation absorbed dose) (11)   The amount of radiation absorbed by a living tissue, regardless of the type of radiation.

RADIOACTIVITY (3)   The energy released by the nucleus of an atom as it undergoes nuclear decay: for example, alpha, beta, and gamma rays.

RADIOPHARMACEUTICAL THERAPY (11)   A method of radiation therapy in which the isotope is administered orally or intravenously. The isotope then uses normal body pathways to seek its target.

REACTANTS (5)   The starting materials in a chemical reaction.

REDUCING AGENT (5)   A substance that causes something else to be reduced.

REDUCTION (5)   The gain of electrons by a substance undergoing a chemical reaction.

REM (roentgen equivalent man) (11)   A weighted unit of radiation.

RESPIRATION (4)   The process by which living organisms take in oxygen and exhale carbon dioxide.

RIBONUCLEIC ACID (RNA) (16)   One of the two major nucleic acids.

RIBOSOMES (16)   The place in the cell where protein synthesis occurs.

ROENTGEN (11)   A measure of the ionizing ability of x rays and gamma rays.

SALT (8)   A compound composed of the positive ions of a base and the negative ions of an acid.

SAPONINS (18)   Chemical compounds that produce soapy lather.

SATURATED HYDROCARBON (12)   An organic compound that contains only carbon and hydrogen atoms, bonded only by single bonds.

SCIENTIFIC NOTATION (2)   A method of writing very large and very small numbers using powers of 10.

SCINTILLATION COUNTER (11)   A device that counts and determines the energy of a radioactive source.

SCRUBBER (10)   An air pollution device that absorbs gaseous pollutants from a smokestack.

SECOND LAW OF THERMODYNAMICS (6)   This can be paraphrased as follows: There is no way of getting 100% usable energy from any energy source, since some energy is always lost as waste heat.

SECONDARY WASTEWATER TREATMENT (9)   A method of wastewater treatment that involves a biological process to remove dissolved solids from the wastewater.

SETTLING TANK (9)   A large rectangular or circular tank in which the solids in wastewater are allowed to settle out.

SICKLE CELL ANEMIA (1)   A hereditary disease in which the red blood cells are sickle-shaped.

SINGLE BOND (4)   A covalent bond in which each atom donates one electron to form the bond.

SINGLE REPLACEMENT REACTION (5)   A reaction in which an uncombined or free element replaces another element in a compound.

SLUDGE (9)   The mudlike material that forms the solids removed from wastewater during wastewater treatment.

SOAPBERRIES (18)   Plants that can be used to produce lather.

SOLAR FLARES (19)   Large flaming outbursts of gas from the sun.

SOLIDS (7)   One of the states of matter. Solids have definite shape and definite volume.

SOLUBLE (5)   A substance that is capable of dissolving in solution is said to be soluble.

SOLUTE (8)   The substance that is being dissolved in solution.

SOLUTION (2)   A homogeneous mixture of two or more substances.

SOLVENT (8)   The substance that is doing the dissolving in a solution.

SPECTROSCOPE (3)   A device that is able to analyze light and break it into its component colors.

STIMULANT (17)   A drug that has an uplifting effect on the central nervous system.

STRUCTURAL FORMULA (12)   A representation of how the carbon atoms are bonded to each other in an organic compound.

SUBSCRIPT (3)   A number written below and to the side of a symbol; for example, in H_2 the 2 is a subscript.

SUN SPOTS (19)   Dark holes that appear on the face of the sun.

SUPERSCRIPT (3)   A number written above and to the side of a symbol; for example in ^{238}U, the 238 is a superscript.

SURFACTANT (18)   Any substance that stabilizes (or holds together) the suspension of nonpolar substances in water.

SYNDET (18)   The abbreviated name for synthetic detergents.

SYNERGISTIC EFFECT (17)   When the effect of combining two substances is greater than the sum of the effect of each substance taken alone. In other words, there is an enhanced effect by combining the two substances.

TELETHERAPY (11)   A method of radiation therapy in which a high-energy beam of radiation is aimed at cancerous tissue.

TESTOSTERONE (17)   A male hormone secreted by the testes. It is responsible for the development of secondary male sex characteristics.

THERMOPLASTIC POLYMER (13)   A polymer that can be softened by heat and remolded into various forms again and again.

THERMOSETTING POLYMER (13)   A polymer that can't be re-formed once it is made.

TORR (7)   A unit of pressure. One torr is the amount of pressure necessary to raise the level of mercury in a barometer 1 millimeter.

TRANQUILIZER (17)   A pharmaceutical agent that is supposed to calm a person without making him or her tired.

TRICKLING FILTER PROCESS (9)   A secondary method of wastewater treatment in which water is sprayed through a rotating sprinkler device over a bed of stones.

TRIGLYCERIDES (15)   Fats in which all three glycerol groups have undergone esterification.

TRIPLE BOND (4)   A covalent bond in which each atom donates three electrons to form the bond.

TRUE SOLUTION (8)   A solution where the solute particles have dissolved to the point of ions, atoms, or molecules into the solvent.

UNIT CELL (7)   A repetitive unit in a crystal lattice.

UNSATURATED HYDROCARBON (12)   An organic compound that contains only carbon and hydrogen, and that has some carbon-carbon double bonds or carbon-carbon triple bonds.

VISIBLE SPECTRUM (3)   The colors of the rainbow: red, orange, yellow, green, blue, indigo, and violet, which compose white light.

VITAMINS (15)   Biologically active compounds that are required by living organisms in order for them to function properly.

VOLATILE (7)   The property of a substance to be easily vaporized.

VULCANIZATION (13)   A process in which rubber is heated with sulfur in order to form a polymer with cross linking. This gives the rubber better elastic properties.

WASHING SODA (18)   The common name for sodium carbonate. This compound has detergent properties.

WATT (6)   A unit of power in the metric system.

X RAYS (3)   Electromagnetic radiation of extremely short wavelength.

Answers to Selected Exercises

Chapter 1

8. (a) technology   (b) science   (c) technology   (d) technology
(e) science
11. the Egyptians
13. the Greeks
22. earth and fire

Chapter 2

2. (a) mixture   (b) mixture   (c) element   (d) element   (e) mixture
(f) compound   (g) mixture   (h) compound   (i) element   (j) mixture
3. (a) atom   (b) molecule
5. (a) 2   (b) 0.5
6. 412,600 years
7. (a) 160   (b) 54   (c) 98   (d) 174   (e) 149   (f) 132
8. (a) 46 g   (b) 0.46 g   (c) 23 g   (d) 8.0 g
9. (a) 2.0 moles   (b) 0.20 mole   (c) $1\overline{0}$ moles   (d) 1.5 moles
10. formulas weight = 386, 3.86 g = 0.0100 mole
11. 0.050 mole

Chapter 3

2. (a) 2 (b) 5 (c) 3 (d) 7 (e) 1 (f) 6 (g) 4

7. (a) 3 (b) 1 (c) 2 (d) 4

9. (a) 2 (b) 3 (c) 1

12. (a) 3 (b) 2 (c) 1

13. (a) $92p, 92e, 146n$ (b) $1p, 1e, 2n$ (c) $35p, 35e, 46n$ (d) $20p, 20e, 20n$

15. period, group (or family)

16. (a) $^{15}_{7}N$ (b) $^{14}_{6}C$

18. 98 electrons

21. (a) K has 2, L has 4 (b) K has 2, L has 5
    (c) K has 2, L has 6 (d) K has 2, L has 8, M has 8

23. All have six electrons in their outermost energy levels.

Chapter 4

1. (a) Ba· (b) ·F̈: (c) Rb· (d) ·S̈n·

2. (a) :C̈l:$^{-1}$ (b) :S̈:$^{-2}$ (c) :P̈:$^{-3}$

3. eight

4. (a) :B̈r:B̈r: (b) H:B̈r: (c) H:C::C:H
                                    H H

    (d) H:P̈:H (e) :C̈l:C::C:C̈l:
         H

6. octet, two

7. (a) double (b) triple (c) single

8. two, one

9. (a) covalent (b) ionic (c) covalent (d) ionic (e) covalent

10. H_2Te, H_2Se, H_2S, and H_2O is most polar

12. (a) Na_3AsO_4 (b) K_2SO_3 (c) $BaCl_2$ (d) $Fe(NO_3)_2$
    (e) $CuSO_4$ (f) Al_2S_3 (g) Rb_2O (h) $Co(NO_2)_2$
    (i) CCl_4 (j) SO_2 (k) $HC_2H_3O_2$ (l) PCl_5

13. (a) silver bromide (b) copper(I) sulfide
    (c) iron(II) nitrate (d) iron(III) nitrate
    (e) calcium hydroxide (f) aluminum sulfite
    (g) zinc phosphide (h) copper(II) chromate
    (i) mercury(II) phosphate (j) cobalt(II) chloride
    (k) osmium(VIII) oxide (l) copper(I) nitride

14. (a) $+2$ (b) $+8$ (c) $+5$ (d) $+6$

15. (a) polar, polar (b) polar, polar (c) nonpolar, nonpolar
    (d) polar, polar

16. (a) Na_2SO_3 (b) K_3AsO_4 (c) $Ba(NO_3)_2$ (d) $FeCl_2$
    (e) CuS (f) $Al_2(SO_4)_3$ (g) $RbCl$ (h) CoO
    (i) CO_2 (j) SO_3 (k) UO_3 (l) PCl_3

17. (a) silver sulfide   (b) copper(I) bromide

(c) mercury(II) nitrate   (d) iron(III) nitrite

(e) calcium sulfite   (f) aluminum hydroxide

(g) zinc chromate   (h) copper(II) phosphide

(i) mercury(II) sulfide   (j) cobalt(II) phosphate

(k) copper(I) phosphide   (l) vanadium(V) oxide

Chapter 5

1. (a) $2\,K + 2\,H_2O \rightarrow 2\,KOH + H_2$ (b) $Mg + Cu(NO_3)_2 \rightarrow Mg(NO_3)_2 + Cu$

(c) $3\,H_2SO_4 + 2\,Fe(OH)_3 \rightarrow Fe_2(SO_4)_3 + 6\,H_2O$ (d) $2\,Na + Br_2 \rightarrow 2\,NaBr$

(e) $2\,Al + 3\,Hg(C_2H_3O_2)_2 \rightarrow 2\,Al(C_2H_3O_2)_3 + 3\,Hg$

2. (a) single replacement   (b) single replacement

(c) double replacement   (d) combination   (e) single replacement

3. (a) $2\,Mg + O_2 \rightarrow 2\,MgO$ (b) $H_2 + I_2 \rightarrow 2\,HI$

(c) $SO_3 + H_2O \rightarrow H_2SO_4$ (d) $N_2 + 3\,H_2 \rightarrow 2\,NH_3$

4. (a) $2\,H_2O \rightarrow 2\,H_2 + O_2$ (b) $2\,HgO \rightarrow 2\,Hg + O_2$

(c) $MgCO_3 \rightarrow MgO + CO_2$ (d) $2\,NaCl \rightarrow 2\,Na + Cl_2$

5. (a) $Mg + 2\,HCl \rightarrow MgCl_2 + H_2$ (b) $Cu + H_2SO_4 \rightarrow$ no reaction

(c) $Zn + Ni(NO_3)_2 \rightarrow Zn(NO_3)_2 + Ni$

(d) $3\,Mg + 2\,H_3PO_4 \rightarrow Mg_3(PO_4)_2 + 3\,H_2$

6. (a) $2\,HBr + Mg(OH)_2 \rightarrow MgBr_2 + 2\,H_2O$

(b) $3\,H_2SO_4 + 2\,Fe(OH)_3 \rightarrow Fe_2(SO_4)_3 + 6\,H_2O$

(c) $NaCl(aq) + Ba(NO_3)_2(aq) \rightarrow$ no reaction

(d) $BaCl_2(aq) + Na_2SO_4(aq) \rightarrow BaSO_4(s) + 2\,NaCl(aq)$

7. (a) Magnesium is oxidized and hydrogen is reduced.
(b) Sodium is oxidized and hydrogen is reduced.
(c) Mercury is oxidized and oxygen is reduced.
(d) Aluminum is oxidized and iron is reduced.

9. (a) single replacement   (b) single replacement
(c) combination   (d) single replacement

10. (a) $2\,K + 2\,H_2O \rightarrow 2\,KOH + H_2$ (b) $H_2 + Br_2 \rightarrow 2\,HBr$

(c) $2\,LiI \rightarrow 2\,Li + I_2$ (d) $Zn + Cu(NO_3)_2 \rightarrow Zn(NO_3)_2 + Cu$

(e) $3\,CuCl_2(aq) + 2\,Na_3PO_4(aq) \rightarrow 6\,NaCl(aq) + Cu_3(PO_4)_2(s)$

(f) $Ag + ZnCl_2 \rightarrow$ no reaction   (g) $2\,H_2SO_3 + O_2 \rightarrow 2\,H_2SO_4$

(h) $SrCl_2(aq) + KNO_3(aq) \rightarrow$ no reaction

Chapter 6

2. matter and energy
3. exothermic
4. endothermic
6. $2\overline{0},000$ cal or $2\overline{0}$ kcal
9. $1\overline{00}$ watts
10. 1195 cal/sec or 1.195 kcal/sec
13. (a) 5   (b) 3   (c) 6   (d) 1   (e) 4   (f) 2

Chapter 7

4. molecular solid
5. (a) increase   (b) decrease
6. The melting point of a substance is not affected very much by changes in atmospheric pressure.
10. decreases
11. increases
12. Henry's law
13. Boyle's law
14. (a) 0.500 atm   (b) 0.200 atm   (c) 2.50 atm   (d) 6.00 atm
15. (a) 1140 torr   (b) 76 torr   (c) 570 torr   (d) 2660 torr
16. (a) $298°K$ (b) $310°K$ (c) $248°K$ (d) $0°K$
17. (a) $100°C$ (b) $27°C$ (c) $-273°C$ (d) $-173°C$
18. $V_f = 57,528$ liters (or about 57,500 liters to three significant figures)
19. $V_f = 217$ ml
20. $V_f = 144$ ml
21. partial pressure of $CO_2 = 38$ torr; partial pressure of $O_2 = 722$ torr

Chapter 8

2. (a) suspension   (b) true solution
3. (a) 12.5%   (b) 30%   (c) 2%
4. 75 g of KCl
5. (a) 2.00 M (b) 0.400 M (c) 1.0 M
6. 88.8 g (or about 89 g to two significant figures)
7. 2.0 liters
8. $2\overline{0}$ mg
12. $HCl + H_2O \rightarrow H_3O^+ + Cl^-$
13. (a) $HCl + NaOH \rightarrow NaCl + H_2O$

(b) $H_2SO_4 + 2\,KOH \rightarrow K_2SO_4 + 2\,H_2O$

(c) $2\,HC_2H_3O_2 + Ca(OH)_2 \rightarrow Ca(C_2H_3O_2)_2 + 2\,H_2O$

(d) $2\,H_3PO_4 + 3\,Mg(OH)_2 \rightarrow Mg_3(PO_4)_2 + 6\,H_2O$

17. (a) $10^{-8} M$ (b) $10^{-5} M$ (c) $10^{-13} M$

18. (a) 6 (b) 9 (c) 1

19. (a) 0 (b) 2 (c) 12 (d) 10

Chapter 9

7. (a) $2 H_2O \xrightarrow{electricity} 2 H_2 + O_2$ (b) $SO_2 + H_2O \rightarrow H_2SO_3$

(c) $Na_2O + H_2O \rightarrow 2 NaOH$ (d) $CaSO_4 \cdot 2 H_2O \xrightarrow{heat} CaSO_4 + 2 H_2O$

10. Aerobic microorganisms require free oxygen in the water to survive and multiply. They indicate the presence of a healthy stream. Anaerobic microorganisms dominate a stream in the absence of free oxygen. They indicate an unhealthy condition in the stream.

12. Chlorination is the most important process in potable water treatment.

15. activated sludge and trickling filter

17. (a) 2 (b) 3 (c) 1

Chapter 10

2. 13.3% O_2, 26.7% N_2, 40% CO_2, 20% H_2

3. 20 ppm

4. 15 ppm

5. 200 $\mu g/m^3$

11. hydrocarbons and nitrogen oxides

14. NO and NO_2

18. mercury, beryllium, and asbestos

23. (a) $C + O_2 \rightarrow CO_2$ (b) $S + O_2 \rightarrow SO_2$

(c) $2 SO_2 + O_2 \rightarrow 2 SO_3$ (d) $SO_3 + H_2O \rightarrow H_2SO_4$

26. scrubber

27. electrostatic precipitator

28. (a) $2 NO + O_2 \rightarrow 2 NO_2$ (b) $NO_2 + sunlight \rightarrow NO + O$

Chapter 11

2. (a) ^4_2He (b) $^0_{-1}e$

5. (a) $^{14}_6C \rightarrow ^0_{-1}e + ^{14}_7N$ (b) $^{24}_{11}Na \rightarrow ^0_{-1}e + ^{24}_{12}Mg$

(c) $^{238}_{92}U \rightarrow ^4_2He + ^{234}_{90}Th$ (d) $^{87}_{36}Kr \rightarrow ^1_0n + ^{86}_{36}Kr$

e) $^{212}_{84}Po \rightarrow ^4_2He + ^{208}_{82}Pb$

6. 12.5 mg

7. 84 years

11. (a) stage 1   (b) stage 2   (c) stage 1

12. bone marrow and reproductive organs

15. 1.1×10^{14} dps

16. 400 rads

17. 0.0004 rad

Chapter 12

3. (a) The corners of the tetrahedron touch.

(b) There is a common edge between the two tetrahedrons.

(c) There is a common side between the two tetrahedrons.

7. (a) C—C—C—C—C—C—C

(b)
$$\begin{array}{c} \text{C} \\ | \\ \text{C—C—C—C—C} \\ | \\ \text{C} \end{array}$$

(c)
$$\begin{array}{c} \text{C—C—C—C—C—C} \\ | \\ \text{C} \end{array}$$

(d)
$$\begin{array}{c} \text{C} \\ | \\ \text{C—C—C—C—C} \\ | \\ \text{C} \end{array}$$

(e)
$$\begin{array}{c} \text{C—C—C—C—C—C} \\ | \\ \text{C} \end{array}$$

(f)
$$\begin{array}{c} \text{C—C—C—C—C} \\ | \\ \text{C} \\ | \\ \text{C} \end{array}$$

(g)
$$\begin{array}{c} \text{C—C—C—C—C} \\ | \quad | \\ \text{C} \quad \text{C} \end{array}$$

(h)
$$\begin{array}{c} \text{C—C—C—C—C} \\ | \quad \quad | \\ \text{C} \quad \quad \text{C} \end{array}$$

(i)
$$\begin{array}{c} \text{C} \\ | \\ \text{C—C—C—C} \\ | \quad | \\ \text{C} \quad \text{C} \end{array}$$

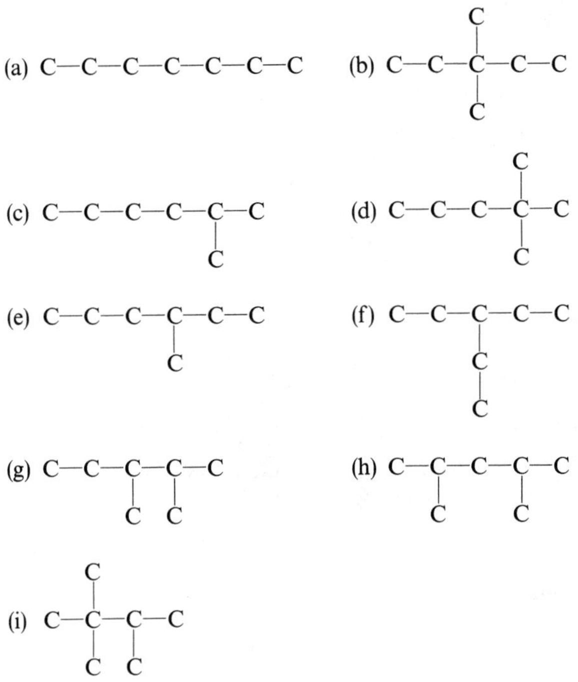

9. (a) cycloheptane   (b) cyclooctyne   (c) cyclononane   (d) cyclobutene

10. (a) $2\,C_8H_{18} + 25\,O_2 \rightarrow 16\,CO_2 + 18\,H_2O$

(b) $CH_3CH_2{-}CH{=}CHCH_3 + H_2 \rightarrow CH_3CH_2CH_2CH_2CH_3$

(c) $CH{\equiv}C{-}CH_2CH_2CH_3 + 2\,H_2 \rightarrow CH_3CH_2CH_2CH_2CH_3$

12. (a) $2\,CH_3CH_2{-}OH \xrightarrow{H_2SO_4} CH_3CH_2{-}O{-}CH_2CH_3 + H_2O$

13. (a) $CH_3{-}\underset{\underset{O}{\|}}{C}H + H_2 \xrightarrow{Pt} CH_3CH_2OH$

14. (a) $CH_3CH_2-\overset{\displaystyle O}{\overset{\displaystyle \|}{C}}-OH + HO-CH_2CH_3$

$\xrightarrow{H_2SO_4} CH_3CH_2-\overset{\displaystyle O}{\underset{\displaystyle \|}{C}}-O-CH_2CH_3 + H_2O$

15. (a) $CH_3-\overset{\displaystyle O}{\underset{\displaystyle \|}{C}}-OH + NH_3 \rightarrow CH_3-\overset{\displaystyle O}{\underset{\displaystyle \|}{C}}-NH_2 + H_2O$

16. (a) primary (b) tertiary (c) primary (d) secondary

Chapter 13

11. $CH_2{=}\overset{\displaystyle CH_3}{\overset{\displaystyle |}{C}}-CH{=}CH_2$
Isoprene

16. $xCH_2{=}CH_2 \xrightarrow{catalyst} \ +CH_2-CH_2-CH_2-CH_2+_x$

Chapter 14

3. $N_2 + 3\,H_2 \xrightarrow[\text{high pressure}]{\text{high temperature}} 2\,NH_3$

5. $C_6H_{12}O_6 + 6\,O_2 \longrightarrow 6\,CO_2 + 6\,H_2O + energy$

8. 20% nitrogen, 10% phosphorus, and 5% potassium
12. chlorobenzene and chloral hydrate

Chapter 15

2. monosaccharides, disaccharides, and polysaccharides
4. 2000 kcal
9. 450 kcal
10. amino acids
18. (a) diglyceride (b) triglyceride (c) monoglyceride
19. (a) unsaturated (b) saturated (c) unsaturated

Chapter 16

2. $H_2N-\overset{\displaystyle R}{\underset{\displaystyle H}{\overset{\displaystyle |}{\underset{\displaystyle |}{C}}}}-\overset{\displaystyle O}{\underset{\displaystyle \|}{C}}-OH$

5. (a) primary   (b) tertiary   (c) secondary   (d) quaternary
9. (a) sucrase   (b) maltase   (c) lactase   (d) lipase   (e) protease

13. ES $\rightleftarrows$ EP

16. Ribose is found in ribonucleic acid and deoxyribose is found in deoxyribonucleic acid.

19. (a) thymine   (b) cytosine   (c) guanine   (d) adenine
20. (a) uracil   (b) cytosine   (c) guanine   (d) adenine
22. (a) 2   (b) 1   (c) 3
24. DNA

Chapter 17

2. Darvon is the brand name and propoxyphene is the generic name.
8. ringing in the ears

13. $OH^- + H^+ \rightarrow H_2O$

$CO_3^{-2} + 2 H^+ \rightarrow H_2O + CO_2$

25. (a) 2   (b) 1   (c) 4   (d) 3

Chapter 18

1. soapberries and soapworts
4. hard water or acidic water

Chapter 19

2. $^2_1H + ^2_1H \xrightarrow[\text{millions of degrees}]{\text{temperature in the}} \, ^3_2He + ^1_0n + \text{energy}$

$^2_1H + ^3_1H \xrightarrow[\text{millions of degrees}]{\text{temperature in the}} \, ^4_2He + ^1_0n + \text{energy}$

6. (a) 5   (b) 3   (c) 1   (d) 6   (e) 2   (f) 8   (g) 4   (h) 7

Appendix A

1. (a) 53   (b) -77 (c) 39   (d) 11   (e) -11
2. (a) -17 (b) -51 (c) -79 (d) 49   (e) 49
3. (a) 9   (b) 64   (c) 6   (d) 1/81   (e) 1/16 or 0.063   (f) 16
4. (a) $(2)^4$ (b) $(2)^2$ (c) $(2)^{-2}$ (d) $(2)^{-4}$
5. (a) $(2)^{-3}$ (b) $(2)^{13}$ (c) $(2)^{-13}$ (d) $(2)^3$
6. (a) $(3)^{10}$ (b) $(4)^3$ (c) $(2)^{-3}$ (d) y^{-5}
7. (a) 7.89×10^{14} (b) 1.003×10^{-7} (c) 3.05×10^6 (d) 2.1×10^{-3}

8. (a) 6×10^8 (b) 6×10^2 (c) 6×10^{-2} (d) 6×10^{-8}

    (e) 4×10^4 (f) 4×10^{14} (g) 4×10^{-14} (h) 4×10^{-4}

9. (a) 16 quarts   (b) 4 miles   (c) 75 feet, 900 inches   (d) 5 m

10. (a) $x = 7$ (b) $a = 6$ (c) $c = 3$ (d) $m = 50$

11. (a) $P = \dfrac{nRT}{V}$ (b) $V = \dfrac{nRT}{P}$ (c) $n = \dfrac{PV}{RT}$.

    (d) $R = \dfrac{PV}{nT}$ (e) $T = \dfrac{PV}{nR}$

Appendix B

1. (a) 85.5 cm   (b) 8.55 dm   (c) 0.855 m   (d) 0.000855 km
2. (a) 4380 ml   (b) 438 cl   (c) 43.8 dl   (d) 0.00438 kl
3. (a) 10 m^2 (b) $1 \times 10^5 \text{ cm}^2$
4. (a) 125 cm^3 (b) $1.25 \times 10^5 \text{ mm}^3$
5. (a) 1.8 m   (b) 180 cm   (c) 1800 mm
6. (a) 1890 ml   (b) 1.89 liters
7. (a) 2.12 quarts   (b) 67.8 fluid ounces
8. (a) $1.54 \times 10^5 \text{ g}$ (b) 340 pounds
9. 0.71 g/ml
10. 200 ml
11. (a) $10°C$ (b) $85°C$ (c) $-40°C$
12. (a) $446°F$ (b) $914°F$ (c) $-234°F$
13. 15 kcal
14. $\Delta H = -1402$ kcal/mole
15. 6.01 kcal/g
16. 327.5 kcal/mole

Index

Alphabetical list of the elements

Name of element	Symbol	Atomic number	Atomic weight
Actinium	Ac	89	(227)
Aluminum	Al	13	26.9815
Americium	Am	95	(243)
Antimony	Sb	51	121.75
Argon	Ar	18	39.948
Arsenic	As	33	74.9216
Astatine	At	85	(210)
Barium	Ba	56	137.34
Berkelium	Bk	97	(247)
Beryllium	Be	4	9.0122
Bismuth	Bi	83	208.980
Boron	B	5	10.811
Bromine	Br	35	79.904
Cadmium	Cd	48	112.40
Calcium	Ca	20	40.08
Californium	Cf	98	(251)
Carbon	C	6	12.01115
Cerium	Ce	58	140.12
Cesium	Cs	55	132.905
Chlorine	Cl	17	35.453
Chromium	Cr	24	51.996
Cobalt	Co	27	58.9332
Copper	Cu	29	63.546
Curium	Cm	96	(247)
Dysprosium	Dy	66	162.50
Einsteinium	Es	99	(254)
Erbium	Er	68	167.26
Europium	Eu	63	151.96
Fermium	Fm	100	253
Fluorine	F	9	18.9984
Francium	Fr	87	(223)
Gadolinium	Gd	64	157.25
Gallium	Ga	31	69.72
Germanium	Ge	32	72.59
Gold	Au	79	196.967
Hafnium	Hf	72	178.49
Hahnium*	Ha	105	(260)
Helium	He	2	4.0026
Holmium	Ho	67	164.930
Hydrogen	H	1	1.00797
Indium	In	49	114.82
Iodine	I	53	126.9044
Iridium	Ir	77	192.2
Iron	Fe	26	55.847
Krypton	Kr	36	83.80
Lanthanum	La	57	138.91
Lawrencium	Lr	103	(257)
Lead	Pb	82	207.19
Lithium	Li	3	6.939
Lutetium	Lu	71	174.97
Magnesium	Mg	12	24.312
Manganese	Mn	25	54.9380
Mendelevium	Md	101	(256)
Mercury	Hg	80	200.59